Graduate Texts in Mathematics 115

Graduate Texts in Mathematics

1 TAKEUTI/ZARING. Introduction to Axiomatic Set Theory. 2nd ed.
2 OXTOBY. Measure and Category. 2nd ed.
3 SCHAEFFER. Topological Vector Spaces.
4 HILTON/STAMMBACH. A Course in Homological Algebra.
5 MACLANE. Categories for the Working Mathematician.
6 HUGHES/PIPER. Projective Planes.
7 SERRE. A Course in Arithmetic.
8 TAKEUTI/ZARING. Axiomatic Set Theory.
9 HUMPHREYS. Introduction to Lie Algebras and Representation Theory.
10 COHEN. A Course in Simple Homotopy Theory.
11 CONWAY. Functions of One Complex Variable. 2nd ed.
12 BEALS. Advanced Mathematical Analysis.
13 ANDERSON/FULLER. Rings and Categories of Modules.
14 GOLUBITSKY/GUILLEMIN. Stable Mappings and Their Singularities.
15 BERBERIAN. Lectures in Functional Analysis and Operator Theory.
16 WINTER. The Structure of Fields.
17 ROSENBLATT. Random Processes. 2nd ed.
18 HALMOS. Measure Theory.
19 HALMOS. A Hilbert Space Problem Book. 2nd ed., revised.
20 HUSEMOLLER. Fibre Bundles. 2nd ed.
21 HUMPHREYS. Linear Algebraic Groups.
22 BARNES/MACK. An Algebraic Introduction to Mathematical Logic.
23 GREUB. Linear Algebra. 4th ed.
24 HOLMES. Geometric Functional Analysis and its Applications.
25 HEWITT/STROMBERG. Real and Abstract Analysis.
26 MANES. Algebraic Theories.
27 KELLEY. General Topology.
28 ZARISKI/SAMUEL. Commutative Algebra. Vol. I.
29 ZARISKI/SAMUEL. Commutative Algebra. Vol. II.
30 JACOBSON. Lectures in Abstract Algebra I: Basic Concepts.
31 JACOBSON. Lectures in Abstract Algebra II: Linear Algebra.
32 JACOBSON. Lectures in Abstract Algebra III: Theory of Fields and Galois Theory.
33 HIRSCH. Differential Topology.
34 SPITZER. Principles of Random Walk. 2nd ed.
35 WERMER. Banach Algebras and Several Complex Variables. 2nd ed.
36 KELLEY/NAMIOKA et al. Linear Topological Spaces.
37 MONK. Mathematical Logic.
38 GRAUERT/FRITZSCHE. Several Complex Variables.
39 ARVESON. An Invitation to C^*-Algebras.
40 KEMENY/SNELL/KNAPP. Denumerable Markov Chains. 2nd ed.
41 APOSTOL. Modular Functions and Dirichlet Series in Number Theory.
42 SERRE. Linear Representations of Finite Groups.
43 GILLMAN/JERISON. Rings of Continuous Functions.
44 KENDIG. Elementary Algebraic Geometry.
45 LOÈVE. Probability Theory I. 4th ed.
46 LOÈVE. Probability Theory II. 4th ed.
47 MOISE. Geometric Topology in Dimensions 2 and 3.

continued after Index

Marcel Berger   Bernard Gostiaux

Differential Geometry:
Manifolds, Curves, and Surfaces

Translated from the French
by Silvio Levy

With 249 Illustrations

Springer-Verlag
New York Berlin Heidelberg
London Paris Tokyo

Marcel Berger
I.H.E.S.
91440 Bures-sur-Yvette
France

Bernard Gostiaux
94170 Le Perreux
France

Translator
Silvio Levy
Department of Mathematics
Princeton University
Princeton, NJ 08544
USA

Editorial Board
F.W. Gehring
Department of Mathematics
University of Michigan
Ann Arbor, MI 48109
USA

P.R. Halmos
Department of Mathematics
Santa Clara University
Santa Clara, CA 95053
USA

AMS Classification: 53–01

Library of Congress Cataloging-in-Publication Data
Berger, Marcel, 1927–
   Differential geometry.
   (Graduate texts in mathematics ; 115)
   Translation of: Géométrie différentielle.
   Bibliography: p.
   Includes indexes.
   1. Geometry, Differential. I. Gostiaux, Bernard.
II. Title. III. Series.
QA641.B4713   1988   516.3'6   87-27507

This is a translation of *Géométric Différentielle: variétés, courbes et surfaces,*
Presses Universitaires, de France, 1987

Text prepared in camera-ready form using T_EX.
Printed and bound by R.R. Donnelley & Sons, Harrisonburg, Virginia.
Printed in the United States of America.

9 8 7 6 5 4 3 2 1

ISBN 0-387-96626-9 Springer-Verlag New York Berlin Heidelberg
ISBN 3-540-96626-9 Springer-Verlag Berlin Heidelberg New York

Preface

This book consists of two parts, different in form but similar in spirit. The first, which comprises chapters 0 through 9, is a revised and somewhat enlarged version of the 1972 book *Géométrie Différentielle*. The second part, chapters 10 and 11, is an attempt to remedy the notorious absence in the original book of any treatment of surfaces in three-space, an omission all the more unforgivable in that surfaces are some of the most common geometrical objects, not only in mathematics but in many branches of physics.

Géométrie Différentielle was based on a course I taught in Paris in 1969–70 and again in 1970–71. In designing this course I was decisively influenced by a conversation with Serge Lang, and I let myself be guided by three general ideas. First, to avoid making the statement and proof of Stokes' formula the climax of the course and running out of time before any of its applications could be discussed. Second, to illustrate each new notion with non-trivial examples, as soon as possible after its introduction. And finally, to familiarize geometry-oriented students with analysis and analysis-oriented students with geometry, at least in what concerns manifolds.

To achieve all of this in a reasonable amount of time, I had to leave out a detailed review of differential calculus. The reader of this book should have a good calculus background, including multivariable calculus and some knowledge of forms in R^n (corresponding to pages 1–85 of [Spi65], for example). A little integration theory also helps. For more details, see chapter 0, where all of the necessary notions and results from calculus, exterior algebra and integration theory have been collected for the reader's convenience.

I confess that, in choosing the contents and style of *Géométrie Différen-tielle*, I emphasized the esthetic side, trying to attract the reader with theorems that are natural and simple to state, instead of providing an exhaustive exposition of the fundamentals of differentiable manifolds. I also decided to include a larger number of global results, rather than giving detailed proofs of local results.

More specifically, here are some of the contents of chapters 1 through 9:

—We start with a somewhat detailed treatment of differential equations, not only because they are used in several parts of the book, but because they tend to be given less an less weight in the curriculum, at least in France.

—Submanifolds of $\mathbf{R}^n$, although sometimes included in calculus courses, are then presented in detail, to pave the way for abstract manifolds.

—Next we define abstract (differentiable) manifolds; they are the basic stuff of differential geometry, and everything else in the book is built on them.

—Five examples of manifolds are then given and resurface several times along the book, thus serving as unifying threads: spheres, real projective spaces, tori, tubular neighborhoods of submanifolds of $\mathbf{R}^n$, and one-dimensional manifolds, i.e., curves. Tubular neighborhoods and normal bundles, in particular, form a class of examples whose study is non-trivial and illustrates a number of more or less refined techniques (chapters 2, 6, 7 and 9).

—Several important topics, for example, Morse theory and the classification of compact surfaces, are discussed without proofs. These "cultural digressions" are meant to give the reader a more complete picture of differential geometry and how it relates with other subjects.

—Two chapters are devoted to curves; this is, in my opinion, justified, because curves are the simplest of manifolds and the ones for which we have the most complete results.

—The exercises consist of fairly concrete examples, except for a few that ask the reader to prove an easy result stated in the text. They range from very easy to very difficult. They are in large measure original, or at least have not appeared in French books. To tackle the more difficult exercises the reader can refer to [Spi79, vol. I] or [Die69].

*   *   *

In deciding to add to the original book a treatment of surfaces, I faced a dilemma: if I were to maintain the leisurely style of the first nine chapters, I would have to limit myself to the basics or make the book far too long. This is especially true because one cannot talk about surfaces in depth without distinguishing between their intrinsic and extrinsic geometries. Once again the desire to give the reader a global view prevailed, and the solution I chose was to be much more terse and write only a kind of "travel guide," or extended cultural digression, omitting details and proofs. Given the abundance of good works on surfaces (see the introduction to chapter 10) and the great number of references sprinkled throughout our material, I feel that the interested reader will have no difficulty in filling in the picture.

Chapter 10, then, covers the local theory of surfaces in $\mathbf{R}^3$, both intrinsic (the metric) and extrinsic (the embedding in space). The intrinsic geometry of surfaces, of course, is the simplest manifestation of riemannian geometry, but I have resisted the temptation to talk about riemannian geometry in higher dimension, even though the field has witnessed spectacular advances in recent years.

Chapter 11 covers global properties of surfaces. In particular, we discuss the Gauss–Bonnet formula, surfaces of constant or bounded curvature, closed geodesics and the cut locus (part I, intrinsic questions); minimal surfaces, surfaces of constant mean curvature and Weingarten surfaces (part II, extrinsic questions).

*   *   *

The contents of this book can serve as a basis for several different courses: a one-year junior- or senior-level course, a one-semester honors course with emphasis on forms, a survey course on surfaces, or yet an elementary course emphasizing chapters 8 and 9 on curves, which can stand more or less on their own, together with section 7.6.

The reader who wants to go beyond the contents of this book will find a number of references inside, especially in chapters 10 and 11, but here are some general ones: [Mil63] is elementary, but a pleasure to read, as is [Mil69], which covers not only Morse theory but many deep applications to differential geometry; [Die69], [Ste64], [Hic65] and [Hu69] cover much of the same ground as as this book, with differences in emphasis; [War71] has a good treatment of Lie groups, which are only mentioned in this work; [Spi79], whose first volume largely overlaps with our chapters 1 to 9, goes on for four more and is especially lucid in offering different approaches to riemannian geometry and expounding its historical development; and [KN69] is the ultimate reference work.

I would like to thank Serge Lang for help in planning the contents of chapters 0 to 9, the students and teaching assistants of the 1969–1970 and 1970–1971 courses for their criticism, corrections and suggestions, F. Jabœuf for writing up sections 7.7 and 9.8, J. Lafontaine for writing up numerous exercises and for the proof of the lemma in 9.5. For feedback on the two new chapters I'm indebted to thank D. Bacry, J.-P. Bourguignon, J. Lafontaine and J. Ferrand.

Finally, I would like to thank Silvio Levy for his accurate and quick translation, and for pointing out several errors in the original. I would also like to thank Springer-Verlag for taking up the translation and the publication of this book.

Marcel Berger
I.H.E.S, 1987

Contents

Preface   v

Chapter 0. **Background**   1

    0.0 Notation and Recap   2
    0.1 Exterior Algebra   3
    0.2 Differential Calculus   9
    0.3 Differential Forms   17
    0.4 Integration   25
    0.5 Exercises   28

Chapter 1. **Differential Equations**   30

    1.1 Generalities   31
    1.2 Equations with Constant Coefficients. Existence of Local
        Solutions   33
    1.3 Global Uniqueness and Global Flows   38
    1.4 Time- and Parameter-Dependent Vector Fields   41
    1.5 Time-Dependent Vector Fields: Uniqueness And Global Flow   43
    1.6 Cultural Digression   44

Chapter 2. **Differentiable Manifolds**   47

    2.1 Submanifolds of $\mathbf{R}^n$ 48
    2.2 Abstract Manifolds   54
    2.3 Differentiable Maps   61
    2.4 Covering Maps and Quotients   67
    2.5 Tangent Spaces   74

2.6 Submanifolds, Immersions, Submersions and Embeddings   85
2.7 Normal Bundles and Tubular Neighborhoods   90
2.8 Exercises   96

Chapter 3. **Partitions of Unity, Densities and Curves**   103

3.1 Embeddings of Compact Manifolds   104
3.2 Partitions of Unity   106
3.3 Densities   109
3.4 Classification of Connected One-Dimensional Manifolds   115
3.5 Vector Fields and Differential Equations on Manifolds   119
3.6 Exercises   126

Chapter 4. **Critical Points**   128

4.1 Definitions and Examples   129
4.2 Non-Degenerate Critical Points   132
4.3 Sard's Theorem   142
4.4 Exercises   144

Chapter 5. **Differential Forms**   146

5.1 The Bundle $\Lambda^r T^* X$ 147
5.2 Differential Forms on a Manifold   148
5.3 Volume Forms and Orientation   155
5.4 De Rham Groups   168
5.5 Lie Derivatives   172
5.6 Star-shaped Sets and Poincaré's Lemma   176
5.7 De Rham Groups of Spheres and Projective Spaces   178
5.8 De Rham Groups of Tori   182
5.9 Exercises   184

Chapter 6. **Integration of Differential Forms**   188

6.1 Integrating Forms of Maximal Degree   189
6.2 Stokes' Theorem   195
6.3 First Applications of Stokes' Theorem   199
6.4 Canonical Volume Forms   203
6.5 Volume of a Submanifold of Euclidean Space   207
6.6 Canonical Density on a Submanifold of Euclidean Space   214
6.7 Volume of Tubes I   219
6.8 Volume of Tubes II   227
6.9 Volume of Tubes III   233
6.10 Exercises   238

Chapter 7. **Degree Theory**   244

7.1 Preliminary Lemmas   245
7.2 Calculation of $R^d(X)$ 251

7.3 The Degree of a Map   253
7.4 Invariance under Homotopy. Applications   256
7.5 Volume of Tubes and the Gauss–Bonnet Formula   262
7.6 Self-Maps of the Circle   267
7.7 Index of Vector Fields on Abstract Manifolds   270
7.8 Exercises   273

Chapter 8. **Curves: The Local Theory**   277

8.0 Introduction   278
8.1 Definitions   279
8.2 Affine Invariants: Tangent, Osculating Plan, Concavity   283
8.3 Arclength   288
8.4 Curvature   290
8.5 Signed Curvature of a Plane Curve   294
8.6 Torsion of Three-Dimensional Curves   297
8.7 Exercises   304

Chapter 9. **Plane Curves: The Global Theory**   312

9.1 Definitions   313
9.2 Jordan's Theorem   316
9.3 The Isoperimetric Inequality   322
9.4 The Turning Number   324
9.5 The Turning Tangent Theorem   328
9.6 Global Convexity   331
9.7 The Four-Vertex Theorem   334
9.8 The Fabricius–Bjerre–Halpern Formula   338
9.9 Exercises   344

Chapter 10. **A Guide to the Local Theory of Surfaces in $\mathbf{R}^3$**   346

10.1 Definitions   348
10.2 Examples   348
10.3 The Two Fundamental Forms   369
10.4 What the First Fundamental Form Is Good For   371
10.5 Gaussian Curvature   382
10.6 What the Second Fundamental Form Is Good For   388
10.7 Links Between the two Fundamental Forms   401
10.8 A Word about Hypersurfaces in $\mathbf{R}^{n+1}$ 402

Chapter 11. **A Guide to the Global Theory of Surfaces**   403
Part I: Intrinsic Surfaces
11.1 Shortest Paths   405
11.2 Surfaces of Constant Curvature   407
11.3 The Two Variation Formulas   409
11.4 Shortest Paths and the Injectivity Radius   410

11.5 Manifolds with Curvature Bounded Below   414
11.6 Manifolds with Curvature Bounded Above   416
11.7 The Gauss–Bonnet and Hopf Formulas   417
11.8 The Isoperimetric Inequality on Surfaces   419
11.9 Closed Geodesics and Isosystolic Inequalities   420
11.10 Surfaces All of Whose Geodesics Are Closed   422
11.11 Transition: Embedding and Immersion Problems   423

Part II: Surfaces in $\mathbf{R}^3$

11.12 Surfaces of Zero Curvature   425
11.13 Surfaces of Non-Negative Curvature   425
11.14 Uniqueness and Rigidity Results   427
11.15 Surfaces of Negative Curvature   428
11.16 Minimal Surfaces   429
11.17 Surfaces of Constant Mean Curvature, or Soap Bubbles   431
11.18 Weingarten Surfaces   433
11.19 Envelopes of Families of Planes   435
11.20 Isoperimetric Inequalities for Surfaces   437
11.21 A Pot-pourri of Characteristic Properties   438

Bibliography   443

Index of Symbols and Notations   453

Index   456

CHAPTER 0

Background

This chapter contains fundamental results from exterior algebra, differential calculus and integration theory that will be used in the sequel. The statements of these results have been collected here so that the reader won't have to hunt for them in other books. Proofs are generally omitted; the reader is referred to [Car71], [Dix68] or [Gui69].

0.0. Notation and Recap

0.0.1. Notation

0.0.2. Let X be a topological space. We denote by $O(X)$ the set of open subsets of X; by $O_x(X)$ the set of open subsets of X containing a point $x \in X$; and by $O_A(X)$ the set of open subsets of X containing a subset $A \subset X$.

0.0.3. If X is a metric space, we let $B(a,r)$ and $\overline{B}(a,r)$ be the open and closed balls of radius r and center a. When $X = \mathbf{R}^d$ we write $B_d(0,1)$ instead of $B(0,1)$.

0.0.4. If E and F are vector spaces over the same field, we let $L(E;F)$ be the vector space of continuous linear maps from E into F (if E and F have finite dimension every linear map is continuous). If $F = \mathbf{R}$ we write E^* instead of $L(E;\mathbf{R})$; this space is called the *dual* of E and its elements are continuous *linear forms* on E.

0.0.5. If X and Y are topological spaces we let $C^0(X;Y)$ be the set of continuous maps from X into Y.

0.0.6. The algebra of continuous functions from X into $\mathbf{R}$ is denoted by $C^0(X)$.

0.0.7. Recap

0.0.8. If X is a compact topological space, $C^0(X)$, with the norm of uniform convergence, is a complete topological space [Car71, I.1.2, example 2].

0.0.9. A finite-dimensional vector or affine space over $\mathbf{R}$ has a canonical topology, given by a norm. All norms are equivalent; in particular, we can take any Euclidean norm [Car71, I.1.6.2].

0.0.10. Example. If E and F are finite-dimensional vector spaces, so is $L(E;F)$: its dimension is equal to $\dim(E) \cdot \dim(F)$.

If E and F are normed vector spaces, $L(E;F)$ has a canonical norm, defined by

$$\|f\| = \sup \big\{ \|f(x)\| : \|x\| = 1 \big\}.$$

Then $\|f \circ g\| \leq \|f\| \cdot \|g\|$ [Car71, equation I.1.5.1], and $L(E;F)$ is a Banach space if F is [Car71, theorem I.1.4.2].

0.0.11. If E and F are isomorphic vector spaces, denote by $\mathrm{Isom}(E;F)$ the set of isomorphisms from E to F. Then

0.0.12 $$\phi : \mathrm{Isom}(E;F) \ni f \mapsto f^{-1} \in \mathrm{Isom}(F;E)$$

is continuous for the norm defined in 0.0.10, as the reader should check [Car71, theorem I.1.7.3].

0.0.13. Lipschitz and contracting maps [Car71, I.4.4.1]

0.0.13.1. Definition. Let X and Y be metric spaces. A map $f : X \to Y$ is a k-*Lipschitz* map if there exists $k \in \mathbf{R}$ such that

$$d\big(f(x), f(y)\big) \le k\, d(x, y)$$

for every $x, y \in X$.

A map $f : X \to Y$ is *locally Lipschitz* if for every $x \in X$ there exists $V \in O_x(X)$ such that $f|_V$ is Lipschitz. A map $f : X \to Y$ is *contracting* if it is k-Lipschitz with $k < 1$.

0.0.13.2. Theorem. *If X is a complete metric space and $t : X \to X$ is contracting, t has a unique fixed point, that is, there exists a unique z such that $t(z) = z$. In addition, $z = \lim_{n\to\infty} t^n(x)$ for every $x \in X$.* $\square$

0.1. Exterior Algebra

Let E be a vector space and $E^* = L(E; \mathbf{R})$ its dual.

0.1.1. We denote by $\Lambda^r E^*$ the vector space of *alternating r-linear forms* on E, that is, continuous maps $\alpha : E^r \to \mathbf{R}$ linear in each variable and satisfying

$$\alpha(\dots, x_i, \dots, x_j, \dots) = -\alpha(\dots, x_j, \dots, x_i, \dots)$$

for every $1 \le i \le j \le r$. One has $\Lambda^1 E^* = E^*$; by convention, $\Lambda^0 E^* = \mathbf{R}$. If E is n-dimensional, $\Lambda^r E^*$ has dimension $\binom{n}{r}$ if $r \le n$ and dimension 0 if $r > n$ [Dix68, 37.1.11].

Recall that, if $f_1, \dots, f_r$ are linear forms on E, we define $f_1 \wedge \cdots \wedge f_r \in \Lambda^r E^*$ by

0.1.2 $$(f_1 \wedge \cdots \wedge f_r)(x_1, \dots, x_r) = \sum_{\sigma \in S_r} \varepsilon_\sigma f_1(x_{\sigma(1)}) \dots f_r(x_{\sigma(r)}),$$

where S_r is the symmetric group on r elements and $\varepsilon_\sigma = \pm 1$ depending on whether σ is an even or odd permutation.

0.1.3. Basis for $\Lambda^r E^*$. Let $\{e_1, \ldots, e_n\}$ be a basis for E and $\{e_1^*, \ldots, e_n^*\}$ the dual basis for E^*. Let $I = (i_1, \ldots, i_r)$ be an r-tuple such that

$$1 \le i_1 < i_2 < \cdots < i_r \le n.$$

The forms $e_I^* = e_{i_1}^* \wedge \cdots \wedge e_{i_r}^*$, as I ranges over all such n-tuples, form a basis for $\Lambda^r E^*$ [Dix68, 37.1.9].

0.1.4. Exterior product of alternating forms. Consider $\alpha \in \Lambda^p E^*$ and $\beta \in \Lambda^q E^*$. The *exterior product* $\alpha \wedge \beta$, an alternating $(p+q)$-linear form, is defined as follows: let A be the subset of S_{p+q} consisting of permutations σ such that

$$\sigma(1) < \sigma(2) < \cdots < \sigma(p) \qquad \text{and} \qquad \sigma(p+1) < \cdots < \sigma(p+q).$$

Then
0.1.5
$$(\alpha \wedge \beta)(x_1, \ldots, x_{p+q}) = \sum_{\sigma \in A} \varepsilon_\sigma \alpha(x_{\sigma(1)}, \ldots, x_{\sigma(p)}) \beta(x_{\sigma(p+1)}, \ldots, x_{\sigma(p+q)})$$

[Dix68, 37.2.5–11]. The exterior product is associative.

0.1.6. If $\alpha \in \Lambda^r E^*$, we say that r is the *degree* of α, and write $\deg \alpha = r$. If $\alpha \in \Lambda^r E^*$ and $\beta \in \Lambda^s E^*$ we have

0.1.7
$$\beta \wedge \alpha = (-1)^{\deg \alpha \, \deg \beta} \alpha \wedge \beta.$$

Thus the exterior product makes the vector space

$$\Lambda E^* = \bigoplus_{r=0}^{\dim E} \Lambda^r E^*$$

into an associative and anticommutative algebra.

0.1.8. Pullbacks. For $f \in L(E; F)$ we define $f^* \in L(\Lambda^r F^*; \Lambda^r E^*)$ by

0.1.9
$$f^* \beta(u_1, \ldots, u_r) = \beta(f(u_1), \ldots, f(u_r))$$

for every $\beta \in \Lambda^r E^*$ and every $u_1, \ldots, u_r \in E$. One immediately sees that

0.1.10
$$f^*(\alpha \wedge \beta) = f^*(\alpha) \wedge f^*(\beta).$$

If $f \in L(E; F)$ and $g \in L(F; G)$ we have

0.1.11
$$(g \circ f)^* = f^* \circ g^*.$$

0.1.12. For $f \in L(E; E)$ and $\beta \in \Lambda^n E^*$, where n is the (finite) dimension of E, we have

0.1.12.1
$$f^* \beta = (\det f)\beta.$$

In fact, $\Lambda^n E^*$ has dimension one, so f^* is multiplication by a constant. If $(e_1, \ldots, e_n)$ is a basis for E and β is of form $e_1^* \wedge \cdots \wedge e_n^*$ (the associated), we have

$$(f^*\beta)(e_1, \ldots, e_n) = \beta(f(e_1), \ldots, f(e_n)) = \det f.$$

Since $f^*\beta = k\beta$, the factor k must be equal to $\det f$.

0.1.13. Orientation. If E has dimension n, the real vector space $\Lambda^n E^*$ has dimension one, so $\Lambda^n E^* \setminus 0$ has two connected components. An *orientation* for E is the choice of one of these two components.

Alternatively, consider on $\Lambda^n E^* \setminus 0$ the equivalence relation $\sim$ given by "$\alpha \sim \beta$ if there exists a strictly positive number k such that $\alpha = k\beta$." The set $O(E) = (\Lambda^n E^* \setminus 0)/\sim$ has two elements, and choosing an orientation for E is the same as choosing one of these elements.

0.1.14. Definition. An n-form $\alpha \in \Lambda^n E^* \setminus 0$ is called *positive* if it belongs to the element of $O(E)$ chosen as the orientation. A basis $\{e_1, \ldots, e_n\}$ for E is called *positive* if for some (hence any) positive $\alpha \in \Lambda^n E^* \setminus 0$ we have $\alpha(e_1, \ldots, e_n) > 0$.

Let E and F be oriented n-dimensional vector spaces, and consider $f \in \text{Isom}(E; F)$. We say that f *preserves orientation* if, for some (hence all) positive $\beta \in \Lambda^n E^* \setminus 0$, we have $f^*\beta$ positive.

If $E = F$, saying that f preserves orientation is the same as saying that $\det f > 0$; this follows from 0.1.12.1 and 0.1.13.

0.1.15. Exterior algebra over a Euclidean space

0.1.15.1. Let E be a Euclidean space, whose scalar product and norm we denote by $(\cdot \mid \cdot)$ and $\|\cdot\|$, respectively. We know that the dual E^* of E is canonically isomorphic to E via the map $\flat : x \mapsto \{y \mapsto (x \mid y)\} \in E^*$ and its inverse $\sharp : E^* \to E$ [Dix68, 35.4.6]. Thus the Euclidean structure of E gives rise to a canonical Euclidean structure on E^*. The spaces $\Lambda^p E^*$ also inherit canonical Euclidean structures [Bou74, III.7, prop. 7]; in the cases that will be treated in this book, namely, $p = 2$ and $p = d = \dim E$, that structure is explicitly defined as follows:

0.1.15.2. $p = 2$. It suffices to define the norm of products $\alpha \wedge \beta$, where $\alpha, \beta \in E^*$. Set

$$\|\alpha \wedge \beta\|^2 = \|\alpha\|^2 \|\beta\|^2 - (\alpha \mid \beta)^2.$$

If $\{e_i\}$ is an orthonormal basis for E, the dual basis $\{e_i^*\}$ of E^* is also orthonormal, and, if

$$\alpha = \sum_i \alpha_i e_i^*, \qquad \beta = \sum_i \beta_i e_i^*,$$

we have

$$\|\alpha \wedge \beta\|^2 = \sum_{i<j} (\alpha_i \beta_j - \alpha_j \beta_i)^2.$$

0.1.15.3. $p = d$. Let $\{e_i\}$ be an orthonormal basis for E; every $\alpha \in \Lambda^d E^*$ can be written as $k\, e_1^* \wedge \cdots \wedge e_d^*$. We define $\|\alpha\| = |k|$. We have to show that $|k|$ does not depend on the chosen orthonormal basis; but this follows from 0.1.12.1 ant the fact that the determinant of an orthogonal transformation is equal to ± 1.

0.1.15.4. We deduce from the previous paragraph that an oriented Euclidean space E of dimension d has a canonical volume element $\lambda_E \in \Lambda^d E^*$, namely, the element of norm 1 belonging to the chosen connected component of $\Lambda^d E^* \setminus 0$.

0.1.15.5. Definition. The form λ_E is called the *canonical volume form* of E.

Notice that λ_E is also defined by the condition that $\lambda_E(e_1, \ldots, e_d) = 1$ for every positive orthonormal basis $\{e_1, \ldots, e_d\}$.

0.1.15.6. Lemma. *If $\{a_i\}_{i=1,\ldots,d}$ is an arbitrary positive basis for E, we have*

$$\lambda_E(a_1, \ldots, a_d) = \sqrt{\det\big((a_i \mid a_j)\big)}.$$

Proof. Let $\{e_i\}_{i=1,\ldots,d}$ be an *orthonormal* positive basis for E, and let A be the matrix whose column vectors are the a_i's in the basis $\{e_i\}$. The definition of matrix multiplication shows that ${}^tAA$, where ${}^tA$ denotes the transpose of A, is just the matrix of scalar products $\big((a_i \mid a_j)\big)$. Thus

$$\det\big((a_i \mid a_j)\big) = \det({}^tAA) = \det{}^tA \det A = (\det A)^2.$$

But

$$\lambda_E(a_1, \ldots, a_d) = \lambda_E(Ae_1, \ldots, Ae_d) = \det A\, \lambda_E(e_1, \ldots, e_d) = \det A,$$

as we wished to prove. $\square$

0.1.15.7. One can also define spaces $\Lambda^p E$, called the *exterior powers* of a vector space [Bou74, III.7.4]. In this book we will just need a skew-symmetric map $\Lambda : E \times E \to \mathbf{R}$. We set, for $x, y \in E$,

$$x \wedge y = x^\flat \wedge y^\flat \in \Lambda^2 E^*,$$

and define Λ by $\Lambda(x, y) = \|x \wedge y\|$, using 0.1.15.2. For example, $\|x \wedge y\| = 1$ if $\{x, y\}$ is an orthonormal basis; in general,

$$\|x \wedge y\|^2 = \|x\|^2 \|y\|^2 - (x \mid y)^2 = \sum_{i<j}(x_i y_j - x_j y_i)^2$$

in an arbitrary orthonormal basis.

0.1.16. Now assume that E is Euclidean, oriented, and three-dimensional. Then λ_E is the *mixed product* of three vectors, written just $(x, y, z) = \lambda_E(x, y, z)$. By lemma 0.1.23, λ_E determines an isomorphism σ between

$\Lambda^2 E^*$ and E; in the notation of 0.1.15.7 this gives rise to a map $E \times E \to E$ defined by

0.1.17 $$(x, y) \mapsto \sigma(x^\flat \wedge y^\flat).$$

This map is called the *cross product* of two vectors $x, y \in E$, and denoted by $x \times y$.

0.1.18. Contractions. Let E be a vector space and ξ an element of E. For every $r \geq 1$ we define a linear map $\text{cont}(\xi) : \Lambda^r E^* \to \Lambda^{r-1} E^*$, called a *contraction* (by ξ), as follows:

0.1.19 $$\big(\text{cont}(\xi)(\alpha)\big)(\xi_1, \ldots, \xi_{r-1}) = \alpha(\xi, \xi_1, \ldots, \xi_{r-1})$$

for every $\alpha \in \Lambda^r E^*$ and $\xi_1, \ldots, \xi_{r-1} \in E$. It is easily checked that $\text{cont}(\xi)$ is an *antiderivation* of ΛE^* of degree -1, that is, for all $\alpha, \beta \in \Lambda E^*$ we have

0.1.20 $$\text{cont}(\xi)(\alpha \wedge \beta) = \big(\text{cont}(\xi)(\alpha)\big) \wedge \beta + (-1)^{\deg \alpha} \alpha \wedge \big(\text{cont}(\xi)(\beta)\big).$$

0.1.21. Use of coordinates. Let E have dimension d, and fix a basis $\{e_1, \ldots, e_d\}$ for E. Take $\alpha \in \Lambda^d E^*$ and an element $\xi = \sum_{j=1}^d x_j e_j$ of E. We have

$$\big(\text{cont}(\xi)(\alpha)\big)(e_1, \ldots, \hat{e}_i, \ldots, e_d) = \alpha\left(\sum_{j=1}^d x_j e_j, e_1, \ldots, \hat{e}_i, \ldots, e_d\right)$$

$$= \sum_{i=1}^d (-1)^{i-1} x_i \alpha(e_1, \ldots, e_i, \ldots, e_d)$$

where $\hat{e}_i$ means that e_i is omitted. Since $\alpha \in \Lambda^d E^*$, there exists a scalar a such that $\alpha = a(e_1^* \wedge \cdots \wedge e_d^*)$, and we have

0.1.22 $$\text{cont}\left(\sum_{i=1}^d \xi_i e_i\right)(\alpha) = \sum_{i=1}^d (-1)^{i-1} a x_i \, e_1^* \wedge \cdots \wedge \hat{e}_i^* \wedge \cdots \wedge e_d^*.$$

Since the forms $e_1^* \wedge \cdots \wedge \hat{e}_i^* \wedge \cdots \wedge e_d^*$ $(i = 1, \ldots, d)$ form a basis for $\Lambda^{d-1} E^*$ (cf. 0.1.3), we deduce that:

0.1.23. Lemma. *If $\alpha \in \Lambda^d E^*$ is non-zero, the map $\xi \mapsto \text{cont}(\xi)(\alpha)$ is an isomorphism between E and $\Lambda^{d-1} E^*$.* $\square$

0.1.24. Densities

0.1.25. Definition. A *density* on a d-dimensional vector space E is a map $\delta : E^d \to \mathbf{R}$ such that $\delta = |\alpha|$ for some $\alpha \in \Lambda^d E^* \setminus 0$.

0.1.26. Example. If $E = \mathbf{R}^d$, the density $\delta_0 = |\lambda_E| = |\det(\cdot)|$ is called the *canonical density* in $\mathbf{R}^d$. More generally, every Euclidean space E admits a canonical density, denoted by μ_E and defined by $\mu_E = |\lambda_E|$, where λ_E is the canonical volume form for an arbitrary orientation of E. By 0.1.15.6 we have

0.1.27 $$\mu_E(a_1, \ldots, a_d) = \sqrt{\det\big((a_i \mid a_j)\big)}$$

for any basis $\{a_1, \ldots, a_d\}$ of E.

0.1.28. The set of densities on E will be denoted by $\mathrm{Dens}(E)$.

0.1.29. Elementary properties of densities

0.1.29.1. *If δ and δ' are densities on E, there exists a constant $k > 0$ such that $\delta' = k\delta$.* $\qquad\square$

0.1.29.2. *If δ, δ' are densities on E and k, k' are non-negative constants not both of which are zero, $k\delta + k'\delta'$ is a density on E.* $\qquad\square$

0.1.29.3. *Let E and F be vector spaces of same dimension d. Let $\delta \in \mathrm{Dens}(F)$ and $f \in \mathrm{Isom}(E; F)$. The map $f^*\delta : E^d \to \mathbf{R}$, defined by*

$$(f^*\delta)(x_1, \ldots, x_d) = \delta\big(f(x_1, \ldots, x_d)\big)$$

for every $x_1, \ldots, x_d \in E$, is a density on E.

Proof. If $\alpha \in \Lambda^d F^* \setminus 0$ is such that $|\alpha| = \delta$, we have

$$(f^*\delta)(x_1, \ldots, x_d) = \delta\big(f(x_1, \ldots, x_d)\big)$$
$$\big|\alpha\big(f(x_1, \ldots, x_d)\big)\big| = \big|(f^*\alpha)(x_1, \ldots, x_d)\big|,$$

so that $f^*\delta$ is the density on E associated with $f^*\alpha \in \Lambda^d E^* \setminus 0$. $\qquad\square$

0.1.29.4. *Let E, F and G be vector spaces of same dimension, $f : E \to F$ and $g : F \to G$ isomorphisms. If δ is a density on G, we have*

$$(g \circ f)^*(\delta) = (f^* \circ g^*)(\delta).$$
$\qquad\square$

From 0.1.12.1 we deduce that

0.1.29.5. *For $f \in \mathrm{Isom}(E; E)$ and $\delta \in \mathrm{Dens}(E)$ we have $f^*(\delta) = |\det(f)|\delta$.* $\qquad\square$

0.1.29.6. *For $\dim(E) = 1$ densities are the same as norms.*

Proof. A density is a map from E into $\mathbf{R}$ such that $\delta = |\alpha|$ for some non-zero $\alpha \in \Lambda^1 E^* = E^*$. Thus

$$\delta(x) \geq 0 \qquad \text{and} \qquad \delta(x) = 0 \Leftrightarrow x = 0$$

(since $\alpha \neq 0$ implies that α is an isomorphism in dimension 1);

$$\delta(\lambda x) = \big|\alpha(\lambda x)\big| = |\lambda|\big|\alpha(x)\big| = |\lambda|\delta(x);$$
$$\delta(x+y) = \big|\alpha(x+y)\big| = \big|\alpha(x) + \alpha(y)\big| \le \big|\alpha(x)\big| + \big|\alpha(y)\big| = \delta(x) + \delta(y). \ \square$$

0.2. Differential Calculus

0.2.1. Definition. Let E and F be Banach spaces and $U \subset E$ open. A map $f : U \to F$ is called *differentiable* at $x \in U$ if there exists a linear map $f'(x) \in L(E; F)$ such that

$$\big\| f(x+h) - f(x) - f'(x)(h) \big\| = o\big(\|h\|\big)$$

(where the notation $o\big(\|h\|\big)$ means that the left-hand side approaches zero faster than $\|h\|$.) If f is differentiable at every $x \in U$ we say that f is differentiable in U.

0.2.2. The map $f'(x)$ is called the *derivative* of f at x.

0.2.3. The map $f' : U \mapsto L(E; F)$ is called the *derivative* of f.

0.2.4. Remark. In the case of a function of a single real variable we recover the elementary notion of the derivative: $L(\mathbf{R}; F)$ is canonically isomorphic to F via the map $\theta \mapsto \theta(1)$, and consider $f'(x)(1)$ is the ordinary derivative.

0.2.5. Definition. Let E and F be Banach spaces and $U \subset E$ open. A map $f : U \to F$ is called *continuously differentiable* if it is differentiable and its derivative f' belongs to $C^0\big(U; L(E; F)\big)$.

We also say that f is (*of class*) C^1. We denote by $C^1(U; F)$ the set of C^1 maps on U, and we set $C^1(U) = C^1(U; \mathbf{R})$.

0.2.6. Theorem. *Let U be a convex open subset of a Banach space E, and $f : U \to F$ a differentiable map such that $\big\| f'(x) \big\| \le k$ for every $x \in U$. Then f is k-Lipschitz* (0.0.13.1).

Proof. See [Dix67, p. 351]. $\square$

0.2.7. Corollary. *Any $f \in C^1(U; F)$ is locally Lipschitz.*

Proof. U is locally convex and f', being continuous, is locally bounded. $\square$

0.2.8. Operations on C^1 maps

0.2.8.1. Theorem. *Let E, F and G be Banach spaces, $U \subset E$ and $V \subset F$ open sets and $f \in C^1(U; F)$ and $g \in C^1(V; G)$ maps with $f(U) \subset V$. Then $g \circ f \in C^1(U; G)$, and, for every $x \in U$, we have*

$$(g \circ f)'(x) = g'\big(f(x)\big) \circ f'(x).$$

Proof. See [Dix68, 47.3.1] or [Car71, theorem I.2.2.1]. □

0.2.8.2. *If f and g are C^1 maps and $\lambda \in \mathbf{R}$ is a constant, $f + g$ and λf are C^1 maps. If multiplication makes sense in F, so is fg.* □

For example, every polynomial function is C^1.

0.2.8.3. *Any linear map $f \in L(E; F)$ is C^1, and satisfies $f'(x) = f'$ for every $x \in E$. If we denote by $L(E, F; G)$ the space of continuous bilinear maps from $E \times F$ into G, we have $L(E, F; G) \subset C^1(E \times F; G)$, and $f'(x, y)(u, v) = f(x, v) + f(u, y)$ for every $x, u \in E$ and $y, v \in F$ [Car71, theorem I.2.4.3].* □

0.2.8.4. *Let $F_1, \ldots, F_n$ be Banach spaces and p_i the projection from $F_1 \times F_2 \times \cdots \times F_n$ into F_i. Then $f \in C^1(U; F_1 \times \cdots \times F_n)$ if and only if $p_i \circ f \in C^1(U; F_i)$ for every i. In addition we have $(p_i \circ f)'(x) = p_i \circ \big(f'(x)\big)$ for every i [Car71, theorem I.2.5.1].* □

0.2.8.5. Let $E_1, \ldots, E_m$ and F be Banach spaces. Consider an open set $U \in O(E_1 \times \cdots \times E_m)$ and a map $f : U \to F$. If

$$\big(\{x_1\} \times \cdots \times \{x_{i-1}\} \times E_i \times \{x_{i+1}\} \times \cdots \times \{x_m\}\big) \cap U$$

is a *section* of U parallel to E_i, we identify the restriction of f to this section (where only the i-th variable varies) with a map defined on a subset of E_i. If the derivative of that restriction with respect to x_i exists, we denote it by $\partial f / \partial x_i$ (or f'_{E_i}, or f'_{x_i}, or $D_i f$). Thus

$$\frac{\partial f}{\partial x_i} \in L(E_i; F),$$

and we have the following result:

0.2.8.6. Proposition. *The map f is C^1 if and only if $\partial f / \partial x_i$ exists and is continuous for all i. In addition,*

0.2.8.7
$$f'(a)(h_1, \ldots, h_m) = \sum_{i=1}^{m} \frac{\partial f}{\partial x_i}(a) \, h_i.$$

Proof. See [Car71, proposition I.2.6.1]. □

0.2.8.8. Particular case. Take $E = \mathbf{R}^m$, $F = \mathbf{R}^n$, $U \in O(E)$ and $f \in C^1(U; F)$ with components $f_1, \ldots, f_n$, where each f_i is a function of the m variables $x_1, \ldots, x_m$. Denoting by $\partial f_i / \partial x_j$ the partial derivatives (in the usual sense) of the components of f, we define the *jacobian matrix* of f at

a to be the matrix

$$\begin{pmatrix} \dfrac{\partial f_1}{\partial x_1}(a) & \cdots & \dfrac{\partial f_1}{\partial x_m}(a) \\ \vdots & \ddots & \vdots \\ \dfrac{\partial f_n}{\partial x_1}(a) & \cdots & \dfrac{\partial f_n}{\partial x_m}(a) \end{pmatrix}.$$

The jacobian matrix is sometimes denoted by $f'(a)$ by abuse of notation.

In this particular case $f \in C^1(U; F)$ if and only if $\partial f_i/\partial x_j \in C^0(U; \mathbf{R})$ for every i and j.

0.2.8.9. Definition and notation. For $f \in C^1(U; E)$ and $U \in O(E)$ the *jacobian* of f, denoted by $J(f)$, is the map

$$J(f) : U \ni x \mapsto \det\big(f'(x)\big) \in \mathbf{R}.$$

For $E = \mathbf{R}^m$ we have $J(f)(a) = \det\big(f'(a)\big)$ (cf. 0.2.8.8).

0.2.9. Examples

0.2.9.1. Definition. A *curve* in $U \in O(E)$ is a pair (I, ϕ), where $I \subset \mathbf{R}$ is an interval and $\phi \in C^1(I; U)$. The *velocity* of ϕ at $t \in I$ is the vector $\phi'(t) \in E$ (cf. 0.2.4).

Now take $U \in O(E)$ and $f \in C^1(U; F)$. Given $x \in U$ and $y \in E$, we can calculate $f'(x)(y)$ by using the velocity of a curve. Choose a curve (I, ϕ) in U such that $0 \in I$, $\phi(0) = x$ and $\phi'(0) = y$. By 0.2.8.1 we have $(f \circ \phi)'(0) = f'(\phi(0)) \circ \phi'(0) = f'(x)(y)$, that is, $f'(x)(y)$ is equal to the velocity of the curve $(I, f \circ \phi)$ at 0.

More rigorously, we should have written (cf. 0.2.4) $\phi'(0)(1) = y$ and

$$(f \circ \phi)'(0)(1) = \big(f'(\phi(0)) \circ \phi'(0)\big)(1) = f'(x)(y).$$

0.2.9.2. Proposition. *Let E and F be isomorphic Banach spaces, and $\phi :$ $\mathrm{Isom}(E; F) \to \mathrm{Isom}(F; E)$ the map given by $\phi(f) = f^{-1}$. The map ϕ is of class C^1 and we have*

$$\phi'(f)(u) = -f^{-1} \circ u \circ f^{-1}.$$

Proof. We must first show that $\mathrm{Isom}(E; F) \in O\big(L(E; F)\big)$. In finite dimension this is obvious since $\mathrm{Isom}(E; F) = \det^{-1}(\mathbf{R} \setminus 0)$ and the map $f \mapsto \det(f)$ is continuous for a fixed choice of bases.

In infinite dimension we must show that for $u_0 \in \mathrm{Isom}(E; F)$ and $u \in L(E; F)$ close enough to u_0 we have $u \in \mathrm{Isom}(E; F)$, which is equivalent to showing that $u_0^{-1}u \in \mathrm{Isom}(E; E)$.

If $f \in L(E; E)$ satisfies $\|f\| < 1$, the map $1 - f$ is invertible (its inverse is $\sum_{n=0}^{\infty} f^n$). Setting $u_0^{-1}u = 1 - f$ we get $f = u_0^{-1}u_0 - u_0^{-1}u$, whence

$$\|f\| \le \|u_0^{-1}\|\|u_0 - u\|,$$

showing that $u_0^{-1}u$ (hence u) is invertible for $\|u_0 - u\| < 1/\|u_0^{-1}\|$ [Car71, theorem I.7.3].

To show differentiability, one can use the explicit formula for the inverse of a matrix in finite dimension (cf. 0.2.8.2), or proceed as follows in arbitrary dimension:

$$\phi(f + u) - \phi(f) + f^{-1} \circ u \circ f^{-1} = (f + u)^{-1} - f^{-1} + f^{-1} \circ u \circ f^{-1}$$
$$= (f + u)^{-1}(f + u)\big((f + u)^{-1} - f^{-1} + f^{-1} \circ u \circ f^{-1}\big)$$
$$= (f + u)^{-1}\big(1 - 1 - u \circ f^{-1} + u \circ f^{-1} + u \circ f^{-1} \circ u \circ f^{-1}\big)$$
$$= (f + u)^{-1}\big(u \circ f^{-1} \circ u \circ f^{-1}\big),$$

whence

$$\big\|\phi(f + u) - \phi(f) + f^{-1} \circ u \circ f^{-1}\big\| \leq \big\|(f + u)^{-1}\big\| \|u\|^2 \|f^{-1}\|^2$$

(cf. 0.0.10). But $\big\|(f + u)^{-1}\big\| \|f^{-1}\|^2$ is bounded for $\|u\|$ small enough, so we get

$$\big\|\phi(f + u) - \phi(f) + f^{-1} \circ u \circ f^{-1}\big\| = o(\|u\|). \qquad \square$$

0.2.10. Higher differentiability class. If f is C^1 on an open set $U \subset E$ and $f' : U \to L(E; F)$ is its derivative, it makes sense to ask whether f' is differentiable, since $L(E; F)$ is a Banach space (0.0.10).

0.2.11. Definition. If $(f')'(x) \in L\big(E; L(E; F)\big)$ exists for all $x \in U$, we say that f is *twice differentiable* and set $f''(x) = (f')'(x)$. We say that f is (of class)C^2 if $f'' \in C^0\big(U; L(E; L(E; F))\big)$.

0.2.12. *Let E, F and G be Banach spaces. The space $L(E, F; G)$ of continuous bilinear maps from $E \times F$ into G is isomorphic to $L\big(E; L(F; G)\big)$* [Car71, I.1.9]. $\qquad \square$

This allows us to state the following result (see [Dix67, p. 356], or [Car71, theorem I.5.1.1]):

0.2.13. Theorem (Schwarz). *If $f : U \to F$ is twice differentiable at a point a, the second derivative $f''(a) \in L(E, E; F)$ is a symmetric bilinear map, that is, for every $h, k \in E$ we have*

$$\big(f''(a)h\big)k = \big(f''(a)k\big)h. \qquad \square$$

0.2.13.1. Second derivative of a composition. The second derivative of a composition of maps $h \circ g$ is given by

$$(h \circ g)''(z) = h''\big(g(z)\big) \circ \big(g'(z), g'(z)\big) + h'\big(g(z)\big) \circ g''(z).$$

This follows from 0.2.8.1 and 0.2.8.3 [Car71, equation I.7.5.1].

0.2.14. We define $C^p(U; F)$ analogously, as the set of p-times differentiable maps, or maps *of class C^p*. We also let

$$C^\infty(U; F) = \bigcap_{p=1}^\infty C^p(U; F)$$

be the set of maps of class C^∞, or differentiable infinitely often.

0.2.15. Properties of maps of class C^p. This section generalizes 0.2.9.

0.2.15.1. A composition of maps of class C^p is of class C^p.

0.2.15.2. If $f, g \in C^p(U; F)$ and $\lambda \in \mathbf{R}$, the functions $f + g$, λg and (when it makes sense) fg are of class C^p. Every polynomial map is C^∞.

0.2.15.3. The space $L(E_1, \ldots, E_n; F)$ of continuous n-linear functions is contained in $C^\infty(E_1 \times \cdots \times E_n; F)$.

0.2.15.4. A map $f : U \to F_1 \times \cdots \times F_n$ is of class C^p if and only if each component $f_i = p_i \circ f$ is.

0.2.15.5. A map $f : U \to F$, where $U \in O(E_1 \times \cdots \times E_n)$, is of class C^p if and only if all its p-th order partial derivatives exist and are continuous.

0.2.15.6. The map $\phi : \operatorname{Isom}(E; F) \to \operatorname{Isom}(F; E)$ defined by $\phi(u) = u^{-1}$ is of class C^∞.

> Throughout this book objects will be of class C^p, for $p \geq 1$, but the value of p won't always be explicitly mentioned.

0.2.16. Example: bump functions

0.2.16.1. Proposition. *For every integer n and every real number $\delta > 0$ there exist maps $\psi \in C^\infty(\mathbf{R}^n; \mathbf{R})$ which equal 1 in $B(0, 1)$ and vanish in $\mathbf{R}^n \setminus B(0, 1 + \delta)$.*

Proof. Consider the function $\phi : \mathbf{R} \to \mathbf{R}$ defined by

$$\phi(t) = \begin{cases} \exp\left(\dfrac{-1}{(t-a)(b-t)}\right) & \text{if } a < t < b, \\ 0 & \text{otherwise.} \end{cases}$$

It is well known (and the reader should check) that $\phi \in C^\infty(\mathbf{R}; \mathbf{R})$. Integrating ϕ and normalizing we get a function $\theta \in C^\infty(\mathbf{R})$ defined by

$$\theta(t) = \frac{\int_{-\infty}^t \phi(s)\, ds}{\int_{-\infty}^{+\infty} \phi(s)\, ds};$$

it is clear that $\theta(t) = 0$ for $t \leq a$ and $\theta(t) = 1$ for $t \geq b$. Now take $a = 1$ and $b = (1 + \delta)^2$; the function $\eta(t) = 1 - \theta(t)$ is C^∞, equal to zero for

$t \geq (1 + \delta)^2$ and equal to 1 for $t \leq 1$. Finally set $\psi(x) = \eta(\|x\|^2)$. Since $x \mapsto \|x\|^2$ is C^∞, the function ψ satisfies the desired conditions.

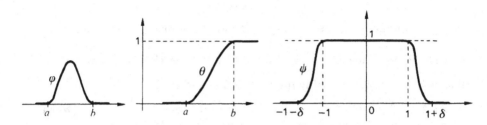

Figure 0.2.16

0.2.17. Diffeomorphisms and the inverse function theorem. The proofs of the results quoted here can be found in [Dix68, §47.4 and 47.5], except for 0.2.22, which is in [Car71, I.4.2.1].

0.2.18. Definition. Let E and F be Banach spaces, $U \subset E$ and $V \subset F$ open sets. A map $f : U \to V$ is called a C^p diffeomorphism $(p \geq 1)$ if f is bijective and both f and f^{-1} are of class C^p.

0.2.19. Proposition. *If $f : U \to V$ is a C^p diffeomorphism, we have $f'(x) \in \mathrm{Isom}(E; F)$ and $\left(f'(x)\right)^{-1} = (f^{-1})'\left(f(x)\right)$ for every $x \in U$.*

Proof. Just differentiate $f^{-1} \circ f = \mathrm{Id}_E$ and $f \circ f^{-1} = \mathrm{Id}_F$, to get

$$(f^{-1})'\left(f(x)\right) \circ f'(x) = \mathrm{Id}_E \qquad \text{and} \qquad f'(x) \circ (f^{-1})'\left(f(x)\right) = \mathrm{Id}_F . \quad \square$$

0.2.20. Definition. A map $f : U \to V$ (of class C^p for $p \geq 1$) is *regular* at x if $f'(x) \in \mathrm{Isom}(E; F)$. It is *regular in U* if it is regular for every $x \in U$.

0.2.21. Example. The map $f : \mathbf{R}^* \times \mathbf{R} \to \mathbf{R}^2$ defined by

$$f(\rho, \theta) = (\rho \cos \theta, \rho \sin \theta)$$

(polar coordinates) is regular. Its jacobian matrix

$$f'(\rho, \theta) = \begin{pmatrix} \cos \theta & -\rho \sin \theta \\ \sin \theta & \rho \cos \theta \end{pmatrix}$$

has determinant $\rho \neq 0$. The map f is not a diffeomorphism (since it is periodic in θ), but its restriction to $\mathbf{R}^* \times \,]0, 2\pi[$ is.

More generally, diffeomorphisms are regular, and regular maps are locally diffeomorphisms:

0.2.22. Inverse function theorem [Car71, I.4.2.1]. *Let U and V be open subsets of Banach spaces E and F, respectively, and $f \in C^p(U; V)$ a map regular at $x_0 \in U$. There exists an open neighborhood $U' \subset U$ of x_0 such that the restriction of f to U' is a C^p diffeomorphism from U' onto $f(U')$.*
$$\square$$

0.2.22.1. Even if f is everywhere regular it need not be injective (example 0.2.21).

0.2.23. Definition. Let E and F be Banach spaces and U an open subset of E. A C^p map $f : U \to F$ is called an *immersion* at x if $f'(x)$ is injective, and a *submersion* if $f'(x)$ is surjective.

The two fundamental theorems below express the fact that submersions and immersions are *locally*, and up to diffeomorphisms of the domain or the range, equivalent to surjective or injective linear maps. In other words, the local behavior of the function is governed by its derivative.

0.2.24. Theorem [Dix68, 47.5.3]. *Let $U \subset \mathbf{R}^m$ be an open set and f : $U \to \mathbf{R}^n$ a map of class C^p, and assume f is an immersion at x. There exist open sets $V \in O_{f(x)}(\mathbf{R}^n)$ and $U' \in O_x(U)$ and a C^p diffeomorphism $g : V \to g(V)$, where $g(V) \subset \mathbf{R}^n$ is open, such that $f(U') \subset V$ and $g \circ f|_{U'}$ coincides with the restriction to U' of the canonical injection $\mathbf{R}^m \cong \mathbf{R}^m \times \{0\}^{n-m} \to \mathbf{R}^n$.* □

0.2.25. Example. For $m = 1$ and $n = 2$ we have an arc of curve in $\mathbf{R}^2$:

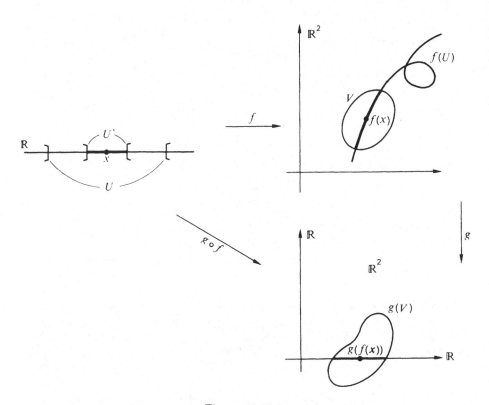

Figure 0.2.25.1

0.2.25.1. Remark. The local character of this statement, that is, the need to restrict the domain, can be clearly seen in the figure on the right: if there is a double point and U' is too big, the composition $g \circ f$ cannot be one-to-one.

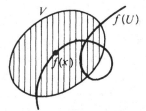

Figure 0.2.25.2

0.2.26. Theorem [Dix68, 47.5.4]. *Let $U \subset \mathbf{R}^m$ be an open set and $f : U \to \mathbf{R}^n$ a map of class C^p, and assume f is a submersion at x. There exist an open set $U' \in O_x(U)$ and a C^p diffeomorphism $g : U' \to g(U')$, where $g(U') \subset \mathbf{R}^n$ is open, such that $f|_{U'} = \pi \circ g|_{U'}$, where $\pi : \mathbf{R}^n \to \mathbf{R}^m$ is the canonical projection.*

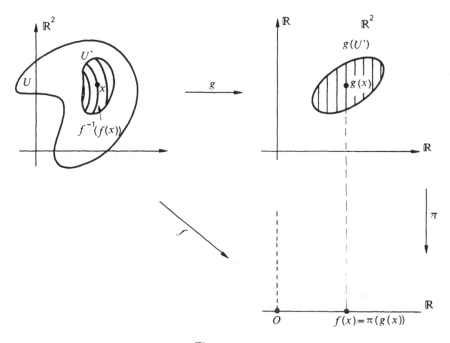

Figure 0.2.26

Theorem 0.2.26 allows one to solve the equation $f(z) = f(x)$ in U'. The solution is $z \in g^{-1}(\pi^{-1}(f(x)))$; but $\pi^{-1}(f(x))$ is the intersection with $g(U')$ of an $(n-m)$-dimensional affine subspace of $\mathbf{R}^n$, and $g^{-1}(\pi^{-1}(f(x)))$ is the image of this subspace (intersect $g(U')$) under the diffeomorphism g^{-1}. This is the so-called *implicit function theorem* [Car71, I.4.7.1].

0.3. Differential Forms

The definitions and notations in this section will be slightly modified in chapter 5 (see 5.2.7).

0.3.1. Definition. Let E be an n-dimensional vector space, where n is finite, and U an open subset of E. A C^p *differential form of degree* r, or r-*form*, on U is a C^p map $\alpha : U \to \Lambda^r E^*$. We denote by $\underline{\Omega}^r_p(U) =$

$C^p(U; \Lambda^r E^*)$ the vector space of C^p differential forms of degree r on U; we also write $\underline{\Omega}^r(U)$ when the differentiability class is not specified.

0.3.2. The vector space

$$\underline{\Omega}_p^*(U) = \bigoplus_{r=0}^{n} \underline{\Omega}_p^r(U)$$

is an associate, anticommutative algebra with the product defined by

$$(\alpha \wedge \beta)(x) = \alpha(x) \wedge \beta(x)$$

for every $x \in U$.

0.3.3. Remark. We have $\underline{\Omega}_p^0(U) = C^p(U) = C^p(U; \mathbf{R})$, since $\Lambda^0 E^* = \mathbf{R}$.

0.3.4. Example. Let $U \subset E$ be an open set, where E is an n-dimensional vector space, and fix a basis $\{e_1, \ldots, e_n\}$ for E and the dual basis $\{e_1^*, \ldots, e_n^*\}$ for E^*. Take $f \in C^p(U)$ and a point $x = (x_1, \ldots, x_n) \in U$. The map

$$x \mapsto f'(x) = \sum_{i=1}^{n} \frac{\partial f}{\partial x_i} e_i^*$$

from U into $\Lambda^1 E^*$ is of class C^{p-1}, so it belongs to $\underline{\Omega}_{p-1}^1(E)$.

0.3.5. Expression in a basis. Consider a form $\alpha \in \underline{\Omega}_p^r(U)$. Since $\alpha(x) \in \Lambda^r E^*$ for $x \in U$ and the e_I^* form a basis for $\Lambda^r E^*$ (0.1.3), there exist scalars $\alpha_{i_1 \ldots i_r}(x) = \alpha_I(x)$ such that

$$\alpha(x) = \sum_{1 \leq i_1 < \cdots < i_r \leq n} \alpha_{i_1 \ldots i_r}(x) \, e_{i_1}^* \wedge \cdots \wedge e_{i_r}^*.$$

0.3.5.1. Let's define $e_I^* = e_{i_1}^* \wedge \cdots \wedge e_{i_r}^* \in \underline{\Omega}_\infty^r(\mathbf{R}^n)$ (by abuse of notation) as the constant map $x \mapsto e_{i_1}^* \wedge \cdots \wedge e_{i_r}^*$. Then we can write

0.3.6 $$\alpha = \sum_I \alpha_I e_I^* = \sum_{i_1 < \cdots < i_r} \alpha_{i_1 \ldots i_r} \, e_{i_1}^* \wedge \cdots \wedge e_{i_r}^*,$$

and $\alpha \in \underline{\Omega}_p^r(U)$ if and only if $\alpha_I \in C^p(U)$ for every I.

0.3.7. Pullbacks

0.3.7.1. Proposition. *Let $U \subset E$ and $V \subset F$ be open sets, $f \in C^p(U; V)$ with $p \geq 1$ a map and $\beta \in \underline{\Omega}_{p-1}^r(V)$ a form on V. The map $f^*\beta$ defined on U by*

$$(f^*\beta)(x) = \big(f'(x)\big)^* \big(\beta(f(x))\big)$$

for $x \in U$ is an r-form of class $p - 1$. The map $f^* : \underline{\Omega}_{p-1}^r(V) \to \underline{\Omega}_{p-1}^r(U)$ is linear.

Proof. One writes $f^*\beta : U \to \Lambda^r E^*$ as the appropriate composition of maps [Car70, I.2.8]; in particular, the map in 0.1.8 gives a map

$$L(E; F) \ni f \mapsto f^* \in L(\Lambda^r F^*; \Lambda^r E^*)$$

which is polynomial, hence C^∞.   □

0.3.7.2. Another proof consists in calculating in coordinates; this gives a practical way to compute $f^*\beta$.

Let $\{f_1, \ldots, f_m\}$ be a basis of F. We have

$$\beta(y) = \sum_I \beta_I(y) f_I^*$$

for every $y \in V$, where $\beta_I \in C^{p-1}(V)$. Thus, for $x \in U$, we have

$$\left(f'(x)\right)^* \beta\left(f(x)\right) = \sum_I (\beta_I \circ f)(x) \left(f'(x)\right)^* f_I^*.$$

If $f_I^* = f_{i_1}^* \wedge \cdots \wedge f_{i_r}^*$ we have, for $u_1, \ldots, u_r \in E$:

$$\begin{aligned}
\left(f'(x)\right)&(f_{i_1}^* \wedge \cdots \wedge f_{i_r}^*)(u_1, \ldots, u_r) \\
&= (f_{i_1}^* \wedge \cdots \wedge f_{i_r}^*)\left(f'(x)(u_1), \ldots, f'(x)(u_r)\right) \\
&= \left((f_{i_1}^* \circ f')(x) \wedge \cdots \wedge (f_{i_r}^* \circ f')(x)\right)(u_1, \ldots, u_r).
\end{aligned}$$

Each $f_{i_k}^* \circ f : U \to \mathbf{R}$ satisfies

$$(f_{i_k}^* \circ f)'(x) = f_{i_k}^*{}'\left(f'(x)\right),$$

and, since $f_{i_k}^*$ is linear and thus equal to its derivative, we get the formula

0.3.8 $\quad \left(f'(x)\right)^* \beta(f)(x) = \displaystyle\sum_{i_1 < \cdots < i_r} (\beta_{i_1 \ldots i_r} \circ f)(x)(f_{i_1}^* \circ f)' \wedge \cdots \wedge (f_{i_r}^* \circ f)'.$

0.3.9. We have

$$\begin{aligned}
f^*(\alpha + \beta) &= f^*\alpha + f^*\beta, \\
f^*(\alpha \wedge \beta) &= f^*\alpha \wedge f^*\beta.
\end{aligned}$$

Thus f^* is an algebra homomorphism.

0.3.10. Remarks

0.3.10.1. If $\beta \in \Omega_{p-1}^0(V)$ we have $f^*(\beta) = \beta \circ f$.

0.3.10.2. If $E = F$ and $\beta(x) = b(x) e_1^* \wedge \cdots \wedge e_n^*$, where n is the dimension of E, we have

$$f^*\beta(y) = a(y) e_1^* \wedge \cdots \wedge e_n^*$$

for $a = J(f)(b \circ f)$, where $J(f)$ is the jacobian of $f \in C^p(U; E)$ (cf. 0.2.8.9). This follows from 0.1.12.1. In other words, setting

0.3.10.3 $\qquad\qquad\qquad \omega_0 = e_1^* \wedge \cdots \wedge e_n^*$

(cf. 0.3.5), we get

0.3.10.4 $f^*(b\omega_0) = (b \circ f) J(f)\omega_0.$

If E, F and G are finite-dimensional vector spaces, $U \subset E$, $V \subset F$ and $W \subset G$ open sets and $f : U \to V$ and $g : V \to W$ maps of class C^p, we have

0.3.10.5 $(g \circ f)^* = f^* \circ g^*.$

0.3.11. Densities on an open set. Notice that, if E is a finite-dimensional vector space, $\mathrm{Dens}(E)$ is an open half-line; indeed, if we fix $\delta_0 \in \mathrm{Dens}(E)$, we have $\mathrm{Dens}(E) = \mathbf{R}_+^* \delta_0$, by 0.1.29.1. Thus the following definition makes sense:

0.3.11.1. Definition. A *density of class* C^p *on* $U \in O(E)$ is a map $\delta \in C^p(U; \mathrm{Dens}(E))$. The set of such densities will be denoted by $\underline{\Delta}_p(U)$.

Once we've fixed $\delta_0 \in \mathrm{Dens}(E)$, giving a density δ is the same as giving $f \in C^p(U; \mathbf{R}_+^*)$ such that $\delta = f\delta_0$. For example, if $U \in O(\mathbf{R}^d)$, we define (and still denote by δ_0) the *canonical density*

$$U \ni x \mapsto \delta_0(x) = \delta_0 \in \mathrm{Dens}(\mathbf{R}^d)$$

(see 0.2.16). And every $\delta \in \underline{\Delta}_p(U)$ will be of the form $f\delta_0$, with $f \in C^p(U; \mathbf{R}_+^*)$.

Following 0.1.29.3, 0.3.7 and 0.3.10.4, we define, for every $f \in C^p(U; V)$ and $\delta \in \underline{\Delta}_{p-1}(V)$, where $U \subset E$ and $V \subset F$ are open, the pullback

0.3.11.2 $f^*\delta \in \underline{\Delta}_{p-1}(U),$

provided that f is regular. If $E = F = \mathbf{R}^d$, we have the formula

0.3.11.3 $f^*(b\delta_0) = (b \circ f)|J(f)|\delta_0.$

We also have

$$(g \circ f)^* = f^* \circ g^*.$$

0.3.12. Exterior differentiation

0.3.12.0. Theorem. *Let E be an n-dimensional real vector space and $U \subset E$ an open set. There exists a unique operator $d : \underline{\Omega}_p^r(U) \to \underline{\Omega}_{p-1}^{r+1}(U)$, for $r = 0, 1, \ldots, n - 1$, such that:*

(i)   *d is additive;*
(ii)   *$d(\alpha \wedge \beta) = d\alpha \wedge \beta + (-1)^{\deg \alpha}\alpha \wedge d\beta$;*
(iii) *$d(d\alpha) = 0$;*
(iv)   *$df = f'$ for every $f \in \underline{\Omega}_p^0(U)$.*

This operation is called exterior differentiation, *and $d\alpha$ is called the* exterior derivative *of α.*

Proof. We just have to calculate in coordinates as in 0.3.6. Any $\alpha \in \underline{\Omega}_p^r(U)$ can be written $\alpha = \sum_I \alpha_I e_I^*$, with $\alpha_I \in C^p(U)$. If d is additive and satisfies (ii) we must have

$$d\alpha = \sum_I d\alpha_I \wedge e_I^* + \sum_I \alpha_I \, de_I^*.$$

Now consider $e_I^* = e_{i_1}^* \wedge \cdots \wedge e_{i_r}^*$, where $I = (i_1, \ldots, i_r)$. Since $e_{i_1}^*$ denotes the i-th coordinate function on E in the basis $\{e_1, \ldots, e_n\}$, we get $(e_{i_1}^*)' = e_{i_1}^*$ (0.2.8.3), whence $e_{i_1}^* = de_{i_1}^*$, by (iv) and because the restriction of $e_{i_1}^*$ to U belongs to $\underline{\Omega}_p^0(U)$. Then $de_{i_1}^* = 0$ by (iii), and we're left with

0.3.12.1 $d\left(\sum_I \alpha_i e_I^*\right) = \sum_I d\alpha_i \wedge e_I^* = \sum_I \alpha_I' \wedge e_I^*,$

where α_I' is defined as in 0.3.4.

This takes care of uniqueness. One can check directly that 0.3.12.1 satisfies (i), (ii) and (iv). As to (iii), it suffices to show that if $f \in \underline{\Omega}_p^0(U)$ (with $p \geq 2$) we have $d(df) = 0$. But

$$d\left(\sum_{i=1}^n \frac{\partial f}{\partial x_i} e_i^*\right) = \sum_{i=1}^n \left(\sum_{j=1}^n \frac{\partial^2 f}{\partial x_i \partial x_j} e_j^*\right) \wedge e_i^* = \sum_{i=1}^n \sum_{j=1}^n \frac{\partial^2 f}{\partial x_i \partial x_j} e_j^* \wedge e_i^*,$$

and this is zero because $e_i^* \wedge e_i^* = 0$, $e_j^* \wedge e_i^* = -e_i^* \wedge e_j^*$, and, by Schwarz's theorem (0.2.13),

$$\frac{\partial^2 f}{\partial x_i \partial x_j} = \frac{\partial^2 f}{\partial x_j \partial x_i}. \qquad \square$$

The operator d satisfies $d \circ f^* = f^* \circ d$, that is, the following diagram commutes:

0.3.13

$$\begin{array}{ccc} \underline{\Omega}_p^r(U) & \xleftarrow{\;f^*\;} & \underline{\Omega}_p^r(V) \\ {\scriptstyle d}\downarrow & & \downarrow{\scriptstyle d} \\ \underline{\Omega}_{p-1}^{r+1}(U) & \xleftarrow{\;f^*\;} & \underline{\Omega}_{p-1}^{r+1}(V) \end{array}$$

The expression given here for the exterior derivative resorts to the canonical basis for $\Lambda^r E^*$. One can instead use the following intrinsic formula, which is taken as a definition in [Car70, I.2.3.1]:

0.3.14. Proposition. If $\alpha \in \underline{\Omega}_p^r(U)$ and $\xi_0, \ldots, \xi_r$ are elements of E, we have, for any $x \in U$:

$$d\alpha(x)(\xi_0, \ldots, \xi_r) = \sum_{i=0}^r (-1)^i \alpha'(x)(\xi_i)(\xi_0, \ldots, \hat{\xi}_i, \ldots, \xi_r),$$

where $\alpha'(x)$ denotes the derivative of $\alpha : U \to \Lambda^r E^*$ and $(\xi_0, \ldots, \hat{\xi}_i, \ldots, \xi_r)$ stands for $(\xi_0, \ldots, \xi_{i-1}, \xi_{i+1}, \xi_r)$.

In fact, take $\alpha = \sum_I \alpha_I e_I^*$; the map $\alpha : U \to \Lambda^r E^*$ has the α_I's for coordinate functions, hence its derivative $\alpha'(x)$ is the linear map $E \to \Lambda^r E^*$ having for coordinate functions

$$x \mapsto \alpha_I'(x) = \sum_{k=1}^n \frac{\partial \alpha_I}{\partial x_k}(x) e_k^*.$$

Thus, for $u \in E$, we have

$$\alpha'(x)(u) = \sum_I \alpha_I'(x)(u) e_I^* = \sum_I \left(\sum_{k=1}^n \frac{\partial \alpha_I}{\partial x_k}(x) e_k^*(u) \right) e_I^*;$$

in particular, since $\alpha'(x)(u) \in \Lambda^r E^*$, we have

$$\alpha'(x)(\xi_i)(\xi_0, \ldots, \hat{\xi}_i, \ldots, \xi_r) = \sum_I \left(\sum_{k=1}^n \frac{\partial \alpha_I(x)}{\partial x_k} e_k^*(\xi_i) \right) e_I^*(\xi_0, \ldots, \hat{\xi}_i, \ldots, \xi_r).$$

On the other hand, consider $d\alpha(x)(\xi_0, \ldots, \xi_r)$. By 0.3.12.1 we have

$$d\alpha(x)(\xi_0, \ldots, \xi_r) = \sum_I \alpha_I'(x) \wedge e_I^*(\xi_0, \ldots, \xi_r)$$

$$= \sum_i \left(\left(\sum_{k=1}^n \frac{\partial \alpha_I(x)}{\partial x_k} e_k^* \right) \wedge e_I^* \right)(\xi_0, \ldots, \xi_r)$$

$$= \sum_i \left(\left(\sum_{k=1}^n \frac{\partial \alpha_I(x)}{\partial x_k} e_k^* \wedge e_I^* \right)(\xi_0, \ldots, \xi_r) \right).$$

Now, if $e_I^* = e_{i_1}^* \wedge \cdots \wedge e_{i_r}^*$, we have (0.1.2):

$$(e_k^* \wedge e_{i_1}^* \wedge \cdots \wedge e_{i_r}^*)(\xi_0, \ldots, \xi_r) = \sum_{\sigma \in S_{r+1}} \varepsilon_\sigma e_k^*(\xi_{\sigma(0)}) \ldots e_{i_r}^*(\xi_{\sigma(r)}).$$

Grouping together terms with same $\sigma(0) = i$, we get

$$(e_k^* \wedge e_I^*)(\xi_0, \ldots, \xi_r) = \sum_{i=0}^r e_k^*(\xi_i) \left(\sum_{\substack{\sigma \in S_{r+1} \\ \sigma(0) = i}} \varepsilon_\sigma e_{i_1}^*(\xi_{\sigma(1)}) \ldots e_{i_r}^*(\xi_{\sigma(r)}) \right).$$

Since $\sigma(0) = i$, the permutation σ maps $\{1, \ldots, r\}$ onto $\{0, \ldots, i-1, i+1, \ldots, r\}$. Consider the map $\tau \in S_{r+1}$ defined by

$$\tau(j) = \begin{cases} i & \text{if } j = 0, \\ j - 1 & \text{if } 1 \leq j \leq i, \\ j & \text{if } i + 1 \leq j \leq r. \end{cases}$$

We have $(\sigma \circ \tau^{-1})(i) = \sigma(0) = i$, so that $\sigma \circ \tau^{-1} = \sigma^{-1}$ leaves i fixed and permutes the other indexes. Furthermore

$$e_{i_1}^*(\xi_{\sigma(1)}) \ldots e_{i_r}^*(\xi_{\sigma(r)}) = e_{i_1}^*(\xi_{\sigma'(0)}) \ldots e_{i_r}^*(\xi_{\sigma'(r)}),$$

where $\sigma'(i)$ does not appear on the right-hand side.

Since $\varepsilon_{\sigma\sigma\tau^{-1}} = \varepsilon_{\sigma'} = \varepsilon_\sigma \varepsilon_{\tau^{-1}}$ and $\varepsilon_{\tau^{-1}} = (-1)^i$ (there being i transpositions), we get

$$\sum_{\substack{\sigma\in S_{r+1} \\ \sigma(0)=i}} \varepsilon_\sigma e_{i_1}^*(\xi_{\sigma(1)})\ldots e_{i_r}^*(\xi_{\sigma(r)}) = (-1)^i \sum_{\sigma'\in S_r} \varepsilon_{\sigma'} e_{i_1}^*(\xi_{\sigma'(0)})\ldots e_{i_r}^*(\xi_{\sigma'(r)}),$$

and

$$(e_k^* \wedge e_I^*)(\xi_0,\ldots,\xi_r) = \sum_{i=0}^r (-1)^i e_k^*(\xi_i) e_I^*(\xi_0,\ldots,\hat{\xi}_i,\ldots,\xi_r),$$

whence the equality

$$d\alpha(x)(\xi_0,\ldots,\xi_r) = \sum_{i=0}^r (-1)^i \left(\sum_I \left(\sum_{k=1}^n \frac{\partial\alpha_I}{\partial x_k} e_k^*(\xi_i) \right) e_I^*(\xi_0,\ldots,\hat{\xi}_i,\ldots,\xi_r) \right)$$
$$= \sum_{i=0}^r (-1)^i \alpha'(x)(\xi_i)(\xi_0,\ldots,\hat{\xi}_i,\ldots,\xi_r).$$

0.3.15. Continuous families of differential forms

0.3.15.1. Definition. A *continuous, one-parameter family* of r-forms of class C^p on $U \in O(E)$ is a continuous map $\alpha : J \times U \to \Lambda^r E^*$, where $J \subset \mathbf{R}$ is a (not necessarily open) interval, satisfying the following conditions: for every $t \in J$, the map $x \mapsto \alpha(t,x)$ is in $C^p(U; \Lambda^r E^*)$; and the p-the derivative of $x \mapsto \alpha(t,x)$ is continuous on $J \times U$.

This implies that the restriction $\alpha_t = \alpha|_{\{t\}\times U}$, for every $t \in J$, belongs to $\Omega_p^r(U)$.

0.3.15.2. Example. The definition is satisfied if $\alpha \in C^p(J \times U; \Lambda^r E^*)$.

Now let α be a continuous, one-parameter family of r-forms of class C^p on U, defined for some interval $J \subset \mathbf{R}$. Let a and b be in J, and $a < b$. Since, for every $x \in U$, the restriction $\alpha|_{I\times\{x\}}$ is continuous, we can define

0.3.15.3
$$\int_a^b \alpha(t,u)\,dt$$

as the ordinary integral of a function of one real variable with values in a finite-dimensional vector space (0.4.7; here the range is $\Lambda^r E^*$). Thus we can consider the map

0.3.15.4
$$u \mapsto \int_a^b \alpha(t,u)\,dt$$

from U into $\Lambda^r E^*$; this map is denoted by $\int_a^b \alpha_t\,dt$.

0.3.15.5. Proposition. *The map $\int_a^b \alpha_t\,dt$ taking $u \in U$ into $\int_a^b \alpha(t,u)\,dt$ belongs to $\Omega_p^r(U)$.*

Proof. This follows by differentianting under the integral sign (see 0.4.8).

$\square$

0.3.15.6. Lemma. *Let α be a continuous, one-parameter family of r-forms of class C^1, where r is less than the dimension of E. For every $a, b \in J$ we have*

$$d\left(\int_a^b dt\right) = \int_a^b d\alpha_t\, dt.$$

This equality makes sense because, since $\alpha_t \in \underline{\Omega}_1^t(U)$, the exterior derivative $d(\alpha_t) \in \underline{\Omega}_0^{r+1}(U)$ is defined. Similarly, by 0.3.15.5, the map $\int_a^b \alpha_t\, dt$ is in $\underline{\Omega}_1^r(U)$, so $d\left(\int_a^b dt\right)$ is also defined and belongs to $\underline{\Omega}_0^{r+1}(U)$.

Proof. Let $\xi_0, \ldots, \xi_r$ be elements of E. By 0.3.14, we have

$$D = \left(d\left(\int_a^b \alpha\, dt\right)\right)(x)(\xi_0, \ldots, \xi_r)$$

$$= \sum_{i=0}^r (-1)^i \left(\int_a^b \alpha(t, x)\, dt\right)'_x (\xi_i)(\xi_0, \ldots, \hat{\xi}_i, \ldots, \xi_r),$$

where $\left(\int_a^b \alpha(t, x)\, dt\right)'_x$ is the derivative of

$$x \mapsto \int_a^b \alpha(t, x)\, dt$$

with respect to x. By 0.4.8 and 0.4.7, we obtain

$$D = \sum_{i=0}^r (-1)^i \left(\int_a^b \frac{\partial \alpha}{\partial x}(t, x)(\xi_i)\, dt\right)(\xi_0, \ldots, \hat{\xi}_i, \ldots, \xi_r)$$

$$= \sum_{i=0}^r (-1)^i \left(\int_a^b \frac{\partial \alpha}{\partial x}(t, x)(\xi_i)(\xi_0, \ldots, \hat{\xi}_i, \ldots, \xi_r)\, dt\right)$$

$$= \int_a^b \left(\sum_{i=0}^r (-1)^i \frac{\partial \alpha}{\partial x}(t, x)(\xi_i)(\xi_0, \ldots, \hat{\xi}_i, \ldots, \xi_r)\right) dt;$$

applying 0.3.14 and again 0.4.7, we get

$$D = \int_a^b \left(d\alpha_t(x)(\xi_0, \ldots, \xi_r)\right) dt = \left(\int_a^b d\alpha_t\, dt\right)(x)(\xi_0, \ldots, \xi_r),$$

concluding the proof. $\square$

0.4. Integration

A systematic reference for the whole of this section in [Gui69].

The theory that we'll need for manifolds is that of Radon measures. This theory works for locally compact topological spaces X which are countable unions of compact spaces. Some texts also require X to be metrizable, in order for a certain lemma [Gui69, p. 37] to be true; but this lemma is automatically true for manifolds (cf. 3.3.11.1).

We denote by $K(X)$ the space of functions $f \in C^0(X)$ having compact support. A *(Radon) measure* on X is a positive linear form μ on $K(X)$ [Gui69, 1.12.3]. The domain of definition of this form can be extended to a space $L^1(X) \supset K(X)$, called the space of functions on X *integrable for* μ. This space will be denoted by

0.4.1
$$L^1(X) = C_\mu^{\text{int}}(X).$$

For $f \in C_\mu^{\text{int}}(X)$ we write

0.4.2
$$\mu(f) = \int_X f\mu.$$

0.4.3.1. On $\mathbf{R}^n$ there is a canonical measure, called the *Lebesgue measure* μ_0 [Gui69, example on p. 10]. For $f \in K(\mathbf{R}^n)$ the integral $\mu_0(f)$ coincides with the ordinary (Riemann) integral. The Lebesgue measure is also defined for $U \in O(\mathbf{R}^n)$.

0.4.3.2. If μ is a measure on X and $a \in C^0(X; \mathbf{R}_+)$, we can define a measure $a\mu$ by $(a\mu)(f) = \mu(af)$. If $f \in C_{a\mu}^{\text{int}}(X)$ we have $af \in C_\mu^{\text{int}}(X)$ [Gui69, 1.11.1], and

$$\int_X f(a\mu) = \int_X (af)\mu.$$

0.4.4. Sets of measure zero. If μ is a measure on X, one has the notion of a subset of X of measure zero [Gui69, p. 10]. For the Lebesgue measure, one can take the following criterion as a definition:

0.4.4.0. Definition. A set in $\mathbf{R}^n$ has *zero Lebesgue measure* if it can be covered by a countable family of cubes whose volumes add up to less than ε, for ε arbitrarily small.

0.4.4.1. Proposition [Gui69, p. 11]. *A countable union of sets of measure zero has measure zero as well.* □

0.4.4.2. Proposition. *The set $\mathbf{R}^m = \mathbf{R}^m \times \{0\} \subset \mathbf{R}^n$, for $m < n$, has Lebesgue measure zero in $\mathbf{R}^n$. In particular, $U \cap \mathbf{R}^m$ has measure zero for any $U \in O(\mathbf{R}^n)$.* □

0.4.4.3. Proposition. *Let a be a positive function on X and μ a measure on X. If A has μ-measure zero, it has $a\mu$-measure zero.*

Proof. Write X as a countable union of compacts and apply 0.4.4.1 and [Gui69, definition on p. 10], together with the fact that continuous functions are bounded on compact sets. □

0.4.4.4. A property is said to hold μ-*almost everywhere* (or just almost everywhere) if it holds for all but a set of measure zero of points. We'll also talk about functions defined almost everywhere.

0.4.4.5. Proposition. *Let $U \in O(\mathbf{R}^n)$ and $f \in C^1(U; \mathbf{R}^n)$. If $A \subset U$ has Lebesgue measure zero, so does $f(A)$.*

Proof. By 0.4.4.1 we can assume that A is contained in $U' \subset U$, where $\overline{U'}$ is compact and U' is convex. Let k be an upper bound for $\|f'\|$ in U'. By 0.2.6, f is k-Lipschitz; in particular, the image under f of a cube of volume α in $\mathbf{R}^n$ will be contained in a cube of volume $k^n\alpha$, which proves the result by 0.4.4.0. □

0.4.4.6. In particular, if $U \in O(\mathbf{R}^n)$, $f \in C^1(U; \mathbf{R}^n)$ and $n > m$, the image $f(U)$ has Lebesgue measure zero in $\mathbf{R}^n$. It suffices to consider the map $\hat{f} : U \times \mathbf{R}^{n-m} \to \mathbf{R}^n$ defined by $\hat{f}(x, y) = f(x)$, since $U \times \{0\} \subset \mathbf{R}^m \times \mathbf{R}^{n-m}$ has measure zero.

0.4.5. If X and Y are spaces with measures μ and ν, respectively, we define on $X \times Y$ a canonical *product measure* $\mu \otimes \nu$ [Gui69, 1.7]. For instance, if μ_n is the Lebesgue measure on $\mathbf{R}^n$, we have $\mu_{m+n} = \mu_m \otimes \mu_n$ [Gui69, example on page 19]. Product measures satisfy Fubini's theorem:

0.4.5.1. Fubini's theorem. *If $f \in C^{\mathrm{int}}_{\mu\otimes\nu}(X \times Y)$ we have, for ν-almost every $y \in Y$,*

$$\{x \mapsto f(x,y)\} \in C^{\mathrm{int}}_\mu(X).$$

Moreover, the function defined ν-almost everywhere by $y \mapsto \int_X f(x,y)\mu$ is in $C^{\mathrm{int}}_\nu(Y)$, and we have

$$\int_{X\times Y} f\,\mu \otimes \nu = \int_Y \left(\int_X f(x,y)\mu \right)\nu. \qquad \square$$

0.4.6. Change of variable formula. *Consider U, $V \in O(\mathbf{R}^n)$ and a diffeomorphism $f : U \to V$ (0.2.18) Let $J(f)$ be as in 0.2.8.9, and let μ_0 be the Lebesgue measure on $\mathbf{R}^n$. If $a \in C^{\mathrm{int}}_{\mu_0}(V)$, we have*

$$(a \circ f)|J(f)| \in C^{\mathrm{int}}_{\mu_0}(U),$$

and

$$\int_U (a \circ f)|J(f)|\mu_0 = \int_V f\mu_0.$$

Proof. See [Gui69, p. 33]. □

0.4.7. Vector-valued integrals. All of the above holds without change for functions with values in a finite-dimensional vector space E. Let μ be a measure on the domain X, and E^* the dual of E. We define $C_\mu^{\text{int}}(X; E)$ to be the space of $f : X \to E$ such that

0.4.7.1 $$\xi \circ f \in C_\mu^{\text{int}}(X) \qquad \text{for every } \xi \in E^*.$$

If $f \in C_\mu^{\text{int}}(X; E)$ we define $\int_X f\mu \in E$ by

0.4.7.2 $$\xi\left(\int_X f\mu\right) = \int_X (\xi \circ f)\mu$$

for all $\xi \in E^*$.

0.4.7.3. If $\{e_i\}_{i=1,\ldots,n}$ is a basis for E and $f = (f_1, \ldots, f_n)$ in that basis, we have

$$\int_X f\mu = \left(\int_X f_1\mu, \ldots, \int_X f_n\mu\right).$$

0.4.8. Differentiation under the integral sign

0.4.8.0. Theorem. *Consider open sets $U \in O(\mathbf{R}^n)$ and $\Lambda \in O(\mathbf{R}^s)$, and a map $U \times \Lambda \to E$ into a finite-dimensional normed vector space E. Let μ be the Lebesgue measure on $\mathbf{R}^n$, and assume that f satisfies the following conditions:*

(i) *for any $\lambda \in \Lambda$, the map $x \mapsto f(x, \lambda)$ belongs to $C_\mu^{\text{int}}(U; E)$;*
(ii) *for any $x \in U$, the map $\lambda \mapsto f(x, \lambda)$ is differentiable and its derivative, denoted by $\frac{\partial f}{\partial \lambda}$, is continuous on $U \times \Lambda$;*
(iii) *there exists $h \in C_\mu^{\text{int}}(U)$ such that*

$$\left\|\frac{\partial f}{\partial \lambda}(x, \lambda)\right\| \leq h(x)$$

for every λ.

Then:

(a) *the map $x \mapsto \frac{\partial f}{\partial \lambda}(x, \lambda)$ belongs to $C_\mu^{\text{int}}(U; L(\mathbf{R}^s; E))$;*
(b) *the map $\lambda \mapsto F(\lambda) = \int_U f(x, \lambda)\mu$ is differentiable;*
(c) *differentiation under the integral sign is allowed:*

$$\frac{\partial F}{\partial \lambda} = \int_U \frac{\partial f}{\partial \lambda}(x, \lambda)\mu.$$

Proof. This follows from [Gui69, p. 26] by applying 0.2.8.6 and 0.2.8.7. $\square$

0.4.8.1. Remark. Conditions (i) and (iii) are satisfied if, for instance, the support of $x \mapsto f(x, \lambda)$ is contained in a compact subset of U independent of λ.

0.4.8.2. Theorem 0.4.8.0 gives rise, by recurrence, to similar results in class C^p. There is also a result in class C^0.

0.5. Exercises

0.5.1. Let E be a d-dimensional oriented Euclidean vector space. Show that, for every p $(0 \leq p \leq d)$, there exists a map

$$* : \Lambda^p E^* \to \Lambda^{d-p} E^*$$

characterized by the condition that

$$(*\alpha)(x_{p+1}, \ldots, x_d) = \alpha(x_1, \ldots, x_p)$$

for any positive orthonormal basis $\{e_i\}_{i=1,\ldots,d}$ and any $\alpha \in \Lambda^p E^*$. Calculate $* \circ *$ as a function of d and p.

0.5.2. Let E be a Euclidean vector space and $(\cdot \mid \cdot)$ the canonical scalar product on E^*. Show that, for every p $(0 \leq p \leq d)$, the formula

$$\|\alpha_1 \wedge \cdots \wedge \alpha_p\|^2 = \big(\det((\alpha_i \mid \alpha_j))\big)^2$$

defines a Euclidean structure on $\Lambda^p E^*$, where $\det\big((\alpha_i \mid \alpha_j)\big)$ indicates the determinant of the matrix whose elements are the $(\alpha_i \mid \alpha_j)$.

0.5.3. Liouville's theorem. The purpose of this exercise is to characterize the differentiable maps of $\mathbf{R}^n$ $(n \geq 3)$ that are *conformal*, that is, whose derivative is, at every point, an angle-preserving linear map.

0.5.3.1. Definitions. A linear map $A : \mathbf{R}^n \to \mathbf{R}^n$ is called a *similarity* if $\|Ax\| = k\|x\|$ for some real number $k \neq 0$ and all $x \in \mathbf{R}^n$; it is easy to see that A is a similarity if and only if A preserves angles. A differentiable map $f : U \to \mathbf{R}^n$, where U is an open subset of $\mathbf{R}^n$, is *conformal* if $f'(x)$ is a similarity for every $x \in U$. It is an *inversion* if there exists a point $c \in \mathbf{R}^n \setminus U$ and a real number $\alpha \neq 0$ such that

$$f(x) = c + \frac{\alpha}{\|x - c\|^2} \cdot (x - c)$$

for $x \in U$; c and α are called the *pole* and *power*, respectively, of the inversion [Ber87, section 10.8]. Finally, f is a *hyperplane reflection* if there exists a hyperplane $H \subset \mathbf{R}^n$ such that $f(x) = 2p(x) - x$, where $p(x)$ is the unique point in H whose distance to x is minimal.

0.5.3.2. Now assume that $n \geq 3$ and that $f : U \to \mathbf{R}^n$ is of class C^3 at least. Show that f is a similiarity composed with one of: (a) a translation; (b) a hyperplane reflection; (c) an inversion. Work in the following way (for

details see [Ber87, 9.5.4]): show first that the function $u(x) = \left\| f'(x) \right\|^{-1}$
satisfies

$$\frac{\partial^2 u}{\partial x_i \partial x_j} = 0 \quad \text{for } i \neq j, \qquad \frac{\partial^2 u}{\partial x_i^2} = \rho, \qquad \sum_i \left(\frac{\partial u}{\partial x_i} \right)^2 = 2\rho u$$

for some constant ρ (the first two formulas say that $\operatorname{Hess} u = \rho \| \cdot \|^2$,
cf. 4.2.2, and the last that $\|\nabla u\|^2 = 2\rho u$). Deduce from this that, if u is
not a constant, it is of the form

$$u = \frac{\rho}{2} \sum_i (x_i - a_i)^2,$$

where the a_i are constants. If u is a constant, show that we're in case (a)
or (b); otherwise show that we're in case (c).

CHAPTER 1

Differential Equations

Apart from their intrinsic interest and their relevance to mechanics and physics, differential equations are also studied as an essential tool in differential geometry (see 7.2.3 and 8.6.13, for example). We start by defining the notion of a differential equation and that of a solution, and by reformulating these concepts in terms of vector fields and integral curves. In 1.2.6 we prove the local existence and uniqueness of integral curves. We also discuss the problem of extending an integral curve into a maximal one (section 1.3).

We continue by studying the behavior of solutions as a function of the initial condition or of parameters appearing in the equation (1.2.7 and 1.4.7). This is carried out in two steps: first we discuss vector fields, that is, differential equations $x' = f(x)$ independent of time (section 1.2). Then we use a technical trick to generalize to the case of equations $x' = f(x, t)$ (section 1.4).

In section 1.6 we discuss linear equations, which enjoy the important property that their solutions exist over the whole interval of definition of the equation. We also state without proof some results which, although not used in the sequel, are so fundamental that we feel we should include them, for the sake of readers with no background in differential equations.

1.1. Generalities

Let E be a real Banach space and ϕ a map from an open subset of $\mathbf{R}$ into E. If ϕ is differentiable, its derivative $\phi'(t)$ at t is a linear map from $\mathbf{R}$ into E, and thus of the form $\lambda \mapsto \lambda V$ for some vector $V \subset E$ (namely $V = \phi'(t)(1)$, $1 \in \mathbf{R}$). In this chapter we will identify the derivative $\phi'(t)$ with the corresponding vector V, that is, we will consider $\phi'(E)$ as an element of E, and the map ϕ' as having values in E instead of $L(\mathbf{R}; E)$ (cf. 0.2.4).

1.1.1. Definition. Let $U \subset \mathbf{R} \times E$ be open and $f : U \to E$ a continuous map. A *solution of the first-order differential equation*

$$\frac{dx}{dt} = f(x, t)$$

is any map $\phi : I \to E$, where $I \subset \mathbf{R}$ is an interval, such that ϕ is of class C^1 and for every $t \in I$ we have $\big(t, \phi(t)\big) \in U$ and $\phi'(t) = f\big(t, \phi(t)\big)$.

In fact it is enough to assume that ϕ is continuous, for then ϕ', being a composition of continuous maps, will also be continuous.

In the case that $E = E_1 \times \cdots \times E_n$ is a product of real Banach spaces, U is an open subset of $\mathbf{R} \times E_1 \times \cdots \times E_n$ and f is a continuous map from U into E, with components $f_i : U \to E_i$, a solution of the equation $dx/dt = f(t, x)$ is an n-tuple of continuous maps $\phi_i : I \to E_i$, where $I \subset \mathbf{R}$ is an interval, such that for every $t \in I$ we have $\big(t, \phi_1(t), \ldots, \phi_n(t)\big) \in U$ and

$$\phi_i'(t) = f_i\big(t, \phi_i(t), \ldots, \phi_n(t)\big).$$

Thus we have a *system of n first-order equations* in n unknowns, often written

$$\frac{dx_i}{dt} = f_i(t, x_1, \ldots, x_n) \qquad \text{for } 1 \le i \le n.$$

1.1.2. Higher-order differential equations. Let E be a Banach space, $U \subset \mathbf{R} \times E^n$ an open set and $f : U \to E$ a continuous map. An *n-th order differential equation* is an equation of the form

$$\frac{d^n x}{dt^n} = f\left(t, x, \frac{dx}{dt}, \ldots, \frac{d^{n-1}x}{dt^{n-1}}\right).$$

A *solution* of such an equation is a map $\phi : I \to E$, where $I \subset \mathbf{R}$ is an interval, such that for every $t \in I$ we have $\big(t, \phi(t), \ldots, \phi^{(n-1)}(t)\big) \in U$ and

$$\phi^{(n)}(t) = f\big(t, \phi(t), \ldots, \phi^{(n-1)}(t)\big).$$

Here again we have identified derivatives with vectors in E.

1.1.3. Proposition. *Every n-th order differential equation can be reduced to a first-order equation.*

Proof. Let $\dfrac{d^n x}{dt^n} = f\left(t, x, \dfrac{dx}{dt}, \ldots, \dfrac{d^{n-1}x}{dt^{t-1}}\right)$ be an n-th order differential equation, and set

$$\frac{dx}{dt} = x_1,$$

$$\frac{dx_1}{dt} = x_2 = \frac{d^2 x}{dt^2}$$

$$\vdots$$

$$\frac{dx_{n-2}}{dt} = x_{n-1} = \frac{d^{n-1}x}{dt^{n-1}}.$$

Solving the given differential equation is the same as determining C^1 maps $\phi, \phi_1, \ldots, \phi_{n-1}$ from an interval $I \subset \mathbf{R}$ into E such that

$$\phi'(t) = \phi_1(t)$$
$$\phi_1'(t) = \phi_2(t)$$

$$\vdots$$

$$\phi_{n-2}'(t) = \phi_{n-1}(t)$$
$$\phi_{n-1}'(t) = f\big(t, \phi(t), \phi_1(t), \ldots, \phi_{n-1}(t)\big).$$

Calling $F : U \to E^n$ the map with coordinate functions

$$f_1(t, x_1, \ldots, x_n) = x_1$$

$$\vdots$$

$$f_{n-1}(t, x_1, \ldots, x_n) = x_{n-1}$$
$$f_n(t, x_1, \ldots, x_n) = f(t, x_1, \ldots, x_n)$$

and setting $X(t) = (x_1, \ldots, x_n)$, we have reduced the problem to solving the first-order differential equation

$$\frac{dX}{dt} = F\big(t, X(t)\big). \qquad \Box$$

Thus, from the theoretical point-of-view, only first-order equations need concern us.

1.2. Equations with Constant Coefficients. Existence of Local Solutions

Here we discuss differential equations of the form $dx/dt = f(x)$, where f is a continuous map from an open subset $U \subset E$ into E.

1.2.1. Definition. Let $U \subset E$ be open. A *vector field* on U is a map $f : U \to E$.

In practice we represent vectors in a vector field as arrows from each point x to the point $x + f(x)$. This makes geometric sense, especially in view of the notion of flows (see figure 1.2.2).

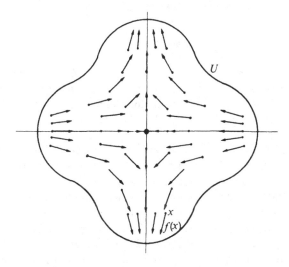

Figure 1.2.1

From now on we assume E is finite-dimensional.

If the vector field f is continuous, we can associate to it the differential equation $x' = f(x)$.

1.2.2. Definition. A C^p *integral curve* of a vector field f is a C^p curve (J, α) in U (0.2.9.1) such that $0 \in J$ and $\alpha'(t) = f(\alpha(t))$ for every $t \in J$. An integral curve α is said to have *initial condition* x_0 if $\alpha(0) = x_0$.

1.2.2.1. Remark. We require that $0 \in J$ just for convenience in the statement of initial conditions, but this requirement is not essential. It's possible to work with arbitrary J and talk about an initial condition $\alpha(t) = x_0$ for $t \in J$ and $x_0 \in U$.

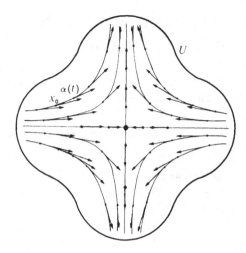

Figure 1.2.2

1.2.3. Definition. Let f be a vector field on U. A *local flow* of f at x_0 consists of a neighborhood $U' \subset U$ of x_0, an open interval J containing 0 and a map $\alpha : J \times U' \to U$ such that, for every $x \in U'$, the restriction of α to $J \times \{x\}$ is an integral curve with initial condition x.

1.2.4. Example. Let $E = \mathbf{R}^2$ and let U be the open triangle determined by the points $O(0,0)$, $A(8,0)$ and $B(4,4)$. Consider the constant map $f : x \mapsto e_1$ taking every $x \in U$ into $e_1 = (0,1)$.

The differential equation $dx/dt = e_1$ can easily be integrated; the integral curve initial condition x_0 is given by $x = te_1 + x_0$, but the values of t must be such that the vector $te_1 + x_0$ is in U. Thus the integral curve with initial condition $x_0 = (4,2)$ is the map $\alpha : \,]-2, 2[\, \to U$ given by $\alpha(t) = te_1 + x_0$. If the initial condition is $x_1 = (2,1)$, the interval of definition of the integral curve is $\,]-1, 5[$. It is clear that for any point in U it is possible to find an integral curve having that point as initial condition.

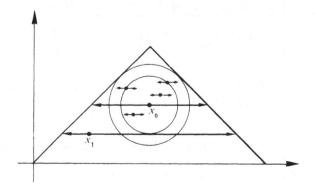

Figure 1.2.4

Consider again the point $x_0 = (4, 2)$. Finding a local flow at x_0 is the same as finding a neighborhood U' of x_0 such that, for any $y \in U'$, the integral curve with initial condition y is defined on some interval J independent of y.

For example, if we take u' to be the open ball of radius 1 centered at x_0, there exists a number $b > 0$ such that every integral curve with initial condition in $B(x_0, 1)$ is defined at least on $]{-b}, b[= J$. On the other hand the ball $U' = B(x_0, \sqrt{2})$ is no good, since it contains points arbitrarily close to the frontier of U, points whose integral curves are only defined on intervals of the form $]-\varepsilon, t_1[$ or $]-t_2, \varepsilon[$, for ε arbitrarily small.

1.2.5. Remark. Clearly the existence of a local flow at x_0 defined on an interval J does not prevent integral curves through points $x \in U'$ from being defined on intervals bigger than J. We thus have a problem of extensibility: see 1.6.1.

1.2.6. Theorem (existence and uniqueness of local flows). *Let f be a k-Lipschitz vector field on U, with $k > 0$. Let x_0 be a point in U, let $a > 0$ be a number such that $\overline{B}(x_0, 2a) \subset U$, and set $l = \sup_{x \in \overline{B}(x_0, 2a)} \|f(x)\|$. For every $b < \inf\left(\frac{a}{l}, \frac{1}{k}\right)$ there exists a unique local flow α at x_0 defined on $]{-b}, b[\times B(x_0, a)$ and continuous on the same set.*

If $k = 0$ the theorem is still valid (for $b < a/l$), but trivial. Indeed, k-Lipschitz means (cf. 0.0.13.1) that for every $x, x' \in U$ we have

$$\|f(x) - f(x')\| \le k\|x - x'\|;$$

if $k > 0$, we get $f(x) = \text{constant} = v$. We're back to example 1.2.4: integral curves in $B(x_0, a)$ are given by $t \mapsto tv + x_0$, for any t such that $\|tv + x_0 - x_0\| < a$, that is, $|t| < a/\|v\|$. But here $\|v\| = l$.

Proof. We're looking for a map $\alpha(t, x)$ such that $\alpha'_t(t, x) = f(\alpha(t, x))$ and $\alpha(0, x) = x$. This is equivalent to having

1.2.6.1
$$\alpha(t, x) = \int_0^t f(\alpha(u, x)) \, du + x.$$

We're thus led to considering the map S_x that associates to α the function $S_x(\alpha)$ given by

1.2.6.2
$$S_x(\alpha)(t) = x + \int_0^t f(\alpha(u)) \, du.$$

Solutions of 1.2.6.1 are fixed points of S_x, that is, maps α such that $S_x(\alpha) = \alpha$. Thus $S_x(\alpha)$ must have values in U, since so does α. Consider $x \in B(x_0, a)$ and the space $\mathcal{M}_x$ of continuous functions $\alpha : [-b, b] \to \overline{B}(x_0, 2a)$ such that $\alpha(0) = x$. We will show that b can be chosen in such a way that S_x maps $\mathcal{M}_x$ into $\mathcal{M}_x$; thus $S_x(\alpha)$, for α in $\mathcal{M}_x$, will have image in $\overline{B}(x_0, 2a)$, hence in U.

Give $\mathcal{M}_x$ the norm of uniform convergence (this is why we're working in a compact interval $[-b, b]$—cf. 0.0.8). Since $S_x(\alpha)$ is continuous on $[-b, b]$, there remains to show that $\left\| S_x(\alpha)(t) - x_0 \right\| \leq 2a$. We have

$$\left\| S_x(\alpha)(t) - x_0 \right\| \leq \left\| x - x_0 \right\| + \left\| \int_0^t f(\alpha(u))\, du \right\| < a + \left| \int_0^t \left\| f(\alpha(u)) \right\| du \right|.$$

Now $u \in [-b, b]$ implies $\alpha(u) \in \overline{B}(x_0, 2a)$, whence $\left\| f(\alpha(u)) \right\| \leq l$, so that $\left\| S_x(\alpha)(t) - x_0 \right\| < a + bl$. If we choose b such that

1.2.6.3
$$b \leq \frac{a}{l},$$

we will indeed have $\left\| S_x(\alpha)(t) - x_0 \right\| \leq 2a$, that is, $S_x(\alpha) \in \mathcal{M}_x$. (If $l = 0$, there is no condition on b.)

Let us try to make S_x contracting. To do this we must find an upper bound for

$$\left\| S_x(\alpha) - S_x(\beta) \right\| = \sup_{|t| \leq b} \left\| \int_0^t f(\alpha(u)) - f(\beta(u))\, du \right\|$$

$$\leq \sup_{|t| \leq b} \left| \int_0^t \left\| f(\alpha(u)) - f(\beta(u)) \right\| du \right|.$$

By assumption, f is a k-Lipschitz field, with $k > 0$, so we have

$$\left\| S_x(\alpha) - S_x(\beta) \right\| \leq \sup_{|t| \leq b} \left| \int_0^t k \left\| \alpha(u) - \beta(u) \right\| du \right|.$$

Since $\left\| \alpha(u) - \beta(u) \right\| \leq \sup_{|u| \leq b} \left\| \alpha(u) - \beta(u) \right\| = \left\| \alpha - \beta \right\|$, we have

$$\left\| S_x(\alpha) - S_x(\beta) \right\| \leq \sup_{|t| \leq b} \left| \int_0^t k \left\| \alpha - \beta \right\| du \right| - kb \left\| \alpha - \beta \right\|.$$

Then S_x will be contracting, and we will be able to apply theorem 0.0.13.2, if b satisfies

1.2.6.4
$$kb < 1.$$

Finally, considering conditions 1.2.6.3 and 1.2.6.4, we conclude that, for $b < \inf\left(\frac{a}{l}, \frac{1}{k}\right)$, there corresponds to each $x \in B(x, a)$ a contracting map $S_x : \mathcal{M}_x \to \mathcal{M}_x$, and $\mathcal{M}_x$ is a complete metric space, since $[-b, b]$ is compact and E is complete (0.0.8). Thus we can associate to each x the fixed point of S_x; this gives a map $\alpha_x : [-b, b] \to \overline{B}(x_0, 2a)$ such that $\alpha_x(0) = x$ and

$$\frac{d\alpha_x(t)}{dt} = f(\alpha_x(t))$$

in $]-b, b[$. We finally define a map $\alpha :]-b, b[\times B(x_0, a) \to \overline{B}(x_0, 2a) \subset U$ by setting $\alpha(t, x) = \alpha_x(t)$. The restriction of α to $]-b, b[\times \{x\}$ is the desired integral curve α_x. This gives a local flow at x_0.

We still have to check that this flow is continuous. For fixed x the map $\alpha(t, x)$ is continuous with respect to t, but we have to study its continuity with respect to t and x simultaneously.

Take $x, y \in B(x_0, a)$ and $t, s \in [-b, b]$. We will show that if (t, x) is close to (s, y) the number $\|\alpha(t, x) - \alpha(s, y)\|$ can be made arbitrarily small. We have

$$\|\alpha(t, x) - \alpha(s, y)\| \leq \|\alpha(t, x) - \alpha(s, x)\| + \|\alpha(s, x) - \alpha(s, y)\|.$$

Now $\|\alpha'_t(t, x)\| = \|f(\alpha(t, x))\| \leq l$ implies

$$\|\alpha(t, x) - \alpha(s, x)\| \leq l|t - s|$$

(theorem 0.2.6). As for the other term, we have

$$\|\alpha(s, x) - \alpha(s, y)\| = \left\| x + \int_0^s f(\alpha(u, x)) \, du - y - \int_0^s f(\alpha(u, y)) \, du \right\|$$

$$\leq \|x - y\| + \left| \int_0^s \|f(\alpha(u, x)) - f(\alpha(u, y))\| \, du \right|.$$

Set $\|\alpha_x - \alpha_y\| = \sup_{|u| \leq b} \|\alpha(u, x) - \alpha(u, y)\|$; since f is k-Lipschitz and $|s| \leq b$, we get

$$\|\alpha(s, x) - \alpha(s, y)\| \leq \|x - y\| + kb\|\alpha_x - \alpha_y\|.$$

But this is true for every $s \in \,]-b, b[$; thus

$$\|\alpha_x = \alpha_y\| \leq \|x - y\| + kb\|\alpha_x - \alpha_y\|,$$
$$(1 - kb)\|\alpha_x = \alpha_y\| \leq \|x - y\|.$$

Since $kb < 1$, we get $\|\alpha_x - \alpha_y\| \leq \|x - y\|/(1 - kb)$.

This completes the proof of continuity: for any $x, y \in B(x_0, a)$ and $s, t \in \,]-b, b[$ we have

$$\|\alpha(t, x) - \alpha(s, y)\| \leq l|t - s| + \frac{1}{1 - kb}\|x - y\|. \qquad \square$$

1.2.7. Theorem. *Any vector field f of class C^p admits a unique local flow, of class C^p, at each point of its domain.*

Proof. Class C^p implies differentiable and locally Lipschitz. Apply theorem 1.2.6 for the existence of a continuous local flow at every point. We will not prove that this flow is of class C^p: see [Lan69, chapter VI, §4]. $\square$

1.3. Global Uniqueness and Global Flows

Theorems 1.2.6 and 1.2.7 guarantee the existence of integral curves with
given initial condition, under certain circumstances. It makes sense to
ask whether two integral curves with same initial condition coincide where
both are defined, and whether they can be extended to larger intervals of
definition. Uniqueness is assured by the next proposition; for extensibility,
see 1.6.1.

1.3.1. Proposition. *Let f be a C^p (or k-Lipschitz) vector field defined
on an open set $U \subset E$. Two integral curves $\alpha_1 : J_1 \to E$ and $\alpha_2 : J_2 \to E$
(where J_1 and J_2 are intervals containing 0) with same initial condition
coincide on $J_1 \cap J_2$.*

This can be interpreted as saying that integral curves do not fork (see
3.5.5).

Proof. Let $Q = \left\{ t \in J_1 \cap J_2 : \alpha_1(t) = \alpha_2(t) \right\}$. We have $Q \subset J_1 \cap J_2$ and
$Q \neq \emptyset$ (since $0 \in Q$). We also have Q closed, since

$$Q = (\alpha_1 - \alpha_2)^{-1}(0) \cap (J_1 \cap J_2).$$

Thus, if we show that Q is open in $J_1 \cap J_2$, it will follow by connectedness
that $Q = J_1 \cap J_2$.

Take $b \in Q$, and consider the maps β_1 and β_2 given by $\beta_i(t) = \alpha_i(t + b)$
$(i = 1, 2)$. We have

$$\beta_i'(t) = \alpha_i'(t + b) = f(\alpha_i(t + b)) = f(\beta_i(t)),$$

so β_1 and β_2 are integral curves for t in the intervals $J_1 - b$ and $J_2 - b$,
respectively. Now $\beta_1(0) = \alpha_1(b) = \alpha_2(b) = \beta_2(0) = x_0$; by local uniqueness
(theorem 1.2.6), this implies that, on an open interval J_k containing 0, the
integral curves β_1 and β_2 coincide. Then α_1 and α_2 coincide on $J_k + b$,
which is an open interval around b. Since this open interval is in Q, we
have shown that Q is open.   □

1.3.1.1. Now suppose that the set of integral curves with initial condition
x is $\left\{ (\alpha_k, J_k) \right\}_{k \in K}$, where J_k is the domain of definition of α_k and K is a
set of indices. Set $J(x) = \bigcup_{k \in K} J_k$ (an open interval), and define $\alpha(t)$, for
$t \in J(x)$, to be equal to $\alpha_k(t)$ for any k such that $t \in J_k$. This is well-defined
by 1.3.1, and $J(x)$ is the largest open interval on which the integral curve
with initial condition x is defined. The integral curve $(J(x), \alpha)$ is denoted
by α_x and called the *maximal integral curve* with initial condition x.

The following problem then arises: for $x_0 \in U$, consider the maximal
integral curve $x \mapsto \alpha(t, x_0)$ with initial condition x_0, and a time $t_1 \in J(x_0)$.
Since $x_1 = \alpha(t_1, x_0) \in U$, there exists an integral curve β with initial
condition x_1, defined on a maximal interval $J(x_1)$. What is the relation
between α and β, and between $J(x_0)$ and $J(x_1)$?

For instance, returning to example 1.2.4 with $x_0 = (3,2)$, we have an integral curve $\alpha(t, x_0) = x_0 + te_1$, with $t \in J(x_0) =]-1, 3[$. Put $t_1 = 2 \in J(x_0)$ and $x_1 = \alpha(2, x_0) = (5, 2)$. The integral curve with initial condition x_1 is given by $\beta(\tau) = x_1 + \tau e_1$, with $\tau =]-3, 1[$. Moreover, $\beta(\tau) = x_0 + 2e_1 + \tau e_1 = x_0 + (2 + \tau)e_1$. We see that $J(x_1)$ is obtained by translating $J(x_0)$ by $-t_1$ (that is, by -2), and that $\beta(\tau) = \alpha(\tau + t_1, x_0)$ for $\tau \in J(x_1)$. This situation is general:

1.3.2. Theorem. *Let f be a vector field of class C^p on an open set $U \subset E$, and take $x_0 \in U$ and $t_1 \in J(x_0)$. The integral curve β with initial condition $x_1 = \alpha(t_1, x_0)$ has maximal interval of definition $J(x_1) = J(x_0) - t_1$, and, for every $t \in J(x_0) - t_1$, we have $\beta(t) = \alpha(t + t_1, x_0)$.*

Proof. Consider the function β given by $\beta(t) = \alpha(t + t_1, x_0)$, which is defined on $J(x_0) - t_1$. We have

$$\beta'(t) = \alpha'(t + t_1, x_0) = f\big(\alpha(t + t_1, x_0)\big) = f\big(\beta(t)\big)$$

and $\beta(0) = \alpha(t_1, x_0) = x_1$, so β is an integral curve with initial condition x_1 defined on $J(x_0) - t_1$. This is the maximal interval of definition for β, otherwise the integral curve with initial condition x_0 would have a larger interval of definition than $J(x_0)$.   □

1.3.3. Definition. Let f be a vector field of class C^p. We set $\mathcal{D}(f) = \{(t, x) \in \mathbf{R} \times U : t \in J(x)\}$. The map $\alpha : \mathcal{D}(f) \to U$ given by $\alpha(t, x) = \alpha_x(t)$, where α_x is the integral curve of f with initial condition x, is called the *global flow* of f, and $\mathcal{D}(f)$ is called its *domain of definition*.

We can write

$$\mathcal{D}(f) = \bigcup_{x \in U} \big(J(x) \times \{x\}\big);$$

we necessarily have $\{0\} \times U \subset \mathcal{D}(f)$, but in general there is no open interval $J \ni 0$ such that $J \times U \subset \mathcal{D}(f)$ (see example 1.2.4, but also theorem 1.3.6).

A nice way of formulating the relation $\beta(t) = \alpha(t + t_1, x_0)$ of theorem 1.3.2 is as follows: for every $(t, x) \in \mathcal{D}(f)$, set

1.3.4 $G_t x = \alpha(t, x).$

We can say that G_t is a "local" map from U into itself. (If $\{t\} \times U \subset \mathcal{D}(f)$ the map G_t is really defined on the whole of U.) With this notation the following equation is true, whenever it makes sense:

1.3.5 $G_t(G_{t_1} x_0) = (G_t \circ G_{t_1}) x_0 = G_{t + t_1} x_0,$

because of 1.3.2. If $\mathbf{R} \times U \subset \mathcal{D}(f)$, formula 1.3.5 becomes $G_t \circ G_s = G_{t+s}$ for every $t, s \in \mathbf{R}$, and it expresses the fact that $t \mapsto G_t$ is a homomorphism from the additive group $\mathbf{R}$ into the group of homeomorphisms of U. We say that the G_t ($t \in \mathbf{R}$) form a *one-parameter group*. In general, for arbitrary $\mathcal{D}(f)$, the G_t form a *semigroup of local homeomorphisms* of U.

Intuitively speaking, 1.3.5 says that walking along α for t_1 seconds, then for t seconds, is the same as walking along α for $t + t_1$ seconds.

1.3.6. Theorem. *If f is a C^p vector field defined on an open set U, then*
(i) *$\mathcal{D}(f)$ is open in $\mathbf{R} \times U$, and*
(ii) *$\alpha \in C^p(\mathcal{D}(f); U)$.*

Proof. We will show that every point $(t_0, x_0) \in \mathcal{D}(f)$ has a neighborhood contained in $\mathcal{D}(f)$ on which the flow α is of class C^p. Since f is of class C^p, so are the local flows (1.2.7), so we just have to demonstrate the existence of a neighborhood of (t_0, x_0) contained in $\mathcal{D}(f)$ and admitting a local flow. The existence of a local flow at x_0 implies that we can find such a neighborhood for points $(0, x_0) \in \mathcal{D}(f)$. Thus the problem is to pass to points $(t_0, x_0) \in \mathcal{D}(f)$, with $t_0 \in J(x_0)$ arbitrary.

We will say that $s \in \mathbf{R}$ satisfies property P if there exists an interval $J \subset \mathbf{R}$ containing s and an open set $U' \subset U$ containing x_0 such that $J \times U' \subset \mathcal{D}(f)$ and $\alpha|_{J \times U'}$ is of class C^p. Let Q_{x_0} be the set of $t \in J(x_0)$ such that every $s \in [0, t]$ satisfies property P. We will show that Q_{x_0} and $J(x_0)$ have the same supremum. An analogous reasoning will show that $J(x_0)$ and $P_{x_0} = \{t \in J(x_0) : \text{every } s \in [t, 0] \text{ satisfies } P\}$ have the same infimum, and consequently that $J(x_0) = Q_{x_0} \cup P_{x_0}$, since Q_{x_0} and P_{x_0} are closed in $J(x_0)$. This will imply the theorem, since then every $(t_0, x_0) \in \mathcal{D}(f)$ has a neighborhood contained in $\mathcal{D}(f)$ on which α is of class C^p.

Let b be the supremum of Q_{x_0}. We can assume that $b < +\infty$, otherwise there is nothing to show (since $Q_{x_0} \subset J(x_0)$). Assume by contradiction that $b < \sup(J(x_0))$; we want to find $t \geq b$ satisfying property P. Thus we're looking for an interval $J \supset b$ and an open set $U' \supset x_0$ such that $J \times U' \subset \mathcal{D}(f)$ and $\alpha|_{J \times U'}$ is of class C^p. Every $t_1 < b$ satisfies property P; thus we should try to consider an integral curve with initial condition $x_0' = \alpha(b, x_0)$, and match it with an integral curve through x_0. Now $b < \sup(J(x_0))$ implies $b \in J(x_0)$; thus there exists an open set $U'' \subset U$ containing x_0, an interval $K =]-a, a[$ and a map $\beta : K \times U'' \to U$ that is a local flow at x_0'.

We will match this flow with a flow whose interval contains 0, by starting from $t_1 < b$, so that $t_1 \in Q_{x_0}$. For that it is necessary that the solution α defined at t_1 take values in U'', the domain of β. But $\alpha(t, x_0)$ is defined on $J(x_0)$, in particular at $t = b$, and is continuous there; since U'' is a neighborhood of $x_0' = \alpha(b, x_0)$, there exists η such that $b - \eta < t_1 < b$ implies $x_1 = \alpha(t_1, x_0) \in U''$.

Since the purpose of the matching is to extend the interval of definition of α to an interval containing $t_1 + K$, so as to go beyond b, we must take t_1 such that $t_1 + a > b$, that is, $t_1 > b - a$. So take t_1 such that

$$\sup((b - a), (b - \eta)) < t_1 < b.$$

Since $t_1 \in Q_{x_0}$, there exists an interval $J_1 \supset t_1$ and an open set $U' \subset U$ containing x_0 such that $J_1 \times U' \subset \mathcal{D}(f)$ and $\alpha|_{J_1 \times U'}$ is of class C^p.

Since $\alpha(t_1, x_0) = x_1$ is in the open set U'' and $x \mapsto \alpha(t_1, x)$ is continuous, we can restrict J_1 and U' and suppose that $\alpha(\{t_1\} \times U') \subset U''$. Take $t \in t_1 + K$ and $x \in U'$. Since $t_1 \in J_1$, we have $\alpha(t_1, x) \in U''$, and there exists an integral curve with initial condition $\alpha(t_1, x)$ defined on K by $\beta(\tau, \alpha(t_1, x))$ for every $\tau \in K$.

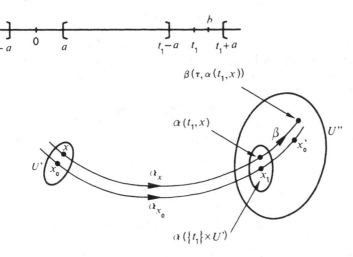

Figure 1.3.6

Thus for every $t \in t_1 + K$, that is $t - t_1 \in K$, the map $t \mapsto \beta(t - t_1, \alpha(t_1, x))$ defines an integral curve which goes through $\alpha(t_1, x)$ at time t_1; by local uniqueness, this says that α is defined for $t \in t_1 + K$ and $x \in U'$.

We have found an interval $I_1 = t_1 + K$ containing b (since $t_1 > b - a$) and an open set $U' \subset U$ containing x_0 and such that, for every $t \in I_1$ and $x \in U'$, the flow $\alpha(t, x)$ exists (since $I_1 \times U' \subset \mathcal{D}(f)$). Furthermore, $\alpha|_{I_1 \times U'}$ coincides with the local flow β, so it is of class C^p. We conclude that b satisfies property P, as do all points in $]b, t_1 + a[$. This contradicts the definition of b if $b < \sup(J(x_0))$, as we wished to show. $\qquad\square$

1.4. Time- and Parameter-Dependent Vector Fields

We consider from now on differential equations of the form $x' = f(t, x)$. We rephrase section 1.1:

1.4.1. Definition. Let $U \subset E$ be open. A *time-dependent vector field* on U is a map $f : J \times U \to E$, where $J \subset \mathbf{R}$ is an open interval containing 0.

An *integral curve* of f with initial condition x_0 is a function $\alpha : K \to U$, defined on an open interval $K \subset \mathbf{R}$ containing 0, and such that $\alpha(0) = x$ and $\alpha'(t) = f(t, \alpha(t))$ for every $t \in K$.

To show the existence of integral curves we will reduce to the case of a time-independent vector field, by considering a vector field defined on an open set of $\mathbf{R} \times E$.

Starting from $f : J \times U \to E$, we define a map $\overline{f} : J \times U \to \mathbf{R} \times E$ as follows:

1.4.2
$$\overline{f}(t, x) = \big(1, f(t, x)\big).$$

Then $\overline{f}$ is a vector field on the open subset $J \times U$ of $\mathbf{R} \times E$. Furthermore, $\overline{f}$ has the same differentiability class as f. If we assume f to be of class C^p, the new vector field $\overline{f}$ will have unique local flows of class C^p everywhere, by theorem 1.2.7.

Take $(s, x) \in J \times U$, and let $\overline{\alpha}$ be the integral curve of $\overline{f}$ with initial condition (s, x), that is, a map $\overline{\alpha} : \overline{J}(s, x) \to J \times U$ (where $\overline{J}(s, x)$ is an interval), whose components we call $\overline{\alpha}_1$ and $\overline{\alpha}_2$:

$$\overline{\alpha}\big(t; (s, x)\big) = \big(\overline{\alpha}_1(t; (s, x)), \alpha_2(t; (s, x))\big).$$

By definition of integral curve we have $\overline{\alpha}\big(0; (s, x)\big) = (s, x)$ and

$$\frac{d\overline{\alpha}}{dt} = \left(\frac{d\overline{\alpha}_1}{dt}, \frac{d\overline{\alpha}_2}{dt}\right) = \overline{f}(\overline{\alpha}_1, \overline{\alpha}_2) = \big(1, f(\overline{\alpha}_1, \overline{\alpha}_2)\big).$$

Thus $\overline{\alpha}_1$ and $\overline{\alpha}_2$ satisfy $\overline{\alpha}_1\big(0; (s, x)\big) = s$, $\overline{\alpha}_2\big(0; (s, x)\big) = x$,

$$\frac{d\overline{\alpha}_1}{dt} = 1 \quad \text{and} \quad \frac{d\overline{\alpha}_2}{dt} = f(\overline{\alpha}_1, \overline{\alpha}_2).$$

Since $\overline{\alpha}_1$ has values in the interval $J \subset \mathbf{R}$, we have

1.4.3
$$\overline{\alpha}_1\big(t; (s, x)\big) = t + s.$$

We then have

$$\frac{d\overline{\alpha}_2}{dt}\big(t; (s, x)\big) = f\big(t + s, \overline{\alpha}_2(t; (s, x))\big).$$

The solutions of $d\alpha/dt = f(t, \alpha)$ can thus be obtained for $s = 0$; they are the functions β defined by

1.4.4
$$\beta(t, x) = \overline{\alpha}_2\big(t; (0, x)\big).$$

This shows the existence and uniqueness of integral curves:

1.4.5. Theorem. *A C^p vector field $f : J \times U \to E$, where $J \subset \mathbf{R}$ is an interval and $U \subset E$ is open, has unique local flows of class C^p.* $\square$

Now consider another Banach space F, and an open set $V \subset F$. We can define a vector field on an open subset $U \subset E$ dependent on a parameter $\lambda \in F$ as a map f from $V \times U$ (or from $J \times V \times U$ if f is also time-dependent) into E. For each $\lambda \in V$ we consider the differential equation

$dx/dt = f(\lambda, x)$, and we would like to know how the local flows depend on λ. More generally, we could consider an equation of the form

$$\frac{dx}{dt} = f(t, \lambda, x).$$

We reduce again to the situation of section 1.2 by defining a vector field on $F \times E$ (or on $\mathbf{R} \times F \times E$). We define $\overline{f} : V \times U \to F \times E$ by setting

1.4.6 $$\overline{f}(\lambda, x) = \big(0, f(\lambda, x)\big).$$

If f is of class C^p, we have defined a new C^p field on $V \times U$; by 1.2.7, $\overline{f}$ admits a local flow $\overline{\alpha}$, with components $\overline{\alpha}_1$ and $\overline{\alpha}_2$, corresponding to an initial condition (λ, x).

This new vector field satisfies $\overline{\alpha}_1\big(0; (\lambda, x)\big) = \lambda$, $\overline{\alpha}_2\big(0; (\lambda, x)\big) = x$,

$$\frac{d\overline{\alpha}_1}{dt}\big(t; (\lambda, x)\big) = 0 \quad \text{and} \quad \frac{d\overline{\alpha}_2}{dt}\big(t; (\lambda, x)\big) = f(\overline{\alpha}_1, \overline{\alpha}_2).$$

It follows that $\overline{\alpha}_1\big(t; (\lambda, x)\big) = \lambda$, and consequently that

$$\frac{d\overline{\alpha}_2}{dt}\big(t; (\lambda, x)\big) = f\big(\lambda, \overline{\alpha}_2(t; (\lambda, x))\big).$$

Thus the map β defined by $\beta(t, x) = \overline{\alpha}_2\big(t; (\lambda, x)\big)$ satisfies

$$\frac{d\beta}{dt}(t, x) = f\big(\lambda; \beta(t, x)\big),$$

that is, β is an integral curve with initial condition x.

1.4.7. Theorem. *Let $V \subset F$ and $U \subset E$ be open sets. Any $f \in C^p(V \times U; U)$ admits a unique local flow $\beta \in C^p(J \times V \times U; U)$.* $\qquad\square$

In particular, this shows that local flows vary C^p-differentiably with the parameter.

1.5. Time-Dependent Vector Fields: Uniqueness And Global Flow

1.5.1. Let $f \in C^p(J \times U; E)$ be a time-dependent vector field. By theorem 1.4.5, proposition 1.3.1 holds for f without change. Thus we can still define for $x \in U$ a *maximal interval* $J(x) \subset J$ on which is defined the *maximal integral curve* $\big(J(x), \alpha_x\big)$ of f with initial condition x. We also set

$$\mathcal{D}(f) = \big\{(t, x) \in J \times U : t \in J(x)\big\}$$

and $\alpha : \mathcal{D}(f) \to U$ with $\alpha(t, x) = \alpha_x(t)$. Theorem 1.3.6 is still valid.

1.5.2. On the other hand, theorem 1.3.2 and formula 1.3.5 no longer hold. This is because, if $t \mapsto \alpha(t)$ is an integral curve, the maps $t \mapsto \alpha(t, s)$, for fixed s, no longer have to be integral curves. To fix 1.3.5, we notice that if $\alpha'(u) = f(u, \alpha(u))$ and we put $\beta(t) = \alpha(t + s)$, we have

$$\beta'(t) = \alpha'(t + s) = f(t + s, \alpha(t + s)) = f(t + s, \beta(t));$$

in other words, β is a solution for the new differential equation

$$\frac{dg}{dt} = g(t, x),$$

where $g(t, x) = f(t + s, x)$. Or again: β is an integral curve of the vector field

$$(t, x) \mapsto g(t, x) = f(t + s, x).$$

1.5.3. Notation. We denote by G_t^s the local map analogous to G_t (defined in 1.3.4) but relative to the vector field $(t, x) \mapsto f(t + s, x)$. In particular, $G_t^0 = G_t$.

Then we have, whenever all operations involved are defined:

1.5.4 $G_{t+s}^r = G_t^{r+s} \circ G_s^r.$

In particular, $G_{t+s} = G_t^s \circ G_s$.

To show this, suppose that, for $x \in U$, we have $r + s + t \in J(x)$; and let $\alpha = \alpha_x$ be the maximal integral curve for f with initial condition x. By 1.5.2 and uniqueness, the map $u \mapsto \alpha(r + u)$ is *the* integral curve for $(u, y) \mapsto f(u + r, y)$ having initial condition $x_1 = \alpha(r)$; thus, if we put $x_2 = \alpha(r+s)$, we have $x_2 = G_s^r x_1$ by definition. Similarly, if $x_3 = \alpha(r+s+t)$ we get $x_3 = G_{s+t}^r x_1$; but on the other hand $x_3 = G_t^{r+s} x_2$, as we wished to prove.

1.6. Cultural Digression

This chapter gives only a partial and somewhat biased treatment of differential equations, but it wouldn't be complete without brief mention of some important topics, even if they won't be necessary in later chapters. We omit proofs, except for theorem 1.6.6.

1.6.0. Comparison tests. If $dx/dt = f(t, x)$ is a differential equation admitting a Lipschitz constant for f and an upper bound for $\|f\|$, one can estimate an upper bound for a given solution of the equation, compare two solutions, or, better yet, compare two approximate solutions [Car71, equation II.1.5.3]. A particular case of this occurs in lemma 1.6.7 below.

1.6.1. Extending a solution. It can be shown that there are only two obstructions to extending a solution of a differential equation to the whole of **R**. Either the integral curve reaches the frontier of U (figure 1.2.4), or the solution is unbounded. A classical example of the latter situation is the equation $x' = -x^2$, whose solution with initial condition x is

$$t \mapsto \frac{x}{tx+1},$$

so that $J(x) =]-1/x, +\infty[$ for $x > 0$. For a precise result, see [Die69, vol. I, 10.5.5] or [Lan68, p. 382, th. 5] (the proof uses 1.6.0). The proof of 1.6.6 gives a particular case of extensions.

1.6.2. Derivatives. When one shows that solutions are differentiable, whether with respect to initial conditions or to parameters (1.2.7 and 1.4.7), one proves at the same time a formula expressing the derivative as a solution of a new differential equation, which has the advantage of being linear (cf. 1.6.4). See [Lan69, p. 135], [Die69, vol. I, 10.7.3.1 and 10.8.4.1], or [Car71, II.3.4 and II.3.6].

1.6.3. Linear equations with constant coefficients. An equation of the form $x^{(n)} + a_1 x^{(n-1)} + \cdots + a_{n-1} x' + a_n x = 0$, with $a_1, \ldots, a_n \in$ **R**, can be explicitly solved. The solutions form a vector space having a basis whose elements are linear combinations of functions of the form $x^{k_i} e^{\lambda_i x}$, where the $k_i \in$ **N** and the λ_i are the roots of the algebraic equation $\xi^n + a_1 \xi^{n-1} + \cdots + a_{n-1} \xi + a_n = 0$. See [Car71, II.2.9].

1.6.4. Linear differential equations. A differential equation $dx/dt = f(t, x)$ is called *linear* if it is of the form

1.6.5 $$\frac{dx}{dt} = A(t)x + b(t),$$

where $A \in C^0\big(J; L(E; E)\big)$ and $b \in C^0(J; E)$, and $J \subset$ **R** is an interval. Thus $U = E$. The important result about linear equations is the following:

1.6.6. Theorem. *For an equation of the form* 1.6.5, *we have* $D(f) = J \times E$; *in other words,* $J(x) = J$ *for every* $x \in E$.

1.6.7. Lemma. *Consider* $w \in C^0\big([t_0, b[; \mathbf{R}_+\big)$ *and assume there exist constants* C *and* K *such that* $g(t) \le C + K \int_{t_0}^t g(u)\, du$ *for every* $t \in [t_0, b[$. *Then* $g(t) \le C e^{k(t-t_0)}$ *for every* t; *in particular,* g *is bounded on* $[t_0, b[$.

Proof. Fix $b' < b$ for the time being, and let B be an upper bound for g in $[t_0, b']$. From the hypothesis of the lemma we deduce, by applying induction and integrating from t_0 to t, that

$$g(t) \le C \left(1 + \frac{K(t-t_0)}{1!} + \cdots + \frac{K^{n-1}(t-t_0)^{n-1}}{(n-1)!} \right) + \frac{BK^n(t-t_0)^n}{n!}$$

for every $t \in [t_0, b']$ and every positive integer n. As n grows we get $g(t) \le Ce^{K(t-t_0)}$ for every $t \in [b_0, b']$. Since this takes place for every $b' < b$, the lemma is proved. $\square$

1.6.8. Lemma. *Assume* $[a, b] \subset J$. *Any solution* α *of 1.6.5 defined on* $]a, b[$ *is bounded on* $]a, b[$.

Proof. Fix $t_0 \in]a, b[$. By 1.6.5, we have

$$\alpha(t) - \alpha(t_0) = \int_{t_0}^{t} \big(A(u)\alpha(u) + b(u)\big)\, du,$$

whence

1.6.9 $$\|\alpha(t) - \alpha(t_0)\| \le k \left| \int_{t_0}^{t} \|\alpha(u)\|\, du \right| + h|t - t_0|,$$

where k and h are upper bounds for $\|A\|$ and $\|b\|$, respectively, over $[a, b]$. Now apply lemma 1.6.7. $\square$

Proof of 1.6.6. Take $x \in E$ and $t_0 \in J$, and let $J(x)$ be the maximal interval of definition for the solution α of 1.6.5 such that $\alpha(t_0) = x$ (cf. 1.2.2.1). Say $J(x) =]t_1, t_2[$; we must show that $t_1 \notin J$ and $t_2 \notin J$. Let's work with t_2.

Let k and h be upper bounds for $\|A\|$ and $\|b\|$ on $[t_0, t_2]$, and j an upper bound for $\|\alpha\|$ on $[t_0, t_2]$ (lemma 1.6.8). Formula 1.6.9 implies that

$$\|\alpha(t) - \alpha(s)\| \le (kj + h)|t - s|$$

for every $s, t \in [t_0, t_2]$. This shows that the $\alpha(t)$ satisfy Cauchy's criterion, and, since E is complete, implies that there exists $x_1 \in E$ such that $x_1 = \lim_{t \to t_2} \alpha(t)$. By local existence we can find a solution β of 1.6.5 defined on an open interval $L \supset t_2$ and satisfying $\beta(t_2) = x_1$. But we have $\alpha(t) = \int_{t_0}^{t} f\big(u, \alpha(u)\big)\, du$ for every $u < t_2$, so, by continuity, $\alpha(t_2) = \int_{t_0}^{t_2} f\big(u, \alpha(u)\big)\, du$. Thus α is a solution for 1.6.5 on $[t_0, t_2]$, which by uniqueness coincides with β on $L \cap [t_0, t_2]$. This means that 1.6.5 has a solution on $[t_0, t_2] \cup L$, obtained by combining α and β. This contradicts the definition of $t_2 = \sup\big(J(x)\big)$. $\square$

Theorem 1.6.6 is but the first stepping-stone in the theory of linear equations. For more, see [Car71, II.2.2 to II.2.9].

CHAPTER 2

Differentiable Manifolds

This chapter forms the backbone of this book, and it is the one with the greatest number of exercises. We first define and give examples of submanifolds of $\mathbf{R}^n$ (section 2.1), the right concrete objects for the study of differential geometry. Next we define parametrizations of submanifolds; coordinate changes from one parametrization to another are the essential ingredients in the definition of abstract manifolds (section 2.2), which are the right objects for the study of abstract (and sometimes even concrete) differential geometry.

We next study mathematical objects associated with manifolds: morphisms, tangent spaces, tangent bundle, submanifolds, immersions, submersions, embeddings. Both as illustrations and for future need, we discuss in detail two general examples: covering spaces (section 2.4) and the normal bundle of a submanifold of a Euclidean space (section 2.7). Another simple result is also proved: the diffeomorphism group of a connected manifold is transitive (2.3.7).

We also give more particular examples (2.1.6.2 and 2.4.12): spheres, real projective spaces and tori. They will be "test objects" later on, when we introduce de Rham groups (sections 5.7 and 5.8) and volume (6.5.5).

2.1. Submanifolds of $\mathbf{R}^n$

For $d \leq n$ the *canonical inclusion* $\mathbf{R}^d \subset \mathbf{R}^n$ is defined as the map i :
$(x_1, \ldots, x_d) \mapsto (x_1, \ldots, x_d, 0, \ldots, 0)$. Similarly, the *canonical isomorphism*
is $\mathbf{R}^n = \mathbf{R}^d \times \mathbf{R}^{n-d}$.

2.1.1. Definition. Let V be a subset of $\mathbf{R}^n$. We say that V is a d-
dimensional C^p *submanifold* of $\mathbf{R}^n$ if, for every $x \in V$, there exists an open
neighborhood $U \subset \mathbf{R}^n$ of x and a map $f : U \to \mathbf{R}^n$ such that $f(U) \subset \mathbf{R}^n$ is
open, f is a C^p diffeomorphism onto its image and $f(U \cap V) = f(U) \cap \mathbf{R}^n$.
The *codimension* of V is $n - d$.

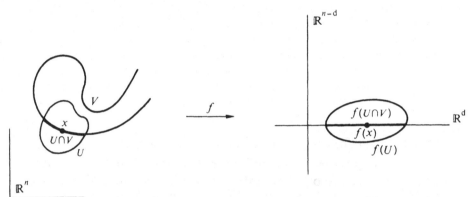

Figure 2.1.1

2.1.2. Theorem. *Let V be a subset of $\mathbf{R}^n$. The following properties are
equivalent:*

(i)   *V is a d-dimensional C^p submanifold of $\mathbf{R}^n$.*

(ii) *For every $x \in V$ there exists an open neighborhood $U \subset \mathbf{R}^n$ of x and
      C^p functions $f_i : U \to \mathbf{R}$ $(i = 1, \ldots, n-d)$ such that the linear forms
      $f_i'(x)$ are linearly independent and*

$$V \cap U = \bigcap_{i=1}^{n-d} f_i^{-1}(0).$$

(iii) *For every $x \in V$ there exists an open neighborhood $U \subset \mathbf{R}^n$ of x and
      a submersion $f : U \to \mathbf{R}^{n-d}$ such that $U \cap V = f^{-1}(0)$.*

(iv) *For every $x \in V$ there exists an open neighborhood $U \subset \mathbf{R}^n$ of $x =
      (\xi_1, \ldots, \xi_n)$, an open neighborhood U' of $x = (\xi_1, \ldots, \xi_d)$ in $\mathbf{R}^d$ and
      C^p functions $h_i : U' \to \mathbf{R}$ $(i = 1, \ldots, n-d)$ such that, possibly after a
      permutation of coordinates, the intersection $V \cap U$ is the graph of the
      map $(h_1, \ldots, h_{n-d}) : U' \to \mathbf{R}^{n-d}$ (under the canonical isomorphism
      $\mathbf{R}^d \times \mathbf{R}^{n-d} = \mathbf{R}^n$).*

(v) *For every $x \in V$ there exists an open neighborhood $U \subset \mathbf{R}^n$ of x, an open neighborhood $\Omega \subset \mathbf{R}^d$ of 0 and a map $g : \Omega \to \mathbf{R}^n$ such that $g'(0)$ is injective, $g(0) = x$, and g is a C^p homeomorphism between Ω and $V \cap U$ (with the topology induced from $\mathbf{R}^n$).*

2.1.3. Remarks. Equivalence between (ii) and (iii) is easy: clearly (iii) implies (ii), and if (ii) holds the map $f : U \to \mathbf{R}^{n-d}$ with components f_i is a submersion at x, and remains a submersion on a neighborhood of x, since the determinant is a continuous map. By restricting U if necessary we obtain (ii).

Submanifolds are a generalization of the notion of plane curves (case $n = 2$ and $d = 1$), thought of as graphs (iv) or as images of functions (v). Space curves, considered as intersections of two surfaces, also fit this picture ((ii) and (iii) with $n = 3$ and $d = 1$).

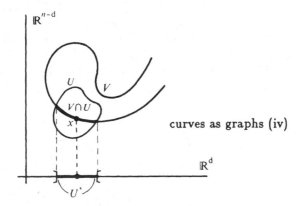

curves as graphs (iv)

Figure 2.1.3.1

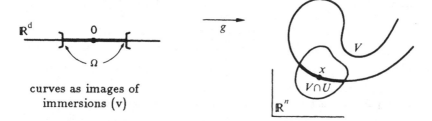

curves as images of immersions (v)

Figure 2.1.3.2

2.1.3.1. In (ii) we think of a submanifold as an intersection of hypersurfaces defined by local equations (cf. 2.1.6.5), in (iii) as the zero-set of a submersion, in (iv) as a graph, and in (v) as the image of an immersion (because if $g'(0)$ is injective, so is $g'(x)$ for x close enough to zero). All of these are *local* descriptions.

2.1.4. Proof. We already know that (ii) $\Leftrightarrow$ (iii); we now show that (iii) $\Rightarrow$ (i) $\Rightarrow$ (v) $\Rightarrow$ (iv) $\Rightarrow$ (ii).

(iii) $\Rightarrow$ (i). This is just theorem 0.2.26.

(i) $\Rightarrow$ (v). Apply definition 2.1.1, using a translation, if necessary, to make $f(x) = 0$. Take $\Omega = f(U \cap V)$ and $g = f^{-1} \circ i$, where $i : \mathbf{R}^d \to \mathbf{R}^n$ is the canonical inclusion. By the definition of induced topology, g is a homeomorphism.

(v) $\Rightarrow$ (iv). After permuting indices if necessary, we can assume that $g'(0)(\mathbf{R}^d) \cap \mathbf{R}^{n-d} = 0$, where $\mathbf{R}^n = \mathbf{R}^d \times \mathbf{R}^{n-d}$. Let $p : \mathbf{R}^n \to \mathbf{R}^d$ be the projection onto the first factor. From $g'(0)(\mathbf{R}^d) \cap \mathbf{R}^{n-d} = 0$ we deduce that $(p \circ g)'(0)(\mathbf{R}^d) = \mathbf{R}^d$; in other words, $p \circ g$ is regular at 0. By theorem 0.2.22 there exists $\Omega' \in O_0(\Omega)$ such that $p \circ g$ is a diffeormorphism between Ω' and $U' = p(g(\Omega')) \in O(\mathbf{R}^d)$. Thus (iv) is satisfied if we take this U' and $h_1, \dots, h_{n-d}$ equal to the $n - d$ last coordinate functions of the map $h = g \circ (p \circ g)^{-1} \in C^p(U', \mathbf{R}^n)$. In fact, $h(U') = g(\Omega')$ by assumption, so, by the definition of induced topology, there exists $U'' \in O_U(\mathbf{R}^n)$ such that $g(\Omega') = h(U') = U'' \cap V$. Thus $U'' \cap V$ is the graph of $(h_1, \dots, h_{n-d}) = h$, as we wished to show.

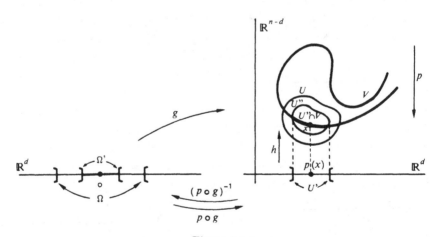

Figure 2.1.4

(iv) $\Rightarrow$ (ii). Just set

$$f_i(x_1, \dots, x_n) = h_i(x_1, \dots, x_d) - x_{i+d}$$

for $i = 1, \dots, n - d$.   □

2.1.5. Remark. In (v) the condition that g be a homeomorphism and an injection at 0 is essential. For example, if $\Omega = \mathbf{R}$ and $g(t) = (t^2, t^3) \in \mathbf{R}^2$,

Figure 2.1.5.1

the map g is C^∞ and a homeomorphism from $\mathbf{R}$ onto $g(\mathbf{R})$, but $g(\mathbf{R})$ is not a submanifold of $\mathbf{R}^2$, since definition 2.1.1 is not satisfied at $x \in (0,0) \in \mathbf{R}^2$ (see exercise 2.8.1); this is because $g'(0) = 0$. If we take $\Omega = \mathbf{R}$ and $g \in C^\infty(\mathbf{R}; \mathbf{R}^2)$ as in figure 2.1.5.1, where the arrows indicate that $\lim_{t\to+\infty} g(t) = \lim_{t\to-\infty} g(t) = g(0)$, definition 2.1.1 is not satisfied at $x = g(0)$ since, for every open set $V \subset \mathbf{R}^2$ containing $g(0)$, the inverse image $g^{-1}(V)$ contains not only an interval around 0 but also two intervals of the form $]-\infty, b[$ and $]c, +\infty[$ (see exercise 2.8.1). See also exercise 2.8.4 for another very important counterexample.

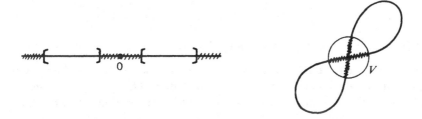

Figure 2.1.5.2

2.1.6. Examples of submanifolds of $\mathbf{R}^n$

2.1.6.1. Proposition. *Let V be a d-dimensional C^p submanifold of $\mathbf{R}^n$ and W an e-dimensional C^p submanifold of $\mathbf{R}^m$. Then $V \times W$ is a $(d + e)$-dimensional C^p submanifold of $\mathbf{R}^{n+m}$.*

Proof. Theorem 2.1.2(ii), applied to $x \in V$ and $y \in W$, gives $n + m - (d + e)$ functions f_i defined on an open neighborhood $U = U_1 \times U_2 \subset \mathbf{R}^{n+m}$ of (x, y) and satisfying (ii) for $V \times W$. See also exercise 2.8.3. □

2.1.6.2. The sphere. The *sphere* $S^d = \{x \in \mathbf{R}^{d+1} : \|x\| = 1\}$ is a compact, d-dimensional, C^∞ submanifold of $\mathbf{R}^{d+1}$. (We call S^1 a *circle*; S^0 is equal to two points.)

To see this, write

$$S^d = \{x = (\xi_1, \ldots, \xi_{d+1}) : \xi_1^2 + \cdots + \xi_{d+1}^2 - 1 = 0\}.$$

Thus S^d is the zero-set of the map $f(\xi_1, \ldots, \xi_{d+1}) = \xi_1^2 + \cdots + \xi_{d+1}^2 - 1$, which is C^∞; furthermore, since

$$f'(x) = (2\xi_1, \ldots, 2\xi_{d+1}),$$

f has non-zero derivative whenever $x = (\xi_1, \ldots, \xi_{d+1})$ is on S^d. Now apply 2.1.2(ii).

Without using coordinates we can also write $f = \|\cdot\|^2 - 1$, and the derivative at x is $f'(x) = 2(x \mid \cdot)$ (cf. 0.1.15.1), which is only zero when $x = 0$.   □

2.1.6.3. The torus, first recipe (see also 2.4.12.1). Let $d \geq 1$ be an integer and $S^1(d^{-1/2}) \subset \mathbf{R}^2$ the circle of center 0 and radius $d^{-1/2}$. Set

$$T^d = \left(S^1(d^{-1/2})\right)^d = \underbrace{S^1(d^{-1/2}) \times \cdots \times S^1(d^{-1/2})}_{d \text{ times}} \subset \mathbf{R}^2 \times \cdots \times \mathbf{R}^2 = \mathbf{R}^{2d}.$$

By 2.1.6.1 and 2.1.6.2, T^d is a d-dimensional C^∞ submanifold of $\mathbf{R}^{2d}$; we call it a *d-torus*. We can also say, in the language of 2.1.2(ii), that T^d is defined by the equations

$$x_1^2 + x_2^2 = \frac{1}{d}, \quad x_3^2 + x_4^2 = \frac{1}{d}, \quad \ldots, \quad x_{2d-1}^2 + x_{2d}^2 = \frac{1}{d}.$$

Notice that T^d is compact and that $T^d \subset S^{2d-1}$.

2.1.6.4. The orthogonal group $O(n)$, defined as the set of invertible $n \times n$ square matrices A such that ${}^t A = A^{-1}$ [Dix68, 35.11.1], is a C^∞ submanifold of $\subset \mathbf{R}^{n^2}$ of dimension $n(n-1)/2$. To see this, show that the map $A \mapsto {}^t A A$ from $\mathbf{R}^{n^2}$ into the set of symmetric matrices, which can be identified with $\mathbf{R}^{n(n+1)/2}$, is surjective (exercise 2.8.10); then apply 2.1.2(iii).

2.1.6.5. Definition. A C^p *hypersurface* in $\mathbf{R}^{d+1}$ is a codimension-one (that is, d-dimensional) C^p submanifold of $\mathbf{R}^{d+1}$.

Theorem 2.1.2(ii) shows that a hypersurface divides $\mathbf{R}^{d+1}$ locally into two regions, namely, $f^{-1}(\mathbf{R}_+^*)$ and $f^{-1}(\mathbf{R}_-^*)$, where $f = f_1$ is a function satisfying 2.1.2(ii). In general it is not possible to distinguish between the two regions, but see section 6.4. One problem of the type "passing from local to global" consists in knowing whether a compact hypersurface separates $\mathbf{R}^{d+1}$ into two parts: see 9.2 and 3.5.2.

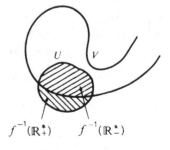

Figure 2.1.6.5

2.1.6.6. Codimension-zero submanifolds are the same as open sets in $\mathbf{R}^d$.

2.1.6.7. Zero-dimensional submanifolds are sets of isolated points in $\mathbf{R}^d$.

2.1.6.8. Veronese's surface. The image of the two-sphere $S^2 = \{(x, y, z) : x^2 + y^2 + z^2 = 1\}$ under the map

2.1.6.9 $\qquad f : (x, y, z) \mapsto (x^2, y^2, z^2, \sqrt{2}yz, \sqrt{2}zx, \sqrt{2}xy)$

is a two-dimensional submanifold of $\mathbf{R}^6$, called Veronese's surface (exercise 2.8.5). See also 2.4.12.2 and 2.6.13.2.

2.1.6.10. See exercises 2.8.1, 2.8.3, 2.8.4 and 2.8.5 for other examples or counterexamples of submanifolds.

2.1.6.11. One-dimensional submanifolds, or curves, will be studied in detail in chapters 8 and 9, especially the cases $d = 1$, $n = 2$ (section 8.5 and the whole of chapter 9) and $d = 1$, $n = 3$ (section 8.6). Chapters 10 and 11 are devoted to the case $d = 2$, $n = 3$ (surfaces in $\mathbf{R}^3$).

2.1.7. Coordinate changes. Theorem 2.1.9 below is the essential link between submanifolds and abstract manifolds.

2.1.8. Definition. A *coordinate system* on a d-dimensional C^p submanifold V of $\mathbf{R}^n$ is a pair (Ω, g) consisting of an open set $\Omega \in O(\mathbf{R}^d)$ and an immersion $g \in C^p(\Omega; \mathbf{R}^n)$ such that $g(\Omega)$ is open in V and g induces a homeomorphism between Ω and $g(\Omega)$.

2.1.9. Theorem. *Let V be a d-dimensional C^p submanifold of $\mathbf{R}^n$ and U_1, U_2 open subsets of $\mathbf{R}^n$ containing a point $x \in V$. If (Ω_1, g_1) and (Ω_2, g_2) are coordinate systems on V such that $g_1(\Omega_1) = V \cap U_1$ and $g_2(\Omega_2) = V \cap U_2$, respectively, the coordinate change $g_2^{-1} \circ g_1$ belongs to $C^p(\Omega_1 \cap g_1^{-1}(U_2); \mathbf{R}^d)$.*

We can express this by saying that g_1 and g_2 are C^p-compatible.

Proof. By definition 2.1.1 there exist an open neighborhood S of x in $\mathbf{R}^n$ and a diffeormorphism $f : S \to f(S)$ such that $f(S \cap V) = f(S) \cap \mathbf{R}^d$. Since $g_1(\Omega_1) = V \cap U_1$ and S is open, we can find a smaller open neighborhood $W_1 \subset \Omega_1$ of $g_1^{-1}(x)$ such that $g_1(W_1) \subset V \cap S$, and consequently that $(f \circ g_1)(W_1) \subset \mathbf{R}^d$. Similarly, there exists a subneighborhood $W_2 \subset \Omega_2$ of $g_2^{-1}(x)$ such that $(f \circ g_2)(W_2) \subset \mathbf{R}^d$. On $W_1 \cap (g_2^{-1} \circ g_1)(W_2)$ we have

$$g_2^{-1} \circ g_1 = g_2^{-1} \circ f^{-1} \circ f \circ g_1 = (f \circ g_2)^{-1} \circ (f \circ g_1),$$

and $f \circ g_1$ and $(f \circ g_2)^{-1}$ are of class C^p (the latter by 0.2.22, since $f \circ g_2$ is bijective and $(f \circ g_2)' = f' \circ g_2'$ is an isomorphism). This shows that $g_2^{-1} \circ g_1$ is C^p on a neighborhood of $g_1^{-1}(x)$; by making x range over $g_1(\Omega_1) \cap g_2(\Omega_2)$ we conclude that $g_2^{-1} \circ g_1$ is of class C^p where defined. $\qquad\square$

For concrete examples of coordinate changes, see exercises 2.8.2 and 2.8.7.

2.2. Abstract Manifolds

2.2.1. Definition. Let X be a set and $p \geq 1$ an integer. A *d-dimensional atlas* of class C^p on X is a set of pairs $\{(U_i, \phi_i)\}_{i \in I}$ satisfying the following axioms:

(AT1) Each U_i is a subset of X and the U_i cover X.

(AT2) Each ϕ_i is a bijection from U_i into an open subset $\phi_i(U_i)$ of $\mathbf{R}^n$, and $\phi_i(U_i \cap U_j) \subset \mathbf{R}^n$ is open for every i and j.

(AT3) For every pair (i, j) the map $\phi_j \circ \phi_i^{-1} : \phi_i(U_i \cap U_j) \rightarrow \phi_j(U_i \cap U_j)$ is a C^p diffeomorphism.

2.2.2. Remarks. In condition (AT3) it should be noticed that ϕ_i^{-1} is defined on $\phi_i(U_i)$, but $\phi_j \circ \phi_i^{-1}$ will only be defined at $x \in \phi_i(U_i)$ if $\phi_i^{-1}(x) \in U_j$, that is, if $x \in \phi_i(U_j)$. Thus the domain of $\phi_j \circ \phi_i^{-1}$ is $\phi_i(U_i \cap U_j)$, and its range is $\phi_j(U_i \cap U_j)$.

One can replace $\mathbf{R}^n$ in the definition by a (perhaps infinite-dimensional) Banach space E. This generalizes the notion of a manifold, but it also complicates it, since linear maps between Banach spaces are not necessarily continuous. See [Lan69, p. 421].

2.2.3. Definition. The pairs (U_i, ϕ_i) are called the *charts* of the atlas $\{(U_i, \phi_i)\}$. A chart *at* or *around* $x \in X$ is one whose domain contains x, and a chart *centered at* x is one mapping x to the origin in $\mathbf{R}^d$. The *local coordinates* associated with a chart (U_i, ϕ_i) are the functions $\phi_{i,k} : U_i \rightarrow \mathbf{R}$ ($1 \leq k \leq d$) such that $\phi_i(x) = (\phi_{i,1}(x), \ldots, \phi_{i,d}(x))$.

2.2.4. Definition. Let $\{(U_i, \phi_i)\}_{i \in I}$ be an atlas on X, let U be a subset of X and $\phi : U \rightarrow \mathbf{R}^d$ a bijection onto an open subset of $\mathbf{R}^d$. The pair (U, ϕ) is said to be a chart *compatible* with the atals $\{(U_i, \phi_i)\}_{i \in I}$ if the union $\{(U, \phi)\} \cup \{(U_i, \phi_i)\}_{i \in I}$ is still an atlas. Two atlases (of same dimension and differentiability class) are *compatible* if their union is still an atlas.

In order for (U, ϕ) to be compatible with an atlas $\{(U_i, \phi_i)\}_{i \in I}$ it is necessary that each $\phi(U \cap U_i)$ and $\phi_i(U \cap U_i)$ be an open subset of $\mathbf{R}^d$ and that the maps $\phi \circ \phi_i^{-1}$ and $\phi^{-1} \circ \phi_i$ be of class C^p on their domains of definition.

Compatibility is an equivalence relation. Thus we arrive at the definition of a manifold:

2.2.5. Definition. A C^p *differentiable structure* ($p \geq 1$) on a set X is an equivalence class of d-dimensional atlases of class C^p on X. A *d-dimensional manifold* of class C^p is a set X endowed with a C^p differentiable structure. A *chart* on X is any chart belonging to any atlas in the differentiable structure of X.

One also says that X is a *differentiable manifold*, as opposed to *topological*, or C^0, *manifolds*, which shall not concern us.

In practice one defines a manifold X by means of a single atlas, whose equivalence class then determines the differentiable structure.

We will now define on a manifold X a canonical topology, one that only depends on the differentiable structure. We could also have started from a topological space X and required that the domains U_i of the charts be open in X and that the maps $\phi_i : U_i \to \phi_i(U_i)$ be homeomorphisms.

2.2.6. Theorem. *Let X be a d-dimensional manifold of class C^p. The set of unions of domains of charts of X forms a topology on X. This topology is called canonical.*

Proof. Let O be the set thus defined; we have to show that O satisfies the two axioms for a topology:

(O1) every union of elements of O is an element of O; and

(O2) every finite intersection of elements of O is an element of O.

Clearly (O1) is satisfied, since a set is in O if and only if it is a union of domains of charts. To show (O2), we just have to consider the intersection of two elements of O. Let them be $A = \bigcup_{j \in J} U_j$ and $B = \bigcup_{k \in K} U_k$; then

$$A \cap B = \bigcup_{(j,k) \in J \times K} (U_j \cap U_k).$$

We have to show that each intersection $U_j \cap U_k$ can be taken as the domain of a chart compatible with the differentiable structure of X. Let (U_j, ϕ_j) be a chart in the differentiable structure of X, and set $\psi = \phi_j|_{U_j \cap U_k}$; we claim that $(U_j \cap U_k, \psi)$ is the desired chart. Clearly $\psi(U_j \cap U_k) = \phi_j(U_j \cap U_k)$ is open in $\mathbf{R}^d$. If (U, ϕ) is any chart in the differentiable structure of X, the composition $\phi \circ \phi_j^{-1}$ is a C^p diffeomorphism between $\phi_j(U \cap U_j)$ and $\phi(U \cap U_j)$, so

$$\phi \circ \psi^{-1} = \phi \circ \phi_j^{-1}|_{\phi_j(U_j \cap U_k \cap U)}$$

is a C^p diffeomorphism between $\psi(U \cap (U_j \cap U_k))$ and $\phi(U \cap (U_j \cap U_k))$. Similarly, $\psi \circ \phi^{-1}$ is a C^p diffeomorphism between $\phi(U \cap (U_j \cap U_k))$ and $\psi(U \cap (U_j \cap U_k))$. This proves compatibility. $\square$

Sometimes it is desirable to characterize the open sets in the canonical topology of X in terms of a single atlas.

2.2.7. Theorem. *Let $\{(V_i, \phi_i)\}_{i \in I}$ be an atlas on a d-dimensional manifold X. A subset U of X is open if and only if $\phi_i(U \cap V_i) \subset \mathbf{R}^d$ is open for every chart (V_i, ϕ_i).*

Proof. We first show that the set $\mathcal{U}$ of subsets of X satisfying the condition of the statement is a topology; then we show that this topology is the same as the one given by the set O in 2.2.6.

Let $(U_\alpha)_{\alpha \in A}$ be a family of elements of $\mathcal{U}$, and consider $U = \bigcup_{\alpha \in A} U_\alpha$. Let (V_i, ϕ_i) be a chart in the atlas. Since $U \cap V_i = \bigcup_{\alpha \in A}(U_\alpha \cap V_i)$ and ϕ_i is a bijection onto its image, we have

$$\phi_i(U \cap V_i) = \phi_i \left(\bigcup_{\alpha \in A} (U_\alpha \cap V_i) \right) = \bigcup_{\alpha \in A} \phi_i(U_\alpha \cap V_i).$$

Thus $\phi_i(U \cap V_i)$ is a union of open sets in $\mathbf{R}^d$, showing that $U \in \mathcal{U}$. A similar reasoning shows that a finite intersection of elements of $\mathcal{U}$ is also in $\mathcal{U}$. It follows that $\mathcal{U}$ is a topology.

To prove that $\mathcal{U} \subset O$, pick $U \in \mathcal{U}$. Since the V_i cover X, we have $U = \bigcup_{i \in I}(U \cap V_i)$; if we show that each $U \cap V_i$ is the domain of a chart compatible with $(V_i, \phi_i)_{i \in I}$ it will follow that $U \in O$.

The desired chart is (W_i, ψ_i), where $W_i = U \cap V_i$ and $\psi_i = \phi_i|_{U \cap V_i}$. First we must show that for any j the domain of the coordinate change $\phi_j \circ \psi_i^{-1}$ is open. We have

$$\psi_i(W_i \cap V_j) = \psi_i(U \cap V_i \cap V_j) = \phi_i(U \cap V_i \cap V_j) = \phi_i(U \cap V_i) \cap \phi_i(V_i \cap V_j),$$

since ϕ_i is bijective. But $\phi_i(U \cap V_i)$ is open because $U \in \mathcal{U}$, and $\phi_i(V_i \cap V_j)$ is open by (AT2) applied to (V_i, ϕ_i) and (V_j, ϕ_j), so we're done. Next we must show that $\phi_j \circ \psi_i^{-1}$ is a C^p diffeomorphism onto its image; but this is clear because $\phi_j \circ \psi_i^{-1}$ is the restriction of $\phi_j \circ \phi_i^{-1}$ to an open set. The proof that $\psi_i^{-1} \circ \phi_j$ is a C^p diffeomorphism is similar.

There remains to prove that $O \subset \mathcal{U}$. This is certainly true if $\mathcal{U}$ is the topology associated with the *maximal* atlas on X, that is, the atlas containing all the charts in the differentiable structure of X. Indeed, any $U \in O$ can be written as a union $\bigcup_{i \in I} U_i$ of domains of charts on X. If (U_k, ϕ_k) is any other chart on X, we have

$$\phi_k(U \cap U_k) = \phi_k \left(\left(\bigcup_{i \in I} U_i \right) \cap U_k \right) = \phi_k \left(\bigcup_{i \in I}(U_i \cap U_k) \right) = \bigcup_{i \in I} \phi_k(U_i \cap U_k),$$

because ϕ_k is bijective. By (AT2) this means that $\phi_k(U \cap U_k)$ is a union of open subsets of $\mathbf{R}^d$, so $U \in \mathcal{U}$ by the definition of $\mathcal{U}$.

The proof of 2.2.7 will be completed if we show the following lemma:

2.2.8. Lemma. *Let $(V_i, \phi_i)_{i \in I}$ and $(W_j, \psi_j)_{j \in J}$ be two atlases on a manifold X. A subset $U \subset X$ is open in the topology $\mathcal{U}$ defined by $(V_i, \phi_i)_{i \in I}$ if and only if it is open in the topology $\mathcal{U}'$ defined by $(W_j, \psi_j)_{j \in J}$.*

Proof. Take $U \in \mathcal{U}$ and let (W_j, ψ_j) be a chart in the second atlas. We must show that $\psi_j(U \cap W_j)$ is open in $\mathbf{R}^d$, in other words, that it is a neighborhood of each of its points. Take $y \in \psi_j(U \cap W_j)$ and x such that $y = \psi_j(x)$. Since X is covered by the V_i, there is i_0 such that $x \in V_{i_0}$. Then $y \in \psi_j(U \cap V_{i_0} \cap W_j)$. But the chart $(U \cap V_{i_0}, \phi_{i_0}|_{U \cap V_{i_0}})$ is compatible with the first atlas (by the proof of $\mathcal{U} \subset O$ in 2.2.7), and in particular $\psi_j(U \cap V_{i_0} \cap W_j)$ is open in $\mathbf{R}^d$ and contained in $\psi_j(U \cap W_j)$. $\qquad\square$

Theorem 2.2.7 provides another way of defining the topology of a manifold.

2.2.9. Theorem. *For every chart (U, ϕ) on a manifold X, considered with its canonical topology, the map $\phi : U \to \phi(U)$ is a homeomorphism.*

Proof. We know that ϕ is a bijection. To prove ϕ is continuous, we take an open set $V \subset \mathbf{R}^d$ and show that $(\phi^{-1}(V), \psi)$, where $\psi = \phi|_{\phi^{-1}(V)}$, is a chart on X, and consequently that $\phi^{-1}(V)$ is open in the canonical topology.

Set $\psi = \phi|_{\phi^{-1}(V)}$, and let (V_i, ϕ_i) be any chart on X. Since $\phi : \psi^{-1}(V) \to V \cap \phi(U)$ is a bijection, we have

$$V \cap \phi(U) \cap \phi(U_i) = V \cap \phi(U \cap U_i),$$

and $\phi_i \circ \psi^{-1}$ is the restriction of $\phi_i \circ \phi^{-1}$ to an open set. It follows that $\phi_i \circ \psi^{-1}$ is a C^p diffeomorphism between $V \cap \phi(U \cap U_i) = \psi(\phi^{-1}(V) \cap U_i)$ and $\phi_i(\phi^{-1}(V) \cap U_i)$.

The map $\psi \circ \phi_i^{-1}$ is defined on $\phi_i(U_i \cap \phi^{-1}(V))$, and equals the restriction of $\phi \circ \phi_i^{-1}$ to

$$\phi_i(U_i \cap \phi^{-1}(V)) = \phi_i(U_i \cap U \cap \phi^{-1}(V)).$$

Thus the image of $\psi \circ \phi_i^{-1}$ is $\phi(U_i \cap U) \cap V$, which is open in $\mathbf{R}^d$ by (AT2). By (AT3) it follows that $\phi_i(U_i \cap \phi^{-1}(V))$ is open in $\mathbf{R}^d$ and that $\psi \circ \phi_i^{-1}$ is a C^p diffeomorphism.

There remains to show that $\phi^{-1} : \phi(U) \to U$ is continuous, that is, that for every open set $S \subset X$ the image $\phi(S)$ is open. Since ϕ is only defined on U we actually have $\phi(S) = \phi(S \cap U)$, and, if $(U_i, \phi_i)_{i \in I}$ is an atlas for X, we have $S \cap U = S \cap U \cap \bigcup_{i \in I} U_i$. Thus

$$\phi(S \cap U) = \phi\left(\bigcup_{i \in I} (S \cap U \cap U_i) \right) = \bigcup_{i \in I} \phi(S \cap U \cap U_i),$$

since ϕ is bijective. But S is open in X, and $(U \cap U_i, \phi|_{U \cap U_i})$ is a chart compatible with the atlas, so it follows from 2.2.7 that $\phi(S \cap U \cap U_i)$ is open in $\mathbf{R}^d$, and so is $\phi(S \cap U)$. $\qquad\square$

2.2.10. Examples of manifolds. To begin with, $\mathbf{R}^n$ is canonically a manifold. It has an atlas with a single chart, $(\mathbf{R}^n, \mathrm{Id}_{\mathbf{R}^n})$. Whenever we discuss $\mathbf{R}^n$ we will be implicitly considering it with this differentiable structure.

2.2.10.1. Theorem. *Any submanifold of $\mathbf{R}^n$ has a canonical differentiable structure, and its canonical topology as a manifold coincides with the topology induced from $\mathbf{R}^n$.*

Proof. Let V be a d-dimensional submanifold of $\mathbf{R}^n$ of class C^p. By theorem 2.1.2(v) there exists for every $x \in V$ an open neighborhood U of

x in $\mathbf{R}^n$, an open set $\Omega_x \subset \mathbf{R}^d$ and a C^p immersion $g_x : \Omega_x \to \mathbf{R}^n$ that is a homeomorphism between Ω_x and $V \cap U$ (cf. 2.1.3.1).

Consider the sets $U_x = g_x(\Omega_x) \subset V$ and the maps $g_x^{-1} : U_x \to \Omega_x$. The pairs $(U_x, g_x^{-1})_x$ form an atlas:

(AT1) $\bigcup_{x \in V} U_x = V$;

(AT2) $g_x^{-1}(U_x) = \Omega_x$ is open in $\mathbf{R}^d$, and $g_x^{-1}(U_x \cap U_y)$ is also open because g_x^{-1} is a homeomorphism, by 2.1.2(v), and $U_x \cap U_y = g_x(\Omega_x) \cap g_y(\Omega_y) \subset \mathbf{R}^n$ is open;

(AT3) follows from theorem 2.1.9.   □

We shall see below (3.1.5) that every abstract manifold is in some sense equivalent to a submanifold of $\mathbf{R}^n$.

2.2.10.2. Proposition. *An open subset U of a manifold X, with the topology induced from the canonical topology of X, is canonically a manifold.*

Proof. Let X be a d-dimensional manifold and $\{(U_i, \phi_i)\}_{i \in I}$ an atlas of X. The pairs $(U \cap U_i, \phi_i|_{U \cap U_i})_{i \in I}$ form an atlas on $U \subset X$:

(AT1) the $U \cap U_i$ cover U because the U_i cover X;

(AT2) setting $\psi_i = \phi_i|_{U \cap U_i}$, we have $\psi_i(U \cap U_i) = \phi_i(U \cap U_i) \in O(\mathbf{R}^d)$ by 2.2.7, and $\psi_i(U \cap U_i \cap U_j) \in O(\mathbf{R}^d)$ because $U \cap U_j$ is open in X;

(AT3) finally, $\psi_j \circ \psi_i^{-1}$ is the restriction of $\phi_j \circ \phi_i^{-1}$ to $\phi_i(U \cap U_i \cap U_j)$; it is thus a C^p diffeomorphism between $\phi_i(U \cap U_i \cap U_j)$ and $\phi_j(U \cap U_i \cap U_j)$.   □

2.2.10.3. Proposition. *Let X and Y be C^p manifolds of dimension d and e and having atlases $(U_i, \phi_i)_{i \in I}$ and $(V_j, \psi_j)_{j \in J}$, respectively. The atlas $(U_i \times V_j, \phi_i \times \psi_j)_{(i,j) \in I \times J}$, where*

$$\phi_i \times \psi_j : (x, y) \mapsto \big(\phi_i(x), \psi_j(y)\big) \in \mathbf{R}^d \times \mathbf{R}^e = \mathbf{R}^{d+e},$$

makes $X \times Y$ into a $(d + e)$-dimensional C^p manifold.

Proof. This is immediate.   □

Notice that if X is of class C^p and Y is of class C^q, the product $X \times Y$ is of class $C^{\inf(p,q)}$. Check also that compatible atlases on X and Y give compatible atlases on $X \times Y$, that is, the differentiable structure obtained for $X \times Y$ depends only on the differentiable structures of X and Y. We can thus say that $X \times Y$ is the *product manifold* of X and Y.

2.2.10.4. A non-Hausdorff manifold. Consider in $\mathbf{R}^2$ the sets $E_1 = \{(x, 0) : x \in \mathbf{R}\}$ and $E_2 = \{(x, 1) : x \in \mathbf{R}\}$. Let $\sim$ be the equivalence relation on $F = E_1 \cup E_2$ defined by

$$\begin{cases} (x, 0) \sim (y, 0) \Leftrightarrow x = y, \\ (x, 1) \sim (y, 1) \Leftrightarrow x = y, \\ (x, 0) \sim (y, 1) \Leftrightarrow x = y < 0. \end{cases}$$

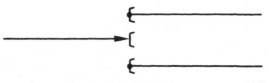

Figure 2.2.10.4

The classes of the quotient set $X = F/\sim$ are represented by elements $(x, 0)$ for $x < 0$ and by elements $(x, 0)$ and $(x, 1)$ for $x \geq 0$, as shown in figure 2.2.10.4.

We now provide X with the manifold structure given by the following two charts:

$$U_1 = \{(x, 0) : x \in \mathbf{R}\}, \quad \phi_1((x, 0)) = x$$

and

$$U_2 = \{(x, 0) : x < 0\} \cup \{(y, 1) : y \geq 0\}, \quad \phi_2((x, 0)) = x, \ \phi_2((y, 1)) = y.$$

Axioms (AT1) and (AT2) are clearly satisfied $\big($notice that $\phi_1(U_1 \cap U_2) = \phi_2(U_1 \cap U_2) = \{x \in \mathbf{R} : x < 0\}\big)$. Let's check (AT3): $\phi_1 \circ \phi_2^{-1}$ is the map $x \mapsto (x, 0) \mapsto x$ from $\phi_2(U_1 \cap U_2) = \,]-\infty, 0[$ onto $\phi_1(U_1 \cap U_2) = \,]-\infty, 0[$, and this is trivially C^∞. Thus X has been given a C^∞ structure.

The topology of X, however, is not Hausdorff. The points $(0, 1)$ and $(0, 0)$ are distinct but have no disjoint neighborhoods: if $U \subset X$ is an open set containing $(0, 0)$, the image $\phi_1(U \cap U_1)$ is open in $\mathbf{R}$ and contains 0, so $]-\alpha, \alpha[\subset \phi_1(U \cap U_1)$ for some $\alpha > 0$, and U contains a set of the form $\{(x, 0) : -\alpha < x < 0\}$. Similarly, an open neighborhood V of $(0, 1)$ in X must contain a set of the form $\{(x, 0) : -\beta < x < 0\}$, and consequently cannot be disjoint from U.

Later on we will disallow this kind of example.

2.2.10.5. Proposition. *If X is a manifold and E is an arbitrary set, $X \times E$ can be given a manifold structure.*

Proof. Let $(U_i, \phi_i)_{i \in I}$ be an atlas on X; then $\big\{(U_i \times \{e\}, \psi_{i,e})\big\}_{i \in I, e \in E}$, where $\psi_{i,e}(u, e) = \phi_i(u)$, is an atlas on $X \times E$. Indeed, intersections of the form $U_i \times \{e\} \cap U_j \times \{F\}$ are empty for $e \neq f$, and for $e = f$ they equal $(U_i \cap U_j) \times \{e\}$, which trivially implies axioms (AT2) and (AT3). $\square$

If E is uncountable, $X \times E$ is an *unreasonable* manifold, because it is too big; this will be made more precise in 3.1.3. Even if E is countable, one may object that $X \times E$ is not connected. But here is an example to show that connected manifolds, too, can be unreasonable:

2.2.10.6. The long line. Let Ω be an uncountable well-ordered set such that every initial segment $H_x = \{y \in \Omega : y < x\}$ is countable (such sets exist; for more details see [Spi79, vol. I, A.1 to A.12]). Let ω be the least element of Ω, and set

$$X = \Omega \times [0, 1[\setminus \{(\omega, 0)\};$$

one can say that X is a collection of intervals indexed by Ω, the first being $]0, 1[$ and the others $[0, 1[$. Order X lexicographically, and give it the associated topology. It can be shown that X becomes in this way a connected one-dimensional C^∞ manifold.

See exercise 3.6.4 for an unreasonable two-dimensional manifold that's easier to construct. In 3.1.6 we will disallow such manifolds, but right now it is enough to agree that from now on

2.2.10.7 $\boxed{\text{every manifold will be assumed Hausdorff.}}$

Of course Hausdorff here refers to the canonical topology. Here are some more properties of the topology of a manifold:

2.2.11. Theorem. *Manifolds are locally compact topological spaces.*

Proof. We must show that every point $x \in X$ has a compact neighborhood. Take $x \in X$ and let (U, ϕ) be a chart at x. By 2.2.9, $\phi : U \to \phi(U)$ is a homeomorphism, and $\phi(U)$ is a neighborhood of $\phi(x)$ in $\mathbf{R}^d$, hence locally compact. Take a compact neighborhood K of $\mathbf{R}^d$ such that $\phi(x) \in K \subset \phi(U)$; since ϕ^{-1} is continuous and X is Hausdorff, $\phi^{-1}(K)$ is a compact neighborhood of x. $\square$

2.2.12. Theorem. *Manifolds are locally connected topological spaces.*

Proof. The proof is analogous: $\phi(U)$, being a neighborhood of $\phi(x)$ in $\mathbf{R}^d$, contains a conneted neighborhood C of $\phi(x)$, so $\phi^{-1}(C)$ is a connected neighborhood of X containing x. And we can choose $\phi^{-1}(C)$ contained in any fixed open neighborhood of x by restricting the domain of the chart. $\square$

2.2.13. Theorem. *A manifold is connected if and only if it is path-connected. connected equiv to path-connected*

Proof. Path-connectedness implies connectedness for any topological space. To show the converse, assume X is connected, pick $x \in X$ and let Q be the set of $y \in X$ that can be joined to x by a path in X. We must prove that $Q = X$, and this we achieve by showing that Q is non-empty, open and closed. Non-emptiness is obvious since $x \in Q$.

Openness: take $y \in Q$ and let (U, ϕ) be a chart at y. Since $\phi(U)$ is open in $\mathbf{R}^d$, there exists $\varepsilon > 0$ such that $B(\phi(y), \varepsilon) \subset \phi(U)$. Take $z \in B(\phi(y), \varepsilon)$. There exists a continuous map $\gamma : [0, 1] \to \mathbf{R}^d$ such that $\gamma(0) = \phi(y)$, $\gamma(1) = z$ and $\gamma([0, 1]) \subset B(\phi(y), \varepsilon)$ (take γ affine linear, for example). The map

$$\phi^{-1} \circ \gamma : [0, 1] \to \phi^{-1}\big(B(\phi(\gamma), \varepsilon)\big)$$

is a path joining y and $\phi^{-1}(z)$. Since we can join x and y, we can equally well join x and $\phi^{-1}(z)$, for any $z \in B(\phi(y), \varepsilon)$. Since $\phi : U \to \phi(U)$ is a homeomorphism, $\phi^{-1}\big(B(\phi(y), \varepsilon)\big)$ is an open neighborhood of y contained in Q, showing that Q is open.

Closedness: take $y \in \overline{Q}$ and let (U, ϕ) be a chart at y. Since $\phi(y) \in \phi(U)$ and $\phi(U)$ is open, there exists $\varepsilon > 0$ such that $B\big(\phi(y), \varepsilon\big) \subset \phi(U)$, and $\phi^{-1}\big(B(\phi(y), \varepsilon)\big) \subset U$. But $y \in \overline{Q}$, and $\phi^{-1}\big(B(\phi(y), \varepsilon)\big)$ is an open set containing y, so there is a point $z \in Q \cap \phi^{-1}\big(B(\phi(y), \varepsilon)\big)$. Since $\phi(z) \in B\big(\phi(y), \varepsilon\big) \subset \phi(U)$, there is a path in $\mathbf{R}^d$ joining $\phi(y)$ and $\phi(z)$, thus also a path in X joining y and z. But $z \in Q$, so there is a path in X joining x and z, hence also one joining x and y, and $y \in Q$. This shows that $\overline{Q} = Q$. $\square$

2.2.14. Important examples of manifolds are the curves, classified in section 3.4, and the compact surfaces (see 4.2.25).

2.3. Differentiable Maps

2.3.1. Definition. Let X and Y be manifolds, of dimension d and e and class C^q and C^r, respectively. Let $p \leq \inf(q, r)$. We say that a continuous map $f : X \to Y$ is *of class C^p*, or C^p *differentiable*, or a C^p *morphism*, if, for every chart (U, ϕ) at $x \in X$ and every chart (V, ψ) at $f(x) \in Y$, the map $\psi \circ f \circ \phi^{-1} : \phi\big(U \cap f^{-1}(V)\big) \to \mathbf{R}^e$ is of class C^p. We will denote by $C^p(X; Y)$ the set of C^p differentiable maps from X into Y.

We sometimes say that f is *differentiable*, without stating the class.

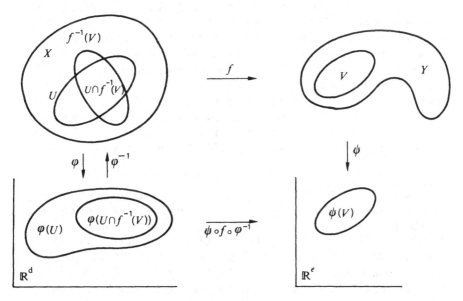

Figure 2.3.1

One needs to go through charts to be able to talk about differentiability. Specifically, the domain of definition of $\psi \circ f \circ \phi^{-1}$ should be open for the definition to make sense. Let's check this (figure 2.3.1): for $\psi \circ f \circ \phi^{-1}(y)$ to be defined, $y \in \phi(U)$ must be such that $f(\phi^{-1}(y))$ is in the domain V of ψ, or equivalently that $\phi^{-1}(y) \in f^{-1}(V)$. Since $\phi : U \to \phi(U)$ is a bijection, this means $y \in \phi(U \cap f^{-1}(V))$.

Now we've assumed that f is continuous, and V is open in Y. Thus $U \cap f^{-1}(V)$ is open in X, and, since ϕ is a homeomorphism, $\phi(U \cap f^{-1}(V))$ is open in $\mathbf{R}^d$. Hence we can talk about $\psi \circ f \circ \phi^{-1}$ being differentiable.

This definition, involving as it does all possible charts at x and $f(x)$, is not always convenient to use. The next theorem helps:

2.3.2. Theorem. *Let X and Y be manifolds of dimension d and e, respectively, and class $\geq p$. Let $f : X \to Y$ be a continuous map. The following conditions are equivalent:*

(i)   *f is C^p differentiable;*
(ii) *for every $x \in X$, every chart (U, ϕ) at x and every chart (V, ψ) at $f(x)$ such that $f(U) \subset V$, the composition $\psi \circ f \circ \phi^{-1} : \phi(U) \to \mathbf{R}^e$ is of class C^p;*
(iii) *for every $x \in X$, there exists a chart (U, ϕ) at x and a chart (V, ψ) at $f(x)$ such that $f(U) \subset V$ and $\psi \circ f \circ \phi^{-1} \subset C^p(\phi(U); \mathbf{R}^e)$.*

Proof. (i) $\Rightarrow$ (ii) is immediate from the definition; just notice that $f(U) \subset V$ implies $U \cap f^{-1}(V) = U$.

(ii) $\Rightarrow$ (iii). Let (V, ψ) be a chart at $f(x)$. Since f is continuous, $f^{-1}(V)$ is open in X and contains x; by the definition of canonical topology (2.2.6) there exists a chart (U, ϕ) at x such that $U \subset f^{-1}(V)$, whence $f(U) \subset V$. If (ii) is true it follows that $\psi \circ f \circ \phi^{-1}$ is of class C^p from $\phi(U)$ into $\mathbf{R}^e$.

(iii) $\Rightarrow$ (i). Let (S, α) be a chart at $x \in X$ and (T, β) one at $f(x) \in Y$. We must show that the map $\beta \circ f \circ \alpha^{-1}$, from the open subset $\alpha(S \cap f^{-1}(T))$ of $\mathbf{R}^d$ into $\mathbf{R}^e$, is of class C^p. It is enough to show that it is C^p on a neighborhood of each point of its domain.

Take $u \in \alpha(S \cap f^{-1}(T))$ and $x' = \alpha^{-1}(u) \in S$. Property (iii), applied to x', gives a chart (U, ϕ) at x' and a chart (V, ψ) at $f(x')$ such that $f(U) \subset V$ and that $\psi \circ f \circ \phi^{-1}$ is of class C^p on $\phi(U)$. Now we can write

$$\beta \circ f \circ \alpha^{-1} = (\beta \circ \psi^{-1}) \circ (\psi \circ f \circ \phi^{-1}) \circ (\phi \circ \alpha^{-1}),$$

with the understanding that this only makes sense if each step in the composition is defined. If we can prove that each step is defined and C^p on a neighborhood of the image of u by the previous steps, we will have shown that $\beta \circ f \circ \alpha^{-1}$ is C^p on a neighborhood of u, and we'll be done.

The coordinate change $\phi \circ \alpha^{-1} : \alpha(S \cap U) \to \phi(S \cap U)$ is of class C^p, and its domain contains $u = \alpha(x')$. Next, $\psi \circ f \circ \phi^{-1}$ is of class C^p on $\phi(U)$, and its domain contains $\phi(x')$, the image of u under $\phi \circ \alpha^{-1}$, by the very choice of U, so $\psi \circ f \circ \phi^{-1}$ is of class C^p on a neighborhood of $\phi(x')$.

Finally, $\beta \circ \psi^{-1}$ is a C^p diffeomorphism between $\psi(T \cap V)$ and $\beta(T \cap V)$. Its domain $\psi(T \cap V)$ contains the image $\psi(f(x'))$ of u under the composition so far, since $f(x') \in V$ by our choice of V and $x' \in f^{-1}(T)$ as the image of $u \in \alpha(S \cap f^{-1}(T))$ under α^{-1}. Thus $\beta \circ \psi^{-1}$ is C^p on a neighborhood of $\psi(f(x'))$, concluding the proof that $\beta \circ f \circ \alpha^{-1}$ is C^p on a neighborhood of u.

2.3.3. Examples of differentiable maps

2.3.3.1. Proposition. *Let X and Y be manifolds. The canonical projections $p : X \times Y \to X$ and $q : X \times Y \to Y$ are differentiable.*

Proof. We prove the result for p. By 2.3.2(iii), it suffices to show that, for every $(x, y) \in X \times Y$, there exists a chart $(U \times V, \phi \times \psi)$ at (x, y) and a chart (W, θ) at $x \in X$ such that $p(U \times V) \subset W$ and $\theta \circ p \circ (\phi \circ \psi)^{-1} :$ $(\phi \times \psi)(U \times V) \to \mathbf{R}^d$ (where d is the dimension of X) is of class C^∞.

Let $(U \times V, \phi \times \psi)$ be a product of charts, as in 2.2.10.3, at the point (x, y). For (W, θ) we take the chart (U, ϕ) at x. We have $p(U \times V) = U$, and the map $\phi \circ p \circ (\phi \times \psi)^{-1}$ is defined on $(\phi \times \psi)(U \times V)$ by

$$(s, t) \mapsto \underbrace{(\phi^{-1}(s), \psi^{-1}(t))}_{\in U \times V} \overset{p}{\mapsto} \phi^{-1}(s) \overset{\phi}{\mapsto} s,$$

which is of class C^∞. $\qquad\qquad\qquad\qquad\qquad\qquad\qquad\qquad\qquad\qquad\square$

2.3.3.2. Proposition. *Let X, Y and Z be manifolds. If $f \in C^p(X; Y)$ and $g \in C^p(Y; Z)$ we have $g \circ f \in C^p(X; Z)$.*

Proof. Take $x \in X$, $y = f(x)$ and $z = g(f(x))$. By assumption and 2.3.2(iii) applied to g, there exists a chart (V, ψ) at $y \in Y$ and a chart (W, θ) at $z \in Z$ such that $g(V) \subset W$ and $\theta \circ g \circ \psi^{-1} \in C^p(\psi(V); \theta(W))$. By assumption and 2.3.1 applied to f, we have $\psi \circ f \circ \phi^{-1} \in C^p(\phi(U \cap f^{-1}(V)); \psi(V))$ for any chart (U, ϕ) at x. Thus 2.3.2(iii) is satisfied for $g \circ f$ and the charts $(U \cap f^{-1}(V), \phi)$ and (W, θ), by 0.2.15.1 applied to

$$\theta \circ (g \circ f) \circ \phi^{-1} = (\theta \circ g \circ \psi^{-1}) \circ (\psi \circ f \circ \phi^{-1}). \qquad\qquad\square$$

2.3.3.3. Proposition. *Let X, Y, Z be C^r manifolds and p, q the canonical projections from $X \times Y$ onto X, Y, respectively. A map $f : Z \to X \times Y$ is of class C^r if and only if the coordinate functions $p \circ f : Z \to X$ and $q \circ f : Z \to Y$ are.*

Proof. If f is of class C^r it follows from 2.3.3.1 and 2.3.3.2 that the compositions $p \circ f$ and $p \circ g$ are also of class C^r.

Conversely, assume $p \circ f$ and $q \circ f$ to be of class C^r. Take $z \in Z$, and consider $f(z) = (x, y) = (p(f(z)), q(f(z)))$. We know that there exists a chart (W_1, θ_1) at z and a chart (U, ϕ) at $x = (p \circ f)(z)$ such that $(p \circ f)(W_1) \subset U$ and $\phi \circ (p \circ f) \circ \theta_1^{-1} : \theta_1(W_1) \to \mathbf{R}^d$, where d is the dimension of X, is of class C^r. Similarly, there exists a chart (W_2, θ_2) at z and a chart (V, ψ) at $y = (q \circ f)(z)$ such that $(q \circ f)(W_2) \subset V$ and

$$\psi \circ (q \circ f) \circ \theta_2^{-1} \in C^r\big(\theta_2(W_2); \mathbf{R}^e\big),$$

where e is the dimension of Y.

Take $W = W_1 \cap W_2$ and $\theta = \theta_1|_W$, and consider the chart (W, θ) at z and the chart $(U \times V, \phi \times \psi)$ at $(x, y) = \big((p \circ f)(z), (q \circ f)(z)\big) \in X \times Y$. We have

$$f(W) \subset (p \circ f)(W) \times (q \circ f)(W) \subset U \times V.$$

To show that the map

$$(\phi \times \psi) \circ f \circ \theta^{-1} = (\phi \circ p \circ f \circ \theta^{-1}) \times (\psi \circ q \circ f \circ \theta^{-1}) : \theta(W) \to \mathbf{R}^{d+e}$$

is of class C^p we must show that its components into $\mathbf{R}^d$ and $\mathbf{R}^e$ are. But

$$\phi \circ p \circ f \circ \theta^{-1} = \phi \circ (p \circ f) \circ \theta_1^{-1} \circ (\theta_1 \circ \theta^{-1}),$$

and $\theta_1 \circ \theta^{-1}$ is a coordinate change, hence C^r-differentiable; $\psi \circ (p \circ f) \circ \theta_1^{-1}$ is C^r by assumption. The proof that $\psi \circ q \circ f \circ \theta^{-1}$ is C^r is similar. $\quad\square$

2.3.3.4. Proposition. *Let Y be a manifold and V a submanifold of $\mathbf{R}^n$. If a map $f : \mathbf{R}^n \to Y$ is of class C^p (where $\mathbf{R}^n$ is considered with its canonical n-dimensional C^∞ manifold structure), the restriction $f|_V : V \to Y$ is of class C^p.*

Proof. Take $x \in V$ and $f(x) = y \in Y$. By definition of the manifold structure on $\mathbf{R}^n$, all pairs (U, Id_U), where $U \subset \mathbf{R}^n$ is open and $\mathrm{Id}_U : U \to U$ is the identity map, are charts on $\mathbf{R}^n$. Since f is of class C^p, there exists an open set $U_1 \subset \mathbf{R}^n$ containing x and a chart (W, ϕ) at $y \in Y$ such that $f(U_1) \subset W$ and $\phi \circ f : U_1 \to \mathbf{R}^r$ is of class C^p, where r is the dimension of Y.

Since V is a submanifold of $\mathbf{R}^n$, there exist by 2.1.2(v) an open neighborhood U_2 of x in $\mathbf{R}^n$, an open set $\Omega \subset \mathbf{R}^d$ (where d is the dimension of V) and an immersive C^p diffeomorphism $g : \Omega \to V \cap U_2$. Furthermore, by example 2.2.10.1, the pair $\big(g(\Omega), g^{-1}\big)$ is a chart at $x \in V$. The set $U = U_1 \cap U_2$ is open in $\mathbf{R}^n$, contains x and satisfies $f(U) \subset W$. Consider $\Omega_1 = g^{-1}(V \cap U)$ and $g_1 = g|_{\Omega_1}$: the pair $\big(g_1(\Omega_1), g_1^{-1}\big)$ is a chart at x because $g_1 : \Omega_1 \to V \cap U$ is a homeomorphism of class C^p. We also have

$$f\big(g_1(\Omega_1)\big) = f(V \cap U) \subset f(U) \subset W,$$

and $\phi \circ f \circ (g_1^{-1})^{-1} = \phi \circ f \circ g_1$ is defined on Ω_1 because $g_1(\Omega_1) \subset V \cap U$ and the domain of $\phi \circ f$ is U. Finally, this composition is of class C^p, since g_1 and $\phi \circ f$ are. This shows that $f|_V \in C^p(V; Y)$. $\quad\square$

2.3.3.5. Proposition. *Let X be a manifold and W a submanifold of $\mathbf{R}^n$. If $f \in C^p(X; \mathbf{R}^n)$ maps X into W we have $f \in C^p(X; W)$.*

Proof. Left to the reader; after detailing several such proofs in this and the previous section, to drill in the technique, we will only outline them from now on, or leave them as exercises. □

2.3.4. Theorem. *Let X be a C^p manifold. The set $C^p(X; \mathbf{R})$, where $\mathbf{R}$ is considered as a manifold (2.2.10) is an $\mathbf{R}$-algebra under the natural operations $f + g$, λf and fg.*

Proof. Let f_1 and f_2 be C^p maps from X into $\mathbf{R}$, and λ_1, λ_2 real numbers. Take $x \in X$. There exists a chart (U, ϕ) at x such that $f_1 \circ \phi^{-1}$ and $f_2 \circ \phi^{-1}$ are of class C^p on $\phi(U)$ (apply 2.3.1 to the chart $(\mathbf{R}, \mathrm{Id}_{\mathbf{R}})$ on $\mathbf{R}$). The map

$$(\lambda_1 f_1 + \lambda_2 f_2) \circ \phi^{-1} = \lambda_1 (f_1 \circ \phi^{-1}) + \lambda_2 (f_2 \circ \phi^{-1})$$

is of class C^p on $\phi(U)$, which shows that $\lambda_1 f_1 + \lambda_2 f_2$ is of class C^p by theorem 2.3.2. The proof for $f_1 f_2$ is analogous. □

2.3.5. Definition. Let X and Y be manifolds. We say that $f : X \to Y$ is a C^p diffeomorphism if f is bijective, $f \in C^p(X; Y)$ and $f^{-1} \in C^p(Y; X)$. We denote by $\mathrm{Diff}(X; Y)$ the set of diffeomorphisms from X into Y, and by $\mathrm{Diff}(X)$ the set of diffeomorphisms from X into itself. We also write $\mathrm{Diff}^p(X; Y)$ and $\mathrm{Diff}^p(X)$.

2.3.6. Remarks

2.3.6.1. It follows from example 2.3.3.2 that $\mathrm{Diff}(X)$ is a group under composition of maps.

2.3.6.2. If (U, ϕ) is a chart on X, the map ϕ is a diffeomorphism onto the open set $\phi(U) \subset \mathbf{R}^n$ with its canonical manifold structure (2.2.10.2).

2.3.7. Theorem. *Let X be a connected manifold. The group $\mathrm{Diff}(X)$ acts transitively on X.*

This means that for any two points $x, y \in X$ there exists a diffeomorphism $f \in \mathrm{Diff}(X)$ such that $f(x) = y$.

Proof. We first study the case $X = \mathbf{R}$, then $X = \mathbf{R}^d$. For arbitrary X we assume at first that x and y are close, so we can use a chart to reduce to the case of $\mathbf{R}^d$.

2.3.7.1. First step. We construct a diffeomorphism of $\mathbf{R}$ that takes 0 to a nearby point but leaves everything outside a neighborhood of 0 fixed. Let $\phi(x) : \mathbf{R} \to [0, 1]$ be a C^∞ bump function (0.2.16) such that $\phi(x) = 1$ on $[-1, 1]$ and $\phi(x) = 0$ if $|x| \geq 2$. Consider the function $g(x) = \varepsilon \phi(x) + x$, where ε is a constant to be determined. We have $g'(x) = \varepsilon \phi'(x) + 1$. Now ϕ' is bounded because it has compact support; thus there exists ε_0 such that $\varepsilon_0 \|\phi'\| < 1$. If we take $\varepsilon < \varepsilon_0$, we have $g'(x) > 0$ for every 0. Thus $g(x)$ is C^∞ and strictly monotonic; by the inverse function theorem it is a C^∞ diffeomorphism. Clearly $g(0) = \varepsilon$ and $g(x) = x$ for $|x| \geq 2$.

2.3.7.2. Second step. We now define a map $h : \mathbf{R}^d \to \mathbf{R}^d$ as follows:

$$h(x_1, \ldots, x_d) = \big(x_1 + \varepsilon\phi(x_1)\phi(x_2^2 + \cdots + x_d^2), x_2, \ldots, x_d\big),$$

where ϕ and ε are as above. The map h is C^∞ because ϕ is; and h is the identity outside $B(0, 2) \times B(0, \sqrt{2}) \subset \mathbf{R} \times \mathbf{R}^{d-1} = \mathbf{R}^d$. Furthermore h is injective (why?) and regular, since its jacobian matrix is upper triangular with all diagonal entries but the first equal to 1, so the determinant is $\partial h_1/\partial x_1 = 1 + \varepsilon\phi'(x_1)\phi(x_2^2 + \cdots + x_d^2)$, which is non-zero by the choice of ε and because $\phi(x_2^2 + \cdots + x_d^2) \leq 1$. Finally, $h(0, \ldots, 0) = (\varepsilon, 0, \ldots, 0)$. By composing h with rotations of $\mathbf{R}^d$ (which are C^∞ diffeomorphisms), we can take 0 to any point $x \in \mathbf{R}^d$ such that $\|x\| \leq \varepsilon$. We can also compose with a homothety to shrink $B(0, 2) \times B(0, \sqrt{2})$ into $B(0, 1)$.

Thus there exists $\varepsilon > 0$ such that, for every $y \in B(0, \varepsilon)$, we can find $h \in \mathrm{Diff}(\mathbf{R}^d)$ such that $h(0) = y$ and that h is the identity outside $B(0, 1)$.

2.3.7.3. Third step. Now we consider an arbitrary manifold X. Let (V, ϕ_1) be a chart at $x \in X$. Since $\phi_1(V)$ is open in $\mathbf{R}^d$, there exists a diffeomorphism $\psi : \mathbf{R}^d \to \mathbf{R}^d$ (a composition of rotations and homotheties) such that $\overline{B}(0, 1) \subset \psi\big(\phi_1(V)\big)$. Set $\phi = \psi \circ \phi_1$; the pair (V, ϕ) is then a chart at $x \in X$ (compatibility is obvious, since ψ is C^∞). We can also assume that $\phi(x) = 0$.

Since $\overline{B}(0, 1) \subset \phi(V)$ and $\phi : V \to \phi(V)$ is a diffeomorphism, the set $U = \phi^{-1}(B(0, \varepsilon))$ is open in X and contains $x = \phi^{-1}(0)$. Now for every $y \in U$ there exists $h \in \mathrm{Diff}(\mathbf{R}^d)$ such that $h(0) = \phi(y)$. Define $f : X \to X$ by

$$f = \begin{cases} \phi^{-1} \circ h \circ \phi & \text{on } \phi^{-1}\big(B(0, 1)\big), \\ \mathrm{Id} & \text{elsewhere.} \end{cases}$$

Clearly f is a bijection, being defined by two bijections on complementary subsets of X. It is also of class C^p, since it equals the identity on $X \setminus \phi^{-1}\big(\overline{B}(0, 1)\big)$ and $\phi^{-1} \circ h \circ \phi$ on V (notice that the two maps coincide where they overlap), and these two sets form an open cover for X. (Why is $X \setminus \phi^{-1}\big(\overline{B}(0, 1)\big)$ open? Because, as in the proof of 2.2.11, $\phi^{-1}\big(\overline{B}(0, 1)\big)$ is the continuous image of a compact set into a Hausdorff space.) Finally, we have $f(x) = y$.

What we have shown is that every $x \in X$ has an open neighborhood U such that, for any $y \in U$, there exists $f \in \mathrm{Diff}(X)$ taking x to y.

2.3.7.4. Fourth step. Let $x \in X$ be arbitrary, and call Q the set of points of X which are the images of x under some diffeomorphism. Since $x \in Q$, we must show that Q is open and closed in X. We prove openness using step 3: if $f(x) = y$ for $f \in \mathrm{Diff}(X)$ and $U \in O_y(X)$ is such that every $z \in U$ is the image of y under some diffeomorphism h, we have $z = h(y) = h\big(f(x)\big) = (h \circ f)(x)$ with $h \circ f \in \mathrm{Diff}(X)$, hence $U \subset Q$. Closedness is proved in the same way (compare the proof of 2.2.13). $\qquad\square$

2.4. Covering Maps and Quotients

Here we give the definition and properties of covering maps, and an explicit way to construct them (2.4.5 and 2.4.9).

2.4.1. Definition. Let X and Y be manifolds of class C^s. A map $p : X \to Y$ is said to be a *covering map* if it satisfies the following properties:

(i) p is surjective and differentiable of class C^s;
(ii) for every $y \in Y$ there exists an open subset $V \subset Y$ containing y such that $p^{-1}(V)$ is of the form

$$p^{-1}(V) = \bigcup_{i \in I} U_i,$$

where the $U_i \subset X$ are pairwise disjoint open sets and $p|U_i : U_i \to V$ is a diffeomorphism for each i.

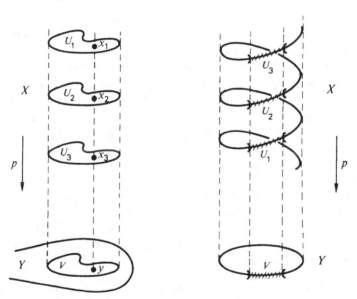

Figure 2.4.1

2.4.2. Remark. The set I may depend on y; but see 2.4.4.

2.4.3. Example. The map $p : t \mapsto (\cos 2\pi t, \sin 2\pi t)$ from $\mathbf{R}$ into the circle S^1 is a covering map.

By 2.1.6.2 and 2.2.10.2 we know that S^1 and $\mathbf{R}$ are both one-dimensional manifolds of class C^∞. Now $t \mapsto (\cos 2\pi t, \sin 2\pi t)$ is C^∞ as a map $\mathbf{R} \to \mathbf{R}^2$; by 2.3.3.5, it is also C^∞ as a map $\mathbf{R} \to S^1$. Let $(x_0, y_0) \in S^1$ be a point, and t_0 a real number such that $p(t_0) = (x_0, y_0)$. Take $\alpha \in]0, \frac{1}{2}[$. The set

$V = p(]t_0 - \alpha, t_0 + \alpha[)$ is open in V and contains (x_0, y_0); its inverse image is

$$p^{-1}(V) = \bigcup_{k \in \mathbf{Z}}]t_0 - \alpha + k, t_0 + \alpha + k[.$$

Finally, the restriction of p to $]t_0 - \alpha + k, t_0 + \alpha + k[$ is a diffeomorphism onto V.

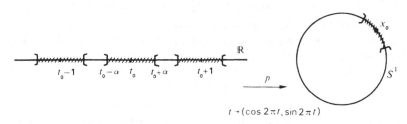

Figure 2.4.3

2.4.4. Theorem. *Let X and Y be manifolds, and $p : X \to Y$ a covering map. The cardinality of $p^{-1}(y)$ is locally constant. If Y is connected, the cardinality of $p^{-1}(y)$ is constant, and called the multiplicity of the covering. If the multiplicity k of p is finite, we say that p (or X) is a k-fold covering.*

Proof. Take $y \in Y$. There exists an open neighborhood $V \subset Y$ of y such that $p^{-1}(V) = \bigcup_{i \in I} U_i$ is a partition of $p^{-1}(V)$ satisfying the conditions of the definition. In particular, for every $i \in I$, there exists a unique $x_i \in U_i$ such that $p(x_i) = y$. Since $p^{-1}(y) \subset p^{-1}(V)$, it follows that $p^{-1}(y)$ and I have the same cardinality. For any other $z \in V$ the inverse image $p^{-1}(z)$ also has the same cardinality as I; this says that the cardinality of $p^{-1}(z)$ is locally constant.

Suppose in addition that X is connected. The technique used to prove 2.2.13 shows that two arbitrary points $x, y \in X$ can be joined by a finite chain of open sets $(U_i)_{i=1,\ldots,k}$ such that $x \in U_1$, $y \in U_k$, $U_i \cap U_{i+1} \neq \emptyset$ for every $i = 1, \ldots, k - 1$, and $\#p^{-1}(z)$ is constant on U_i for every i. This shows that $\#p^{-1}(x) = \#p^{-1}(y)$. $\qquad \square$

2.4.5. Definition. Let X be a manifold and G a subgroup of $\mathrm{Diff}(X)$. We say that G acts *properly discontinuously without fixed points* if the following to conditions are satisfied:

(i) For every $x, y \in X$ such that y is not in the orbit $G(x)$ of x there exist neighborhoods $U \ni x$ and $V \ni y$ such that, for every $g \in G$, the intersection $g(U) \cap V$ is empty.

(ii) For every $x \in X$ there exists an open neighborhood $U \subset X$ of x such that, for every $g \in G$ distinct from the identity, the intersection $g(U) \cap U$ is empty.

We recall that the *orbit* $G(x)$ of a point $x \in X$ under the action of G is the set $\{g(x) : g \in G\}$.

2.4.6. Remarks. Condition (i) can be stated in the following equivalent way:

(i′) For every $x, y \in X$ such that y is not in the orbit $G(x)$ of x there exist neighborhoods $U \ni x$ and $V \ni y$ such that, for every $g, h \in G$, the intersection $g(U) \cap h(V)$ is empty.

Indeed, $g(U) \cap h(V) = \emptyset$ if and only if $(h^{-1} \circ g)(U) \cap (V) = \emptyset$, and g and h arbitrary in G means $h^{-1} \circ g$ arbitrary in G.

Condition (ii) explains the phrase "without fixed points": if x were a fixed point, that is, if there were $g \in G$ distinct from the identity such that $g(x) = x$, we would have $x \in g(U) \cap U$ for every neighborhood U of x. This condition also expresses the fact that elements of G are not "close" to one another; hence the word "discontinuous".

2.4.7. Examples

2.4.7.1. Take $X = \mathbf{R}^n$ (cf. 2.2.10) and $G \cong \mathbf{Z}^n$ consisting of translations of the form $x \mapsto g_k(x) = x + k$, where

$$k \in \mathbf{Z}^n = \big\{(k_1, \ldots, k_n) : k_i \in \mathbf{Z} \text{ for all } i\big\}.$$

We have $G \subset \mathrm{Diff}^\infty(\mathbf{R}^n)$. Let's show that G acts properly discontinuously without fixed points.

Take first x and y such that $y \notin G(x)$. We have $\inf\big\{d(y, z) : z \in G(x)\big\} = \varepsilon > 0$, since there are only a finite number of points of $G(x)$ in any compact subset of $\mathbf{R}^n$. The sets $U = B(x, \varepsilon/2)$ and $V = B(y, \varepsilon/2)$ clearly satisfy (i). As for (ii), take $U = B(x, \frac{1}{2})$ for any fixed $x \in \mathbf{R}^n$.

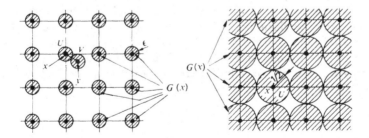

Figure 2.4.7.1

2.4.7.2. Consider S^d, a submanifold of $\mathbf{R}^{d+1}$, and take for G the subgroup of $\mathrm{Diff}(S^d)$ formed by the two maps Id_{S^d} and $-\mathrm{Id}_{S^d} : x \mapsto -x$ (the *antipodal* map). We claim that G acts properly discontinuously without fixed points. Just take $U = B(x, \varepsilon/2)$ and $V = B(y, \varepsilon/2)$, where $x \neq \pm y$ and $\varepsilon = \inf\big\{d(x, y), d(x, -y)\big\}$, to satisfy (i), and $U = B(x, \sqrt{2})$ to satisfy (ii). (Here we work with the metric induced from $\mathbf{R}^{d+1}$; if we want to use the intrinsic metric of the sphere instead—where distances are measured along arcs of great circle—we can replace $\sqrt{2}$ by $\pi/2$.)

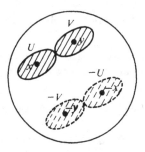 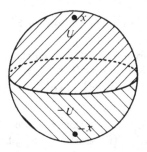

Figure 2.4.7.2

2.4.8. Proposition. *Let X be a manifold and G a subgroup of $\mathrm{Diff}(X)$. The relation $\sim$ on X defined by "$x \sim y$ if and only if there exists $g \in G$ such that $g(x) = y$" is an equivalence relation.*

Proof. This is obvious. □

We can consider the *quotient* $X/\sim$, also denoted by X/G, and the canonical surjection $p : X \to X/G$.

2.4.9. Theorem. *Let X be a d-dimensional C^p manifold and G a subgroup of $\mathrm{Diff}(X)$ acting properly discontinuously without fixed points. There exists on X/G a unique d-dimensional C^p manifold structure such that the canonical map $p : X \to X/G$ is a covering map.*

Proof. Take $y \in Y = X/G$, and let x be a representative of the class y. Since G acts properly discontinuously without fixed points, there exists an open neighborhood $U \subset X$ of x such that $g(U) \cap U = \emptyset$ for every $g \in G$. We can assume by shrinking U that U is the domain of a chart (U, ϕ). Set $V = p(U)$, and consider $p^{-1}(V) = \{z \in X : p(z) \in V\}$. A point z belongs to $p^{-1}(V)$ if and only if there exists $u \in U$ such that $p(z) = p(u)$, if and only if $z \sim u$, if and only if there exists $u \in U$ and $g \in G$ such that $z = g(u)$. Thus

$$p^{-1}(V) = \bigcup_{g \in G} g(U).$$

The choice of U guarantees that the $g(U)$ are pairwise disjoint. Thus we can define $\psi : V \to \mathbf{R}^d$ as follows: for $v \in V = p(U)$ there exists a unique representative u of v in U, which we write (by abuse of notation) $u = p^{-1}(v)$. Then we set $\psi = \phi \circ p^{-1}$.

We assert that the pairs (V, ψ) thus defined form a d-dimensional atlas of class C^p on $X/G = Y$. We check the axioms:

(AT1) Since $y \in Y$ was arbitrary, the domains V cover X/G.

(AT2) The map p^{-1} (in the sense above) is a bijection from V onto U, and ϕ is a bijection from U onto $\phi(U)$. Thus ψ is a bijection from V onto $\phi(U)$, which is open in $\mathbf{R}^n$. Consider two charts (V, ψ) and (V', ψ')

on X/G, deriving from charts (U, ϕ) and (U', ϕ') on X. If $v \in V \cap V'$, take $x \in U$ such that $p(x) = v$ and $x' \in U'$ such that $p(x') = v'$. Since $p(x) = p(x')$ there exists $g \in G$ such that $g(x') = x$; thus $\big(g(U'), \phi' \circ g^{-1}\big)$ is still a chart on X, since $g \in \text{Diff}(x)$. Finally,

$$(\phi \circ p^{-1})(V \cap V') = \phi\big(U \cap p^{-1}(V \cap V')\big) = \phi\bigg(U \cap \bigg(\bigcup_{h \in G} h(U')\bigg)\bigg)$$

is open in $\mathbf{R}^d$ because $U \cap \big(\bigcup_{h \in G} h(U')\big)$ is open in X, each $h \in G$ being continuous.

(AT3) Here's where we use the auxiliary chart $\big(g(U'), \phi' \circ g^{-1}\big)$. Class C^p being a local property, it is enough to show that $\psi' \circ \psi^{-1}$ is C^p in a neighborhood of $\psi(y) = \phi(x)$. On the open set $\phi\big(U \cap g(U')\big)$ we have

$$\psi' \circ \psi^{-1} = \big(\phi' \circ g^{-1} \circ p^{-1}\big) \circ \big(\phi \circ p^{-1}\big)^{-1} = \big(\phi' \circ g^{-1}\big) \circ \phi^{-1},$$

which is of class C^p by (AT3) applied to X.

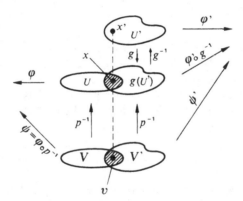

Figure 2.4.9

All that remains to show is that X/G is Hausdorff (cf. 2.2.10.7), and that p is a covering map. Let y and y' be points of X/G belonging to $p^{-1}(y)$ and $p^{-1}(y')$, respectively. Again because the action is properly discontinuous without fixed points, there exist open neighborhoods U and U' of x and x', respectively, such that $g(U) \cap U' = \emptyset$ for every $g \in G$. We can in fact assume that U and U' are domains of charts. Then $V = p(U)$ and $V' = p(U')$ are domains of charts in X/G, hence open in X/G (2.2.6), and they contain y and y' respectively. They are also disjoint: $z \in V \cap V'$ would imply $p^{-1}(V \cap V') \neq \emptyset$, that is, $p^{-1}(V) \cap p^{-1}(V') \neq \emptyset$, or again

$$\bigg(\bigcup_{g \in G} g(U)\bigg) \cap \bigg(\bigcup_{g' \in G} g'(U')\bigg) \neq \emptyset,$$

and there would exist $g, g' \in G$ such that $g(U) \cap g'(U') \neq \emptyset$ and $(g'^{-1} \circ g)(U) \cap U' \neq \emptyset$, contradicting the choice of U and U'.

Now to check that p is a covering map. If $V = p(U)$ is the domain of a chart on X/G, we have $p^{-1}(V) = \bigcup_{g \in G} g(U)$, where the $g(U)$ are open in X because $G \subset \text{Diff}(X)$. Thus $p^{-1}(V)$ is open and p is continuous. If $x \in X$ and (U, ϕ) is a chart at x, let (V, ψ) be the associated chart at $p(x) = y \in X/G$. We have $p(U) \subset V$, and $\psi \circ p \circ \phi^{-1} = \phi \circ p^{-1} \circ p \circ \phi^{-1} = \text{Id}_{\phi(U)}$ is of class C^p from $\phi(U)$ into $\mathbf{R}^d$. Thus p is C^p differentiable, by 2.3.2. Finally, if V is the domain of a chart associated with (U, ϕ), we have

$$p^{-1}(V) = \bigcup_{g \in G} g(U),$$

where the $g(U)$ are diffeomorphic to V and form a partition of $p^{-1}(V)$. $\square$

2.4.10. Corollary. *If X is a compact manifold, so is X/G.*

Proof. The canonical map $p : X \to X/G$ is continuous and surjective, and X/G is Hausdorff. $\square$

2.4.11. The following criterion holds (exercise 2.8.12): let X and Y be differentiable manifolds of class C^p, and $G \subset \text{Diff}(X)$ a group acting on X properly discontinuously without fixed points. We consider X/G with its canonical manifold structure, and let $p : X \to X/G$ be the canonical projection. Then a map $f : X/G \to Y$ is of class C^p if and only if $f \circ p : X \to Y$ is.

2.4.12. Examples.

2.4.12.1. The torus, second recipe (cf. 2.1.6.3). By 2.4.7.1 and 2.4.9 the set $Y = \mathbf{R}^n/\mathbf{Z}^n$ is a C^∞ manifold, covered by $\mathbf{R}^n$. As a manifold, Y is the product of n copies of $\mathbf{R}/\mathbf{Z}$ (2.2.10.3). Now the reader can guess that $\mathbf{R}/\mathbf{Z}$, the set of reals modulo 1, is diffeomorphic to the circle S^1; this will be rigorously proved in 2.6.13.1. Thus $Y = \mathbf{R}^n/\mathbf{Z}^n$ is diffeomorphic to $(S^1)^n$, or to the torus $T^n = \left(S^1(n^{-1/2})\right)^n$ introduced in 2.1.6.3. Taking the quotient $Y = \mathbf{R}^n/\mathbf{Z}^n$ is another way to obtain the manifold T^n.

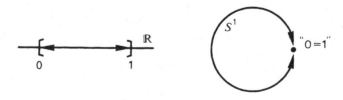

Figure 2.4.12.1

2.4.12.2. Real projective spaces, first recipe (cf. 2.6.13.3). By 2.4.7.2 and 2.4.9 the set $P^d(\mathbf{R}) = S^d/G = S^d/\pm \mathrm{Id}_{S^d}$ is a C^∞ manifold, called d-dimensional real projective space. Since S^d is compact, so is $P^d(\mathbf{R})$. This definition of projective space coincides with the one given in projective geometry, namely the quotient of $\mathbf{R}^{d+1} \setminus 0$ by the equivalence relation "$x \sim y$ if and only if there exists $k \in \mathbf{R}$ such that $y = kx$". Indeed, every $x \in \mathbf{R}^{d+1} \setminus 0$ gives rise to $x/\|x\| \in S^d$, so the quotient set of $\mathbf{R}^{d+1} \setminus 0$ by $\sim$ concides with the quotient of S^d by $\sim|_{S^d}$; but for $x, y \in S^d$ the condition $x \sim y$ is equivalent to $y = \pm y$.

The map introduced in 2.1.6.9 factors:

2.4.12.3

$$
\begin{array}{ccc}
S^2 & \xrightarrow{\ \ p\ \ } & P^2(\mathbf{R}) \\
& \searrow{\scriptstyle f} \quad \swarrow{\scriptstyle \underline{f}} & \\
& \mathbf{R}^6 &
\end{array}
$$

since $f(-x, -y, -z) = f(x, y, z)$. This yields a map $\underline{f} : P^2(\mathbf{R}) \to \mathbf{R}^6$ which, as the reader should verify, is injective.

2.4.12.4. The Klein bottle. Consider the manifold $T^2 = \mathbf{R}^2/\mathbf{Z}^2$, and denote by $\overline{x}$ the class of $x \in \mathbf{R}$ modulo $\mathbf{Z}$. The map $g : T^2 \to T^2$ defined by

$$g(\overline{x}, \overline{y}) = \left(\overline{x + \tfrac{1}{2}}, -\overline{y}\right)$$

is a diffeomorphism of T^2 satisfying $g^2 = \mathrm{Id}_{T^2}$. We show that the group $\{\mathrm{Id}_{T^2}, g\}$ acts properly discontinuously without fixed points.

First take u such that $g(u) \neq u$. Since T^2 is a Hausdorff space we can find disjoint open neighborhoods U_1 and V_1 of u and $g(u)$, respectively. Since g is a diffeomorphism, there exists an open set $U \ni u$ contained in U_1 and such that $g(U) \subset V_1$; thus $g(U) \cap U \subset V_1 \cap U_1 = \emptyset$, showing that G has no fixed points.

Now take $u, v \in T^2$ with $v \notin G(u) = \{u, g(u)\}$. There exist disjoint open neighborhoods U_1 of u and V_1 of v, and disjoint open neighborhoods U_2 of $g(u)$ and V_2 of v. Let Ω_2 be an open neighborhood of u such that $g(\Omega_2) \subset U_2$, and set $U = U_1 \cap \Omega_2$, $V = V_1 \cap V_2$. Then $U \cap V \subset U_1 \cap V_1 = \emptyset$ and

$$g(U) \cap V \subset g(\Omega_2) \cap V_2 \subset U_2 \cap V_2 = \emptyset,$$

and G is properly discontinuous.

The quotient $K = T^2/G$ is called the *Klein bottle*. We see that K is a compact manifold (2.4.10). Let's try to represent K, starting from $\mathbf{R}^2$. Passing from $\mathbf{R}^2$ to $T^2 = \mathbf{R}^2/\mathbf{Z}^2$ is achieved by identifying pairs of points (x, y) and $(x + k, y + l)$, with $(k, l) \in \mathbf{Z}^2$. We can represent T^2 by the square $ABCD$, including the sides DA and AB, but not BC or CD (figure 2.4.12.4). Or, if we include all sides, we can identify DA with CB and AB with DC.

Now identify pairs of points of the form (x, y) and $(x + \tfrac{1}{2}, -y)$; this corresponds to throwing away the rectangle $SCBR$ (not including the side

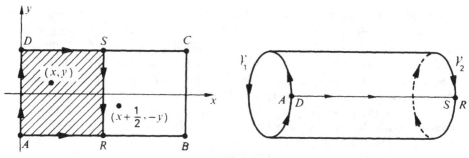

Figure 2.4.12.4

RS), then gluing AR and DS together to get a cylinder, and finally gluing DA and RS together (the two circles on the boundary of the cylinder), with orientation reversed. Like this:

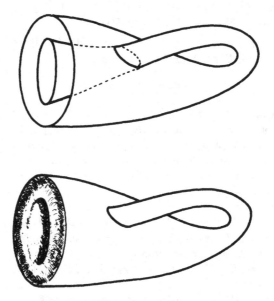

Figure 2.4.12.5

2.5. Tangent Spaces

Before introducing tangent spaces to abstract manifolds, we study the case of submanifolds of $\mathbf{R}^n$.

2.5.1. Definition. Let V be a submanifold of $\mathbf{R}^n$. A vector $z \in \mathbf{R}^n$ is said to be *tangent to* V at x if there exists a C^1 curve $\alpha : I \to V$ (where $I \subset \mathbf{R}$ is an interval containing 0, cf. 0.2.9.1) such that $\alpha(0) = x$ and $\alpha'(0) = z$.

2.5.2. Remarks. Strictly speaking, $\alpha'(0)$ is a linear map from $\mathbf{R}$ into $\mathbf{R}^n$, but we have identified it with the vector $\alpha'(0) \cdot 1 \in \mathbf{R}^n$ (cf. 0.2.4).

The condition $0 \in I$ just lightens the notation somewhat, but we could allow the curve to be defined on an interval I containing some t_0 such that $\alpha(t_0) = x$ and $\alpha'(t_0) = z$.

2.5.3. Theorem. *Let V be a submanifold of $\mathbf{R}^n$ and $x \in V$ a point. The set of tangent vectors to V at x is a vector subspace of $\mathbf{R}^n$, which we call the tangent space to V at x, and denote by $T_x V$.*

We will need two lemmas for the proof:

2.5.4. Lemma. *Let E and F be finite-dimensional vector spaces, $U \subset E$ an open set, $f : U \to F$ a C^1 map, V a submanifold of E contained in U and W a submanifold of F such that $f(V) \subset W$. Take $x \in V$ and set $y = f(x)$. If z is a tangent vector to V at x, the image $f'(x)(z)$ is a tangent vector to W at $y = f(x)$.*

Proof. There exists a curve $\alpha : I \to V$ such that $\alpha(0) = x$ and $\alpha'(0) = z$. The function $\gamma = f \circ \alpha : I \to W$ is also C^1 and satisfies $\gamma(0) = f(x) = y$ and

$$\gamma'(0) = f'\big(\alpha(0)\big)\big(\alpha'(0)\big) = f'(x)(z). \qquad \square$$

2.5.5. Lemma. *Let V be a submanifold of a real vector space E and U an open subset of E. A vector in E is tangent to $V' = V \cap U$ at x if and only if it is tangent to V at x.*

Proof. Assume first $z \in T_x V'$; there exists a curve $\alpha : I \to V'$ such that $\alpha(0) = x$ and $\alpha'(0) = z$. But α is also a curve in V, so $z \in T_x V$. For the converse, take $z \in T_x V$ and let $\alpha : I \to V$ be a curve such that $\alpha(0) = x$ and $\alpha'(0) = z$. Since U is open and α is a continuous map taking 0 inside $V' = V \cap U$, there exists a neighborhood J of 0 contained in I and such that $\alpha(J) \subset U$. Thus $\alpha(J) \subset V \cap U = V'$, and the restriction $\beta = \alpha|_J$ is a curve in V' of class C^1 satisfying $\beta(0) = x$ and $\beta'(0) = z$. $\qquad \square$

Proof of 2.5.3. Let V be a d-dimensional submanifold of $\mathbf{R}^n$ and $x \in V$ a point. By definition 2.1.1 there exists an open neighborhood U of x in $E = \mathbf{R}^n$ and a C^p diffeomorphism $f : U \to f(U)$ such that $f(U \cap V) = f(U) \cap \mathbf{R}^d$. By 2.5.5 we have $z \in T_x V$ if and only if $z \in T_x(U \cap V)$; by 2.5.4 this is equivalent to $f'(x)(z)$ being tangent to $f(U \cap V) = f(U) \cap \mathbf{R}^d$ at $f(x)$, and again to $f'(x)(z) \in T_{f(x)} \mathbf{R}^d$ since $f(U)$ is open. Now $T_{f(x)} \mathbf{R}^d = \mathbf{R}^d$ because any $u \in \mathbf{R}^d$ is the derivative of a curve in $\mathbf{R}^d$, say $\alpha(t) = f(x) + tu$; thus $z \in T_x V$ if and only if $f'(x)(z) \in \mathbf{R}^d$. Since $f'(x)$ is bijective we see that $T_x V = \big(f'(x)\big)^{-1}(\mathbf{R}^d)$ is a subspace of E. $\qquad \square$

2.5.6. Remark. The proof of 2.5.3 shows that the dimension of the tangent space to a submanifold of $\mathbf{R}^n$ is the same as the dimension of the manifold.

The various characterizations of submanifolds of $\mathbf{R}^n$ (theorem 2.1.2) lead to equivalent characterizations of tangent spaces. In particular we have the following result:

2.5.7. Theorem. *Let V be a d-dimensional C^p submanifold of $\mathbf{R}^n$.*

(i) *Take $x \in V$ and let $U \ni x$ be an open subset of $\mathbf{R}^n$ on which are defined real-valued functions $(f_i)_{1 \le i \le n-d}$ of class C^p such that*

$$V \cap U = \bigcap_{i=1}^{n-d} f_i^{-1}(0)$$

and that the vectors $f_i'(x)$, for $1 \le i \le n-d$, are linearly independent. The tangent space to V at x is given by

$$T_x V = \bigcap_{i=1}^{n-d} \left(f_i'(x)\right)^{-1}(0).$$

(ii) *Take $x \in V$ and let $U \ni x$ and $\Omega \ni 0$ be open subsets of $\mathbf{R}^n$ and $\mathbf{R}^d$, respectively, and $g : \Omega \to V \cap U$ a C^p homeomorphism such that $x = g(0)$ and $g'(0)$ is injective. Then $T_x(V) = g'(0)(\mathbf{R}^d)$.*

2.5.7.1. Example. If V has codimension one, that is, if it is locally given by a single equation f, we have $T_x V = \left(f'(x)\right)^{-1}(0)$. For instance, take $V = S^d \subset \mathbf{R}^{d+1}$ and $f = \|\cdot\|^2 - 1$. Then $f'(x) = 2(x \mid \cdot)$ (0.2.8.3), and $T_x S^d$ is the hyperplane orthogonal to x.

2.5.7.2. Proof. (i) Define $F : U \to \mathbf{R}^{n-d}$ as $F = (f_1, \ldots, f_{n-d})$. We have $V \cap U = F^{-1}(0)$ and F is a submersion at x, so

$$F'(x)^{-1}(0) = \bigcap_{i=1}^{n-d} \left(f_i'(x)\right)^{-1}(0)$$

is a vector space of dimension $n - (n - d) = d$. If we can show that $T_x V$ is contained in $\left(F'(x)^{-1}\right)(0)$ it will follow that the two spaces are identical.

Take $z \in T_x V$, and let $\alpha : I \to V$ be a curve such that $\alpha(0) = x$ and $\alpha'(0) = z$. After shrinking I if necessary, we can assume that $\alpha(I) \subset U$. The map $F \circ \alpha = \gamma$ is a curve in $\mathbf{R}^{n-d}$ satisfying $\gamma(I) = F\left(\alpha(I)\right) \subset F(U \cap V) = \{0\}$; in particular, $\gamma'(0) = 0$. But

$$\gamma'(0) = F'\left(\alpha(0)\right)\left(\alpha'(0)\right) = F'(x)(z),$$

which concludes the proof.

(ii) By 2.1.6.6 an open set $\Omega \subset \mathbf{R}^d$ is a d-dimensional submanifold of $\mathbf{R}^d$, so the tangent space to Ω at any point is $\mathbf{R}^d$. Since $g : \Omega \to V \cap U$ is a C^p

homeomorphism, it follows from 2.5.4 and 2.5.5 that

$$g'(0)(\mathbf{R}^d) \subset T_{g(0)}(V \cap U) = T_{g(0)}V.$$

But $g'(0)$ is injective, so $g'(0)(\mathbf{R}^d)$ and $T_{g(0)}V = T_x V$ have the same dimension. Thus they are equal, and we're done.   □

Now let $z \in T_x V$ be a tangent vector to a submanifold $V \subset \mathbf{R}^n$, and consider two compatible coordinate systems (2.1.8) g and h on V satisfying $g(0) = h(0) = x$. By the theorem, $T_x V = g'(0)(\mathbf{R}^d) = h'(0)(\mathbf{R}^d)$, so we can find $u, v \in \mathbf{R}^d$ such that $g'(0)(u) = z = h'(0)(v)$. This means that

$$v = \big((h'(0))^{-1} \circ g'(0)\big)(u) = (h^{-1} \circ g)'(0)(u);$$

in other words, the two triples (Ω_g, g, u) and (Ω_h, h, v) determine the same tangent vector if and only if $v = (h^{-1} \circ g)'(0)(u)$.

This points the way in generalizing the idea of tangent spaces to abstract manifolds: we just have to substitute charts for coordinate systems.

2.5.8. Proposition. *Let X be a d-dimensional manifold of class C^p and $x \in X$ a point. Consider all triples of the form (U, ϕ, u), where (U, ϕ) is a chart at x and u is a vector in $\mathbf{R}^d$. The relation $\sim$ defined by "$(U, \phi, u) \sim (V, \psi, v)$ if and only if $(\psi \circ \phi^{-1})'(\phi(x))(u) = v$" is an equivalence relation.*

Proof. Since $\phi \circ \phi^{-1}$ is the identity on $\phi(U)$, its derivative is everywhere the identity, and $(\phi \circ \phi^{-1})'(\phi(x))(u) = u$; this shows $\sim$ is reflexive.

For symmetry, notice that

$$\big((\psi \circ \phi^{-1})'(\phi(x))\big)^{-1} = (\phi \circ \psi^{-1})'\big((\psi \circ \phi^{-1})(\phi(x))\big) = (\phi \circ \psi^{-1})'(\psi(x)),$$

that is, $(\psi \circ \phi^{-1})'(\phi(x))(u) = v$ implies $u = (\phi \circ \psi^{-1})'(\psi(x))(v)$.

Finally, assume that $(U, \phi, u) \sim (V, \psi, v)$ and $(V, \psi, v) \sim (W, \theta, w)$, and consider the composition $\theta \circ \phi^{-1}$:

$$\begin{aligned}
(\theta \circ \phi^{-1})'(\phi(x))(u) &= \big((\theta \circ \psi^{-1}) \circ (\psi \circ \phi^{-1})\big)'(\phi(x))(u) \\
&= \big(((\theta \circ \psi^{-1})'(\psi \circ \phi^{-1}(\phi(x)))) \circ (\psi \circ \phi^{-1})'\big)(\phi(x))(u) \\
&= (\theta \circ \psi^{-1})'(\psi(x))(v) = w,
\end{aligned}$$

which shows that $(U, \phi, u) \sim (W, \theta, w)$ and that $\sim$ is transitive.   □

2.5.9. Definition. Let X be a manifold and $x \in X$ a point. A *tangent vector* to X at x is a $\sim$-equivalence class of triples (U, ϕ, u). The set of tangent vectors to X at x will be denoted by $T_x X$.

2.5.10. Remark. A chart (U, ϕ) at x determines an associated isomorphism $\theta_x : T_x X \to \mathbf{R}^d$, which takes $z \in T_x X$ to the unique vector $u \in \mathbf{R}^d$ such that $(U, \phi, u) \in z$. Bijectivity follows because the vector $u \in \mathbf{R}^d$ in (U, ϕ, u) is arbitrary.

2.5.11. Theorem. *Let X be a d-dimensional abstract manifold of class C^p. The tangent space $T_x X$ to X at x has a canonical d-dimensional vector space structure.*

Proof. Fix a chart (U, ϕ) at $x \in X$, and consider the associated bijection $\theta_x : T_x X \to \mathbf{R}^d$. The vector space structure on $T_x X$ is defined by pulling back the one on $\mathbf{R}^d$. More precisely,

$$\lambda z + \lambda' z' = \theta_x^{-1}\big(\lambda \theta_x(z) + \lambda' \theta_x(z')\big).$$

This structure does not depend on the choice of (U, ϕ). Indeed, if $\eta_x : T_x X \to \mathbf{R}^d$ is the bijection associated with another chart (V, ψ), definitions 2.5.8 and 2.5.10 show that

2.5.11.1 $\eta_x \circ \theta_x^{-1} = (\psi \circ \phi^{-1})'\big(\phi(x)\big),$

which is linear.   □

2.5.12. Proposition. *Let X be a submanifold of $\mathbf{R}^n$ and $x \in X$ a point. The tangent space E_x to the submanifold X at x (definition 2.5.1) is canonically isomorphic to the tangent space to the abstract manifold X at x (definition 2.5.9).*

Proof. This should be true by the very way we arrived at the notion of tangent vectors for abstract manifolds, but let's check nonetheless. Let (Ω, g) be a coordinate system of X such that $g(0) = x$; the map $g'(0) : \mathbf{R}^d \to E_x$ is an isomorphism. But $(U, \phi) = \big(g(\Omega), g^{-1}\big)$ is a chart on X centered at x, so we can consider the associated isomorphism $\theta_x : T_x X \to \mathbf{R}^d$. Our canonical isomorphism will be

2.5.12.1 $g'(0) \circ \theta_x : T_x X \to E_x.$

This does not depend on (Ω, g): if (Ω', h) is another coordinate system such that $h'(0) = x$, formula 2.5.11.1 says that $\eta_x \circ \theta_x^{-1} = (h^{-1} \circ g)'(0)$, whence $g'(0) \circ \theta_x = h'(0) \circ \eta_x$, as we wished to prove.   □

2.5.12.2. In particular, consider $U \in O(\mathbf{R}^d)$ with its canonical manifold structure given by the chart (U, i), where $i : U \to \mathbf{R}^d$ is the inclusion. In the submanifold sense we have $T_x U = \mathbf{R}^d$, so we deduce from 2.5.10 the existence of a *canonical* isomorphism

2.5.12.3 $\theta_x : T_x U \to \mathbf{R}^d$

associated with x and the chart (U, i). Notice that 2.5.12.3 does not contradict 2.5.12.1, since the natural coordinate system g of U such that $g(0) = x$ is just the translation $g : z \mapsto z + x$, so that $g'(0) = \mathrm{Id}_{\mathbf{R}^d}$.

2.5.12.4. Sometimes it is useful *not* to identify $T_x U$ with $\mathbf{R}^d$; one should rather visualize $T_x U$ as the space $(x, \mathbf{R}^d)$ of bound vectors [Ber87, 2.1.3] based at $x \in \mathbf{R}^d$, and θ_x as the projection $(x, z) \mapsto z$.

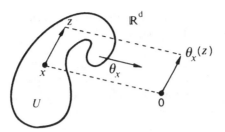

Figure 2.5.12.4

2.5.13. The tangent map. Now consider two manifolds X and Y, and a C^p map $f : X \to Y$. We want to define a linear map $T_x f$ from the tangent space $T_x X$ into the tangent space $T_{f(x)} Y$ in the following fashion: using 2.3.2(iii), take $z \in T_x X$ and charts (U, ϕ) and (V, ψ) at x and $f(x)$, respectively, such that $f(U) \subset V$ and $\psi \circ f \circ \phi^{-1}$ is of class C^p. Let $\theta_x : T_x X \to \mathbf{R}^d$ and $\eta_{f(x)} : T_{f(x)} Y \to \mathbf{R}^e$ be the isomorphisms associated with (U, ϕ) and (V, ψ) (2.5.10). Set

2.5.13.1 $$T_x f = \eta_{f(x)}^{-1} \circ (\psi \circ f \circ \phi^{-1})'(\phi(x)) \circ \theta_x,$$

so that the diagram on the right commutes:

2.5.13.2

$$
\begin{array}{ccc}
U \xrightarrow{\ f\ } V & & T_x X \xrightarrow{\ \ T_x f\ \ } T_{f(x)} Y \\
\phi \downarrow \quad \downarrow \psi & & \theta_x \downarrow \qquad \downarrow \eta_{f(x)} \\
\phi(U) \xrightarrow{\psi \circ f \circ \phi^{-1}} \psi(V) & & \mathbf{R}^d \xrightarrow{(\psi \circ f \circ \phi^{-1})'(\phi(x))} \mathbf{R}^e
\end{array}
$$

This definition will be consistent if it depends only on x, not on (U, ϕ) and (V, ψ). To check that, let (U_1, ϕ_1) and (V_1, ψ_1) be new charts, with associated maps ξ_x and $\varsigma_{f(x)}$. By formula 2.5.11.1 we can write (omitting the points where the derivatives are taken)

$$\varsigma_{f(x)}^{-1} \circ (\psi_1 \circ f \circ \phi_1^{-1})' \circ \xi_x$$

$$= \varsigma_{f(x)}^{-1} \circ ((\psi_1 \circ \psi^{-1}) \circ (\psi \circ f \circ \phi^{-1}) \circ (\phi \circ \phi_1^{-1}))' \circ \xi_x$$

$$= (\varsigma_{f(x)}^{-1} \circ (\psi_1 \circ \psi^{-1})') \circ (\psi \circ f \circ \phi^{-1})' \circ ((\phi \circ \phi_1^{-1})' \circ \xi_x)$$

$$= \eta_{f(x)}^{-1} \circ (\psi \circ f \circ \phi^{-1})' \circ \theta_x,$$

as we wished to show.

2.5.14. Definition. Let X and Y be manifolds and $f : X \to Y$ a C^p map. The *tangent map to f at a point* $x \in X$ is the map $T_x f : T_x X \to T_{f(x)} Y$ defined by 2.5.13.1.

2.5.15. Theorem. *Let X and Y be manifolds and $f : X \to Y$ a C^p map. The tangent map $T_x f : T_x X \to T_{f(x)} Y$ is linear. If Z is another manifold and $g : Y \to Z$ is a C^p differentiable map, we have*

$$T_x(g \circ f) = (T_{f(x)} g) \circ (T_x f).$$

Proof. This follows directly from 2.5.13.1 and from the linearity of θ_x, $\eta_{f(x)}$ (2.5.11) and $(\psi \circ f \circ \phi^{-1})'(\phi(x))$. $\qquad\qquad\qquad\qquad\qquad\qquad\square$

2.5.16. Example. Let X be a manifold and $U \in O(X)$ an open submanifold of X (2.2.10.2). For every $x \in U$, the tangent spaces T_xU and T_xX are canonically isomorphic and will henceforth be identified. The isomorphism is given by T_xi, where $i : U \to X$ is the inclusion.

2.5.17. Curves and velocity

2.5.17.1. Definition. A *curve* of class C^p in a manifold X (of class C^q, $q \geq p$) is a pair (I, α) consisting of an open interval $I \subset \mathbf{R}$, taken with its canonical manifold structure, and a map $\alpha \in C^p(I; X)$.

2.5.17.2. Definition. For $I \in O(\mathbf{R})$ and $t \in I$, we denote by $1_t \in T_tI$ the tangent vector to I at t defined by $\theta_t(1_t) = 1$, where $1 \in \mathbf{R}$ and θ_t is as in 2.5.12.2. The map $t \mapsto 1_t$ is called the *canonical vector field* on I (cf. 3.5.1).

2.5.17.3. Definition. Let (I, α) be a curve in a manifold X. The *velocity* of α at $t \in I$, denoted by $\alpha'(t)$, is the tangent vector to X at $\alpha(t)$ given by

$$\alpha'(t) = (T_t\alpha)(1_t).$$

Notice that this definition does not coincide, for $X = U \in O(\mathbf{R}^n)$, with the one introduced in 0.2.9.1 for $E = \mathbf{R}^n$. But they coincide modulo the canonical isomorphism θ_x. In the present setting the property established in 0.2.9.1 can be reformulated as follows:

2.5.17.4. Proposition. *For $f \in C^p(X;Y)$ and (I, α) a curve in X, we have*

$$(T_{\alpha(t)}f)(\alpha'(t)) = (f \circ \alpha)'(t);$$

in particular, we can calculate T_xf using velocities.

Proof. We have, by 2.5.17.3 and 2.5.15:

$$(f \circ \alpha)'(t) = (T_t(f \circ \alpha))(1_t) = (T_{\alpha(t)}f \circ T_t\alpha)(1_t)$$
$$= (T_{\alpha(t)}f)((T_t\alpha)(1_t)) = (T_{\alpha(t)}f)(\alpha'(t)). \qquad \square$$

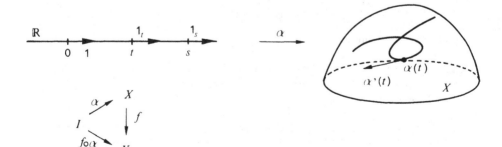

Figure 2.5.17

2.5.18. Proposition. *Let X and Y be C^p manifolds and $p : X \times Y \to X$ and $q : X \times Y \to Y$ the canonical projections. For every $(x, y) \in X \times Y$, the map*

$$T_{(x,y)}p \times T_{(x,y)}q : T_{(x,y)}(X \times Y) \to T_x X \times T_y Y$$

is an isomorphism.

This will allow us to identify $T_{(x,y)}(X \times Y)$ with $T_x X \times T_y Y$.

Proof. By 2.3.6.2 and 2.5.16 we can assume that $X = \mathbf{R}^d$ and $Y = \mathbf{R}^e$. Then the maps 2.5.12.3, here denoted by $\theta_{(x,y)}$, θ_x and θ_y, make the following diagram commute:

2.5.18.1

$$
\begin{array}{ccc}
T_{(x,y)}(\mathbf{R}^d \times \mathbf{R}^e) & \xrightarrow{\ T_{(x,y)}p\ } & T_x \mathbf{R}^d \\
{\scriptstyle \theta_{(x,y)}} \downarrow & & \downarrow {\scriptstyle \theta_x} \\
\mathbf{R}^d \times \mathbf{R}^e & \xrightarrow{\ \ p\ \ } & \mathbf{R}^d.
\end{array}
$$

Thus $T_{(x,y)}p \times T_{(x,y)}q$ is, up to bijections, equal to $p \times q$, which is an isomorphism. □

Here is the counterpart of 0.2.19 and 0.2.22 for manifolds.

2.5.19. Proposition. *If $f : X \to Y$ is a diffeomorphism, $T_x f : T_x X \to T_{f(x)} Y$ is an isomorphism.*

Proof. Since f is a diffeormorphism, f^{-1} exists and $f^{-1} \circ f = \mathrm{Id}_X$. Now $T_x(\mathrm{Id}_X)$ is the identity on $T_x X$, because if z is a tangent vector representing the triple (U, ϕ, u), the image $T_x(\mathrm{Id}_X)(z)$ is the tangent vector represented by the triple

$$(U, \phi, (\phi \circ \mathrm{Id}_X \circ \phi^{-1})'(\phi(x))(u)) = (U, \phi, u).$$

On the other hand, 2.5.15 gives

$$(T_x f^{-1} \circ f) = (T_{f(x)} f^{-1}) \circ (T_x f) = T_x(\mathrm{Id}_X) = \mathrm{Id}_{T_x X}.$$

It follows that $T_x f$ is an isomorphism between $T_x X$ and $T_{f(x)} Y$, and that $(T_x f)^{-1} = T_{f(x)} f^{-1}$. □

2.5.20. Proposition. *Let $f : X \to Y$ be a differentiable map of class C^p such that $T_x f \in \mathrm{Isom}(T_x X; T_{f(x)} Y)$. There exists an open neighborhood $U \subset X$ of x such that $f(U)$ is open in Y and f is a diffeomorphism from U into $f(U)$.*

Proof. We first express $T_x f$ in terms of charts, reducing to the case of open sets in $\mathbf{R}^n$, where we can apply the inverse function theorem (0.2.22). Using the fact that f is of class C^p and 2.3.2(iii) we take charts (U, ϕ) and (V, ψ) at x and $f(x)$, respectively, such that $f(U) \subset V$. We also take the isomorphisms $\theta_x : T_x X \to \mathbf{R}^d$ and $\theta_{f(x)} : T_{f(x)} \to \mathbf{R}^e$ (where d and e are the dimensions of X and Y).

By 2.5.13.1 and our assumptions we have

$$(\psi \circ f \circ \phi^{-1})'(\phi(x)) = \theta_{f(x)} \circ T_x f \circ \theta_x^{-1} \in \mathrm{Isom}(\mathbf{R}^d; \mathbf{R}^e),$$

so $\psi \circ f \circ \phi^{-1}$ is regular at $\phi(x)$. By 0.2.22 there exists $\Omega \in O_{\phi(x)}(\phi(U))$ such that $\psi \circ f \circ \phi^{-1}$ is a diffeomorphism between Ω and $\psi(f(\phi^{-1}(\Omega)))$.

Since ϕ and ψ are diffeormorphisms (see 2.3.6.2), the composition $f = \psi^{-1} \circ (\psi \circ f \circ \phi^{-1}) \circ \phi$ is also one. $\qquad\qquad\qquad\qquad\qquad\qquad\square$

2.5.21. Remark. Covering maps are local diffeomorphisms everywhere, by definition 2.4.1, but the converse is false. For example, take $p : X \to Y$, where $X =]0, 3\pi[$, $Y = S^1$ and $p(t) = (\cos t, \sin t)$. This can't be a covering map because S^1 is connected but the conclusion of 2.4.4 is not satisfied: we have $p^{-1}(p(3\pi/2)) = \{3\pi/2\}$ but $p^{-1}(p(\pi/2)) = \{\pi/2, 5\pi/2\}$.

2.5.22. Canonical isomorphisms. The canonical isomorphism θ_x between $T_x U$ and $\mathbf{R}^d$ defined in 2.5.12.3 for $U \in O(\mathbf{R}^d)$ can be generalized to any open submanifold U of a finite-dimensional real vector space E. By 2.5.16, we can assume $U = E$. Let $f \in L(E; \mathbf{R}^d)$ be arbitrary; for $x \in E$, we define

2.5.22.1 $$\theta_x : T_x E \to E$$

by setting $\theta_x = f^{-1} \circ \eta_x$, where η_x is the isomorphism associated in 2.5.10 with the chart $f : E \to \mathbf{R}^d$ of E. This θ_x is independent of f: if $g \in L(E; \mathbf{R}^d)$ is another linear map, with associated isomorphism $\varsigma_x : T_x E \to \mathbf{R}^d$, we have $(g \circ f^{-1})' = g \circ f^{-1}$ because $g \circ f^{-1}$ is linear (0.2.8.3), and $f^{-1} \circ \eta_x = g^{-1} \circ \varsigma_x$ by 2.5.11.1.

This allows us to introduce the following notion, which lies halfway between tangent maps and derivatives:

2.5.23. Definition. Let X be a manifold, E a finite-dimensional vector space and $f \in C^p(X; E)$. The *differential* of f at E, denoted by $df(x)$, is the map $\theta_{f(x)} \circ T_x f : T_x X \to E$, where $\theta_{f(x)} : T_{f(x)} E \to E$ is the map defined in 2.5.22.1.

2.5.23.1
$$\begin{array}{ccc} T_x X & \xrightarrow{T_x f} & T_{f(x)} E \\ & {\scriptstyle df(x)}\searrow & \downarrow{\scriptstyle \theta_{f(x)}} \\ & & F \end{array}$$

We leave it to the reader to check that if $U \in O(E)$ and $f \in C^p(U; F)$, where E and F are finite-dimensional vector spaces, the diagram below commutes:

2.5.23.2
$$\begin{array}{ccc} T_x U & \xrightarrow{T_x f} & T_{f(x)} E \\ {\scriptstyle \theta_x}\downarrow & {\scriptstyle df(x)}\searrow & \downarrow{\scriptstyle \theta_{f(x)}} \\ E & \xrightarrow[f'(x)]{} & F, \end{array}$$

that is, $T_x f = \theta_{f(x)}^{-1} \circ f'(x) \circ \theta_x$ (cf. 2.5.13.1).

2.5.23.3. If $f \in C^p(X; E)$ and $g \in C^p(Y; X)$ we have $d(f \circ g)(y) = df(g(y)) \circ T_y g$, as a trivial consequence of 2.5.15.

2.5.24. Tangent bundles. Let X be a manifold, and consider the disjoint union $TX = \bigcup_{x \in X} T_x X$ of the tangent spaces to X at all of its points.

2.5.25. Theorem. *If X is a d-dimensional manifold of class C^p, the set TX has canonical manifold structure of class C^{p-1} and dimension $2d$. We call TX the tangent bundle of X.*

Proof. To define an atlas we consider the projection $\pi : TX \to X$ given by $\pi(z) = x$ if $z \in T_x X$. Let (U, ϕ) be a chart on X. To any $z \in \pi^{-1}(U) \subset TX$ we associate the point

$$\left(\phi(\pi(z)), \theta_{\pi(z)}(z)\right) \in \mathbf{R}^{2d} = \mathbf{R}^d \times \mathbf{R}^d,$$

where $\theta_{\pi(z)} : T_{\pi(z)} \to \mathbf{R}^d$ is the isomorphism associated with (U, ϕ) at $\pi(z)$ (2.5.10); this makes sense because $z \in T_{\pi(z)} X$ and $\pi(z) \in U$. We thus have a map

2.5.25.1 $$\tau_\phi = (\phi \circ \pi, \theta_{\pi(\cdot)}) : \pi^{-1}(U) \to \mathbf{R}^d \times \mathbf{R}^d = \mathbf{R}^{2d}.$$

We claim that, as (U, ϕ) ranges over all charts on X, the pairs $\{(\pi^{-1}(U), \tau_\phi)\}$ forms an atlas on TX, and makes it into a (Hausdorff) manifold of class C^{p-1}.

(AT1) Since $\pi : TX \to X$ is surjective and X is covered by the U, so is TX covered by the $\pi^{-1}(U)$.

(AT2) Each τ_ϕ is injective because $\theta_{\pi(\cdot)}$ and ϕ are. The image of τ_ϕ is $\tau_\phi(\pi^{-1}(U))$, which is open in $\mathbf{R}^{2d}$. If $(\pi^{-1}(U), \tau_\phi)$ and $(\pi^{-1}(V), \tau_\psi)$ are charts on TX, we have

$$\tau_\phi(\pi^{-1}(U) \cap \pi^{-1}(V)) = \tau_\phi(\pi^{-1}(U \cap V)),$$

which is open in $\mathbf{R}^{2d}$ because $\phi(U \cap V)$ is open in $\mathbf{R}^d$.

(AT3) Again for two charts $(\pi^{-1}(U), \tau_\phi)$ and $(\pi^{-1}(V), \tau_\psi)$ on TX, we must show that $\tau_\psi \circ \tau_\phi^{-1}$ is of class C^{p-1} from $\tau_\phi(\pi^{-1}(U) \cap \pi^{-1}(V))$ into $\tau_\psi(\pi^{-1}(U) \cap \pi^{-1}(V))$, that is, from $\phi(U \cap V) \times \mathbf{R}^d$ into $\psi(U \cap V) \times \mathbf{R}^d$. If η is the isomorphism associated with (V, ψ) we have

$$\tau_\psi \circ \tau_\phi^{-1} = (\psi \circ \phi^{-1}, \eta \circ \theta^{-1}) = (\psi \circ \phi^{-1}, (\psi \circ \phi^{-1})'),$$

by 2.5.11.1, and this is a C^{p-1} map because $\psi \circ \phi$ is a C^p map.

So much for the differentiable structure. We next show that TX is Hausdorff. The following observation will be useful: if W is open in $\mathbf{R}^d$, $\phi(U) \times W$ is open in $\phi(U) \times \mathbf{R}^d$ and $\tau_\phi^{-1}(\phi(U) \times W)$ is open in TX, since, by 2.2.9, τ_ϕ is a homeomorphism from $\pi^{-1}(U)$ onto its image $\phi(U) \times \mathbf{R}^d$. This said, take distinct points $z_1, z_2 \in TX$.

If $z_1 \in T_{x_1}X$ and $z_2 \in T_{x_2}X$ with $x_1 \neq x_2$, there exist disjoint open sets U_1 and U_2 containing x_1 and x_2, because X is Hausdorff. Since U_1 and U_2 are domains of charts, $\pi^{-1}(U_1)$ and $\pi^{-1}(U_2)$ are open in TX, and they separate z_1 and z_2.

If $\pi(z_1) = \pi(z_2) = x$, take a chart (U, ϕ) at $x \in X$. The vector z_1 is represented by a triple (U, ϕ, u_1) and z_2 by (U, ϕ, u_2), and $u_1 \neq u_2$ because $z_1 \neq z_2$. We can find disjoint open neighborhoods W_1 and W_2 of u_1 and u_2 in $\mathbf{R}^d$, and the inverse images $\tau_\phi^{-1}(\phi(U) \times W_1)$ and $\tau_\phi^{-1}(\phi(U) \times W_2)$ are disjoint open neighborhoods of z_1 and z_2 in TX, by the observation above. $\square$

We now associate to a differentiable map $f : X \to Y$ a map $TX \to TY$.

2.5.26. Theorem. *If $f : X \to Y$ is of class C^p, the map $Tf : TX \to TY$ defined by $Tf|_{T_x X} = T_x f$ for every x is of class C^{p-1}.*

Proof. We use 2.3.2(iii), after having shown that for every $z \in TX$ there exist charts $\left(\pi^{-1}(U), \tau_\phi\right)$ at z and $\left(\pi^{-1}(V), \tau_\psi\right)$ at $Tf(z)$ such that $Tf\left(\pi^{-1}(U)\right) \subset \pi^{-1}(V)$ and that $\tau_\psi \circ Tf \circ \tau_\phi^{-1}$, defined on the open set $\tau_\phi\left(\pi^{-1}(U)\right)$, is of class C^{p-1} (notice that the same letter is being used to denote two different maps $\pi : TX \to X$ and $\pi : TY \to Y$). We also must show that Tf is continuous.

Take $z \in TX$ and set $x = \pi(z)$, so that $z \in T_x X$. By definition, $(Tf)(z) = (T_x f)(z)$ is in $T_{f(x)}Y$. Since f is of class C^p, there exist charts (U, ϕ) at x and (V, ψ) at $y = f(x)$ such that $f(U) \subset V$ and $\psi \circ f \circ \phi^{-1} \in C^p\left(\phi(U), \mathbf{R}^e\right)$, where e is the dimension of Y. The desired charts at $z \in TX$ and $(Tf)(z) \in TY$ will be $\left(\pi^{-1}(U), \tau_\phi\right)$ and $\left(\pi^{-1}(V), \tau_\psi\right)$, respectively.

We have $Tf\left(\pi^{-1}(U)\right) \subset \pi^{-1}(V)$ because any $t_0 \in \pi^{-1}(U)$ is tangent to X at a point x_0 in U, so $(Tf)(t_0) = (T_{x_0}f)(t_0)$ is tangent to Y at $f(x_0)$. But $f(x_0) \in V$ because $f(U) \subset V$, so $(Tf)(t_0) \in \pi^{-1}(V)$.

Formulas 2.5.13.1 and 2.5.25.1 imply that

2.5.26.1 $$\tau_\psi \circ Tf \circ \tau_\phi^{-1} = \left(\psi \circ f \circ \phi^{-1}, (\psi \circ f \circ \phi^{-1})'\right).$$

This shows that $\tau_\psi \circ Tf \circ \tau_\phi^{-1}$ is of class C^{p-1}, since $\psi \circ f \circ \phi^{-1}$ is of class C^p. It also shows that Tf is continuous, because $\tau_\psi \circ Tf \circ \tau_\phi^{-1}$ is and τ_ϕ and τ_ψ are homeomorphisms (2.2.9). $\square$

2.5.27. Proposition. *Let X, Y and Z be manifolds of class C^p at least, and $f : X \to Y$, $g : Y \to Z$ differentiable maps of class C^p. Then $T(g \circ f) = Tg \circ Tf$.*

Proof. This is just a restatement of 2.5.15. $\square$

2.5.28. Velocity map. Let $I \subset \mathbf{R}$ be an interval and $\xi : t \mapsto 1_t$ the map from I into TI defined in 2.5.17.2. We have $\xi \in C^\infty(I; TI)$. Indeed, in the canonical charts for I and TI (the latter being deduced from the former), ξ has the expression $t \mapsto (t, 1)$; now use 2.3.1.

It follows that, if (I, α) is a curve on a manifold X (2.5.17), we can define the *velocity map* $\alpha' : I \to TX$ of α by $\alpha' = T\alpha \circ \xi$, and $\alpha' \in C^{p-1}(I; TX)$ by 2.5.26.

2.5.29. The derivative. Let E be a finite-dimensional vector space, X a manifold and $f : X \to E$ a differentiable map. We associate to f its *derivative* $df : TX \to E$, defined as the collection of the $df(x)$ for $x \in X$ (cf. 2.5.23.1). The following diagram commutes:

$$TX \xrightarrow{Tf} TE$$
$$df \searrow \quad \downarrow \theta$$
$$E,$$

where θ is also defined as the collection of the θ_y for $y \in E$.

If $g \in C^p(X; Y)$ we have $d(f \circ g) = df \circ Tg$ (2.5.23.3).

2.6. Submanifolds, Immersions, Submersions and Embeddings

We now extend the notion of submanifolds, introduced in 2.1, from $\mathbf{R}^n$ to abstract manifolds. The reader in encouraged to draw himself the figures essential to his understanding, as we did in 2.1.

2.6.1. Definition. Let X be a d-dimensional manifold and Y a subset of X. We say that Y is an e-dimensional *submanifold of X* if for every $y \in Y$ there exists a chart (U, ϕ) of X at y such that $\phi(U \cap Y) = \phi(U) \cap \mathbf{R}^e$.

Here we have identified $\mathbf{R}^e$ with $\mathbf{R}^e \subset \mathbf{R}^d$ via the canonical inclusion

$$\mathbf{R}^e \to \mathbf{R}^e \times \{0\} \subset \mathbf{R}^e \times \mathbf{R}^{d-e} = \mathbf{R}^d.$$

2.6.1.1. Remark. Definition 2.6.1 can be reformulated as follows: a subset $Y \subset X$ is a submanifold if the inclusion of Y in X is everywhere locally diffeomorphic to the inclusion of $\mathbf{R}^e$ in $\mathbf{R}^d$.

2.6.2. Theorem. *Let X be a manifold of class C^p and Y a submanifold of X. For (U, ϕ) ranging over all charts satisfying the condition in 2.6.1, the pairs $(U \cap Y, \phi|_{U \cap Y})$ form a C^p atlas for Y. In particular, Y has a canonical manifold structure of class C^p.*

Proof. Axiom (AT1) is trivial. To show (AT2), let (U, ϕ) and (V, ψ) be charts satisfying the condition: we have

$$\phi\big((U \cap Y) \cap (V \cap Y)\big) = \phi(U \cap V) \cap \mathbf{R}^e.$$

As to (AT3), we have

$$(\psi|_{U \cap V \cap Y}) \circ (\phi|_{U \cap V \cap Y})^{-1} = (\psi \circ \phi^{-1})|_{\phi(U \cap V) \cap \mathbf{R}^e},$$

which is clearly of class C^p if $\psi \circ \phi^{-1}|_{\phi(U \cap V)}$ is. $\square$

The next lemma shows that every chart on Y is locally of the type required by 2.6.1:

2.6.3. Lemma. *Let X be a manifold, $Y \subset X$ a submanifold and (W, η) a chart of Y at y. There exists a chart (U, ϕ) of X at y such that the intersection $U \cap Y$ is open in W, its image $\phi(U \cap Y)$ equals $\phi(U) \cap \mathbf{R}^e$, and the restrictions $\phi|_{U \cap Y}$ and $\eta|_{U \cap Y}$ coincide.*

Proof. By definitions 2.2.5 and 2.6.1, there exists a chart (V, ψ) of X at y such that $\psi(V \cap Y) = \psi(V) \cap \mathbf{R}^e$ and that $\eta \circ \psi^{-1} : \psi(V \cap W) \to \eta(V \cap W)$ is a diffeomorphism, where $\psi(V \cap W)$ and $\eta(V \cap W)$ are open in $\mathbf{R}^e \subset \mathbf{R}^d$. By the definition of the product topology on $\mathbf{R}^d = \mathbf{R}^e \times \mathbf{R}^{d-e}$, there exist $Z \in O_0(\mathbf{R}^{d-e})$ and $S \in O_{\psi(y)}\big(\psi(V \cap W)\big)$ such that

$$S \times Z \in O_{\psi(y)}\big(\psi(V)\big).$$

We extend $\eta \circ \psi^{-1}|_{S \times \{0\}}$ into a diffeomorphism σ between $S \times Z$ and an open set in $\mathbf{R}^d$ by setting

$$\sigma(s, z) = \big((\eta \circ \psi^{-1})(s, 0), z\big).$$

The desired chart (U, ϕ) is then $\big(\psi^{-1}(S \times Z), \sigma \circ \psi\big)$. $\square$

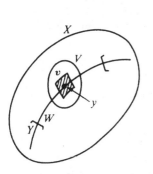 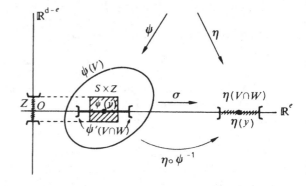

Figure 2.6.3

2.6.4. Proposition. *If X is a manifold and Y is a submanifold of X, the topology of Y as a manifold coincides with the topology induced from X.*

Proof. Applying the definition in 2.2.6 to Y, we see that any open set in the manifold topology of Y is a union of domains of charts of Y, or, by lemma 2.6.3, a union of subsets of Y each of which is the intersection of Y with an open subset of X. Thus such a set is open in the topology induced from X. Conversely, if U is open in the induced topology, we have $U = W \cap Y$ by 2.2.6, where W is a union of domains of charts of X. But for every chart (S, τ) of X such that $S \cap Y \neq \emptyset$, there exists a chart (V, ϕ), with $V \subset S$, of the type required in 2.6.1 (apply definition 2.6.1 to $y \in S \cap Y$ and restrict (V, ϕ) to $(V \cap S, \phi|_S)$ if necessary). □

2.6.5. Proposition. *Let X be a manifold, Y a submanifold of X and Z a submanifold of the manifold Y (2.6.2). Then Z is a submanifold of X.*

Proof. Let d, e and f be the dimensions of X, Y and Z, respectively, and consider the canonical inclusions $\mathbf{R}^f \subset \mathbf{R}^e \subset \mathbf{R}^d$. Definition 2.6.1 shows that, for any point $z \in Z$, there exists a chart (W, η) of Y at z such that $\eta(W \cap Z) = \eta(W) \cap \mathbf{R}^f \subset \mathbf{R}^e$. Apply lemma 2.6.3 to $z \in Y$ and the chart (W, η) of the submanifold $Y \subset X$. After restricting W, if necessary, we can assume that there exists a chart (U, ϕ) of X at z such that $\phi(U \cap Y) = \phi(U) \cap \mathbf{R}^e$ and $W = U \cap Y$. But then

$$\phi(U \cap Z) = \eta(U \cap Z) = \eta(W \cap Z) = \eta(W) \cap \mathbf{R}^f = \phi(U) \cap \mathbf{R}^f. \quad \square$$

2.6.6. Proposition (cf. 2.3.3.4). closure of diff maps under restriction *Let X and Y be manifolds, $f \in C^p(X; Y)$ a differentiable map and Z a submanifold of X. We have $f|_Z \in C^p(Z; Y)$.*

Proof. Apply 2.6.1.1 (cf. exercise 2.8.19). □

2.6.7. Proposition (cf. 2.3.3.5). ditto of range *Let X and Z be manifolds, Y a submanifold of X and $f : Z \to X$ an arbitrary map taking Z inside Y. Let f_Y be f considered as a map from Z into Y. Then $f \in C^p(Z; X)$ is equivalent to $f_Y \in C^p(Z; Y)$.*

Proof. By 2.6.1.1, we can reduce to the case where E, F and G are vector spaces, $Z \in O(E)$, $X = F \times G$, $Y = F \times \{0\} \subset F \times G$ and $f : Z \to F \times G$ is of the form $(f_F, 0)$. By 0.2.8.4, $(f_F, 0)$ is of class C^p if and only if f is. The reader is encouraged to fill in the details (exercise 2.8.19). □

2.6.8. Proposition. inclusion is differentiable *The inclusion $i : Y \to X$ of a submanifold Y into a manifold X is differentiable, and the tangent map $T_y i : T_y Y \to T_y X$ is injective for every $y \in Y$.*

Thus we can identify $T_y Y$ with a vector subspace of $T_y X$, and write $T_y Y \subset T_y X$.

Proof. By 2.6.1.1 we can assume that we're dealing with $i : E \to E \times F$ of the form $i(y) = (y, 0)$, in which case the statement is trivial. $\square$

We now extend to manifolds the notions introduced in 0.2.17, 0.2.20 and 0.2.23, as well as the subsequent results:

2.6.9. Definition. Let X and Y be manifolds and $f : X \to Y$ a differentiable map. We say that f is an *immersion* at x if $T_x f$ is injective, and a *submersion* if $T_x f$ is surjective. We say that f is *regular* at x if $T_x f \in \mathrm{Isom}(T_x X; T_{f(x)} Y)$. The map f is called an *immersion*, a *submersion* or *regular* if the relevant property holds at every $x \in X$. Finally, f is an *embedding* of X in Y if f is an injective immersion and a homeomorphism onto its image $f(X)$.

The results below generalize theorem 2.1.2.

2.6.10. Proposition. *Let Y be a submanifold of a manifold X. The canonical injection $i : Y \to X$ is an embedding of Y in X; in particular, it is an immersion. Conversely, if Z and X are manifolds and $f : Z \to X$ is a differentiable map that is an immersion at $x \in Z$, there exists $U \in O_x(Z)$ such that $f|_U$ is an embedding of U in X. In additon, $f(U)$ is a submanifold of X and f is a diffeomorphism between U and $f(U)$.*

Proof. Since each of these properties is invariant under diffeomorphisms, we can use charts to reduce to the case $Y \in O(E)$ and $X = F$, where E and F are finite-dimensional vector spaces. We then apply theorem 0.2.24, which further reduces the picture locally to the canonical injection $E \to E \times \{0\} \subset E \times F$. In this case $E \times \{0\}$ is indeed a submanifold of $E \times F$, and $E \to E \times \{0\}$ is an embedding, and even a diffeomorphism onto its image. $\square$

2.6.11. Corollary. *If $f : Z \to X$ is an embedding, $f(Z)$ is a submanifold of X and f is a diffeomorphism between Z and $f(Z)$.*

Proof. By 2.6.10, f is everywhere a local diffeomorphism; thus $f^{-1} : f(Z) \to Z$, which exists and is continuous by assumption, is also everywhere locally differentiable. By 2.3.2(iii), f^{-1} is differentiable. $\square$

Watch out: in general, an injective immersion is not an embedding (cf. 2.1.5). But non-compactness is the only obstruction:

2.6.12. Theorem. *An injective immersion $f : Z \to X$ from a compact manifold Z into a manifold X is an embedding; in particular, its image is a submanifold of X.*

Proof. It is a fact from point-set topology that a bijective continuous map from a compact space into a Hausdorff topological space is a homeomorphism. $\square$

2.6.13. Examples

2.6.13.1. The manifold $\mathbf{R}/\mathbf{Z}$ (2.4.12.1) is diffeomorphic to the circle S^1.

Indeed, $\overline{f} : \mathbf{R} \ni t \mapsto (\cos \pi t, \sin \pi t) \in \mathbf{R}^2$ defines a map f on the quotient $\mathbf{R}/\mathbf{Z}$, since $f(t+k) = f(t)$ for all $k \in \mathbf{Z}$:

2.6.13.2

$$\begin{array}{ccc} \mathbf{R} & \xrightarrow{\ \ p\ \ } & \mathbf{R}/\mathbf{Z} \\ & \searrow{\scriptstyle \overline{f}} \quad \swarrow{\scriptstyle f} & \\ & \mathbf{R}^2 & \end{array}$$

The quotient map f is injective and its image is S^1; also, $f \in C^{\infty}(\mathbf{R}/\mathbf{Z}; \mathbf{R}^2)$ by 2.4.11. Since $\mathbf{R}/\mathbf{Z}$ is compact, being the image under p of the compact interval $[0, 1]$, all that is left to show before we can apply 2.6.12 is that f is an immersion. It is enough to check that $\overline{f}$ is an immersion, because p is regular; but this is obvious, since $\overline{f}'(t) = (-\pi \sin \pi t, \pi \cos \pi t) \neq 0$.

2.6.13.3. The projective plane, second recipe. The map $\underline{f}$ in 2.4.12.3 is an embedding of $P^2(\mathbf{R})$ into $\mathbf{R}^6$.

We know that $P^2(\mathbf{R})$ is compact (2.4.12.2) and $\underline{f}$ is injective; the result will follow from 2.6.12 if we can show that $\underline{f}$, or, equivalently, f (since $p : S^2 \to P^2(\mathbf{R})$ is regular) is an immersion. Now $S^2 \subset \mathbf{R}^3$ is a submanifold, and f is the restriction of

$$f : \mathbf{R}^3 \ni (x, y, z) \mapsto (x^2, y^2, z^2, \sqrt{2}\, yz, \sqrt{2}\, zx, \sqrt{2}\, xy) \in \mathbf{R}^6$$

to S^2. Thus it suffices to show that $T_{(x,y,z)}f$, restricted to $T_{(x,y,z)}S^2$, is injective; by definition 2.5.13.1, this amounts to checking that $f'(x, y, z)$ is injective.

The jacobian matrix of f at (x, y, z) is (0.2.8.8)

$$\begin{pmatrix} 2x & 0 & 0 \\ 0 & 2y & 0 \\ 0 & 0 & 2z \\ 0 & \sqrt{2}\, z & \sqrt{2}\, y \\ \sqrt{2}\, z & 0 & \sqrt{2}\, x \\ \sqrt{2}\, y & \sqrt{2}\, x & 0 \end{pmatrix},$$

which has rank three if $(x, y, z) \neq 0$. Thus $f'(x, y, z)$ is injective, and *a fortiori* injective when restricted to $T_{(x,y,z)}S^2$.

2.6.13.4. In chapters 8 and 9 we will study in detail immersions and embeddings in $\mathbf{R}^n$ of the real line $\mathbf{R}$ and of the circle S^1, which are the one-dimensional manifolds: see sections 8.1 and 9.1.

2.6.14. Proposition. *Let X and Y be manifolds and $f : X \to Y$ a submersion at $x \in X$. There exists $U \in O_x(X)$ such that $f^{-1}(f(x)) \cap U$ is a submanifold of X. Conversely, let Y be a submanifold of X and $y \in Y$ a point; there exists $U \in O_x X$ and a submersion $f : U \to \mathbf{R}^{d-e}$ (where $d = \dim X$ and $e = \dim Y$) such that $Y \cap U = f^{-1}(0)$.*

Proof. Apply theorem 0.2.26.   □

Another way to phrase 2.6.14, paralleling 2.1.2(ii), is the following:

2.6.15. Proposition. *Let X be a d-dimensional manifold of class C^p and $x \in X$ a point. Consider k functions f_i of class C^q $(1 \leq q \leq p)$, defined on a neighborhood U of x, taking x to 0 and such that the derivatives $df_i(x)$ $(i = 1, \ldots, k)$ are linearly independent (as linear forms on $T_x X$, cf. 2.5.23). There exists a subneighborhood V of x in U such that*

$$V \cap \bigcap_{i=0}^{k} f_i^{-1}(0)$$

is a $(d-k)$-dimensional submanifold of X, where X is considered as a manifold of class C^q.

Proof. It suffices to consider the differentiable map

$$F : U \ni x \mapsto \big(f_1(x), \ldots, f_k(x)\big) \in \mathbf{R}^k,$$

which is a submersion at x because the $df_i(x)$ $(i = 1, \ldots, k)$ are linearly independent. Proposition 2.6.14 then shows that $F^{-1}\big(F(x)\big) = F^{-1}(0)$ is a submanifold of some open set $V \in O_x(U)$.   □

For a more complete statement, see exercise 2.8.20. The converse of 2.6.15 follows immediately from definition 2.6.1 by transferring to U the $d - e$ coordinate functions $x_{e+1}, \ldots, x_d : \mathbf{R}^d \to \mathbf{R}$.

2.7. Normal Bundles and Tubular Neighborhoods

Let X be an abstract manifold, E a Euclidean space and $f : X \to E$ an immersion; we assume throughout this section that the differentiability class of X and f is at least two. For $x \in X$ we will form the subspace

2.7.1 $N_x X = \big(\theta_{f(x)}(T_x f(T_x X))\big)^{\perp},$

where $\perp$ denotes the *orthogonal complement* in Euclidean space, and θ is as in 2.5.22.1:

2.7.2. Definition. The set

$$N_x X = \big\{ z \in E : (z \mid u) = 0 \text{ for any } u \in \theta_{f(x)}(T_x f(T_x X)) \big\}$$

is called the *normal space* to (X, f) at x (or, if f is understood, the normal space to X at x, or the normal space to $f(X)$ at $f(x)$).

Notice that $\dim N_x X = \dim E - \dim X$.

More generally, we introduce the following notation for subsets of the product $X \times E$:

2.7.3. Notation. We set

$$NX = \{(x, v) \in X \times E : v \in N_x X\},$$
$$N^\varepsilon X = \{(x, v) \in NX : \|v\| < \varepsilon\},$$
$$\overline{N^\varepsilon X} = \{(x, v) \in NX : \|v\| \le \varepsilon\},$$
$$NU^\varepsilon X = \{(x, v) \in NX : \|v\| = \varepsilon\},$$
$$NUX = \{(x, v) \in NX : \|v\| = 1\} = NU^1 X.$$

2.7.4. Definition. The set NX is called the *normal fiber bundle* of (X, f), and NUX the *normal unitary fiber bundle* of (X, f).

We also introduce a *canonical map* can : $NX \to E$ as follows:

2.7.5 $$\text{can}(x, v) = f(x) + v.$$

2.7.6. Definition. The set $\text{Tub}^\varepsilon X = \text{can}(N^\varepsilon X)$ is called the *tubular neighborhood* of radius ε around (X, f).

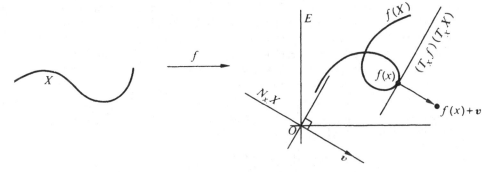

Figure 2.7.6.1

Here's the reason for this term. Assume, for simplicity, that X is a submanifold of E, that is, f is the canonical injection from X into E. We can draw $(T_x X)^\perp$ as the set of bound vectors (2.5.12.4) based at x and orthogonal to $T_x X$, so $\text{can}(x, v) = x + v$ is the tip of such a vector and $\text{Tub}^\varepsilon X$ is the set of tips of all normal vectors at x with norm less than ε, for $x \in X$. If X is a curve in $\mathbf{R}^3$, what we get is a tube of radius ε around X; if X is a curve in $\mathbf{R}^3$, we get a strip of width 2ε. Watch out: it is not true that $\text{Tub}^\varepsilon X$ is the set of points of E whose distance to X is less than ε; see exercise 2.8.29. Notice that $\overline{N^\varepsilon X}$ is the closure and $NU^\varepsilon X$ the frontier of $N^\varepsilon X$.

2.7.7. Theorem. *Let X be a d-dimensional abstract manifold of class C^p and E an n-dimensional Euclidean space. If $f : X \to E$ is an immersion of class C^p, the sets NX, $N^\varepsilon X$ and NUX are manifolds of class C^{p-1} and dimension n, n and $n - 1$, respectively.*

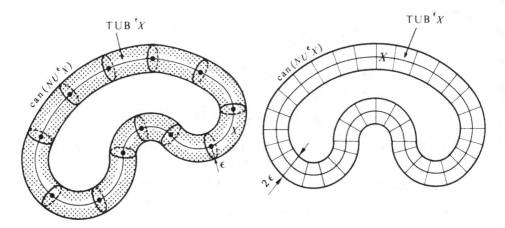

Figure 2.7.6.2

Proof. Since E is a C^∞ manifold, $X \times E$ is a manifold (2.2.10.3). We will show that NX, $N^\epsilon X$ and NUX are submanifolds of $X \times E$ by writing them as zero sets of equations and using 2.6.15.

Consider NX first. Take $(x, v) \in NX$, and let (U, ϕ) be a chart of X at x. We have $(y, w) \in (U \times E) \cap NX$ if and only if $y \in U$ and $w \in N_y X$. We want to make the latter condition explicit.

Introduce the isomorphism θ_y associated with (U, ϕ) and y (2.5.10), and the isomorphism $\theta_{f(y)}$ from 2.5.22.1. For $z \in T_y X$ and $u = \theta_x(z) \in \mathbf{R}^d$ we have, by definition 2.5.13.1:

$$\theta_{f(y)}\big((T_y f)(z)\big) = (f \circ \phi^{-1})'(\phi(y))(u).$$

Thus the subspace $\theta_{f(y)}\big((T_y f)(T_y X)\big)$ of E is spanned by the d vectors $(f \circ \phi^{-1})'(\phi(y))(e_i)$, where $\{e_1, \ldots, e_d\}$ is the canonical basis of $\mathbf{R}^d$. It follows that $w \in N_y X$ if and only if for all $i = 1, \ldots, d$ we have

2.7.8 $\big(w \mid (f \circ \phi^{-1})'(\phi(y))e_i\big) = 0.$

Consider the maps $h_i : U \times E \to \mathbf{R}$ $(i = 1, \ldots, d)$ given by

2.7.9 $h_i(y, w) = \big(w \mid (f \circ \phi^{-1})'(\phi(y))e_i\big).$

We have

$$(U \times E) \cap NX = \bigcap_{i=1}^{d} h_i^{-1}(0),$$

and if we can show that the maps h_i are of class C^{p-1} and that the $dh_i(x, v)$ are linearly independent, it will follow by theorem 2.6.15 that NX is a submanifold of $X \times E$ of class C^{p-1} and dimension $n + d - d = n$. Since the map $(\alpha, \beta) \mapsto (\alpha \mid \beta)$ from $E \times E$ into $\mathbf{R}$ is of class C^∞ and the map

$y \mapsto (f \circ \phi^{-1})(\phi(y))(e_i)$ from U into E is of class C^{p-1}, the h_i are indeed of class C^{p-1}.

Left to study are the maps $dh_i(x, v) = \theta_{h_i(x,v)} \circ T_{(x,v)}h_i : T_{(x,v)}(X \times E) \to \mathbf{R}$, where $\theta_{h_i(x,v)} : T_{h_i(x,y)}\mathbf{R} \to \mathbf{R}$ is the canonical isomorphism associated at $h_i(x, v)$ with the chart $(\mathbf{R}, \mathrm{Id}_{\mathbf{R}})$ of $\mathbf{R}$ (2.5.32.2). But $N_x X$ is a submanifold of $U \times E$ (since it is a vector subspace of E and $\{x\} \times E$ is a submanifold of $X \times E$), and for $(x, v) \in N_x X$ the space $T_v E$ is a subspace of $T_{(x,v)}(X \times E)$ (2.5.18); thus, if we can show that the derivatives $dh_i(x, v)$ $(i = 1, \ldots, d)$, restricted to $T_v E$, are linearly independent, they will necessarily be linearly independent as maps on $T_{(x,v)}(X \times E)$. By applying 2.5.23.3 to the inclusion $\{x\} \times E \subset X \times E$ we get

$$dh_i(x, v)|_{T_v E} = d(h_i|_E)_v,$$

and the $h_i|_E$ are linear forms $\left(\cdot \mid (f \circ \phi^{-1})'(\phi(x))(e_i) \right)$ associated with the fixed vectors $(f \circ \phi^{-1})'(\phi(x))(e_i)$ $(i = 1, \ldots, d)$; thus the differentials of these linear forms are, up to the isomorphism θ_p, equal to themselves (0.2.8.3), and consequently linearly independent. This completes the proof that NX is a submanifold.

It also takes care of $N^\epsilon X$, which is open in NX, being of the form $\{(x, v) \in NX : \|v\| \leq \epsilon\}$: just apply 2.2.10.2.

Finally, $NUX = \{(x, v) \in NX : \|v\| = 1\}$ is also a submanifold of NX, being defined by one equation $f(x, v) = 0$, where $f : NX \to \mathbf{R}$ is given by $f(x, v) = \|v\|^2 - 1$; we just have to show that $df(x, v) \neq 0$ for every point $(x, v) \in NUX$. The technique is the same one we used in the last step of the proof that NX is a manifold: the restriction of f to $\{x\} \times E$ is the quadratic form $\|v\|^2 - 1$, whose derivative at v (up to θ_v) is just the linear form $2(v \mid \cdot)$, which is certainly non-zero if $\|v\| = 1$. This completes the proof of theorem 2.7.7. $\qquad\square$

We now want to show that in certain good cases the map defined in 2.7.5 is an embedding of $N^\epsilon X$ in E. We will need the following result:

2.7.10. Theorem. *For every $x \in X$ the map can is regular at $(x, 0)$.*

Proof. This map is defined by $\mathrm{can}(x, v) = f(x) + v$. It is enough to show that the tangent map $T_{(x,0)}(\mathrm{can}) : T_{(x,0)}NX \to T_{f(x)}E$ is surjective, since $\mathrm{can}(x, 0) = f(x)$ and the spaces NX and E have the same dimension. The subspace $T_x f(T_x X)$ of $T_{f(x)}E$ has as a complementary subspace the image of $\left(\theta_{f(x)}(T_x f(T_x X))\right)^\perp$ under $\theta_{f(x)}^{-1}$. By abuse of notation we shall denote this complementary subspace by $\left(T_x f(T_x X)\right)^\perp$, so that $T_{f(x)}E$ admits the direct sum decomposition

$$T_{f(x)}E = \left(T_x f(T_x X)\right) \oplus \left(T_x f(T_x X)\right)^\perp.$$

To prove that $T_{(x,0)}$ is surjective we just have to show that for every vector $u \in T_x f(T_x X)$ there exists $z \in T_{(x,0)} NX$ such that $(T_{(x,0)}(\text{can}))(z) = u$, and similarly for every $v \in (T_x f(T_x X))^{\perp}$.

First observe that if γ is a curve in NX such that $\gamma'(t) = z$, with $z \in T_{(x,0)} NX$, we have $(T_{(x,0)}(\text{can}))(z) = (\text{can} \circ \gamma)'(t)$ by 2.5.17.4. So take $u \in T_x f(T_x X)$ and $w \in T_x X$ such that $T_x f(w) = (u)$. Let α be a curve of class C^p in X such that $\alpha(0) = x$ and $\alpha'(0) = w$. If we set $\gamma(t) = (\alpha(t), 0)$, where $0 \in E$, we will have a curve of class C^p in NX such that

$$(\text{can} \circ \gamma)'(0) = T_x f(\alpha'(0)) = T_x f(w) = u,$$

and thus a vector $z = \gamma'(0) \in T_{\gamma(0)} NX = T_{(x,0)} NX$ such that

$$T_{(x,0)} \text{can}(z) = (\text{can} \circ \gamma)'(0) = u.$$

Similarly, consider $v \in (T_x f(T_x X))^{\perp}$, and define a curve in NX by $\gamma(t) = (x, tv)$. Then $(\text{can} \circ \gamma)(t) = \text{can}(x, tv) = f(x) + tv$, so $(\text{can} \circ \gamma)'(0) = v$ and the vector $z = \gamma'(1) \in T_{\gamma(0)} NX = T_{(x,0)} NX$ satisfies

$$(\text{can} \circ \gamma)'(0) = T_{(x,0)} \text{can}(z) = v. \qquad \square$$

2.7.11. Remarks. In general, can is not regular everywhere. For instance, take $X = S^1 \subset \mathbf{R}^2$; for every $x \in S^1$ we have $x \in N_x S^1$, so $\text{can}(x, -x) = 0$ (we say that 0 is a *focal point* of S^1).

Furthermore, it is generally not possible to find $\varepsilon > 0$ such that the restriction of can to $N^{\varepsilon} X \to E$ is an embedding, although theorem 2.7.10 implies that can is a local embedding around each point $(x, 0)$ (2.5.20). Figure 2.7.11 shows why: as X tapers off to infinity, points on both sides

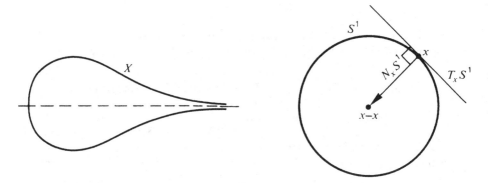

Figure 2.7.11

get closer and closer, and ε must get smaller and smaller. But if X is compact this kind of thing cannot happen:

2.7.12. Theorem. *Let X be a compact manifold, E a Euclidean vector space, $f : X \to E$ an embedding of class C^p. There exists $\varepsilon > 0$ such that the restriction* can $: N^\varepsilon X \to$ Tub$^\varepsilon X$ *is a diffeomorphism.*

Proof. The map can is regular, and by 2.5.20 a local diffeormorphism, at every point of the form $(x, 0)$ for $x \in X$. Thus there exists an open neighborhood $W_x \subset NX$ of $(x, 0)$ such that can$|_{W_x}$ is a diffeomorphism onto its image. Since NX is a submanifold of $X \times E$, we can find a number $\varepsilon_x > 0$ and an open neighborhood $U_x \subset X$ of x such that

$$N^{\varepsilon_x} X \cap (U_x \times E) \subset W_x.$$

Now X is compact and covered by the U_x, so there exists a finite number of points $x_1, \ldots, x_n$ such that $X = U_{x_1} \cup \cdots \cup U_{x_n}$. Set $\varepsilon = \inf_{1 \le i \le n} \varepsilon_{x_i}$; clearly can $: N^\varepsilon X \to$ Tub$^\varepsilon X$ is a local diffeomorphism, because any $(x, v) \in N^\varepsilon X$ satisfies

$$(x, v) \in N^\varepsilon X \cap (U_{x_i} \times E) \subset N^{\varepsilon_{x_i}} X \cap (U_{x_i} \times E) \subset W_{x_i}$$

for some i.

There remains to show that there exists ε such that can$|_{N^\varepsilon X}$ is injective. Assume there is no such ε and take, for each $n \ge 1$, two distinct points $p_n, q_n \in N^{1/n} X$ such that can$(p_n) =$ can(q_n). The sequences (p_i) and (q_i) are contained in a compact set (for example, $X \times \overline{B}(0, 1)$), so we can take subsequences converging to p and q, respectively. Since $p_n \in N^{1/n} X$, we can write $p_n = (x_n, u_n)$ with $\|u_n\| < 1/n$, so in the limit we must have $p = (x, 0)$ and $q = (y, 0)$. Also, by continuity,

$$\text{can}(p) = \lim_{n \to \infty} \text{can}(p_n) = \lim_{n \to \infty} \text{can}(q_n) = \text{can}(q),$$

that is, $f(x) = f(y)$. Since f is injective by assumption, we have $x = y$. But the restriction of can to a neighborhood of $(x, 0) = p = q$ is a diffeomorphism; and for n big enough the points p_n and q_n must lie in such a neighborhood. Thus the assumption can$(p_n) =$ can(q_n) implies $p_n = q_n$, which is absurd.

Thus we have found $\varepsilon > 0$ such that can$|_{N^\varepsilon X}$ is injective and regular. From this we deduce that can$|_{N^{\varepsilon'} X}$, for any $\varepsilon' < \varepsilon$, is an embedding. We use the same technique as in the proof of theorem 2.6.12, by noticing that $\overline{N}^{\varepsilon'} X \subset N^\varepsilon X$ is compact (since it is closed in the compact subset $X \times \overline{B}(0, \varepsilon)$ of $X \times \mathbf{R}^n$), and thus that $(\text{can}|_{\overline{N}^{\varepsilon'} X})^{-1}$ is continuous; in particular, $(\text{can}|_{N^{\varepsilon'} X})^{-1}$ is continuous. $\square$

2.7.13. Examples. We will meet the case $d = 1$, $n = 3$ (tubes) in 10.2.3.2 and the case $d = 2$, $n = 3$ (parallel surfaces) in 10.2.3.11 and 10.6.8.

2.8. Exercises

2.8.1. Which of the sets below are submanifolds of $\mathbf{R}^n$? Of what differentiability class?

$$\{(x, y, z) \in \mathbf{R}^3 : x^3 + y^3 + z^3 - 3xyz = 1\};$$
$$\{(x, y, z) \in \mathbf{R}^3 : x^2 + y^2 + z^2 = 1 \text{ and } x^2 + y^2 - x = 0\};$$
$$\{(t, t^2) : t \in \mathbf{R}_-\} \cup \{(t, -t^2) : t \in \mathbf{R}_+\};$$
$$\{(x, y) \in \mathbf{R}^2 : x = 0 \text{ or } y = 0\};$$
$$\{(t^2, t^3) : t \in \mathbf{R}\};$$
$$\{(x, y) \in \mathbf{R}^2 : y = |x|\};$$
$$\left\{\left(\cos t, \cos \frac{t}{3} + \sin \frac{t}{3}\right) : t \in \left]0, 4\pi + \frac{3\pi}{4}\right[\right\}.$$

2.8.2. Check the compatibility of the various parametrizations of S^2 of the type

$$(x, y) \mapsto (x, y, \sqrt{1 - x^2 - y^2}).$$

2.8.3. Prove in several ways, using the conditions in 2.1.2, that if $V \subset \mathbf{R}^n$ and $W \subset \mathbf{R}^m$ are submanifolds, so is $V \times W \subset \mathbf{R}^n \times \mathbf{R}^m = \mathbf{R}^{n+m}$.

2.8.4. Show that the set

$$\{(\cos(\sqrt{2}\,t)(2 + \cos t), \sin(\sqrt{2}\,t)(2 + \cos t), \sin t) \in \mathbf{R}^3 : t \in \mathbf{R}\}$$

is not a submanifold of $\mathbf{R}^3$. Show that this set is dense in the torus

$$\{(\cos t(2 + \cos s), \sin t(2 + \cos s), \sin t) \in \mathbf{R}^3 : s, t \in \mathbf{R}\}.$$

2.8.5. Veronese's surfaces and generalizations
(a) Using the map $F : \mathbf{R}^6 \setminus 0 \to \mathbf{R}^6$ defined by

$$F(x_1, \ldots, x_6) = (2x_1 x_2 - x_6^2, 2x_2 x_3 - x_4^2, 2x_3 x_1 - x_5^2,$$
$$x_4 x_5 - \sqrt{2}\, x_3 x_6, x_5 x_6 - \sqrt{2}\, x_1 x_4, x_6 x_4 - \sqrt{2}\, x_2 x_5),$$

show that

$$\{(x^2, y^2, z^2, \sqrt{2}\, yz, \sqrt{2}\, zx, \sqrt{2}\, xy) \in \mathbf{R}^6 : x, y, z \in \mathbf{R}\}$$

is a submanifold of $\mathbf{R}^6$.
(b) Deduce that

$$\{(x^2, y^2, z^2, \sqrt{2}\, yz, \sqrt{2}\, zx, \sqrt{2}\, xy) \in \mathbf{R}^6 : x^2 + y^2 + z^2 = 1\}$$

is a submanifold of $\mathbf{R}^6$.

(c) Show that, for every n, the set

$$\{(\underbrace{\ldots, x_i^2, \ldots,}_{n} \underbrace{\ldots, \sqrt{2}\, x_i x_j, \ldots,}_{n(n-1)/2}) \in \mathbf{R}^{n(n+1)/2} : x_1^2 + \cdots + x_n^2 = 1\}$$

is a submanifold of $\mathbf{R}^{n+n(n-1)/2} = \mathbf{R}^{n(n+1)/2}$.

(d) Same question for

$$\{(\underbrace{\ldots, |z_i|^2, \ldots,}_{n} \underbrace{\ldots, \sqrt{2}\, \mathrm{Re}(z_i \bar{z}_j), \ldots,}_{n(n-1)/2} \underbrace{\ldots, \sqrt{2}\, \mathrm{Im}(z_i \bar{z}_j), \ldots}_{n(n-1)/2} \in \mathbf{R}^{n^2} :$$

$$|z_1|^2 + \cdots + |z_n|^2 = 1\}$$

where $|z|$, $\mathrm{Re}(z)$ and $\mathrm{Im}(z)$ denote the absolute value, real part and imaginary part, respectively, of a complex number z.

(e) Same as (d), substituting quaternions for complex numbers.

2.8.6. Prove 2.2.10.3 in detail.

2.8.7. Stereographic projections. Let i_N be the inversion (0.5.3.1) with pole $N = (0, \ldots, 0, 1)$ taking $S^d \setminus N$ onto the plane $\{x_{d+1} = 0\} \subset \mathbf{R}^{d+1}$, and i_S the corresponding inversion with pole $S = (0, \ldots, 0, -1)$. Write formulas for i_N and i_S, and show that together these two maps form a C^∞ atlas on S^d, compatible with the usual submanifold structure of the sphere.

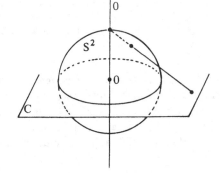

Figure 2.8.7

2.8.8. Grassmannians

(a) Show that the set $V_{k,d}$ of k-tuples of linearly independent vectors in $\mathbf{R}^n$ is an open subset of $(\mathbf{R}^d)^k$.

(b) Let $G_{k,d}$ be the quotient of $V_{k,d}$ by the equivalence relation obtained by identifying k-tuples of vectors that span the same subspace of $\mathbf{R}^d$. Thus, $G_{k,d}$ can be understood as the set of k-dimensional vector subspaces of $\mathbf{R}^d$. Show that $G_{k,d}$, with its quotient structure, is a compact space. (To prove Hausdorffness, show that for every pair (X, Y) of points of $G_{k,d}$ there exists a continuous real-valued function ρ such that $\rho(X) \neq \rho(Y)$. For compactness, consider the subspace $\tilde{V}_{k,d}$ of $V_{k,d}$ consisting of orthonormal sets in k vectors, where $\mathbf{R}^d$ is taken with its canonical Euclidean structure. Show that $\tilde{V}_{k,d}$ is compact, and construct a continuous map $\tilde{V}_{k,d} \to G_{k,d}$.) The spaces $G_{k,d}$ are called *grassmanianns*.

(c) Let X be an element of $G_{k,d}$ and U_X the set of $Y \in G_{k,d}$ such that $Y \cap X^{\perp} = \{0\}$. Define a map $T_X : U_X \to L(X; X^{\perp})$ as follows: For $Y \in U_X$ and $x \in X$, let $T_X(Y)x$ be the $X^{\perp}$-component of the unique vector $z \in Y$ whose X-component is x. Show that T_X is a homeomorphism. (Hint: let $(x^i)_{1 \leq i \leq k}$ be an orthornormal basis for X, and let $y^i(Y)$ be the unique vector in Y whose X-component is x^i. We have $y^i(Y) = x^i + T_X(Y)x^i$; show that each map $Y \to Y^i(Y)$ is continuous.)

(d) Show that $(U_X, T_X)_{X \in G_{k,d}}$ is a C^{∞} atlas on $G_{k,d}$, and that the topology determined on $G_{k,d}$ by this atlas is the same as the quotient topology inherited from $V_{k,d}$.

2.8.9. Let H be a *hyperquadric* in $\mathbf{R}^d$, with equation

$$\sum_{1 \leq i \leq j \leq d} a_{ij} x_i x_j + \sum_{i=1}^{d} b_i x_i = 1,$$

where the symmetric matrix (a_{ij}) is invertible (such a hyperquadric is called *central*). Show that H, which is a $(d-1)$-dimensional C^{∞} submanifold of $\mathbf{R}^d$, is C^{∞} diffeomorphic to $S^k \times \mathbf{R}^{d-k-1}$, where $(k, d-k)$ is the *signature* of the quadratic form associated with the (a_{ij}) (this means that in appropriately chosen coordinates $y_1, \ldots, y_d$ of $\mathbf{R}^d$ the quadratic form has the expression $\sum_{i=1}^{k} y_i^2 - \sum_{i=k+1}^{d} y_i^2$; cf. 4.2.8).

2.8.10. In this exercise and the next E denotes a d-dimensional Euclidean space, $\text{End}(E)$ the set of linear maps from E into itself, $\text{GL}(E) \subset \text{End}(E)$ the set of invertible linear maps, $\text{SL}(E) \subset \text{GL}(E)$ the set of linear maps of determinant one, and $O(E)$ the set of linear maps u such that ${}^t u u = 1$. Show that $\text{GL}(E)$, $\text{SL}(E)$ and $O(E)$ are C^{∞} submanifolds of $\text{End}(E)$, and calculate their dimension.

2.8.11. Show that, for $u \in \text{End}(E)$, the series $\sum_{n=0}^{\infty} \dfrac{u^n}{n!}$ converges in the canonical norm (0.0.10). Its sum is denoted by $\exp u$.

(a) Show that $\det(\exp u) = e^{\text{Tr}(u)}$, and that $\exp(u+v) = \exp u \exp v$ if u and v commute.

(b) Show that $u \mapsto \exp(u)$ is a C^{∞} map from $\text{End}(E)$ into $\text{GL}(E)$.

(c) Using the inverse function theorem, show that there exists an open neighborhood V of 0 in $\text{End}(E)$, an open neighborhood V' of 0 in the vector space $\{u \in \text{End}(E) : \text{Tr}(u) = 0\}$ and an open neighborhood

V'' of 0 in the vector space $\{u \in \text{End}(E) : {}^t u + u = 0\}$ such that $\exp|_V$, $\exp|_{V'}$ and $\exp|_{V''}$ are parametrizations of V, V' and V'' onto neighborhoods of the identity in $\text{GL}(E)$, $\text{SL}(E)$ and $O(E)$, respectively. This is another way to prove the dimensions in 2.8.10.

(d) Let $S(E)$ be the subspace of $\text{End}(E)$ consisting of symmetric linear maps and $S^+(E)$ the subspace consisting of symmetric linear maps whose associated quadratic form is positive definite. Show that $S^+(E)$ is open in $S(E)$, and that the map $u \mapsto u^2$ is a C^∞ diffeomorphism from $S^+(E)$ into itself. Then construct a C^∞ diffeomorphism between $\text{GL}(E)$ and $S^+(E) \times O(E)$.

2.8.12. Let G be a group of diffeomorphisms of X acting properly discontinuously without fixed points, with associated projection $p : X \to X/G$, and let Y be a manifold. Show that $f \in C^p(X/G; Y)$ if and only if $f \circ p \in C^p(X; Y)$:

$$
\begin{array}{ccc}
X & & \\
p \downarrow & \searrow{\scriptstyle f \circ p} & \\
X/G & \xrightarrow{\ f\ } & Y
\end{array}
$$

2.8.13. Identify $\mathbf{R}^2$ with $\mathbf{C}$. Is the map $z \mapsto z^3$ of $\mathbf{C}$ into itself a covering map?

2.8.14. Take $X = \mathbf{R}^2 \setminus 0$ and let G be the order two group generated by $x \mapsto -x$. Show that G acts properly discontinuously without fixed points, and that the quotient X/G is diffeomorphic to X.

2.8.15. Consider the sphere $S^{2d-1} \subset \mathbf{R}^{2d} = \mathbf{C}^d$, and let $q_1, \ldots, q_d$ be integers. Let G be the group generated by the map $g : \mathbf{C}^d \to \mathbf{C}^d$ defined by

$$(z_1, \ldots, z_d) \mapsto (e^{2\pi i / q_1} z_1, \ldots, e^{2\pi i / q_d} z_d).$$

Can the process of theorem 2.4.9 be used to give S^{2d-1}/G a differentiable structure?

2.8.16. Blow-ups. Let U be an open neighborhood of 0 in $\mathbf{R}^d$. In the manifold $U \times P^{d-1}(\mathbf{R})$, let U' be the subset of points (x, z) such that, for some set $(z_1, \ldots, z_d)$ of homogeneous coordinates for z, we have $x_j z_k - x_k z_j$ for every pair of indices (j, k) (this condition clearly does not depend on the choice of homogeneous coordinates).

(a) Show that U' is a closed d-dimensional submanifold of $U \times P^{d-1}(\mathbf{R})$. We say that U' is the manifold obtained from U by *blowing up* at the origin.

(b) Let π_U be the restriction to U' of the projection $U \times P^{d-1}(\mathbf{R}) \to U$. Show that $\pi_U^{-1}(0)$ is a submanifold of U' diffeomorphic to $P^{d-1}(\mathbf{R})$, and that the restriction of π_U to $U' \setminus \pi_U^{-1}(0)$ is a diffeomorphism between this set and $U \setminus 0$.

2.8.17. (You should read this exercise now, but probably won't be able to solve it until after reading 4.2.13 and 5.2.9 below.) If X is a C^∞ manifold, we denote by $C^\infty(X)$ the **R**-algebra of C^∞ real-valued functions on X, and by $V(X)$ the **R**-vector space of C^∞ vector fields on X (that is, C^∞ maps $\xi : X \to TX$ such that $p \circ \xi = \mathrm{Id}_X$; see 3.5.1). A *derivation* on $C^\infty X$ is an **R**-linear map $D : C^\infty(X) \to C^\infty(X)$ such that $D(fg) = D(f)g + fD(g)$ for any $f, g \in C^\infty(X)$. The set of derivations on $C^\infty(X)$ is denoted by $\mathrm{der}\big(C^\infty(X)\big)$.

2.8.17.1. Take $\xi \in V(X)$ and $f \in C^\infty(X)$, and define $df(\xi) = \xi(f) \in C^\infty(X)$ by

$$\big(df(\xi)\big)(x) = df(x)\big(\xi(x)\big)$$

for every $x \in X$. Show that $f \mapsto \xi(f)$ belongs to $\mathrm{der}\big(C^\infty(X)\big)$. We will show that, conversely, every $D \in \mathrm{der}\big(C^\infty(X)\big)$ comes from some $\xi \in V(X)$.

(a) Take $U \in O(\mathbf{R}^d)$ and let $x_1, \ldots, x_d$ be the canonical coordinates in $\mathbf{R}^d$. Given a vector field

$$\xi = \sum_i a_i \frac{\partial}{\partial x_i},$$

where $a_i \in C^\infty(U)$, calculate $\xi(f)$.

(b) If $D \in \mathrm{der}\big(C^\infty(X)\big)$ and f is constant on X we have $Df = 0$.

(c) Let $D \in \mathrm{der}\big(C^\infty(X)\big)$ be a derivation, $U \in O(X)$ an open set and $f, g \in C^\infty(X)$ functions such that $f|_U = g|_U$. Show that $Df|_U = Dg|_U$.

(d) If $U \in O(\mathbf{R}^d)$ is an open set and $x_1, \ldots, x_d$ the canonical coordinates on $\mathbf{R}^d$, show that, for every $D \in \mathrm{der}\big(C^\infty(U)\big)$, $f \in C^\infty(U)$ and $u \in U$ we have

$$(Df)(u) = \sum_i (Dx_i)(u)\frac{\partial f}{\partial x_i}(u).$$

Deduce that given D there exists $\xi \in V(U)$ such that $Df = \xi(f)$ for any $f \in C^\infty(U)$.

(e) Let X be an arbitrary manifold and $D \in \mathrm{der}\big(C^\infty(X)\big)$. Show that there exists $\xi \in V(X)$ such that $Df = \xi(f)$ for every $f \in C^\infty(X)$.

2.8.17.2. The bracket of two vector fields

(a) Let X be a manifold and $\xi, \eta \in V(X)$. Show that the map

$$f \mapsto \xi\big(\eta(f)\big) - \eta\big(\xi(f)\big)$$

belongs to $\mathrm{der}\big(C^\infty(X)\big)$. We will denote by $[\xi, \eta]$ the vector field associated to this map by 2.8.17.1, and call it the *bracket* of ξ and η.

(b) For $\xi, \eta, \varsigma \in V(X)$, show that $[\xi, [\eta, \varsigma]] + [\varsigma, [\xi, \eta]] + [\eta, [\varsigma, \xi]] = 0$.

(c) Let $U \in O(\mathbf{R})$ and

$$\xi = \sum_i a_i \frac{\partial}{\partial x_i}, \qquad \eta = \sum_i b_i \frac{\partial}{\partial x_i},$$

with $a_i, b_i \in C^\infty(U)$, be vector fields in U. Compute the functions c_i such that

$$[\xi, \eta] = \sum_i c_i \frac{\partial}{\partial x_i}.$$

(d) Let Y be a submanifold of X and $\xi, \eta \in V(X)$ vector fields such that $\xi(y)$ and $\eta(y)$ lie in T_yY for every $y \in Y$. Show that $[\xi, \eta](y) \in T_yY$ for every $y \in Y$. See 3.5.15 for the converse (the Frobenius integrabitity theorem.

2.8.18. Let $X \subset Y$ and $X' \subset Y'$ be submanifolds. Show that $X \times X' \subset Y \times Y'$ is also one.

2.8.19. Give detailed proofs for 2.6.6 and 2.6.7.

2.8.20. Let X be a d-dimensional manifold of class C^p and $x \in X$ a point. If $f_1, \ldots, f_k \in C^p(U)$, where U is open in X, are functions whose derivatives at x are linearly independent, show that there exists a chart for X at x whose first k coordinates are the f_i.

2.8.21. In the space of real 2×2 matrices, consider the open set Ω consisting of the invertible matrices, and the set G of those of determinant 1. Prove that G is a C^∞ submanifold of Ω, and that the maps $M \mapsto M^{-1}$ from G into itself and $(M, N) \mapsto MN$ from G^2 into G are differentiable.

2.8.22. The helicoid. Show that the set $V = \{(t \cos z, t \sin z, z) : t, z \in \mathbf{R}\}$ is a submanifold of $\mathbf{R}^3$. Let $p : \mathbf{R}^2 \ni (x, y, z) \mapsto (x, y) \in \mathbf{R}^2$ be the canonical projection. Is $p|_V$ a submersion?

2.8.23. Show that there exists no C^r embedding of S^1 in $\mathbf{R}$, for $r \geq 0$. Is there a C^1 immersion of S^1 in $\mathbf{R}$?

2.8.24. Let X be a d-dimensional connected manifold and Y a closed submanifold of codimension ≥ 2. Show that $X \setminus Y$ is connected.

2.8.25. Show that the formula

$$h(u, v) = \left(2\operatorname{Re}(u\bar{v}), 2\operatorname{Im}(u\bar{v}), |u|^2 - |v|^2\right),$$

where $u, v \in \mathbf{C}$ satisfy $|u|^2 + |v|^2 = 1$, defines a C^∞ map $h : S^3 \to S^2$, called the *Hopf fibration*. Show that every $x \in S^2$ has an open neighborhood U such that there exists a diffeomorphism $\phi : U \times S^1 \to h^{-1}(U)$ satisfying $h(\phi(y, z)) = y$ for all $y \in U$ and $z \in S^1$. In particular, h is a submersion. (Hint: use the stereographic projection, 2.8.7.)

Using quaternions, find a map $h : S^7 \to S^4$ having similar properties.

2.8.26. Define an equivalence relation on $\mathbf{C}^{d-1} \setminus 0$ by "$x \sim y$ if and only if there exists $\lambda \in \mathbf{C}^*$ such that $x = \lambda y$". The quotient of $\mathbf{C}^{d-1} \setminus 0$ by this relation is called the d-dimensional *complex projective space*, and denoted by $P^d(\mathbf{C})$; the canonical projection is denoted by $p : \mathbf{C}^{d+1} \setminus 0 \to P^d(\mathbf{C})$. A set of *homogeneous coordinates* for a point $X \in P^d(\mathbf{C})$ is a $(d+1)$-tuple $x \in \mathbf{C}^{d+1}$ such that $p(x) \in X$.

(a) Show that there exists a C^∞ atlas, defined by means of homogeneous coordinates, that makes $P^d(\mathbf{C})$ into a $2d$-dimensional manifold; the projection p is a submersion for this structure.

(b) Same question for $P^d(\mathbf{H})$, where we have substituted the quaternions $\mathbf{H}$ for $\mathbf{C}$.

2.8.27. Same notation as in 2.8.26.

(a) Let π be the restriction of $p : \mathbf{C}^{d+1} \setminus 0 \to P^d(\mathbf{C})$ to

$$S^{d+1} = \left\{ (u_1, \ldots, u_{d+1}) : \sum_{i=1}^{d+1} |u_i|^2 = 1 \right\}.$$

Show that, for every $X \in P^d(\mathbf{C})$, there exists a neighborhood U of X and a diffeomorphism $\phi : U \times S^1 \to \pi^{-1}(U)$ such that $\pi(\phi(y, z)) = y$ for any $y \in U$ and $z \in S$.

(b) State and prove an analogous property for $P^d(\mathbf{H})$.

2.8.28. Do parts (b), (c) and (d) of exercise 2.8.5 by using 2.8.16 and the technique introduced in 2.6.13.2.

2.8.29. Let $X \subset E$ be a submanifold of a Euclidean space E and set

$$B(X, \varepsilon) = \{ y \in E : d(y, X) < \varepsilon \},$$

where $d(y, X) = \inf \{ \|y - x\| : x \in X \}$. Show that $\mathrm{Tub}^\varepsilon X \subset B(X, \varepsilon)$ for all ε. Show that the two are generally not equal. Can you find conditions for equality to hold?

2.8.30. Let $p : \mathbf{R}^{n+1} \setminus 0 \to P^n(\mathbf{R})$ be the canonical projection (2.4.12.2). Show that the $n + 1$ maps

$$c_i : \mathbf{R}^n \ni (x_1, \ldots, x_n) \mapsto p(x_1, \ldots, x_i, 1, x_{i+1}, \ldots, x_n) \in P^n(\mathbf{R}),$$

for $i = 0, \ldots, n$, define an atlas $\{(c_i(\mathbf{R}^n), c_i^{-1})\}$ for $P^n(\mathbf{R})$ (cf. exercise 2.8.26 for the complex case).

CHAPTER 3

Partitions of Unity, Densities and Curves

In this chapter we develop analytic tools that will be fundamental to our study of manifolds. Partitions of unity (section 3.2) are the most important of these tools, but to prove their existence (3.2.4) we need to impose some restrictions on our manifolds (3.1.6); we provide some motivation for this step in section 3.1.

Using partitions of unity, we define densities and prove their existence (3.3.8). From that we derive a class of measures on a given manifold, called Lebesgue measures (3.3.11). No one Lebesgue measure on a manifold is canonical, but the notion of sets of measure zero is (3.3.13).

As an application of densities, we prove that every connected manifold of dimension one is a line or a circle (3.4.1).

The last tool we introduce is vector fields on manifolds. The compact case is particularly interesting (3.5.13); a typical application of it will be the proof of Moser's theorem (7.2.3).

3.1. Embeddings of Compact Manifolds

3.1.1. Theorem. *Let X be a compact, d-dimensional abstract manifold. For some integer n there exists an embedding of X in $\mathbf{R}^n$.*

Thus every compact manifold is homeomorphic to a submanifold of some $\mathbf{R}^n$.

Proof. By 2.6.12 it is enough to discover an integer n and a map $F : X \to \mathbf{R}^n$ that is at the same time an injection and an immersion. We will need the following lemma:

3.1.2. Lemma. *Let $x \in X$ be a point. There exist open neighborhoods $W_x \subset V_x \subset X$ of x and a C^p-differentiable function $f : X \to \mathbf{R}$ such that $f = 0$ on $X \setminus V_x$, $f = 1$ on W_x and $0 < f(z) < 1$ for $z \in V_x \setminus \overline{W}_x$. In other words, f is a bump function around x.*

Proof. Just use charts to reduce to the case of $\mathbf{R}^d$, where such functions are known to exist. More exactly, take a function $g : \mathbf{R}^d \to \mathbf{R}$ of class C^∞ such that $g = 1$ on $B(0,1)$, $g = 0$ on $\mathbf{R}^d \setminus \overline{B}(0,2)$ and $0 < g(t) < 1$ for $t \in B(0,2) \setminus \overline{B}(0,1)$; such a function exists by 0.2.16. Let (U, ϕ) be a chart at $x \in X$ and λ a positive number such that $B(0,3) \subset \lambda\phi(U)$. Set $\psi(x) = \lambda\phi(x)$ for $x \in U$. The pair (U, ψ) is a chart compatible with the atlas of X and satisfying $B(0,3) \subset \psi(U)$. Since $\psi : U \to \psi(U)$ is a homeomorphism, the sets $V_x = \psi^{-1}\big(B(0,2)\big)$ and $W_x = \psi^{-1}\big(B(0,1)\big)$ are open. Clearly $x \in W_x \subset V_x$.

We define $f : X \to \mathbf{R}$ by $f = g \circ \psi$ on U and $f = 0$ on $X \setminus U$. Already we have $0 \le f \le 1$. Since $W_x \subset U$ we have $f(y) = g\big(\psi(y)\big) = 1$ for $y \in W_x$. Moreover, if $y \in X \setminus V_x$, either $y \notin U$ or $y \in U$ with $\psi(y) \notin B(0,2)$; either way $f(y) = 0$. This also shows that $0 < f(y) < 1$ for $z \in V_x \setminus \overline{W}_x$.

Finally, f is of class C^∞ on $X \setminus \overline{V}_x$ (being identically zero), and of class C^p on U (being the composition of g and ψ). The sets U and $X \setminus \overline{V}_x$ form an open cover for X ($X \setminus \overline{V}_x$ is open because X is Hausdorff—cf. the proof of 2.2.11), implying that f is everywhere of class C^p.   □

Now consider for each $x \in X$ the open sets W_x and V_x given by the lemma. The W_x cover X, so by compactness we can choose a finite number of points $x_1, \ldots, x_n$ such that $X = \bigcup_{k=1}^n W_{x_k}$.

Let f_i be the bump function associated by the lemma with the point x_i, and (U_i, ϕ_i) the corresponding chart. We construct $F : X \to \mathbf{R}^{n(d+1)}$ by setting

$$F(x) = \big(f_1(x), \ldots, f_n(x), f_1(x)\phi_1(x), \ldots, f_n(x)\phi_n(x)\big).$$

First we make sure that F is well-defined: although the ϕ_i are only defined on U_i, the products $f_i\phi_i$ are defined on the whole of X because $\mathrm{supp}(f_i) \subset U_i$ (the proof is identical to that of 3.1.2). Then we must check that F is an injective immersion of class C^p.

Clearly F is injective: if $F(x) = F(y)$ and i_0 is such that $x \in W_{x_{i_0}}$, we have $f_{i_0}(x) = 1 = f_{i_0}(y)$, whence $y \in W_{x_{i_0}}$ because of the way we defined f_{i_0}. Thus x and y are in U_{i_0}, and, since ϕ_{i_0} is injective, the equality $f_{i_0}(x)\phi_{i_0}(x) = f_{i_0}(y)\phi_{i_0}(y)$ forces $x = y$.

There remains to show that F is an immersion, that is, $T_x F$ is injective for any x. We can write $TF = \big(Tf_1, \ldots, Tf_n, T(f_1\phi_1), \ldots, T(f_n\phi_n)\big)$, and $T_x F$ will be injective if any of the $T_x(f_k\phi_k)$ are. Choose i such that $x \in W_{x_i}$; on W_{x_i} we have $f_i\phi_i = \phi_i$, and $T_x\phi_i$ is injective because ϕ_i is a diffeomorphism. This implies that $T_x(f_i\phi_i)$ is injective, as desired.   □

The question arises whether any abstract manifold can be embedded in some $\mathbf{R}^n$. The answer is no. To give a counterexample, we introduce the following notion:

3.1.3. Definition. A topological space T is said to be *separable* if its topology has a countable base (this means that there exists a countable set $\mathcal{B}$ of open subsets of T such that every open subset of T is a union of elements of $\mathcal{B}$).

3.1.4. Examples

3.1.4.1. $\mathbf{R}^d$ is separable, for we can take $\mathcal{B} = \big\{B(x,r) : x \in \mathbf{Q}^d, r \in \mathbf{Q}\big\}$, where $\mathbf{Q}$ is the set of rational numbers.

3.1.4.2. A subspace of a separable topological space is separable.

3.1.4.3. Proposition. *Every submanifold of $\mathbf{R}^n$ is separable.*

Proof. This follows from the previous two statements and the fact that the canonical topology on a submanifold is the induced topology from $\mathbf{R}^n$. □

Thus a non-separable manifold cannot be embedded in $\mathbf{R}^n$, since separability is preserved under diffeomorphisms. This answers our question in the negative, because there exist non-separable manifolds: for example the manifold introduced in 2.2.10.5. It may be argued that this example is not connected; but the long half-line (2.2.10.6) is not separable, either, nor is Prüfer's surface (exercise 3.6.4).

Non-separability is the sole obstruction to embedding in $\mathbf{R}^n$:

3.1.5. Theorem. *Any separable abstract manifold X can be embedded in some $\mathbf{R}^n$.*

The proof, due to Whitney, is more difficult than in the compact case; we refer the reader to [Ste64, p. 63]. One of the building blocks in the proof is the existence of partitions of unity on a separable manifold; this result we will demonstrate, since it will be useful many times in the future.

3.1.6

> From now on all manifolds
> will be assumed Hausdorff and separable.

3.1.7. The tangent bundle. Having said that, we must verify that if X is a manifold according to this convention its tangent bundle TX (2.5.24) also is. This is true because X, and consequently TX, are countable unions of domains of charts, and we have the following result:

3.1.7.1. Proposition. criterion for separability *A manifold that is a countable union of domains of charts is separable.*

Proof. In fact, $\mathbf{R}^n$ is separable, and a countable union of countable sets is countable. □

3.1.7.2. Similarly, it must be checked that if X is a separable manifold and G acts properly discontinuously without fixed points on X (2.4.9) the manifold X/G is separable. We leave this to the reader.

3.2. Partitions of Unity

3.2.1. Definition. Let T be a topological space. An open cover $\mathcal{U} = \{U_i : i \in I\}$ for T is said to be *locally finite* if, for every $x \in T$, there exists a neighborhood U of x in T such that $U \cap U_i = \emptyset$ except for a finite number of indices i.

If $\mathcal{U} = \{U_i : i \in I\}$ and $\mathcal{V} = \{V_j : j \in J\}$ are two open covers, we say that $\mathcal{U}$ is *subordinate* to $\mathcal{V}$ if, for every $i \in I$, there exists $j \in J$ such that $U_i \subset V_j$.

3.2.2. Definition. Let X be a manifold of class C^p. *A partition of unity* on X is a family $\{(U_i, \psi_i) : i \in I\}$, where the $U_i \subset X$ are open and the ψ_i are C^p maps from X into $\mathbf{R}$, satisfying the following conditions:

(PU1) $\psi_i(x) \geq 0$ for every $x \in X$ and every $i \in I$;
(PU2) the support of ψ_i is contained in U_i;
(PU3) $\mathcal{U} = \{U_i : i \in I\}$ is a locally finite cover of X;
(PU4) $\sum_{i \in I} \psi_i(x) = 1$ for every $x \in X$.

3.2.3. Remark. The sum in (PU4) makes sense because for any given x only finitely many of the $\psi_i(x)$ are non-zero: by (PU3) we can find an open neighborhood of U intersecting only finitely many of the U_i, and by (PU2) only the associated ψ_i are non-zero at x.

3.2.4. Theorem. *Let X be a (separable and Hausdorff) manifold. For every open cover $\mathcal{U}$ of X there exists a partition of unity subordinate to $\mathcal{U}$.* existence of ... subordinate to a partition

Notice that we do not assume that $\mathcal{U}$ is locally finite.

Proof. There are several steps.

3.2.5. Lemma. *A (separable) manifold X has a countable base $\mathcal{B} = \{U_n : n \in \mathbf{N}\}$ whose elements are relatively compact (that is, their closure $\overline{U}_n$ in X is compact).*

Proof. By definition X has a countable base $\{W_n : n \in K\}$. Take $x \in X$ and let (V, ψ) be a chart at x. Since $\psi(V)$ is open in $\mathbf{R}^d$ and $\mathbf{R}^d$ is locally compact, there exists a compact neighborhood L of $\psi(x)$ contained in $\psi(V)$. The map ψ is a homeomorphism from V onto its image, so $U = \psi^{-1}(\mathring{L})$ is open in X, and its closure $\overline{U} = \psi^{-1}(L)$ is compact. If we set $\phi = \psi|_U$ the pair (U, ϕ) is a chart at x with relatively compact domain.

Let K' be the set of $n \in K$ such that W_n is contained in relatively compact domains of charts, and set $\mathcal{B}' = \{W_n : n \in K'\}$. For $n \in K'$ we have $\overline{W}_n \subset \overline{U}$ compact. Clearly $\mathcal{B}'$ is countable; we will show that it is a base for the topology of X.

Let $S \subset X$ be open; we want to write S as a union of sets W_n, for $n \in K'$. For $x \in S$ there exists a chart (U, ϕ) of X at x such that $\overline{U}$ is compact. But for any $V \subset U$ the closure $\overline{V}$ is still compact, so, by restricting U if necessary, we can assume that we have a chart (U_x, ϕ) at x whose domain is relatively compact and contained in S.

Since U_x is open in X and $\{W_n : n \in K\}$ is a base for X there exists $K_x \subset K$ such that $U_x = \bigcup_{n_x \in K_x} W_{n_x}$. For such values of n_x we have $W_{n_x} \subset U_x$ relatively compact, so $n_x \in K'$, showing that U_x is a union of elements of $\mathcal{B}'$. Since S is the union of the U_x for all $x \in S$, we have written S as a union of elements of $\mathcal{B}'$, and shown that $\mathcal{B}'$ is a base. $\qquad\square$

3.2.6. Lemma. *Let X be a (separable) manifold. There exists a sequence $\{A_k : k \in \mathbf{N}\}$ of compact subsets of X such that $X = \bigcup_{k \in \mathbf{N}} A_k$, with $A_k \subset \mathring{A}_{k+1}$ for every k. exhaustion by compact subsets*

Proof. By the previous lemma X has a countable cover $\{U_n\}$ consisting of relatively compact open subsets. Set $A_1 = \overline{U}_1$. For $x \in A_1$ take n_x such that U_{n_x} contains x; the U_{n_x} cover A_1 as x ranges over A_1, but in fact finitely many of them are enough, since A_1 is compact. In particular, we have $A_1 \subset \bigcup_{1 \le k \le h_1} U_k$ for some h_1, and this union is relatively compact, so we set $A_2 = \bigcup_{1 \le k \le h_1} \overline{U}_k$. (If $h_1 = 1$, which can happen if $\overline{U}_1 = U_1$, we add 1 to h_1, so our sequence won't be stationary.) Continuing in this fashion we obtain the desired sequence. $\qquad\square$

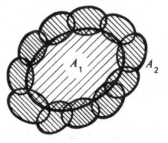

Figure 3.2.6

3.2.7. Lemma. *For every open cover $\mathcal{U}$ of X there exists an atlas $\{(V_k, \phi_k): k \in K\}$ of X such that the open cover $\mathcal{V} = \{V_k : k \in K\}$ is subordinate to $\mathcal{U}$, locally finite, and satisfies the conditions $\phi_k(V_k) = B(0,3)$ and*

$$\bigcup_{k \in K} \phi_k^{-1}(B(0,1)) = X.$$

Proof. Take $x \in X$ and let $W \in \mathcal{U}$ be an element of the cover containing x. Let (V, ψ) be a chart at x such that $V \subset W$ and $B(0,3) \subset \psi(V)$ (use a homothety if necessary). The pair (U, ϕ), where $U = \psi^{-1}(B(0,3))$ and $\phi = \psi|_U$, is a chart at x such that $\phi(U) = B(0,3)$ and $U \subset W$.

Now consider the sequence $\{A_k : k \in K\}$ found in the previous lemma. For each k, the set $A_{k+1} \setminus \mathring{A}_k$ is compact (being closed in A_{k+1}), and $\mathring{A}_{k+2} \setminus A_{k-1}$ is open. For every $x \in A_{k+1} \setminus \mathring{A}_k$ we can find a chart (V_x, ϕ_x) at x such that $\phi_x(V_x) = B(0,3)$ and that V_x is contained in some element of $\mathcal{U}$ and in $\mathring{A}_{k+2} \setminus A_{k-1}$. Since $x \in \phi_x^{-1}(B(0,1))$, we can cover the compact set $A_{k+1} \setminus \mathring{A}_k$ with a finite number of such open sets $\phi_x^{-1}(B(0,1))$.

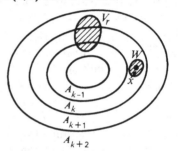

Figure 3.2.7

If we now consider all of the differences $A_{k+1} \setminus \mathring{A}_k$, plus the single compact set A_1 (all these sets together cover X), and put together all the V_x obtained by the process above, we get a countable family which we denote by $\mathcal{V} = \{V_r : r \in \mathbf{N}\}$. We claim that the corresponding charts (V_r, ϕ_r) form an atlas satisfying the conditions of the lemma.

In fact, we have already seen that the V_x cover X, and the (V_r, ϕ_r) are charts of X, so $\mathcal{V}$ is an atlas. We have required that each V_r be contained in an element of $\mathcal{U}$, so the atlas is subordinate to $\mathcal{U}$. Finally, we must show that $\mathcal{V}$ is locally finite. Assume, without loss of generality, that the sequence $\{A_k : k \in K\}$ is strictly increasing, and take $x \in X$. Choose k such that $x \in W = A_{k+1} \setminus \mathring{A}_k$. If $W \cap V_r$ is not empty, V_r must be contained in one of the sets $\mathring{A}_{h+3} \setminus A_h$, $\mathring{A}_{h+2} \setminus A_{h-1}$, $\mathring{A}_{h+1} \setminus A_{h-2}$, since those are the only sets of the form $\mathring{A}_{h+2} \setminus A_{h-1}$ that intersect $A_{k+1} \setminus \mathring{A}_k$. By construction, each of these three difference sets contains only a finite number of the V_r; this shows that W only intersects a finite number of sets V_r. $\square$

We can now conclude the proof of theorem 3.2.4. Let f be a bump function on $\mathbf{R}^d$, that is, a function $f : \mathbf{R}^d \to [0,1]$ of class C^∞ such that $f(x) = 1$ for $x \in B(0,1)$ and $f(x) = 0$ for $x \notin B(0,2)$. Let (V_k, ϕ_k) be one of the charts constructed in the proof of lemma 3.2.7. For each k, define $\theta_k : X \to \mathbf{R}$ as $\theta_k = 0$ on $X \setminus V_k$ and $\theta_k = f \circ \phi_k$ on V_k. Clearly θ_k is of class C^p, because it is a composition of C^p maps on V_k and it vanishes identically on the open set $X \setminus \phi_k^{-1}(B(0,2))$ containing the complement of V_k. Only a finite number of θ_k are non-zero at any $x \in X$, since $x \in V_k$ for only a finite number of indices k. In addition, at least one θ_k is non-zero at any $x \in X$, since the sets $\phi_k^{-1}(B(0,1))$ cover X.

Thus the function $\theta(x) = \sum_k \theta_k(x)$ is well-defined, of class C^p and strictly positive, and we get our partition of unity by setting $\psi_k = \dfrac{\theta_k}{\theta}$ for each k. Namely, each ψ_k is of class C^p, takes values in $[0,1]$, and is contained in some element $U \in \mathcal{U}$ because $\operatorname{supp}\theta_k = \operatorname{supp}\psi_k$ (here we have used lemma 3.2.7). In addition, the covering $\{V_k : k \in \mathbf{N}\}$ is locally finite, and we have $\sum_k \psi_k(x) = 1$. $\square$

3.3. Densities

3.3.1. Let X be a C^p manifold. We denote by $\operatorname{Dens}(X)$ the disjoint union of the sets $\operatorname{Dens}(T_x X)$, for $x \in X$ (0.1.24), and we call this set the density bundle of X. The projection $p : \operatorname{Dens}(X) \to X$ is defined by $p(\operatorname{Dens}(T_x X)) = \{x\}$. A density on X is a map $\delta : X \to \operatorname{Dens}(X)$ such that $p \circ \delta = \operatorname{Id}_X$, or equivalently, $\delta(x) \in \operatorname{Dens}(T_x X)$ for every $x \in X$. We denote by $\Delta(X)$ the set of densities on X.

3.3.2. If X and Y are manifolds, $f : X \to Y$ is of class C^p and regular, and $\delta \in \Delta(Y)$ is a density, we can define the pullback $f^*\delta \in \Delta(X)$ of δ (see 0.1.29.3 and 0.3.11). We have $(g \circ f)^* = f^* \circ g^*$.

3.3.3. If $X = U \in O(E)$, where E is a vector space, the definition above does not coincide with 0.3.11.1. (That's why we used the notation $\underline{\Delta}$ in 0.3.11.1.) The definitions coincide, however, if we identify $T_x U$ with E by means of θ_x (2.5.22.1).

3.3.4. If X is a manifold, (U, ϕ) is a chart on X and $\delta \in \Delta(X)$, we obtain by 3.3.2 and 3.3.3 a density $(\phi^{-1})^*\delta \in \underline{\Delta}(\phi(U))$.

3.3.5. Definition. Let X be a C^p manifold. A density $\delta \in \Delta(X)$ is said to be of class C^q $(0 \le q \le p-1)$ if $(\phi^{-1})^*\delta \in \underline{\Delta}_q(\phi(U))$ for every chart on X.

3.3.6. In practice it suffices to check that $(\phi^{-1})^*\delta \in \underline{\Delta}_q(\phi(U))$ for the charts in a single atlas of X. Indeed, if (U, ϕ) and (V, ψ) are two charts and $V \supset U$, the property $(\psi^{-1})^*\delta \in \underline{\Delta}_q(\psi(V))$ implies $(\phi^{-1})^*\delta \in \underline{\Delta}_q(\phi(U))$. This follows from formula 0.3.11.3 applied to $\psi \circ \phi^{-1}$, because, by 3.3.2,

$$(\phi^{-1})^*\delta = (\psi \circ \phi^{-1})^*\big((\psi^{-1})^*\delta\big);$$

if $(\psi^{-1})^*\delta = b\delta_0$ with b of class C^q $(q \le p-1)$ we have

$$(\phi^{-1})^*\delta = (b \circ f)\big|J(f)\big|\,\delta_0 \in \underline{\Delta}_q(\phi(U))$$

(recall that the jacobian of a map of class C^p is of class C^{p-1}).

3.3.7. Example. If $f : X \to Y$ is a regular differentiable map and $\delta \in \Delta_q(Y)$ we have $f^*\delta \in \Delta_q(X)$ (3.3.2 and 3.3.8). In particular, this is the case if $p : X \to Y$ is a covering map (2.4.1).

3.3.8. Theorem. *Let X be a manifold of class C^p ($p \geq 1$). There exist densities of class C^q on X for every $0 \leq q \leq p-1$; the set of such densities is denoted by $\Delta_q(X)$. If $\delta, \delta' \in \Delta_q(X)$ are densities on X, there exists $g \in C^q(X; \mathbf{R}_+^*)$ such that $\delta' = g\delta$; conversely, if $\delta \in \Delta_q(X)$ and $g \in C^q(X; \mathbf{R}_+^*)$ we have $g\delta \in \Delta_q(X)$.*

Proof. By theorem 3.2.4 we can find a partition of unity $\big\{(U_k, \phi_k, \psi_k)\big\}_{k \in K}$ such that each U_k is the domain of a chart (U_k, ϕ_k) on X. Define $\delta \in \Delta(X)$ by

$$3.3.9 \qquad\qquad \delta = \sum_k \psi_k \phi_k^* \delta_0,$$

where δ_0 is as in 0.3.11 and the identification of 3.3.3 is in effect. We have $\delta \in \Delta(X)$ by 0.1.29.2 and because $\sum_k \psi_k = 1$ implies that for every $x \in X$ there exists $k \in K$ such that $\psi_k(x) > 0$. There remains to show that $\delta \in \Delta_{p-1}(X)$. Apply 3.3.6 to the atlas $\{(U_k, \phi_k)\}$ to obtain, for any h:

$$(\phi_h^{-1})^*\delta = \sum_k (\phi_h^{-1})^*(\psi_k \phi_k^* \delta_0) = \sum_k (\psi_k \circ \phi_h^{-1})\big((\phi_k \circ \phi_h^{-1})^* \delta_0\big)$$

$$= \sum_k (\psi_k \circ \phi_h^{-1}) \big| J(\phi_k \circ \phi_h^{-1}) \big| \delta_0$$

$$= \Big(\sum_k (\psi_k \circ \phi_h^{-1}) \big| J(\phi_k \circ \phi_h^{-1}) \big| \Big) \delta_0.$$

The function that multiplies δ_0 is of class C^{p-1}.

The second part of the theorem is local in nature. Let (U, ϕ) be a chart, and consider the densities $(\phi^{-1})^*\delta'$ and $(\phi^{-1})^*\delta$. Writing $\delta' = g\delta$ with $g : X \to \mathbf{R}_+^*$ (0.1.29.1), we have

$$(\phi^{-1})^*\delta' = a'\delta_0 = (g \circ \phi^{-1})a\delta_0$$

on $\phi(U)$. By assumption, $a, a' \in C^q\big(\phi(U); \mathbf{R}_+^*\big)$, so $a' = (g \circ \phi^{-1})a$ implies $g \circ \phi^{-1} \in \big(\phi(U); \mathbf{R}_+^*\big)$, whence $g|_U \in C^q\big(\phi(U); \mathbf{R}_+^*\big)$. The converse is trivial. $\qquad\square$

3.3.10. Remark. One consequence of 3.3.8 is that densities are not unique, indeed far from it. Abstract manifolds do not have a canonical density. But we will see in section 6.6 that submanifolds of Euclidean spaces do.

The next result is fundamental:

3.3.11. Theorem. *Let X be a manifold of class C^p. A density $\delta \in \Delta_0(X)$ canonically determines a measure on X, denoted by $d\delta$ (or simply δ).*

Proof. We work in several steps:

3.3.11.1. By the remarks at the beginning of section 0.4, it is enough to associate to δ a positive linear form μ on $K(X)$, since X is locally compact by 2.2.11 and a countable union of compact sets by 3.1.6 and 3.2.6. (Strictly speaking, the result we're using, as proved in [Gui69], depends on X being metrizable, but this need not concern us for two reasons: first, [Gui69] only uses metrizability in showing the existence of partitions of unity, and we already have done that; and every manifold is metrizable anyway, either by 3.1.5 or by the argument in exercise 3.6.3. We will not use the metrizability of manifolds in this book.)

3.3.11.2. First define μ on the set of $f \in K(X)$ such that $\operatorname{supp} f \subset U$, where U is the domain of some chart (U, ϕ). If $(\phi^{-1})^*\delta = a\delta_0$, where δ_0 is the canonical density on $\mathbf{R}^d$, set

3.3.11.3
$$\mu(f) = \int_{\phi(U)} (f \circ \phi^{-1}) a\delta_0,$$

which makes sence since $(f \circ \phi^{-1})a$ lies in $K(\phi(U))$ and is continuous (definition 3.3.5). To see that $\mu(f)$ depends only on f and not on the chart (U, ϕ), take another chart (V, ψ) such that $\operatorname{supp} f \subset V$. If $(\psi^{-1})^*\delta = b\delta_0$, we obtain

$$\int_{\phi(U)} (f \circ \phi^{-1}) a\delta_0 = \int_{\psi(V)} (f \circ \psi^{-1}) b\delta_0$$

by applying 0.4.6 and formula 0.3.11.3 to the diffeomorphism $\psi \circ \phi^{-1} : \phi(U) \to \psi(V)$.

3.3.11.4. To extend μ to $K(X)$, take a partition of unity $\{(U_i, \phi_i, \psi_i)\}_{i \in I}$ associated with the charts (U_i, ϕ_i), as in the proof of 3.3.8, and set

3.3.11.5
$$\mu(f) = \sum_{i \in I} \mu(\psi_i f),$$

which makes sense because $\operatorname{supp} f$, being compact, intersects only finitely many domains of charts (3.2.2).

We have to show that 3.3.11.5 does not depend on $\{(U_i, \phi_i, \psi_i)\}_{i \in I}$, so let $\{(V_j, \eta_j, \varsigma_j)\}_{j \in J}$ be another partition. As we have seen, only a finite number of functions $\psi_i f$ and $\varsigma_j f$ are non-zero; since $\sum_i \psi_i = \sum_j \varsigma_j = 1$ and μ is additive, we have

$$\sum_i \mu(\psi_i f) = \sum_i \mu\left(\psi_i\left(\sum_j \varsigma_j f\right)\right) = \sum_i \mu\left(\sum_j \psi_i \varsigma_j f\right) = \sum_{i,j} \mu(\psi_i \varsigma_j f)$$

and

$$\sum_{j} \mu(s_j f) = \sum_{j} \mu\left(s_j\left(\sum_{i} \psi_i f\right)\right) = \sum_{j} \mu\left(\sum_{i} \psi_i s_j f\right) = \sum_{i,j} \mu(\psi_i s_j f).$$

(This calculation is classical in integration theory: see [Gui69, proposition 1.15]).

That μ is positive follows from 3.3.11.3, 3.3.11.5 and the fact that the Lebesgue measure is positive; linearity is trivial.   □

3.3.12. Definition. A *Lebesgue measure* on a manifold X is a measure deduced from a continuous density (one of class C^0) by theorem 3.3.11.

If $d\delta'$ and $d\delta$ are Lebesgue measures on X, we can find, by 3.3.8, a function $g \in C^0(X; \mathbf{R}_+^*)$ such that $d\delta' = g \, d\delta$. Then 0.4.4.3 implies that a set A has $d\delta$-measure zero if and only if it has $d\delta'$-measure zero. Hence the definition:

3.3.13. Definition. A subset A of a manifold X is said to have *measure zero* if it has measure zero for some Lebesgue measure on X.

3.3.14. Notation. Let X be a manifold and $\delta \in \Delta_0(X)$ a density. We denote by $C_\delta^{\text{int}}(X)$ the space of $d\delta$-integrable functions $X \to \mathbf{R}$ (0.4.1). The integral of a function $f \in C_\delta^{\text{int}}$ will be denoted by one of the following:

3.3.15
$$\mu(f) = \int_X f\delta = \int_X f \, d\delta.$$

3.3.16. Proposition. *Let X and Y be manifolds, $f \in \mathrm{Diff}(X; Y)$ a differentiable map and ε a continuous density on Y. Let $\delta = f^*\varepsilon$ be the (continuous) pullback of ε (3.3.7). For any $b \in C_\varepsilon^{\text{int}}(Y)$ we have $b \circ f \in C_\delta^{\text{int}}(X)$ and*
$$\int_X (b \circ f)\delta = \int_Y b\varepsilon.$$

Proof. This follows from 3.3.3 and the proof of 3.3.11.   □

In the language of measure theory [Gui69, 6.2.2] this means that the measure associated to ε is the image under f of the measure associated to $\delta = f^*\varepsilon$.

3.3.17. Examples

3.3.17.1. If $f \in \mathrm{Diff}(X; Y)$ and A has measure zero in X, the image $f(A)$ has measure zero in Y.

3.3.17.2. If $Y \subset X$ is a submanifold of codimension grater than zero, Y has measure zero in X. To see this, cover Y with a countable family of charts (U, ϕ) such that $\phi(U \cap Y) = \phi(U) \cap \mathbf{R}^e$ (definition 2.6.1), where $\mathbf{R}^e \subset \mathbf{R}^d$ and $d = \dim X$, $e = \dim Y$. By 0.4.4.3, $\phi(U) \cap \mathbf{R}^e$ has measure zero in $\phi(U)$ for some measure $a\delta_d$, hence for δ_d itself; we finish off with 0.4.4.1 and 3.3.17.1.

3.3.17.3. Proposition. *Let X and Y be manifolds of the same dimension, and $f \in C^1(X;Y)$. If A has measure zero in X, the image $f(A)$ has measure zero in Y.*

Proof. By 0.4.4.1 we can use charts to reduce to the case $X = U \in O(\mathbf{R}^n)$, $Y = \mathbf{R}^n$, and the Lebesgue measure. We then apply 0.4.4.5. $\square$

3.3.17.4. Corollary. *If $\dim X < \dim Y$ and $f \in C^1(X;Y)$ the image $f(X)$ has measure zero in Y.*

Proof. Extend f to $\overline{f} : X \times \mathbf{R}^{d-e}$ by $\overline{f}(x,y) = f(x)$, where $d = \dim Y$ and $e = \dim X$. Now use the fact that X has measure zero in $X \times \mathbf{R}^{d-e}$ (3.3.17.2). $\square$

3.3.17.5. See exercise 3.6.1 for the behavior of Lebesgue measures under covering maps.

3.3.17.6. Sard's theorem (4.3.1), a fundamental tool in differential geometry, asserts that certain subsets associated with differentiable maps have measure zero.

3.3.18. Product densities. Let E and F be d- and e-dimensional vector spaces, where d and e are finite. For $\delta \in \mathrm{Dens}(E)$ and $\varepsilon \in \mathrm{Dens}(F)$ there exist, by 0.1.25, $\alpha \in \Lambda^d E^*$ and $\beta \in \Lambda^e F^*$ such that $\delta = |\alpha|$ and $\varepsilon = |\beta|$. We have

$$p^* \alpha \wedge q^* \beta \in \Lambda^{d+e} (E \times F)^* \setminus 0.$$

3.3.18.1. Definition. The density $\delta \otimes \varepsilon = |p^* \alpha \wedge p^* \beta|$ on $E \times F$ is called the *product* of δ and ε.

This does not depend on α and β because α and β are determined up to a factor ± 1.

3.3.18.2. Examples. If δ_d and δ_e are the canonical densities on $\mathbf{R}^d$ and $\mathbf{R}^e$, respectively (0.1.26), we have

$$\delta_d \otimes \delta_e = \delta_{d+e},$$

the canonical density on $\mathbf{R}^{d+e} = \mathbf{R}^d \times \mathbf{R}^e$. If $\delta = a\delta_d$ and $\varepsilon = a\delta_e$ we have

$$\delta \otimes \varepsilon = ab\delta_{d+e}.$$

3.3.18.3. Lemma. *If E, E', F, F' are finite-dimensional vector spaces, $f \in$ Isom$(E; E')$ and $g \in Isom(F; F')$ are isomorphisms and δ' and ε' are densities in E' and F', respectively, we have*

$$(f \times g)^*(\delta' \otimes \varepsilon') = f^*\delta' \otimes g^*\varepsilon'.$$

Proof. Use definition 0.1.25 and the following fact, easily derived from the definitions of exterior product and pullback: if $\alpha' \in \Lambda E'^*$, $\beta' \in \Lambda F'^*$ and p, q denote the canonical projections from both $E' \times F'$ and $E \times F$, we have

$$(f \times g)^*(p^*\alpha' \wedge q^*\beta') = p^*(f^*\alpha') \wedge q^*(g^*\beta'). \qquad \square$$

Let X and Y be manifolds and $\delta \in \Delta_0(X)$, $\varepsilon \in \Delta_0(Y)$ densities. For every $(x, y) \in X \times Y$ we use the identification in 2.5.18 to write $\delta(x) \otimes \varepsilon(y) \in \text{Dens}(T_{(x,y)}(X \times Y))$. With this we can define the product density $\delta \otimes \varepsilon \in \Delta(X \times Y)$:

3.3.18.4 $\quad \delta \otimes \varepsilon : X \times Y \ni (x, y) \mapsto \delta(x) \otimes \varepsilon(y) \in \text{Dens}(T_{(x,y)}(X \times Y)).$

3.3.18.5. Proposition. *The map $\delta \otimes \varepsilon$ is a continuous density on $X \times Y$, and the measure on $X \times Y$ associated with $\delta \otimes \varepsilon$ is the product of the measures on X and Y associated with δ and ε, respectively* (0.4.5).

Proof. By 3.3.6 it's enough to check continuity with respect to one particular atlas; and, sure enough, we choose the atlas from definition 2.2.10.3. By 2.5.18, 3.3.18.2 and 3.3.18.3, with the product chart $(U \times V, \phi \times \psi)$, we have

3.3.18.6 $\qquad\qquad ((\phi \times \psi)^{-1})^*(\delta \otimes \varepsilon) = ab\delta_{d+e},$

where $(\phi^{-1})^*\delta = a\delta_d$ and $(\psi^{-1})^*\varepsilon = b\delta_e$. Since a and b are continuous, so is ab.

Now we have to study the measure on $X \times Y$ associated with $\delta \otimes \varepsilon$. By the construction in theorem 3.3.11, it is enough to do so within domains of charts $(U \times V, \phi \times \psi)$. In this case the measure on $\phi(U) \times \psi(V) \subset \mathbf{R}^d \times \mathbf{R}^e = \mathbf{R}^{d+e}$ associated with $((\phi \times \psi)^{-1})^*(\delta \otimes \varepsilon)$ is exactly the one associated with $ab\delta_{d+e}$ (formula 3.3.18.6).

There remains to check that the measure associated with $ab\delta_{d+e}$ is the product of the measures associated with $a\delta_d$ and $b\delta_e$. This follows from the fact that δ_{d+e} is the product $\delta_d \otimes \delta_e$ (0.4.5) and from Fubini's theorem (0.4.5.1):

$$\int_{\phi(U) \times \psi(V)} f(x, y)(a(x)b(y)\delta_{d+e}) = \int_{\phi(U) \times \psi(V)} (f(x, y)a(x)b(y))\delta_{d+e}$$

$$= \int_{\psi(V)} \left(\int_{\phi(U)} f(x, y)a(x)b(y)\delta_d \right) \delta_e$$

$$= \int_{\psi(V)} b(y) \left(\int_{\phi(U)} f(x, y)(a(x)\delta_d) \right) \delta_e.$$

□

3.3.18.7. Corollary. *Let X and Y be manifolds with densities δ and ε, respectively, and $f \in C^{\text{int}}_{\delta \otimes \varepsilon}(X \times Y)$ an integrable map. For ε-almost every $y \in Y$ the map $\{x \mapsto f(x,y)\}$ is δ-integrable. In addition, the function $y \mapsto \int_X f(x,y)\delta$ (defined ε-almost everywhere) is ε-integrable, and*

$$\int_{X \times Y} f(x,y)\delta \otimes \varepsilon = \int_Y \left(\int_X f(x,y)\delta \right) \varepsilon.$$

Proof. This is Fubini's theorem (0.4.5.1). □

3.4. Classification of Connected One-Dimensional Manifolds

In this section we show that there are only two homeomorphism types of one-dimensional manifolds. For a classification of compact two-dimensional manifolds, see 4.2.25.

3.4.1. Theorem. *Every connected one-dimensional manifold is diffeomorphic to $\mathbf{R}$ or to S^1.*

3.4.2. Convention. In this section a *curve* will mean a connected one-dimensional manifold of class C^1. (Later on, in chapters 8 and 9, we will use the word curve in a more precise way.) *Densities* will always be assumed continuous.

Proof. The main idea is this: for one-dimensional submanifolds of $\mathbf{R}^n$, we have a particular parametrization, based on the natural notion of arclength. In the case of abstract manifolds, we must define the arclength to begin with, which is the same as choosing a norm on $T_x X$. Since $T_x X$ has dimension one, norms and densities are the same (by 0.1.29.6).

3.4.3. Definition. Let X be a curve and δ a density on X. A *parametrization by arclength* for X (with respect to δ) is a map $\alpha : I \to X$, where $I \subset \mathbf{R}$ is an interval, such that α is a diffeomorphism onto its image and $\delta(\alpha(t))((T_t\alpha)(1_t)) = 1$ for every $t \in I$.

This makes sense because $\delta(\alpha(t))$ is in fact a norm on $T_{\alpha(t)} X$ and $(T_t\alpha)(1_t)$ lies in $T_{\alpha(t)} X$.

3.4.4. Lemma. *Let X be a curve, δ a density on X and (I,α) a parametrization of X. Then X has a parametrization by arclength (J,β) with respect to δ.*

It follows from this lemma that every curve has parametrizations by arclength, since every curve admits densities (theorem 3.3.8) and parametrizations (inverses of charts, for instance).

Proof. This is analogous to parametrizing a curve in $\mathbf{R}^3$ (in the elementary sense) by its arclength. Let (I, α) be a parametrization and $\theta_t : T_t\mathbf{R} \to \mathbf{R}$, for $t \in I$, the canonical isomorphism (2.5.12.3). The given density δ on X can be pulled back to a density $\alpha^*\delta$ on I, given by

$$\alpha^*\delta(t) = (T_t\alpha \circ \theta_t^{-1})^*\delta(\alpha(t)),$$

and by 0.1.29.1 we can write $\alpha * \delta = g\delta_0$, where δ_0 is the canonical density on $\mathbf{R}$ (0.1.26) and the scalar function $g : I \to \mathbf{R}_+^*$ is continuous (3.3.5). Let t_0 be a fixed point in I, and define a map $\gamma : I \to \mathbf{R}$ by

$$\gamma(t) = \int_{t_0}^{t} g(s)\, ds.$$

We see that γ is defined and differentiable on I, and its derivative is $\gamma'(t) = g(t) > 0$ by the fundamental theorem of calculus; it follows that γ is a bijection between I and an interval $J \subset \mathbf{R}$. Let $\beta : J \to X$ be given by $\beta(s) = \alpha \circ \gamma^{-1}$; we want to show that β is a parametrization by arclength, which is the same as showing that $\beta^*\delta = \delta_0$. But we have

$$\beta^*\delta = (\gamma^{-1})^*\alpha^*\delta = (\gamma^{-1})^*(g\delta_0),$$

and this equals $(1/g) \cdot g\delta_0 = \delta_0$ by 0.3.11.3 and because $J(\gamma^{-1}) = (\gamma^{-1})' = (\gamma')^{-1} = 1/g$.

This also shows that $T_s\beta$ is non-zero, and so establishes an isomorphism between $T_s\mathbf{R}$ and $T_{\beta(s)}X$, the two spaces being one-dimensional. Thus β is regular and a local diffeomorphism by 2.5.20; restricting J if necessary we obtain a parametrization by arclength. $\square$

3.4.5. Lemma. *Let (I, α) and (J, β) be parametrizations of X by arclength (with respect to the same density δ). Assume that $\alpha(I) \cap \beta(J) \neq \emptyset$. Then either $\alpha(I) \cap \beta(J)$ is connected, and there exists a third parametrization by arclength (K, γ) such that $\gamma(K) = \alpha(I) \cup \beta(J)$; or $\alpha(I) \cap \beta(J)$ has two connected components, and X is diffeomorphic to S^1.*

Proof. The map $\alpha^{-1} \circ \beta$ is defined at $s \in J$ if and only if $x = \beta(s) \in \beta(J) \cap \alpha(I)$; in this case set $t = (\alpha^{-1} \circ \beta)(s)$. The map that associates to $x \in \beta(J) \cap \alpha(I)$ the pair (s, t), where $s \in J$ and $t \in I$ are the unique scalars such that $\beta(s) = x$ and $\alpha(t) = x$, is continuous, since α and β are homeomorphisms. Thus the set $\Gamma = \{(s, t) : \alpha(t) = \beta(s) = x \text{ for some } x \in \beta(J) \cap \alpha(I)\}$ has as many connected components as $\beta(J) \cap \alpha(I)$.

Notice that Γ is the graph of the map $s \mapsto t = (\alpha^{-1}\circ\beta)(s)$, which satisfies $|(\alpha^{-1}\circ\beta)'(s)| = 1$, since (I, α) and (J, β) are parametrizations by arclength (formula 0.3.11.3). Thus Γ is made up of segments of slope ± 1, and these segments cannot have endpoints in $J \times I$, because Γ is closed (being the set of pairs (s, t) such that $\beta(s) - \alpha(t) = 0$) and open (its projection on I being $\alpha^{-1}(\alpha(I) \cap \beta(J))$, which is open in I). In addition, for a given value of s there exists at most one value of t such that $\beta(s) = \alpha(t)$, since α and β are bijections; thus there can be at most one segment starting or ending on each side of $J \times I$. In particular, there can't be segments of slope $+1$ and -1 at the same time:

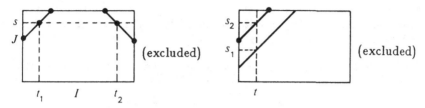

Figure 3.4.5.1

The only possible cases are:

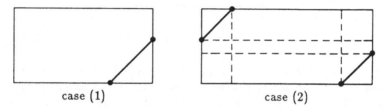

case (1)   case (2)

Figure 3.4.5.2

This already shows that Γ, and consequently $\alpha(I) \cap \beta(J)$, have either one or two connected components.

3.4.5.1. First case. After translating I and J and perhaps switching the sign of t, we can assume that $I = {]b, a[}$ and $J = {]0, c[}$ (possibly with $b = -\infty$, $a = +\infty$ or $c = +\infty$), and that Γ is a single segment of slope $+1$ joining $(0,0)$ and (a,a).

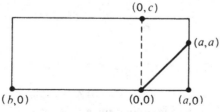

Figure 3.4.5.3

(In the figure we have assumed $a \leq c$; the reader should analyze the case $a > c$.) We must find a parametrization by arclength (K, γ) such that

$\gamma(K) = \alpha(I) \cup \beta(J)$. Take $K =]b, c[$, and define γ by

$$\gamma(t) = \begin{cases} \alpha(t) & \text{for } t \in]b, 0], \\ \alpha(t) = \beta(t) & \text{for } t \in]0, a[, \\ \beta(t) & \text{for } t \in [a, c[. \end{cases}$$

Clearly $\gamma(K) = \alpha(I) \cup \beta(J)$, and γ is a diffeomorphism onto its image because it is that when restricted to $]b, a[$ and to $]0, c[$. The last condition in the definition of a parametrization by arclength is also satisfied, since $(T_t\gamma)(1_t)$ is equal to either $(T_t\alpha)(1_t)$ or $(T_t\beta)(1_t)$.

3.4.5.2. Second case. Since Γ consists of two segments, we can assume that it looks like this:

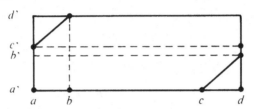

Figure 3.4.5.4

This can only happen if $I =]a, d[$ and $J =]a', d'[$ are bounded; if $d' = +\infty$, for example, there would be a horizontal line intersecting the two segments. After a translation we can assume $a' = c$, whence $b' = d$. Thus

$$a < b \le c < d = b' \le c' < d'.$$

We define a map $\hat{h} :]a, d'[\to X$ as follows:

$$\hat{h}(t) = \begin{cases} \alpha(t) & \text{for } t \in]a, d[, \\ \beta(t) & \text{for } t \in]c, d'[. \end{cases}$$

By construction, α and β coincide on $]c, d[$ and, more importantly, we have $\hat{h}(t + c' - a) = \hat{h}(t)$ for $t \in]a, b[$. Thus we can extend $\hat{h}$ to $\mathbf{R}$ by requiring it to be periodic of period $c' - a$. By section 2.4 and by construction, $\hat{h}$ factors:

$$\begin{array}{ccc} & \mathbf{R} & \\ {\scriptstyle p}\downarrow & & \searrow{\scriptstyle \hat{h}} \\ \mathbf{R}/(c' - a)\mathbf{Z} & \xrightarrow{\ \ h\ \ } & X \end{array}$$

By 2.6.13.1, we know that $Y = \mathbf{R}/(c' - a)\mathbf{Z}$ is diffeomorphic to S^1. By construction, h is regular and injective; since Y is scompact, theorem 2.6.12 implies that Y is diffeomorphic to its image $h(Y)$ in X. But h is surjective, since $h(Y) = \alpha(I) \cup \beta(J)$ is both open and closed (being the continuous image of a compact set) and X is connected. This shows that X and $Y \simeq S^1$ are diffeomorphic. $\square$

We can now conclude the proof of theorem 3.4.1. Assume that X is not diffeomorphic to S^1. It is clear that any linearly ordered set of parametrizations by arclength, under the partial order $(I, \alpha) \leq (J, \beta)$ if $\alpha(I) \subset \beta(J)$, admits an upper bound. By Zorn's lemma, we can take a maximal element (L, ε) in the set of parametrizations by arclength; then $\varepsilon(L)$ is obviously open in X, and also closed. Otherwise we could find $x \in \overline{\varepsilon(L)} \setminus \varepsilon(L)$, and a parametrization by arclength (J, β) around x (3.4.4). By the previous lemma, we could combine (L, ε) and (J, β) into a single parametrization (K, γ) such that $\gamma(K) = \varepsilon(L) \cup \beta(J) \ni x$, contradicting maximality. Since $\varepsilon(L)$ is both open and closed in X and X is connected, we have $\varepsilon(L) = X$. Since ε is a diffeomorphism, we conclude that $\mathbf{R}$ is diffeomorphic to an open interval in $\mathbf{R}$, hence to $\mathbf{R}$. $\square$

3.4.6. Corollary. *If X is a curve and δ is a density on X, one of the following situations obtains:*

(i) *X is diffeomorphic to $\mathbf{R}$, and has a global parametrization by arclength;*

(ii) *X is diffeomorphic to S^1, and has a periodic parametrization by arclength (that is, there exists a map $f : \mathbf{R} \to X$ and a scalar $T \in \mathbf{R}_+^*$ such that $f(t + T) = f(t)$, the restriction of f to $[0, T[$ is injective, and $\delta(f'(t))$ for all t).*

If we require $f(0) = x$, such a parametrization is unique up to a sign flip $t \mapsto -t$. $\square$

3.5. Vector Fields and Differential Equations on Manifolds

> In this section the differentiability class is at least 2.

We will now transport to manifolds the notions, introduced in chapter 1, of vector fields and differential equations. The local theorems remain true because manifolds are locally diffeomorphic to open sets in $\mathbf{R}^n$ and diffeomorphisms preserve integral curves (2.5.17.4). The new facts here are global uniqueness (3.5.4) and the results for compact manifolds (3.5.9 and 3.5.13).

3.5.1. Definition. Let X be a manifold of class C^p. A *vector field* on X is a C^{p-1} map $\xi : X \to TX$ such that $\xi(x) \in T_x X$ for every $x \in X$.

If π denotes the map taking $z \in TX$ to the point $\pi(z) = x \in X$ such that $z \in T_x X$, a vector field is characterized by $\pi \circ \xi = \mathrm{Id}_X$. The canonical vector field on $\mathbf{R}$ (2.5.17.2) is an example of a vector field.

Let (U, ϕ) be a chart on X. If u is a point in the open set $\phi(U) \subset \mathbf{R}^d$, we have $\xi(\phi^{-1}(u)) \in TX$, hence $(T_{\phi^{-1}(u)}\phi)(\xi(\phi^{-1}(u))) \in T_u \mathbf{R}^d$. To obtain a vector field in the sense of 1.2.1, consider the map $f : \phi(U) \to \mathbf{R}^d$ defined by

$$f(u) = \left(\theta_u \circ T_{\phi^{-1}(u)}\phi\right)\left(\xi(\phi^{-1}(u))\right),$$

where $\theta_u : T_u \mathbf{R}^d \to \mathbf{R}^d$ is the isomorphism introduced in 2.5.12.3. By 2.5.26, the map f is of class C^{p-1}.

We have the following commutative diagram:

$$
\begin{array}{ccccc}
TX \supset TU & \xrightarrow{\ T\phi\ } & T(\phi(U)) & \xrightarrow{\ \theta\ } & \mathbf{R}^d \\
\xi \uparrow\uparrow \pi & & \downarrow \pi & & \downarrow \mathrm{Id} \\
U & \xrightarrow[\ \phi\ ]{\phi^{-1}} & \phi(U) & \xrightarrow[\ f\ ]{} & \mathbf{R}^d.
\end{array}
$$

3.5.2. Remark. Vector fields on an open subset W of $\mathbf{R}^d$ considered as a manifold (definition 3.5.1) or as a subset of $\mathbf{R}^d$ (definition 1.2.1) are distinct objects. In the first case they are maps in $C^p(W; TW)$ (such that $\pi \circ \xi = \mathrm{Id}_W$), in the second maps in $C^p(W; \mathbf{R}^d)$. But since we can identify $T_{(\pi \circ \xi)(w)}$ with $\mathbf{R}^d$ via $\theta_{\pi(\xi(w))}$, we will not distinguish between the two notions.

We can now adapt the study carried out in chapter 1.

3.5.3. Definition. An *integral curve* of a vector field ξ on X is a curve (I, α) on X, where $I \subset \mathbf{R}$ is an interval and $\alpha : I \to X$ is such that $\alpha'(t) = \xi(\alpha(t))$ for every $t \in I$ (for the definition of $\alpha'(t)$ see 2.5.17). The *initial condition* of (I, α) is $\alpha(0)$.

We leave to the reader the definition of a local flow and the statement of the theorem of local existence and uniqueness of local flows (cf. 1.2.6).

3.5.4. Theorem (global uniqueness). *Let (J_1, α_1) and (J_2, α_2) be integral curves of a vector field on X, and assume that $\alpha_1(0) = \alpha_2(0)$. Then*

$$\alpha_1|_{J_1 \cap J_2} = \alpha_2|_{J_1 \cap J_2}.$$

Proof. As in the proof of 1.3.1, one shows that

$$Q = \left\{t \in J_1 \cap J_2 : \alpha_1(t) = \alpha_2(t)\right\}$$

is open in $J_1 \cap J_2$. It is also closed because

$$Q = (J_1 \cap J_2) \cap (\alpha_1, \alpha_2)^{-1}(\Delta),$$

where Δ is the diagonal of $X \times X$, a closed subset since we assume X to be Hausdorff (2.2.10.7). $\square$

3.5.5. Remark. This theorem would be false if X were not Hausdorff. For example, we can consider on the manifold X from example 2.2.10.4 the canonical vector field ξ arising from the vector field $t \mapsto 1_t$ on $\mathbf{R}$ (2.5.17.2).

Figure 3.5.5

Then ξ has two integral curves α_1, α_2 which coincide for $t < 0$ but "fork" at $t = 0$ (figure 3.5.5).

For $x \in X$ we can define the maximal interval $J(x)$ and the maximal integral curve $(J(x), \alpha_x)$ with initial value x, as in 1.3.1.1.

3.5.6. Definition. Let ξ be a vector field on X, and set $\mathcal{D}(\xi) = \{(t, x) \in \mathbf{R} \times X : t \in J(x)\}$. The *global flow* of ξ is the pair $(\mathcal{D}(\xi), \alpha)$, where $\alpha : \mathcal{D}(\xi) \to X$ is the map defined by $\alpha(t, x) = \alpha_x(t) = G_t x$.

The notation $G_t x$ is introduced in analogy with 1.3.4 and 1.3.5.

3.5.7. Theorem. *If ξ is a vector field of class C^{p-1} on X the set $\mathcal{D}(\xi)$ is open in $\mathbf{R} \times X$, and $\alpha \in C^{p-1}(\mathcal{D}(\xi); X)$.*

Proof. The same as that of 1.3.6.   □

As before, whenever the maps involved are defined, we have

3.5.8 $G_t(G_{t_1} x_0) = (G_t \circ G_{t_1}) x_0 = G_{t+t_1} x_0.$

The one new result here has to do with compact manifolds:

3.5.9. Theorem. *If ξ is a vector field on a compact manifold X, we have $\mathcal{D}(\xi) = \mathbf{R} \times X$, and $G_t \in \mathrm{Diff}(X)$ for every $t \in \mathbf{R}$. Furthermore, the map $t \mapsto G_t$ is a group homomorphism from the additive group $\mathbf{R}$ into $\mathrm{Diff}(X)$.*

Proof. Take $x \in X$. By the existence of local flows we can find an open neighborhood U_x of x and an interval I_x containing 0 such that, for every $z \in U_x$, there exists an integral curve $\alpha : I_x \to X$ with initial condition z. Thus $I_x \times U_x \subset \mathcal{D}(\xi)$. By compactness, a finite number of neighborhoods U_{x_i} $(1 \le i \le n)$ suffice to cover X. If $\varepsilon > 0$ is such that $]-\varepsilon, \varepsilon[\subset I_{x_i}$ for all i, the integral curve with initial condition $x \in X$ is defined for every $|t| < \varepsilon$, that is, $]-\varepsilon, \varepsilon[\subset J(x)$ for every $x \in X$.

Now consider the integral curve $\alpha_{\alpha_x(t)}$ with initial condition $\alpha_x(t)$. This curve is defined for $|t| < \varepsilon$, and satisfies $\alpha_{\alpha_x(t)}(t) = \alpha_x(t + t)$. This means that the integral curve with initial condition α is actually defined for every t such that $|t| < 2\varepsilon$, that is, $]-2\varepsilon, 2\varepsilon[\subset J(x)$ for every $x \in X$. Repeating the argument we get $]-2^n \varepsilon, 2^n \varepsilon[\subset J(x)$ for every $x \in X$ and $n \in \mathbf{R}$; this implies $J(x) = \mathbf{R}$ and $\mathcal{D}(\xi) = \mathbf{R} \times X$.

This also shows that formula 3.5.8 holds for every $t \in \mathbf{R}$. Applying it to case $t_1 = -t$ we get

$$(G_t \circ G_{-t})(x_0) = G_0(x_0).$$

But by definition 3.5.3, $G_0(x) = \alpha(0, x) = x$. Thus

$$G_t \circ G_{-t} = G_{-t} \circ G_t = G_0 = \mathrm{Id}_x,$$

so that $G_t \in \mathrm{Diff}(t)$ and the map $t \mapsto G_t$ is a group homomorphism. $\qquad \square$

3.5.10. Remark. Theorem 3.5.9 says that a vector field ξ on a compact manifold X gives rise to a one-parameter group of diffeomorphisms of X. If X is arbitrary, the flow α is just a semigroup of local diffeomorphisms (cf. the discussion following 1.3.5).

In order to generalize section 1.4, we have to define time-dependent vector fields:

3.5.11. Definition. Let X be a manifold of class C^p. A *time-dependent vector field*, or *differential equation*, of class C^q on X (where $1 \leq q \leq p-1$) is a map $\xi \in C^p(I \times X; TX)$ such that $\pi \circ \xi = p$, where $I \subset \mathbf{R}$ is an interval containing 0 and $p : I \times X \to X$ is the canonical projection.

A *solution* or *integral curve* of ξ is a pair (I, α) such that $\alpha'(t) = \xi(t, \alpha(t))$ for every $t \in I$.

All the notions from section 1.4 have counterparts here: the local existence and uniqueness of integral curves, local flows, the global uniqueness of integral curves (cf. 3.5.4), the maximal interval of definition $J(x)$ of an integral curve with initial condition x, the global flow α and its domain $\mathcal{D}(\xi) = \{(t, x) \in \mathbf{R} \times X : t \in J(x)\}$.

Let α be the global flow. For every $(t, x) \in \mathcal{D}(\xi)$, set $G_t x = \alpha(t, x)$. The vector field η defined by $\eta(h, y) = \xi(s + h, y)$, where $s \in \mathbf{R}$ is fixed, also has a global flow β, and we set $G_t^s x = \beta(t, z)$ for $\beta \in \mathcal{D}(\eta)$. In particular, $G_t = G_t^0$. With this notation we have, whenever the maps involved are defined:

3.5.12 $$G_{t+s}^r = G_t^{r+s} \circ G_s^r.$$

3.5.13. Theorem. *If X is compact and ξ is a time-dependent vector field defined on $I \times X$ we have $\mathcal{D}(\xi) = I \times X$. In addition, $G_t \in \mathrm{Diff}(X)$ for every $t \in I$.*

Proof. If $\mathcal{D}(\xi) = I \times X$, the map G_t is defined for every $t \in I$. Since $G_t = G_t^0$, it follows from 3.5.12 that $G_{-t}^t \circ G_t^0 = G_0^0 = \mathrm{Id}_X$ and $G_t^0 \circ G_{-t}^t = G_0^t = \mathrm{Id}_X$, so that G_t^0 is a diffeomorphism with inverse G_{-t}^t.

To prove the first assertion, we assume by contradiction that the maximal interval $J(x)$ for some $x \in X$ is smaller than I. This implies either $\sup(J(x)) \in I$ or $\inf(J(x)) \in I$; we can assume without loss of generality that we're in the first case. Set $b = \sup(J(x))$, and take a sequence $\{t_n\}_{n \in \mathbf{N}}$ of points in $[0, b[$ converging to b.

By compactness, the sequence $\{\alpha(t_n, x)\}$
accumulates at some point $y \in X$, so we can
assume by taking a subsequence that $\lim_{n\to\infty}$
$\alpha(t_n, x) = y$. At the point $(b, y) \in I \times X$
there exists a local flow β of the vector field
$\eta : (s, z) \mapsto \xi(s + b, z)$, and this local flow is
defined on $]-\varepsilon, \varepsilon[\times U$, where U is an open
neighborhood of x in U.

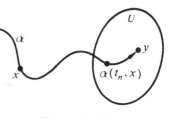

Figure 3.5.13

Since $\lim_{n\to\infty} t_n = b$ and $\lim_{n\to\infty} \alpha(t_n, x) = y$, we can take n large
enough to have $|b - t_n| < \varepsilon$ and $\alpha(t_n, x) \in U$. But

$$\beta(b - t_n, \alpha(t_n, x)) = \alpha(t_n + b - t_n, x) = \alpha(b, x),$$

showing that $b \in J(x)$ and consequently that $]t_n, t_n + \varepsilon[\subset J(x)$, which
contradicts the assumption $b = \sup(J(x))$. $\qquad\square$

3.5.13.1. Note. This gives another proof for the first part of theorem 3.5.9.

3.5.14. Conversely, we can reconstruct vector fields from one-parameter
families of diffeomorphisms. In fact, let $H \in C^p(I \times X; X)$, where $I \subset \mathbf{R}$
is an interval containing 0, be a map such that each section $H_t : x \mapsto$
$H_t(x) = H(t, x)$ is a diffeomorphism, and $H_0 = \mathrm{Id}_x$. A time-dependent
vector field ξ on X is obtained by setting $\xi(t, x) = \beta'(t) \in T_x X$, where β
is the curve $s \mapsto H(s, (H_t)^{-1}(x))$. By construction, the global flow of ξ is
$\alpha(t, x) = H_t x$, for $(t, x) \in \mathcal{D}(\xi) = I \times X$.

The reader should verify that $\xi \in C^{p-1}(I \times X; TX)$ and that ξ is time-
independent if $H_{t+s} = H_t \circ H_s$ (exercise 3.6.2).

**3.5.15. Cultural digression: the bracket of two vector fields and
the Frobenius integrability theorem**

3.5.15.1. In this presentation we omit proofs, which the reader can find in
[Die69, vol. I, 10.9.4]; [War71, p. 42]; [Ste64, p. 130]; [Spi79, vol. I, p. 6–19
and 7–21], and [Cho68, p. 192]. For simplicity, we assume all objects to be
C^∞. Manifolds are not necessarily compact.

3.5.15.2. Let ξ and η be vector fields on a manifold X, and denote their
respective flows by G_t and H_s. Is it true, at least locally, that $G_t \circ H_s =$
$H_s \circ G_t$ for $r, s \in \mathbf{R}$? In general the answer is no, there is no reason why
the two flows should commute. It all depends on the bracket $[\xi, \eta]$, defined
in exercise 2.8.17.2. In order to see why, fix $x \in X$ and consider the curve
originating at x and defined by

$$t \mapsto c(t) = (H_{-t} \circ G_{-t} \circ H_t \circ G_t)(x)$$

(this is defined for t small enough).

The curve c measures how far ξ and η are from commuting. We have

$$c(t) = x + \frac{t^2}{2}([\xi, \eta](x)) + O(t^3),$$

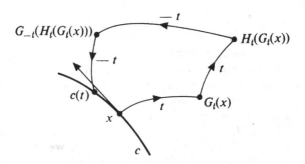

Figure 3.5.15.2

or, in other terms,

$$c'(t) = 0 \quad \text{and} \quad c'(\sqrt{t})(0) = [\xi, \eta](x).$$

In particular, a necessary condition for the flows to commute is that $[\xi, \eta] = 0$ everywhere. This condition is also sufficient: if $[\xi, \eta]$ is the zero vector field, we have $G_t \circ H_s = H_s \circ G_t$ for all $s, t \in \mathbf{R}$ locally, and even globally if X is compact (theorem 3.5.9).

3.5.15.3. Now let's take up the converse of exercise 2.8.17.2(d), which said that if Y is a submanifold of X and ξ, η are vector fields on X such that $\xi(y), \eta(y) \in T_y Y$ for all $y \in Y$, the bracket $[\xi, \eta](y)$ also belongs to $T_y Y$ for all y. First we have to think of a converse: let ξ and η be vector fields on X and assume that $\xi(x)$ and $\eta(x)$ are linearly independent in $T_x X$ for every $x \in X$. Denote by $P(x)$ the vector subspace of $T_x X$ spanned by $\xi(x)$ and $\eta(x)$. Is there a submanifold Y of X such that $T_y Y = P(y)$ for every $y \in Y$? Is there such a submanifold going through every point of X? We have seen that a necessary condition is that $[\xi, \eta](x) \in P(x)$ for every $x \in X$.

This kind of question lies in the realm of partial differential equations (in two variables), whereas the integration of vector fields has to do with ordinary differential equations. The most general answer is provided by the Frobenius integrability theorem:

3.5.15.4. Theorem (Frobenius). *Let X be a n-dimensional manifold and $d \le n$ an integer. Let ξ_i $(i = 1, \ldots, d)$ be vector fields on X such that, for every $x \in X$, the vectors $\xi_1(x), \ldots, \xi_d(x)$ are linearly independent, and denote by $P(x)$ the subspace of $T_x X$ spanned by them. A necessary and sufficient condition for the existence, for every $x \in X$, of a d-dimensional submanifold $Y \ni x$ of X such that $T_y Y = V(y)$ for all $y \in Y$ is that*

$$[\xi_i, \xi_j](x) \in V(x)$$

for every $x \in X$ and every $i, j = 1, \ldots, d$. In addition, for a fixed $x \in X$, such a submanifold is locally unique. □

3.5.15.5. This theorem is of great importance in differential geometry. We do not use it directly in this book, but it will appear in disguise in 10.2.2.7 and 10.4.9.5. Namely, when is a two-parameter family of lines in $\mathbf{R}^3$ the family of normals to a surface? Exactly when the field of planes in $\mathbf{R}^3$ orthogonal to the lines in the family satisfies the conditions in Frobenius' theorem. And if this is the case, we can actually find a whole one-parameter family of parallel surfaces whose family of normals is the original two-parameter family of lines.

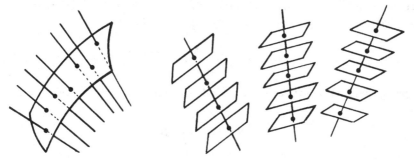

Figure 3.5.15.5

Frobenius' theorem is a key ingredient in the proof of the important theorem 10.7.3. It is also used in the proof of Darboux's theorem on two-forms (cf. chapter 5): if a closed two-form ω has rank r (that is $\omega^r = \omega \wedge \cdots \wedge \omega \neq 0$ and $\omega^{r+1} = 0$), it is locally of the form

$$\omega = dx_1 \wedge dx_2 + dx_3 \wedge dx_4 + \cdots + dx_{2r-1} \wedge dx_{2r}$$

for appropriately chosen coordinates $x_1, \ldots, x_n$ (see [Ste64], p. 140).

In addition, Frobenius' theorem plays a fundamental role in the theory of Lie groups and Lie algebras (see [War71], for example). In riemannian geometry, it shows up in the study of symmetric spaces, of totally geodesic submanifolds and of holonomy groups: see [KN69] and [Bes86, chapter 10], for example.

3.5.15.6. We conclude with the dual version of Frobenius' theorem—an equivalent result, but couched in the language of differential forms (chapter 5). For each $x \in X$, define a d-dimensional subspace $P(x)$ of $T_x X$ as the intersection of the kernel of the $n - d$ differential one-forms $\omega_1, \ldots, \omega_{n-d}$, which we assume everywhere linearly independent. In symbols,

$$P(x) = \bigcap_{i=1}^{n-d} \left(\omega_i(x)\right)^{-1}(0).$$

The statement that parallels 3.5.15.4 is that a local submanifold tangent to $P(x)$ exists for every $x \in X$ if and only if each exterior derivative $d\omega_i(x)$ is a linear combination of exterior products $\omega_i(x) \wedge \omega_j(x)$. The equivalence between the two statements is easy, and the reader should prove it. The proof of Frobenius' theorem itself is much harder.

3.6. Exercises

3.6.1. Let $p : X \to Y$ be a covering map, $\varepsilon \in \Delta_0(Y)$ and $\delta = p^*\varepsilon$ (3.3.7). Assume X and Y are compact. Show that $p^{-1}(y)$ is finite for every $y \in Y$, and that, for every continuous function $f : X \to \mathbf{R}$, we have

$$\int_X f\delta = \int_{y \in Y} \left(\sum_{x \in p^{-1}(y)} f(x) \right) \varepsilon.$$

In particular, if Y is connected, p is a k-fold cover (2.4.4) and g is continuous on X, we have

$$\int_X (g \circ p)\delta = k \int_Y g\varepsilon.$$

3.6.2. Prove the statements at the end of 3.5.14.

3.6.3. We will show, without using 3.1.5, that every manifold is metrizable.

(a) If E is a vector space, denote by $B(E)$ the set of symmetric bilinear forms on E. For $f \in L(E; F)$, define $f^* : B(F) \to B(E)$ by

$$(f^*\alpha)(x, y) = \alpha\big(f(x), f(y)\big).$$

A *riemannian structure* on a manifold X is a map $g : X \to B(TX)$ such that $g(x) \in B(T_x X)$ is positive definite for every $x \in X$ and that, for any chart (U, ϕ) on X, the map

$$U \ni u \mapsto \big((\theta_u \circ T_{\phi(u)})(\phi^{-1})\big)^* \big(g(\phi(u))\big) \in B(\mathbf{R}^n)$$

is differentiable. Show that every manifold (satisfying 3.1.6) admits a riemannian structure. Fix a riemannian structure g for X, and assume that X is connected.

(b) If $\big([a, b], f\big)$ is a curve on X, the *length* of f is defined by $\mathrm{leng}(f) = \int_a^b \sqrt{g(f'(t), f'(t))}\, dt$. The *distance* between two points $x, y \in X$ is the scalar

$$d(x, y) = \inf \big\{\mathrm{leng}(f) : f \text{ is a curve such that } f(a) = x \text{ and } f(b) = y\big\}.$$

Show that $d(x, y) = d(y, x)$ and that $d(x, z) \leq d(x, y) + d(y, z)$ for every $x, y, z \in X$.

(c) Show that $d(x, y) = 0$ implies $x = y$. This shows that the function $d : X \times X \to \mathbf{R}$ is a metric, called the *intrinsic metric* associated with the riemannian manifold X. (Hint: show first that if $d(x, y) = 0$ there exists a chart (U, ϕ) at x such that $y \in U$. Then show that, if g is a riemannian structure on an open subset U of $\mathbf{R}^n$ and $K \subset U$ is compact, there exists $k > 0$ such that $g(x, x) \geq k\|x\|^2$ for every $x \in K$, where $\|\cdot\|$ denotes the standard norm in $\mathbf{R}^n$.)

3.6.4. Prüfer's surface. Let $E = \mathbf{R}^3$, and denote by $\mathbf{R}_a^2$ the plane $z = a$, with its usual topology. Consider E as the disjoint union of the $\mathbf{R}_a^2$, and give it the corresponding topology (so that a subset of E is open if and only if its intersection with each $\mathbf{R}_a^2$ is open). Denoting points of $\mathbf{R}_a^2$ by $(x, y)_a$, form the equivalence relation

$$(x, y)_a \sim (x', y')_b \Leftrightarrow (x, y)_a = (x', y')_b \text{ or } (y = y' > 0 \text{ and } xy + a = x'y' + b).$$

Set $P = E/\sim$, and let $p : E \to P$ be the canonical map.

(a) Show that P is Hausdorff, locally compact (the redundancy is intentional, for pedagogical reasons), and that for every $a \in \mathbf{R}$ the restriction of p to $\mathbf{R}_a^*$ is a homeomorphism onto its image.

(b) If $X = p((x, y)_a)$, set $f_a(X) = (x, y)$. Show that $(p(\mathbf{R}_a^*), f_a)_{a \in \mathbf{R}}$ is a C^∞ atlas, and that the topology underlying the differential structure thus obtained is identical with the quotient topology.

(c) Show that the set $\{p((0, 0)_a) : a \in \mathbf{R}\}$ is discrete in P. Deduce that P is not separable. P is called the *Prüfer surface*.

3.6.5. Let G be a *Lie group*, that is, a C^∞ manifold with a group structure such that the maps $(x, y) \mapsto xy$ and $x \mapsto x^{-1}$ are C^∞. We denote left translations $x \mapsto ax$ by L_a, and the identity element by $e \in G$.

(a) Let $\delta_e \neq 0$ be a density on $T_e(G)$. For $a \in G$, set $\delta_a = (T_a L_{a^{-1}})^* \delta_e$. Show that this defines a density on G.

(b) Let $d\delta$ be the measure associated with the density in (a). Show that, for every continuous function $f : G \to \mathbf{R}$ with compact support and every $a \in G$ we have

$$\int_G f(x) \, d\delta = \int_G f(ax) \, d\delta.$$

We say that $d\delta$ is a *left Haar measure* on G.

(c) Write down $d\delta$ for $G = \mathrm{GL}(2; \mathbf{R}) = \mathrm{Isom}(\mathbf{R}^2; \mathbf{R}^2)$.

CHAPTER 4

Critical Points

After defining critical points and regular values of a differentiable map $f : X \to Y$ between manifolds, we study two particular cases: $Y = \mathbf{R}$ and $\dim X = \dim Y$.

In the case of real-valued functions $f : X \to \mathbf{R}$ we define non-degenerate critical points and the index of f at such a point (4.2.11). We show that around a non-degenerate critical point f looks, up to diffeomorphism, like a non-degenerate quadratic form (4.2.12). This result has important consequences: for example, it governs the behavior of a surface in $\mathbf{R}^3$ with respect to its tangent plane at a non-degenerate point (4.2.20). Another consequence is Morse theory, which is not within the scope of this book but nonetheless deserves a digression (4.2.24) for the reader's information and for future reference (section 7.5).

The case $\dim X = \dim Y$ is the setting for Sard's theorem, which asserts that almost all points of Y are regular values of $f : X \to Y$ (4.3.1). The existence of regular values is used many times in this book (7.3.1, 7.5.4, 9.2.8).

4.1. Definitions and Examples

4.1.1. Definition. Let X and Y be manifolds and $f : X \to Y$ a differentiable map. A point $x \in X$ is said to be *regular* if f is a submersion at x, and *critical* otherwise.

A point $y \in Y$ is called a *regular value* if every $x \in f^{-1}(y)$ is regular, and a *critical value* otherwise. In particular, y is a regular value if $y \notin f(X)$.

In this chapter we will discuss the following two particular cases:

4.1.2. If $Y = \mathbf{R}$ and $f \in C^p(X)$, a point x is critical if and only if $T_x f$ is not surjective. By definition 2.5.23 and since the range space is one-dimensional, this is the same as saying that

$$x \text{ is critical} \Leftrightarrow df(x) = 0.$$

4.1.3. If $\dim X = \dim Y$, a point $x \in X$ is regular if and only if $T_x f \in \mathrm{Isom}(T_x X; T_{f(x)} Y)$; in other words (cf. definition 2.6.9),

$$x \text{ is regular} \Leftrightarrow f \text{ is regular at } x.$$

4.1.4. Examples

4.1.4.1. Let E be a Euclidean space, with inner product $(\cdot \mid \cdot)$, and $u \neq 0$ a vector in E. Let $V \subset E$ be a submanifold, and $f : V \to \mathbf{R}$ the map $f : x \mapsto (u \mid x)$. If $N_x V$ denotes the normal subspace to V at x under the embedding $V \to E$ (2.7.2), we have

4.1.4.2 x is critical $\Leftrightarrow u \in N_x V.$

To prove this, apply 4.1.2. The derivative $df(x)$ is just the linear form $df(x) = (u \mid \theta_x(\cdot))$ on $T_x X$, by 0.2.8.3 and 2.5.23.2, and because f can be considered as the restriction of $(u \mid \cdot)$ to the submanifold V. Thus $(u \mid \theta_x(\cdot))$ vanishes on $T_x X$ if and only if $u \in (\theta_x(T_x X))^\perp = N_x X$.

If u points straight up, $(u \mid \cdot)$ is the height and 4.1.4.2 says that the height function is critical at x if the tangent space is horizontal.

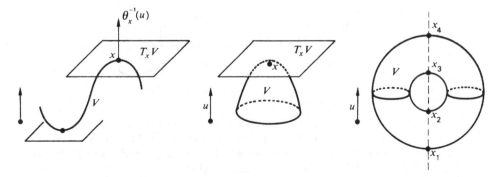

Figure 4.1.4.2

4.1.4.3. Particular case. Take the sphere $S^n \subset \mathbf{R}^{n+1}$. If $\{e_1, \ldots, e_{n+1}\}$ is the canonical basis of $\mathbf{R}^{n+1}$, the map $x \mapsto (x \mid e_i)$ has as critical points $(0, \ldots, 0, \pm 1, 0, \ldots, 0)$, where only the i-th coordinate is non-zero.

4.1.4.4. If X is the torus $V = S^1(1/\sqrt{2}) \times S^1(1/\sqrt{2})$ (2.1.6.3), embedded in $\mathbf{R}^3$ as shown in figure 4.1.4.2, the map $x \mapsto (x \mid e_3)$ has four critical points.

4.1.5. Proposition. *Let X and Y be manifolds of same dimension and $f : X \to Y$ a differentiable map. Assume that X is compact and that $y \in Y$ is a regular value of f. There exists a neighborhood V of y such that the restriction of f to $f^{-1}(V)$, considered as a map from $f^{-1}(V)$ into V, is a covering map.*

We shall see in 4.3.6 that any f has lots of regular values.

Proof. Let $y \in Y$ be a regular value; then every $x \in f^{-1}(y)$ is a regular point, that is, $T_x f \in \mathrm{Isom}(T_x X; T_y Y)$ (4.1.3). By 2.5.20 there exists an open neighborhood U_x of x such that $f|_{U_x}$ is a diffeomorphism between U_x and $f(U_x)$. It follows that $f^{-1}(y)$ is discrete, for if there were a sequence (x_n) of points of $f^{-1}(y)$ accumulating at x we would have $x_n \in U_x$ for n large enough, and $f|_{U_x}$ wouldn't be injective.

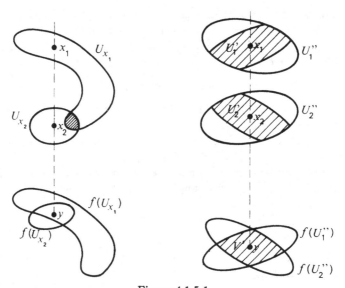

Figure 4.1.5.1

Thus $f^{-1}(y)$ is a finite set, being discrete in a compact set. Let $x_1, \ldots, x_n$ be its points. Each x_i is contained in an open set U_{x_i} restricted to which f is a diffeomorphism onto its image. But as they stand the U_{x_i} do not satisfy definition 2.4.1; we have to modify them.

First we shrink them into pairwise disjoint sets U_i'' (still containing x_i), which is possible because X is Hausdorff and there are only a finite number of x_i. But the images $f(U_i'')$ may not coincide. So we set $V' = \bigcap_{i=1}^n f(U_i'')$ and $U_i' = U_i'' \cap f^{-1}(V')$. Since $f|_{U_i'}$ is bijective, U_i' is open and $f(U_i') = V'$; the restriction of f to U_i' is still a diffeomorphism onto its image. The only condition for a covering map that's left to check is that $f^{-1}(V') = \bigcup_{i=1}^n U_i'$. For now we just have $\bigcup_{i=1}^n U_i' \subset f^{-1}(V')$, since $U_i' \subset f^{-1}(V')$ for all i. What else can be in the inverse image of V'?

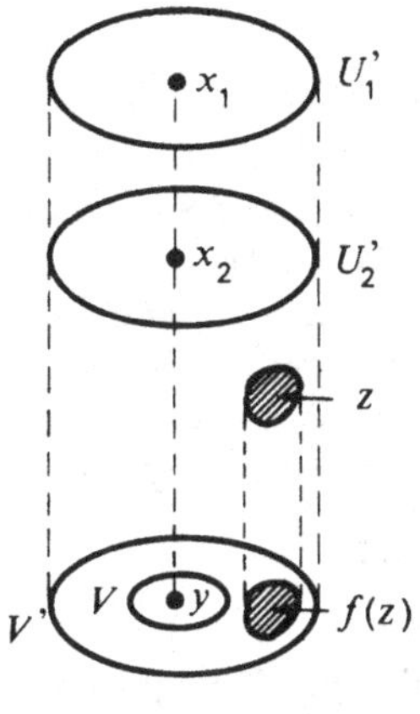

Assume that $\overline{V}'$ is compact; we can get this by restricting V, since Y is locally compact (2.2.11). The set

$$Z = f^{-1}(\overline{V}') \setminus \bigcup_{i=1}^n (U_i').$$

is closed in X, hence compact. Thus $f(Z)$ is compact, hence closed in V', and, since $y \notin f(Z)$, we can take a neighborhood V of y such that $V \cap f(Z) = \emptyset$. Setting $U_i = f^{-1}(V) \cap U_i'$, we finally get $f^{-1}(V) = \bigcup_{i=1}^n U_i$, and the restriction of f to $f^{-1}(V)$ is a covering map. $\qquad\square$

Figure 4.1.5.2

4.1.6. Corollary. *The cardinality of $f^{-1}(y)$ is constant in a neighborhood of a regular value.*

Proof. Use 2.4.4. $\qquad\square$

4.1.7. Proposition 4.1.5 does not hold around critical points, or when X is not compact. Figure 4.1.7 shows counterexamples to corollary 4.1.6: around y_1 and y_2 the cardinality of $f^{-1}(y)$ changes.

4.1.8. See exercises 4.4.3 to 4.4.6 for explicit calculations of critical points.

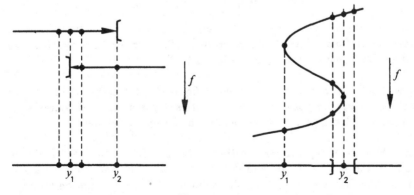

Figure 4.1.7

4.2. Non-Degenerate Critical Points

Let X be a d-dimensional manifold and $f : X \to \mathbf{R}$ a function of class C^p, with $p \geq 2$. For $x \in X$, the derivative $df(x)$ is a linear function on $T_x X$ (2.5.23), and so can be seen as an element of $(T_x X)^*$, the dual of $T_x X$.

If (U, ϕ) is a chart at $x \in X$, the composition $f \circ \phi^{-1}$ is a C^p map from $\phi(U)$ into $\mathbf{R}$, and as such it has a second derivative $(f \circ \phi^{-1})''$ which is a symmetric bilinear form on $\mathbf{R}^d$ (0.2.13). Let $\theta : T_x X \to \mathbf{R}^d$ be the isomorphism associated with (U, ϕ) (2.5.10), and define a symmetric bilinear form on $T_x X$ by

4.2.1 $\qquad\qquad B = (f \circ \phi^{-1})''(\phi(x)) \circ (\theta, \theta),$

that is, $B(u, v) = (f \circ \phi^{-1})''(\phi(x)) \circ (\theta(u), \theta(v))$ for every $u, v \in T_x X$. In principle B might depend on (U, ϕ), but in fact it doesn't:

4.2.2. Proposition and definition. *Let X be a d-dimensional manifold of class C^p, with $p \geq 2$. If $x \in X$ is a critical point of $f \in C^p(X)$, the symmetric bilinear form on $T_x X$ defined by 4.2.1 does not depend on the choice of ϕ. We denote it by $\mathrm{Hess}_x f$, and call it the hessian of f at x.*

4.2.3. Proof. Let (U, ϕ) and (V, ψ) be charts around x, and $\theta : T_x X \to \mathbf{R}^d$ and $\eta : T_x X \to \mathbf{R}^d$ the corresponding isomorphisms (2.5.10). If $df(x) = 0$, the equality

$$(f \circ \phi^{-1})''(\phi(x)) \circ (\theta, \theta) = (f \circ \psi^{-1})''(\psi(x)) \circ (\eta, \eta)$$

follows directly from 2.5.11.1 and 0.2.13.2 applied to $z = \phi(x)$, $h = f \circ \psi^{-1}$ and $g = \psi \circ \phi^{-1}$. $\qquad\qquad\qquad\qquad\qquad\qquad\qquad\qquad\qquad\qquad\square$

4.2.4. Let (I, α) be a curve on X such that $\alpha'(0) = u \in T_x X$ and x is a critical point of f. We have

$$(\mathrm{Hess}_x f)(u, u) = \frac{d^2(f \circ \alpha)}{dt^2}(0),$$

where the expression in the right-hand side is just the ordinary second derivative of $f \circ \alpha : I \to \mathbf{R}$. See exercise 4.4.2. This provides another way to show 4.2.2 (without using formula 0.2.13.2).

4.2.5. Example. If X is the torus in figure 4.1.4.2 and f is the height function, $\mathrm{Hess}_{x_1} f$ is positive definite and $\mathrm{Hess}_{x_4} f$ is negative definite, while $\mathrm{Hess}_{x_2} f$ and $\mathrm{Hess}_{x_3} f$ take both positive and negative values.

4.2.6. Recap on symmetric bilinear forms. Let E be a finite-dimensional vector space over $\mathbf{R}$, and denote by $\mathrm{Bilsym}(E)$ the vector space of symmetric bilinear forms on E. The following statements can be found in [Dix68, pp. 35–37]:

Let $\lambda \in \mathrm{Bilsym}(E)$. Two vectors $x, y \in E$ are called *orthogonal* with respect to λ if $\lambda(x, y) = 0$. It is possible to find a basis $\mathcal{B} = \{e_1, \ldots, e_n\}$ for E whose elements are pairwise orthogonal with respect to λ; for such a basis the matrix $\Lambda_\mathcal{B}$ of λ is diagonal:

$$\Lambda_\mathcal{B} = \begin{pmatrix} k_1 & & 0 \\ & \ddots & \\ 0 & & k_n \end{pmatrix}, \qquad k_i = \lambda(e_i, e_i).$$

The number of strictly positive diagonal entries is independent of the choice of the diagonalizing basis $\mathcal{B}$, and similarly for strictly negative and zero entries.

4.2.7. Definition. We say that λ is *non-degenerate* if all the k_i are non-zero.

4.2.8. Definition. The *index* of λ is defined as the number of strictly negative entries k_i.

4.2.9. Examples

4.2.9.1. Consider again the torus in figure 4.1.4.2. The index is zero at x_1, two at x_4, and one at x_2 and x_3. This is related to the behavior of f with respect to its tangent plane: see 4.2.20.

4.2.9.2. The form $\lambda(x_1, \ldots, x_d) = -x_1^2 - \cdots - x_i^2 + x_{i+1}^2 + \cdots + x_d^2$ on $\mathbf{R}^d$ is non-degenerate and has index i.

4.2.10. See exercises 4.4.3 to 4.4.6 for explicit calculations of indices.

4.2.11. Definition. Let $f : X \to \mathbf{R}$ be of class C^p. A critical point x of f is said to be *non-degenerate* if $\mathrm{Hess}_x f$ is non-degenerate. The *index* of x is by definition the index of $\mathrm{Hess}_x f$; we denote it by $\mathrm{ind}_x f$.

4.2.12. Theorem. *Let X be a C^p manifold, $p \geq 3$, and $f \in C^p(X)$ a function having a non-degenerate critical point x of index i. There exists a chart (U, ϕ) centered at x such that, in the associated coordinate system (2.2.3), f has the form*

$$(f \circ \phi^{-1})(x_1, \ldots, x_d) = -x_1^2 - x_2^2 - \cdots - x_i^2 + x_{i+1}^2 + \cdots + x_d^2.$$

We prove the theorem for $d = 2$. For arbitrary d the argument is analogous and does not present any additional difficulty. We need two lemmas.

4.2.13. Lemma (Morse). *Let $U \subset \mathbf{R}^2$ be open and star-shaped at $(0, 0)$, and $f \in C^p(U; \mathbf{R})$ a function such that $f(0, 0) = 0$. There exist $g, h \in C^{p-1}(U; \mathbf{R})$ such that $f = xg + yh$, where x and y are the coordinate functions.*

Proof. By definition of star-shaped, $(tx, ty) \in U$ for any $t \in [0, 1]$ and $(x, y) \in U$, that is, the segment joining $(0, 0)$ to any point $(x, y) \in U$ lies entirely in U. Thus the function $t \mapsto f(tx, ty)$ is defined on $[0, 1]$, and satisfies

$$\int_0^1 \frac{d}{dt}\big(f(tx, ty)\big)\, dt = f(x, y) - f(0, 0) = f(x, y).$$

By the chain rule,

$$\frac{d}{dt}\big(f(tx, ty)\big) = x\frac{\partial f}{\partial x}(tx, ty) + y\frac{\partial f}{\partial y}(tx, ty),$$

so if we set

$$g = \int_0^1 \frac{\partial f}{\partial x}(tx, ty)\, dt, \qquad h = \int_0^1 \frac{\partial f}{\partial y}(tx, ty)\, dt$$

we get

$$f(x, y) = x\int_0^1 \frac{\partial f}{\partial x}(tx, ty)\, dt + y\int_0^1 \frac{\partial f}{\partial y}(tx, ty)\, dt = xg + yh.$$

Since f is of class C^p, its partial derivatives, and consequently g and h, are of class C^{p-1}, which proves the lemma. $\qquad\square$

4.2.14. Remark. We have

$$g(0, 0) = \int_0^1 \frac{\partial f}{\partial x}(0, 0)\, dt = \frac{\partial f}{\partial x}(0, 0) \qquad \text{and} \qquad h(0, 0) = \frac{\partial f}{\partial y}(0, 0).$$

4.2.15. Lemma. *Let $U \subset \mathbf{R}^2$ be open and start-shaped at $(0, 0)$, and $f : (\xi, \eta) \mapsto f(\xi, \eta)$ a C^p real-valued function on U. If $f(0, 0) = 0$ and $f'(0, 0) = 0$, there exist functions $u, v, w \in C^{p-2}(U; \mathbf{R})$ such that*

$$f = x^2 u + 2xyv + y^2 w.$$

Proof. We have $f'(0, 0) = 0$, that is, $\partial f/\partial x(0, 0) = \partial f/\partial y(0, 0) = 0$. By 4.2.13, there exist $g, h \in C^{p-1}(U; \mathbf{R})$ such that $f = xg + yh$. But $g(0, 0) = \partial f/\partial x(0, 0)$ by 4.2.14, so $g(0, 0) = 0$, and since $g \in C^{p-1}(U; \mathbf{R})$, an application of 4.2.13 to g gives $u, v_1 \in C^{p-2}(U; \mathbf{R})$ such that $g = xu + yv_1$. Similarly, we can find $v_2, w \in C^{p-2}(U, \mathbf{R})$ such that $h = xv_2 + yw$. Setting $v = (v_1 + v_2)/2$, we get

$$f = xg + yh = x^2 u + xy(v_1 + v_2) + y^2 w = x^2 u + 2xyv + y^2 w. \qquad\square$$

4.2.16. Remark. This formula is similar to a Taylor series expansion, but has no higher-order terms.

Proof of theorem 4.2.12. The idea is to imitate the diagonalization of a quadratic form:

$$ax^2 + 2bxy + cy^2 = a\left(x^2 + 2\frac{b}{a}xy + \frac{c}{a}y^2\right) = a\left(x + \frac{b}{a}y\right)^2 + \left(c - \frac{b^2}{a}\right)y^2;$$

but here the coefficients are functions, so we need to take some precautions.

By composing f with a chart centered at x, we reduce to the case of $f \in C^p(U)$, where $U \in O(\mathbf{R}^2)$ can be assumed star-shaped at 0. By assumption, $f(0) = 0$ and 0 is a critical point for f. By lemma 4.2.15, there exist $u, v, w \in C^{p-2}(U)$ such that $f = ux^2 + 2vxy + wy^2$.

We now set

$$r = \frac{\partial^2 f}{\partial x^2}(0,0), \qquad s = \frac{\partial^2 f}{\partial xy}(0,0), \qquad t = \frac{\partial^2 f}{\partial y^2}(0,0),$$

so that $\mathrm{Hess}_0 f = \left(\begin{smallmatrix} r & s \\ s & t \end{smallmatrix}\right)$. Since $\mathrm{Hess}_x f$ is non-degenerate, $rt - s^2 \neq 0$ is non-zero. We can also assume that r is non-zero; otherwise, either t is non-zero and we interchange x and y, or s is non-zero (with $r = t = 0$), and we change to new coordinates $x' = (x + y)/2$, $y' = (x - y)/2$.

Assume $r > 0$, for example. There exists an open set U', contained in U and star-shaped at 0, such that $u(x,y) > 0$ for $x, y \in U'$. On U' we have

4.2.17
$$f = u\left(x + \frac{v}{u}y\right)^2 + y^2\left(w - \frac{v^2}{u}\right).$$

Assuming first that $rt - s^2 > 0$, we can take a smaller open set U'', star-shaped at 0 and such that $w - v^2/u > 0$ on U''. Thus we can write $f = \xi^2 + \eta^2$, where

$$\xi = \sqrt{u}\left(x + \frac{v}{u}y\right) \qquad \text{and} \qquad \eta = y\sqrt{w - \frac{v^2}{u}}.$$

The functions ξ and η are of class C^{p-2} because the radicals do not vanish. There remains to show that the map

$$F : U'' \ni (x,y) \longmapsto \big(\xi(x,y), \eta(x,y)\big) \in \mathbf{R}^2$$

defines a chart at $(0,0)$. By 2.5.20 it is enough to show that F is regular at $(0,0)$. Direct calculation shows that the jacobian of F at $(0,0)$ (0.2.8.8) is

$$\begin{pmatrix} u(0,0) & v(0,0) \\ 0 & \sqrt{w(0,0) - \dfrac{v^2(0,0)}{u(0,0)}} \end{pmatrix},$$

whose determinant is $\sqrt{rt - s^2} \neq 0$.

If $rt - s^2 < 0$, we write $f = \xi^2 - \eta^2$, with

$$\xi = \sqrt{u}\left(x + \frac{v}{u}y\right) \qquad \text{and} \qquad \eta = y\sqrt{\frac{v^2}{u} - w}. \qquad \square$$

4.2.18. Corollary. *Non-degenerate critical points of $f \in C^p(X)$ are isolated in the set of critical points.*

Proof. By theorem 4.2.12, f is locally of the form

$$f \circ \phi^{-1} = -x_1^2 - x_2^2 - \cdots - x_i^2 + x_{i+1}^2 + \cdots + x_d^2.$$

Thus $(f \circ \phi^{-1})'(x_1, \ldots, x_d) = 2(-x_1, \ldots, -x_i, x_{i+1}, \ldots, x_d)$, whose only zero is $(x_1, \ldots, x_d) = 0$.   □

If x is degenerate, there is no reason why it should be isolated. Think of $(x, y) \mapsto x^2$, or, better yet, $(x, y) \mapsto 0$!

4.2.19. Example. If X is compact and $f \in C^p(X)$ only has non-degenerate critical points, there are only finitely many of them.

4.2.20. Application. Now we study the position of a surface S (that is, a two-dimensional submanifold of $\mathbf{R}^3$) with respect to its tangent plane.

By theorem 2.1.2(iv) S can be written locally as the graph of a function $z = f(x, y)$. Let A be the point $(a, b, f(a, b)) \in S$. The tangent plane $T_A S$ has equation

$$(x - a)\frac{\partial f}{\partial x}(a, b) + (y - b)\frac{\partial f}{\partial y}(a, b) + (z - f(a, b))(-1) = 0$$

by 2.5.7.1; thus the height $p(x, y)$ of the point of $T_x S$ with first two coordinates x and y is

$$(x - a)\frac{\partial f}{\partial x}(a, b) + (y - b)\frac{\partial f}{\partial y}(a, b) + f(a, b).$$

We are interested in the sign of

$$h(x, y) = f(x, y) - f(a, b) - (x - a)\frac{\partial f}{\partial x}(a, b) - (y - b)\frac{\partial f}{\partial y}(a, b).$$

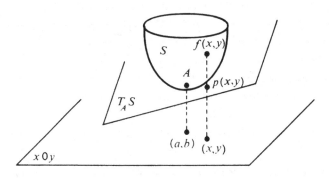

Figure 4.2.20.1

Assume that

$$rt - s^2 = \frac{\partial^2 f}{\partial x^2}(a, b)\frac{\partial^2 f}{\partial y^2}(a, b) - \left(\frac{\partial f}{\partial x \partial y}(a, b)\right)^2 \neq 0;$$

since h and f have the same second derivatives and

$$h(a, b) = \frac{\partial h}{\partial x}(a, b) = \frac{\partial h}{\partial y}(a, b) = 0,$$

it follows from 4.2.12 that there exists an open set $U \subset \mathbf{R}^2$ containing (a, b) and local coordinates $\xi(x, y)$ and $\eta(x, y)$ such that $h|_U = \pm\xi^2 \pm \eta^2$.

If the index of x is 0, we have $h|_U = \xi^2 + \eta^2$: S lies above the tangent plane, and only touches it at A.

If the index is 2, we have $h|_U = -\xi^2 - \eta^2$: S lies below the tangent plane, and only touches it at A.

If the index is 1, we have $h|_U = \xi^2 - \eta^2$, for example. The surface S intersects its tangent plane along the curves $\xi = \eta$ and $\xi = -\eta$. It lies strictly above that plane in two of the regions bounded by these curves, and strictly below in the other two.

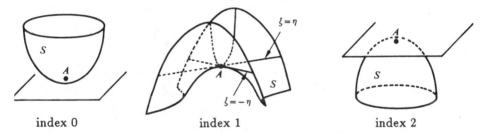

index 0   index 1   index 2

Figure 4.2.20.2

In the case of index 1, the tangents to the intersection curves at A have slopes (in $T_A S$) given by the roots μ of the equation $\mu^2 r - 2\mu s + t$; these roots are distinct.

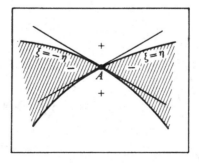

Figure 4.2.20.3

4.2.21. Remark. In order to write S locally as a graph, we've had to choose one particular basis for $\mathbf{R}^3$, and assume that $rt - s^2 \neq 0$ in these coordinates. It can be shown (exercise 4.4.9) that the condition $rt - s^2 \neq 0$ at a point $A \in S$ does not depend on the basis chosen, only on A and S. If this condition is satisfied we say that A is a *non-degenerate* point of S. It can also be shown (exercise 4.4.9) that the sign of $rt - s^2 \neq 0$ does not depend on the basis; by the discussion above the nature of $S \cap T_A S$ is determined if A is a non-degenerate point of S. We say that S has positive or negative (total) curvature at A according to whether $rt - s^2$ is positive or negative. See also 6.9.7 and section 10.5.

4.2.22. Remark. On the other hand, if A is degenerate, nothing can be said about $S \cap T_A S$. The graphs of the following functions $\mathbf{R}^2 \to \mathbf{R}$ all show different behaviors with respect to the xy-plane:

$$f_1(x, y) = x^2,$$
$$f_2(x, y) = x^2 y^2,$$
$$f_3(x, y) = x(x^2 - 3y^2) \quad \text{(monkey saddle)},$$
$$f_4(x, y) = \exp\left(-\frac{1}{x^2 + y^2}\right) \sin \frac{1}{x^2 + y^2}.$$

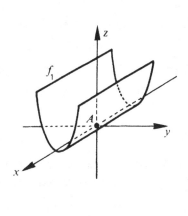

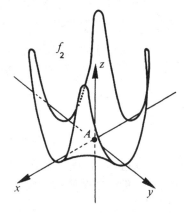

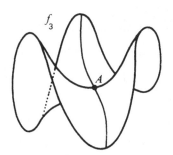

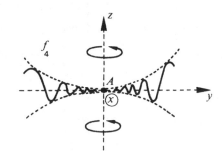

Figure 4.2.22

4.2.23. Remark. For $rt - s^2 > 0$, Taylor's formula yields the results in 4.2.20 in an elementary way.

4.2.24. Cultural digression: Morse theory. Morse theory, whose starting point is theorem 4.2.12, has numerous geometric applications: for example, it can be used to proved the existence of certain geometrical objects. Here we present only some fundamental results; for proofs and applications, see [Mil63].

4.2.24.1. Theorem (see [Mil63, p. 37], and also exercise 4.4.10). *Let X be a compact C^p manifold. Every function $f \in C^p(X)$ can be uniformly approximated by functions $g \in C^p(X)$ having only non-degenerate critical points; in fact, we can require that g approximate not only f but all of its "derivatives" (so to speak) up to order p.* □

4.2.24.2. Definition (see [Gre67, §23]). Every compact d-dimensional manifold X can be canonically assigned certain finite-dimensional real vector spaces $H^k(X)$ $(k = 0, 1, \ldots, d)$, called the *real cohomology groups* of X. The dimension of $H^k(X)$ is called the k-th *Betti number* of X, and denoted by $b_k(X)$. The alternating sum

$$\chi(X) = \sum_{k=0}^{d} (-1)^k b_k(X)$$

is called the *Euler characteristic* of X.

The correspondence $X \to H^k(X)$ is functorial: if $f : X \to Y$ is continuous, there exist group homomorphisms $f^* : H^k(Y) \to H^k(X)$ such that $(f \circ g)^* = g^* \circ f^*$ and $(\mathrm{Id}_X)^* = \mathrm{Id}_{H^k(X)}$ for every X, k, $f \in C^0(X;Y)$ and $g \in C^0(Z;Y)$. Functioriality implies that homeomorphic manifolds have isomorphic cohomology groups, hence same Betti numbers and Euler characteristics. In particular this is true for diffeomorphic manifolds. In 5.4.10 we will discuss *de Rham groups*, which are vector spaces isomorphic to the H^k.

4.2.24.3. Examples (see also sections 5.6 to 5.8)

$X = S^d$ (sphere): all the b_k are zero except $b_0 = b_d = 1$, so $\chi = 1 + (-1)^d$.

$X = P^d(\mathbf{R})$ (real projective space): all the b_k are zero except $b_0 = 1$ and $b_d = 0$ if d is even, 1 if odd.

$X = T^d$ (torus): $b_k = \binom{d}{k}$ for every k, so $\xi = (1 - 1)^d = 0$.

$X = K$ (Klein bottle): $b_0 = b_1 = 1$, $b_2 = 0$, $k = 0$.

$X = T_g$ (surface with g holes, figure 4.2.24.3.1): $b_0 = b_2 = 1$, $b_1 = 2g$, $\chi = 2(1 - g)$.

$X = U_g$, where U_g denotes the connected sum of g copies of the projective plane:

$$U_g = U_{g-1} \# P^2(\mathbf{R}) = \underbrace{P^2(\mathbf{R}) \# \cdots \# P^2(\mathbf{R})}_{k \text{ copies}}.$$

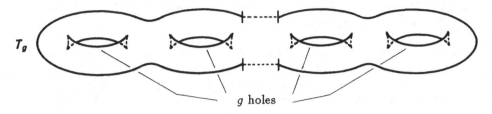

T_g

g holes

Figure 4.2.24.3.1

Figure 4.2.24.3.2

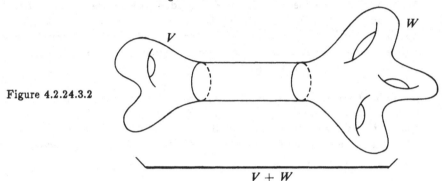

V

W

V + W

(The connected sum of two manifolds of the same dimension is the operation shown in figure 4.2.24.3.2; it is denoted by the operator #.)
Notice that U_2 is the Klein bottle K and $U_3 = U_2 \# U_1 = T_1 \# U_1$. By induction one can show that $U_{2n+1} = T_n \# U_1$ and $U_{2n+2} = T_n \# K$. All the U_g are non-orientable, and their invariants are $b_0 = 1$, $b_1 = g - 1$, $b_2 = 0$ and $\chi = 2 - g$.

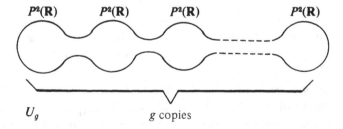

$P^2(\mathbf{R})$ $P^2(\mathbf{R})$ $P^2(\mathbf{R})$ $P^2(\mathbf{R})$

U_g g copies

Figure 4.2.24.3.3

4.2.24.4. The fundamental theorem of Morse theory. *Let X be a compact manifold and $f \in C^1(X)$ a function having only non-degenerate critical points. Denote by $c_k(f)$ the number of critical points of f of index k (this number is finite by 4.2.19). For every k we have $c_k(f) \geq b_k(f)$ and*

$$\chi(X) = \sum_{k=0}^{d} (-1)^d c_k(f).$$ □

The importance of this theorem is that it links the $b_k(X)$, which are invariants of the manifold, with the $c_k(X)$, which are invariants associated with an arbitrary (non-degenerate) function on the manifold. For a proof, see [Mil63, theorem 1.3.5 and §1.5].

4.2.25. Cultural digression: Classification of compact surfaces. A *surface* is a connected, two-dimensional manifold. From examples 4.2.24.3 it is clear that the classification of surfaces is more difficult than that of curves (one-dimensional manifolds), carried out in section 3.4. A modern treatment of the problem, based on Morse theory, can be found in [Gra71]. For a classical, "cut-and-paste" treatment, see [Mas77]. The fundamental result is the following:

4.2.25.1. Theorem. *Every orientable compact surface is diffeomorphic to some T_g. Every non-orientable compact surface is diffeomorphic to some U_g.* □

Notice that this implies that the orientable double cover (5.3.27) of U_g is T_g.

4.2.26. Cultural digression: Other manifolds. The classification of non-compact surfaces is in some sense unachievable, for various reasons: one can remove from a compact surface an arbitrary compact set (possible an unwieldy one, like a Cantor set); also, there exist surfaces with unbounded topology (figure 4.2.26). There exist also surfaces that are "non-orientable at infinity": take an infinite connected sum of projective planes, for instance. See [Mas77, pp. 47–51].

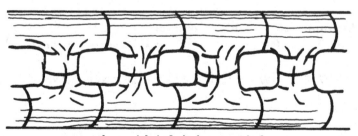

surface with infinitely many holes

Figure 4.2.26

How about higher dimensions? Even in the compact case, the problem is open. We mention a few milestones. First, every finitely generated group can be realized as the fundamental group of a compact manifold of dimension 4 (see [Mas77, p. 143], for example). Since the classification of such groups is logically undecidable, so is the classification of four- (and higher-) dimensional manifolds.

For simply connected manifolds, a general result, using homotopy types and tangent bundles, reduces the classification of *simply connected* compact manifolds of dimension ≥ 5 to an algebraic problem. In dimensions three and four, the problem is still open; see, for example, [Thu82] and [Thu88] (dimension 3) and [Don83] (dimension 4).

4.3. Sard's Theorem

Let X and Y be C^k manifolds, $k \geq 1$, of dimension n and m, respectively. Sard's theorem asserts that, if $f \in C^r(X;Y)$ with $r \geq \sup(n - m + 1, 1)$, the image of the set A of critical points of f has measure zero in Y (3.3.12). Here we deal with the case $n = m$, the only one we shall need. The case $n < m$ is just 3.3.17.3; the proof for $n > m$ is much more difficult. See [Die69, vol. III, 16.23] for a general proof.

Notice that we're saying that f has few *images* of critical points, that is, lots of regular values. Critical *points* can be very numerous—think of a constant map.

4.3.1. Theorem (Sard). *Let X and Y be manifolds of same dimension d, and $f \in C^1(X;Y)$. The set of critical values of f has measure zero in Y.*

Proof. Since X and Y have countable atlases (3.1.6 and 3.2.6), we can consider just the case $X = U \in O(\mathbf{R}^d)$ and $f \in C^1(U; \mathbf{R}^d)$, by 0.4.4.1 and 3.3.17.3.

4.3.2. Lemma. *Let $U \subset \mathbf{R}^d$ be open and $f \in C^1(U; \mathbf{R}^d)$. For every compact set $K \subset U$ there exists $\lambda : \mathbf{R}^+ \to \mathbf{R}^+$ such that $\lim_{t\to 0} \lambda(t) = 0$ and*

$$\|f(x) - f(y) - f'(y)(x - y)\| \leq \lambda(\|x - y\|)\|x - y\|$$

for all $x, y \in K$.

Proof. We have

$$\|f(x) - f(y) - f'(y)(x - y)\|$$

$$= \left\| \int_0^1 \frac{d}{dt}\big(f(y + t(x - y))\big) - f'(y)(x - y)\, dt \right\|$$

$$= \left\| \int_0^1 \big(f'(y + t(x - y))(x - y) - f'(y)(x - y)\big)\, dt \right\|$$

$$\leq \int_0^1 \|\big(f'(y + t(x - y)) - f'(y)\big)(x - y)\|\, dt.$$

For $t \in [0,1]$ and $(x,y) \in K^2$, the norm $\|f'(y+t(x-y)) - f'(y)\|$ is bounded because K is compact; in particular, we can set

$$\lambda(r) = \sup\{\|f'(y+t(x-y)) - f'(y)\| : (x,y) \in K^2, t \in [0,1], \|x-y\| = r\}.$$

Then

$$\|f(x) - f(y) - f'(y)(x-y)\| \le \lambda(\|x-y\|)\|x-y\|$$

by definition, and $\lim_{t\to 0} \lambda(t) = 0$ by the uniform continuity of f' on compact sets. $\square$

4.3.3. We continue the proof of 4.3.1. Since $U \subset \mathbf{R}^d$ is a countable union of compact cubes (3.1.4.1), we can assume by 0.4.4.1 that U contains the cube $[0,1]^d = K$ and prove that the image of the set $A \subset K$ of critical points in K has measure zero. We will use the covering criterion 0.4.4.0.

Set $M = \sup\{\|f'(x)\| : x \in K\}$; by 0.2.6 we have $\|f(x) - f(a)\| \le M\|x-a\|$ for every $x \in K$ and $a \in A$, so

$$f(x) \in B\big(f(a), M\|x-a\|\big).$$

Let a be a critical point of f; for every $z \in \mathbf{R}^d$, the image $f'(a)(z)$ belongs to $f'(a)(\mathbf{R}^d)$, which is a proper subspace of $\mathbf{R}^d$. Fix an affine hyperplane H containing $f(a) + f'(a)(\mathbf{R}^d)$. We have

$$d\big(f(x), H\big) \le \|f(x) - (f(a) + f'(a)(x-a))\| \le \lambda(\|x-a\|)\|x-a\|,$$

the first inequality because $f(a) + f'(x-a) \in H$, and the second by lemma 4.3.2.

If $x \in K$ is within distance η of a, this gives $d\big(f(x), H\big) \le \lambda(\eta)\eta$, and, since $\|f(x) - f(a)\| \le M\eta$, we conclude that $f(x)$ is in the intersection of the ball of radius $M\eta$ centered at $f(a)$ with the slab of thickness $2\lambda(\eta)\eta$ around H. This intersection is contained in a box of height $2\lambda(\eta)\eta$ and base a cube of edge $2M\eta$, so

4.3.4 $$\text{volume}\big(f[B(a,\eta) \cap K]\big) \le 2\lambda(\eta)\eta(2M\eta)^{d-1}.$$

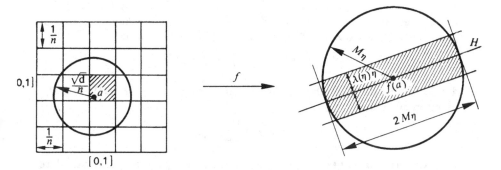

Figure 4.3.5

Now subdivide the cube $K = [0,1]^d$ into n^d cubes of side $1/n$. If such a little cube contains a critical point a, it must lie in the ball $B(a, \sqrt{d}/n)$, so its image will have volume no greater than

$$2^d M^{d-1} \left(\frac{\sqrt{d}}{n}\right)^d \lambda\left(\frac{\sqrt{d}}{n}\right),$$

by 4.3.4. There are at most n^d cubes containing critical points, so we have covered $f(A)$ with cubes whose volumes add up to no more than

4.3.5 $\qquad n^d 2^d M^{d-1} \dfrac{1}{n^d}(\sqrt{d})^d \lambda\left(\dfrac{\sqrt{d}}{n}\right) = 2^d M^{d-1}(\sqrt{d})^d \lambda\left(\dfrac{\sqrt{d}}{n}\right).$

As n increases, this expression approaches zero (lemma 4.3.2), which shows that $f(A)$ has measure zero. $\qquad\qquad\square$

4.3.6. Corollary. *The set of regular values of a differentiable map f : $X \to Y$, with $\dim X = \dim Y$, is dense in Y. In particular, f has regular values.* $\qquad\square$

4.3.7. Corollary. *Let $f : X \to Y$, with $\dim X = \dim Y$, be a differentiable map, and A the set of its critical points. If δ is a density on Y and $h \in C_\delta^{\mathrm{int}}(Y)$ we have*

$$\int_Y h\delta = \int_{Y\backslash f(A)} h\delta. \qquad\qquad\square$$

4.4. Exercises

4.4.1. Let V be a submanifold of $\mathbf{R}^d$ and $p \in \mathbf{R}^d$ a point not in V. Let $f_p : V \to \mathbf{R}$ be the function given by $f_p(x) = \|x - p\|^2$. Characterize the critical points of f_p by means of $T_x(V)$.

4.4.2. Let X be a manifold, $f : X \to \mathbf{R}$ a C^∞ function having a critical point a, and α a curve on X such that $\alpha(0) = a$. Show that

$$\mathrm{Hess}_a\big(\alpha'(0)1, \alpha'(0)1\big) = (f \circ \alpha)''(0).$$

4.4.3. Let $p : S^d \to P^d(\mathbf{R})$ be the canonical map, and $f : P^d(\mathbf{R}) \to \mathbf{R}$ the function such that

$$f\big(p(x)\big) = \sum_{i=1}^{d+1} i x_i^2.$$

Find the critical points of f, show they are non-degenerate, and calculate their indices.

4.4.4. Let $g : T^2 \to T^2$ be the map defined by $(x, y) \mapsto (x + \frac{1}{2}, y)$ and G the group $\{\mathrm{Id}_{T^4}, g\}$. Let $p : T^2 \to K = T^2/G$ be the canonical projection into the Klein bottle, and $h : K \to \mathbf{R}$ the function defined by

$$h\big(p(x, y)\big) = \cos 2\pi x \sin 2\pi y.$$

Find the critical points of f, show they are non-degenerate, and calculate their indices.

4.4.5. Same setting as 4.4.3, but the function now is $\sum_{i=1}^{d+1} a_i x_i^2$. Discuss degeneracy and index as a function of the a_i.

4.4.6. (a) Same question as 4.4.5, but for the $P^d(\mathbf{C})$ (2.8.26) and the function $\sum_{i=1}^{d+1} a_i |z_i|^2$.
(b) Same question for $P^d(\mathbf{H})$ and the same function.
(c) Find $\chi\big(P^d(\mathbf{C})\big)$ and $\chi\big(P^d(\mathbf{H})\big)$ using 4.2.24.4.
(d) Same question as 4.4.3, for the sphere S^d and the function $\sum_{i=1}^{d+1} i x_i^2$.

4.4.7. Prove the result in 4.2.18, for $rt - s^2 > 0$, using only Taylor series.

4.4.8. Determine $S \cap T_A S$ for the examples in 4.2.22. Give other examples showing that $S \cap T_A S$ can be a very complicated set.

4.4.9. Let $S \subset \mathbf{R}^3$ be a surface and $A \in S$ a point. Consider two bases for $\mathbf{R}^3$ such that S is locally (around A) the graph of a function f in the first basis, and the graph of g in the second. Let the expression of A be $\big(a, b, f(a, b)\big)$ in the first basis and $\big(u, v, f(u, v)\big)$ in the second.
    Show that if

$$\frac{\partial^2 f}{\partial x^2}(a, b) \frac{\partial^2 f}{\partial y^2}(a, b) - \left(\frac{\partial^2 f}{\partial x \partial y}(a, b) \right)^2$$

is non-zero, so is

$$\frac{\partial^2 g}{\partial x^2}(u, v) \frac{\partial^2 g}{\partial y^2}(u, v) - \left(\frac{\partial^2 g}{\partial x \partial y}(u, v) \right)^2,$$

and that these two numbers have the same sign.

4.4.10. Let $f \in C^\infty(\mathbf{R}^d; \mathbf{R})$ be such that $f'(0) = 0$. By adding to f an appropriate quadratic form, show that we can approximate f by a function g having a non-degenerate critical point at g, where the approximation is in the sense of uniform convergence of all derivatives up to order k (where $k > 0$ is fixed) on a compact whose interior contains 0.

CHAPTER 5

Differential Forms

This is a fundamental chapter, coming right after chapter 2 in importance. Differential forms are the most commonly encountered mathematical objects when one deals with manifolds: they occur naturally in geometry, physics, mechanics...

After defining differential forms, we introduce exterior differentiation and pullbacks, and study their properties and their expression in local coordinates.

Next we make a systematic study of orientability, based on volume forms (section 5.3).

Section 5.4 introduces de Rham groups, which are vector spaces functiorially associated with manifolds, and having to do with the spaces of differential forms on them. De Rham groups are important in geometry and in other fields as well: in physics, for example, they are connected with integrable vector fields. We calculate the groups of star-shaped plane sets, of spheres, real projective spaces and tori.

In the course of the chapter (5.3.32) we define the important notion of a manifold-with-boundary, which is the setting for Stokes' theorem.

5.1. The Bundle $\Lambda^r T^* X$

5.1.1. We wish to extend to manifolds the notions introduced in section 0.3. Let X be an n-dimensional C^q manifold, where $q \geq 2$. For $x \in X$, let

$$T_x^* X = L(T_x X; \mathbf{R}) = (T_x X)^*$$

be the dual of $T_x X$. For every $0 \leq r \leq n$, let

5.1.2 $$\Lambda^r T^* X = \bigcup_{x \in X} \Lambda^r (T_x^* X)$$

be the disjoint union of the spaces $\Lambda^r (T_x^* X)$, and let $p : \Lambda^r T^* X \to X$ be the map taking $\alpha \in \Lambda^r (T_x^* X)$ to $x = p(\alpha)$. We will make $\Lambda^r T^* X$ into a manifold, as we did with TX in 2.5.24.

5.1.3. Let (U, ϕ) be a chart on X and, for every $x \in U$, consider the canonically associated isomorphism $\theta_x : T_x \to \mathbf{R}^n$ (2.5.10). Define a map

5.1.4 $$\tau_\phi = p^{-1}(U) \ni \alpha \mapsto \left(\phi(p(\alpha)), (\theta_{p(\alpha)}^{-1})^*(\alpha) \right) \in \mathbf{R}^n \times \Lambda^r (\mathbf{R}^n)^*,$$

where $(\theta_{p(\alpha)}^{-1})^*$ is the pull-back of $\theta_{p(\alpha)}^{-1}$ in the sense of 0.1.8.

5.1.5. Theorem. *Let X be an n-dimensional C^q manifold, where $q \geq 2$, and $0 \leq r \leq n$ an integer. The pairs $(p^{-1}(U), \tau_\phi)$, for (U, ϕ) ranging over all possible charts on X, form a C^{q-1} atlas on $\Lambda^r T^* X$. This atlas makes $\Lambda^r T^* X$ into a separable, Hausdorff, C^{q-1} manifold of dimension $n + \binom{n}{r}$.*

Proof. Analogous to 2.5.25; we just have to add separability, as in 3.1.7, and check that the charts are compatible. Let (U, ϕ) and (V, ψ) be charts on X, and θ, η the associated isomorphisms (2.5.10). The corresponding charts on $\Lambda^r T^* X$ are $\tau_\phi = \left((\phi \circ p), (\theta_{p(\cdot)}^{-1})^* \right)$ and $\tau_\psi = \left((\psi \circ p), (\eta_{p(\cdot)}^{-1})^* \right)$. By 0.3.10.5 and 2.5.11.1 we get

$$\tau_\psi \circ \tau_\phi^{-1} = \left(\psi \circ \phi^{-1}, (\eta_{p(\cdot)}^{-1})^* \circ ((\theta_{p(\cdot)}^{-1})^*)^{-1} \right) = \left(\psi \circ \phi^{-1}, (\theta_{p(\cdot)} \circ \eta_{p(\cdot)}^{-1})^* \right)$$

$$= \left(\psi \circ \phi^{-1}, ((\phi \circ \psi^{-1})' \circ \psi \circ \phi^{-1})^* \right).$$

But $\psi \circ \phi^{-1}$ and $(\phi \circ \psi^{-1})'$ are of class C^{q-1} by assumption, and $f \mapsto f^*$ is of class C^∞ by 0.3.7.1. $\qquad\square$

5.1.6. Remark. What we just did was to use the functor $E \mapsto \Lambda^r E^*$ from the category of vector spaces into itself to construct the fiber bundle $\Lambda^r T^* X$ from the tangent bundle TX. This works for any functor on vector spaces; in this way we get bundles $\otimes^r TX$, $\otimes^r T^* X$, $\mathrm{End}(TX)$, and so on.

5.2. Differential Forms on a Manifold

5.2.1. Definition. Let X be an n-dimensional $C^{q'}$ manifold and q, r integers such that $0 \leq q \leq q' - 1$ and $0 \leq r \leq n$. A C^q *differential form of degree* r, or *r-form*, on X is a map $\omega \in C^q(X; \Lambda^r T^* X)$ such that $p \circ \omega = \mathrm{Id}_X$. The space of r-forms of class C^q on X will be denoted by $\Omega_q^r(X)$.

5.2.2. Criterion. Let $\omega : X \to \Lambda^r T^* X$ be a map satisfying $p \circ \alpha = \mathrm{Id}_X$. How does one know if ω is an r-form? Apply 2.3.2: $\omega \in C^q(X; \Lambda^r T^* X)$ if and only if $\tau_\phi \circ \omega \circ \phi^{-1}$ is of class C^q for every chart (U, ϕ) on X (or, better yet, for every chart in a fixed atlas on X). From 5.1.4 and the identity $p \circ \phi = \mathrm{Id}_X$ it follows that

5.2.2.1
$$\tau_\phi \circ \omega \circ \phi^{-1} = \left(\phi \circ p \circ \omega \circ \phi^{-1}, (\theta^{-1})^* (\omega \circ \phi^{-1}) \right) = \left(\mathrm{Id}_{\phi(U)}, (\theta^{-1})^* (\omega \circ \phi^{-1}) \right).$$

This pair is of class C^q if and only if the second element is, since $\mathrm{Id}_{\phi(U)}$ is C^∞; notice that $(\theta^{-1})^* (\omega \circ \phi^{-1}) : \phi(U) \to \Lambda^r (\mathbf{R}^n)^*$ is an r-form on $\phi(U)$. If (U, ϕ) is a chart on X and $\omega : X \to \Lambda^r T^* X$ satisfies $p \circ \omega = \mathrm{Id}_X$, set

5.2.2.2 $\boxed{\phi}\omega = (\theta^{-1})^* (\omega \circ \phi^{-1}) : \phi(U) \to \Lambda^r (\mathbf{R}^n)^*.$

5.2.2.3. Proposition. *A map* $\omega : T \to \Lambda^r T^* X$ *satisfying* $p \circ \omega = \mathrm{Id}_X$ *is an r-form if and only if* $\boxed{\phi}\omega \in \underline{\Omega}_q^r(\phi(U))$ *for all of the charts in some atlas on X.* $\square$

5.2.3. Sum and wedge product. For $\omega, \sigma \in \Omega_q^r(X)$, we define $\omega + \sigma$ by $(\omega + \sigma)(x) = \omega(x) + \sigma(x)$ for every $x \in X$. We define $k\omega$ and $\omega \wedge \tau$ in the same way, for $k \in \mathbf{R}$ and $\tau \in \Omega_q^s(X)$. We have to check that $\omega + \sigma$ and $k\omega$ are r-forms and that $\omega \wedge \tau$ is an $(r+s)$-form; but this follows directly from the formulas in 0.1.9 and 0.1.10:

$$\boxed{\phi}(\omega + \sigma) = \boxed{\phi}\omega + \boxed{\phi}\sigma,$$
$$\boxed{\phi}(k\omega) = k\left(\boxed{\phi}\omega \right),$$
$$\boxed{\phi}(\omega \wedge \tau) = \boxed{\phi}\omega \wedge \boxed{\phi}\tau.$$

5.2.3.1. Proposition. $\Omega_q^r(X)$ *is a real vector space. The direct sum*

$$\Omega_q(X) = \bigoplus_{r=0}^{n} \Omega_q^r(X)$$

is an associative, anticommutative, graded algebra. $\square$

5.2.4. Pullbacks. Let X and Y be manifolds, $f : X \to Y$ a map of class $C^{q'}$ and $\omega \in \Omega_q^r(Y)$ $(q \leq q' - 1)$ an r-form on X. Define $f^* \omega : X \to \Lambda^r T^* X$ by

5.2.4.1 $f^* \omega = (Tf)^* (\omega \circ f),$

that is, $(f^*\omega)(x) = (T_x f)^* \big(\omega(f(x)) \big)$. This map is called the *pullback* of ω by f.

5.2.4.2. Proposition. *We have $f^*\omega \in \Omega_q^r(X)$, and $f^* : \Omega_q(Y) \to \Omega_q(X)$ is an algebra homomorphism.*

Proof. Take charts (U, ϕ) and (V, ψ) on X and Y, respectively, as in 2.3.2(iii). Proposition 5.2.4.2 follows from 5.2.3.1 and the formula

5.2.4.3
$$\boxed{\phi}(f^*\omega) = (\psi \circ f \circ \phi^{-1})^* \left(\boxed{\psi}\omega \right),$$

where $(\psi \circ f \circ \phi^{-1})^*$ denotes the pullback in the sense of 0.3.7, and

$$\boxed{\psi}\omega \in \underline{\Omega}_q^r(\psi(V)).$$

Formula 5.2.4.3 is a consequence of 0.3.10.5, 0.3.7 and 2.5.13.1, where θ and η are associated with (U, ϕ) and (V, ψ), respectively:

$$\begin{aligned}
\boxed{\phi}(f^*\omega) &= (\theta^{-1})^* \big((Tf)^* (\omega \circ f) \circ \phi^{-1} \big) \\
&= \big((\theta^{-1})^* \circ (Tf)^* \big) (\omega \circ f \circ \phi^{-1}) \\
&= (Tf \circ \theta^{-1})^* \big(\omega \circ (f \circ \phi^{-1}) \big) \\
&= \big(\eta^{-1} \circ (\psi \circ f \circ \phi^{-1})' \big)^* \big(\omega \circ (f \circ \phi^{-1}) \big) \\
&= \big((\psi \circ f \circ \phi^{-1})' \big)^* \big((\eta^{-1})^* (\omega \circ (f \circ \phi^{-1})) \big) \\
&= \big((\psi \circ f \circ \phi^{-1})' \big)^* \big((\eta^{-1})^* ((\omega \circ \psi^{-1})(\psi \circ f \circ \phi^{-1})) \big) \\
&= (\psi \circ f \circ \phi^{-1})^* \big((\eta^{-1})^* (\omega \circ \psi^{-1}) \big) \\
&= (\psi \circ f \circ \phi^{-1})^* \left(\boxed{\psi}\omega \right).
\end{aligned}$$

That f^* is an algebra homomorphism is evident. Also, we deduce from 0.1.11 that

5.2.4.4
$$(g \circ f)^* = f^* \circ g^*$$

for $f \in G^{q'}(X; Y)$ and $g \in G^{q'}(Y; Z)$. $\square$

5.2.5. Restrictions. Let X be a manifold, $\omega \in \Omega_q^r(X)$ a form and Y a submanifold of X. Denote the inclusion map by $i : Y \to X$.

5.2.5.1. Definition. The map $i^*\omega = \omega|_Y$ is called the *restriction* of ω to Y.

A common example is $Y = U \in O(X)$.

5.2.5.2. We will often prove that a form is zero by a dimension argument. This means that, if $Y \subset X$ is an m-dimensional submanifold of X and ω is an r-form on X, with $r > m$, the restriction $\omega|_Y$, being still an r-form, must be zero. (Recall that $\Lambda^r E^* = 0$ if $r > \dim E$.)

5.2.6. The derivative of a function. Let X be a C^q manifold and $f \in C^q(X)$ a function. In 2.5.29 we defined the derivative $df : TX \to \mathbf{R}$, from which we can now get a map $X \ni x \mapsto df(x) \in T_x^* X$. This map is still called the *derivative* of f, and denoted by

5.2.6.1 $df : X \to T^* X = \Lambda^1 T^* X.$

5.2.6.2. Proposition. *We have* $df \in \Omega^1_{q-1}(X)$.

Proof. By construction, $p \circ df = \mathrm{Id}_X$. The result will follow from 5.2.2.3 and the next two lemmas.   □

5.2.6.3. Lemma. *If* $f \in C^q(Y)$ *and* $G \in C^q(X;Y)$ *we have* $d(f \circ G) = G^*(df)$.

Proof. By 2.5.29, $d(f \circ G) = df \circ TG$; since df is linear, $G^*(df) = df \circ TG$ by definition 0.1.8.   □

5.2.6.4. Lemma. *If* U *is open in a finite-dimensional vector space* E *and* $f \in C^q(U)$, *we have* $df = \Omega^1_{q-1}(U)$.

Proof. For the chart $(U, \phi) = (U, \mathrm{Id}_U)$ we have, by 2.5.32.2 and 5.2.2.2:

5.2.6.5 $\boxed{\phi}(df) = f'.$ □

5.2.6.6. In particular, $df = 0$ if and only if f is locally constant.

5.2.7. Open subsets of vector spaces. The reader may have noticed that the notion of a differential form introduced in 5.2.1, applied to an open subset of a vector space E considered as a manifold, does not coincide with the one introduced in 0.3.1. This is the reason why the Ω are underlined in section 0.3. But $\Omega(U)$ and $\underline{\Omega}(U)$, for $U \in O(E)$, are identical up to the canonical isomorphisms $\theta : T_x E \to E$, for every $x \in U$.

From now on we'll generally work with the new notion, even for $U \in O(E)$. Formula 0.3.6 will look nicer now, because of the following lemma:

5.2.7.1. Lemma. *Let* U *be an open subset of an* n-*dimensional vector space* E, $\{e_i\}_{i=1,\dots,n}$ *a basis for* E *and* $x_i : U \to \mathbf{R}$ $(i = 1,\dots,n)$ *the associated coordinate functions. Every* $\omega \in \Omega^r_q(U)$ *can be written in a unique way as*

5.2.7.2 $\omega = \sum_I \omega_I \, dx_I = \sum_{i_1 < \dots < i_r} \omega_{i_1 \dots i_r} \, dx_{i_1} \wedge \dots \wedge dx_{i_r},$

where $\omega_I \in C^q(U)$ *for every* I, *and* dx_i *is the derivative of* x_i *for every* $1 \le i \le n$.

Proof. By the proof of 5.2.6.4, the correspondence between $\Omega^q_r(U)$ and $\underline{\Omega}^q_r(U)$ is given by $\boxed{\phi}$, where $\phi = \mathrm{Id}_U$. By 0.3.6, $\boxed{\phi}\omega$ can be written as a sum $\sum_I \alpha_I e_I^*$. Setting $\omega = \sum_I \alpha_I \, dx_I$, with $dx_I = dx_{i_1} \wedge \dots \wedge dx_{i_r}$, and taking into account that $\phi = \text{Id}_U$ and that $\boxed{\phi}$ preserves the algebra structure (5.2.3), we conclude from 5.2.6.5 that

$$\boxed{\phi}\left(\sum_{i_1 < \cdots < i_r} \alpha_{i_1 \ldots i_r} \, dx_{i_1} \wedge \cdots \wedge dx_{i_r} \right)$$

$$= \sum_{i_1 < \cdots < i_r} \alpha_{i_1 \ldots i_r} \, \boxed{\phi}(dx_{i_1}) \wedge \cdots \wedge \boxed{\phi}(dx_{i_r})$$

$$= \sum_{i_1 < \cdots < i_r} \alpha_{i_1 \ldots i_r} (x_{i_1})' \wedge \cdots \wedge (x_{i_r})'.$$

Since x_i is a linear form, $(x_i)' = e_i^*$ (see the proof of 0.3.12), whence

$$\boxed{\phi}\left(\sum_I \alpha_i \, dx_I \right) = \sum_I \alpha_I e_I^*,$$

concluding the proof. $\qquad\square$

5.2.7.3. In particular, for $E = \mathbf{R}^n$, we will make implicit use of 5.2.7.2 with $\{e_i\}_{i=1,\ldots,n}$ equal to the canonical basis.

5.2.8. Local expression in coordinates. Let X be an n-dimensional C^p manifold, (U, ϕ) a chart on X and $\{y_i\}_{i=1,\ldots,n}$ the corresponding local coordinates on U, that is, the functions $y_i = x_i \circ \phi$ (2.2.3).

5.2.8.1. Proposition. *Every form $\omega \in \Omega^r(X)$, after being suitably restricted (cf. 5.2.5.1), can be written in a unique way as*

5.2.8.2 $$\omega|_U = \sum_I \omega_I^U \, dy_I = \sum_{i_1 < \cdots < i_r} \omega_{i_1 \ldots i_r}^U \, dy_{i_1} \wedge \cdots \wedge dy_{i_r}.$$

Given $q \leq p - 1$, an r-form ω is of class C^q if and only if each ω_I^U is a function of class C^q on U, for U ranging over the domains of charts of some atlas on X.

Proof. This follows from 5.2.2.3, 5.2.7.2, 5.2.4.2 and 5.2.6.3, applied to $(\phi^{-1})^* \omega \in \Omega^r(\phi(U))$. $\qquad\square$

Let $f : X \to Y$ be differentiable, and let's calculate the expression in coordinates (5.2.8.2) of the pullback $f^* \omega$ of a form $\omega \in \Omega^r(X)$. If the expression of ω in the local coordinates $\{y_i\}_{i=1,\ldots,n}$ corresponding to a chart (U, ϕ) on Y is

$$\omega|_U = \sum_{i_1 < \cdots < i_r} \omega_{i_1 \ldots i_r}^U \, dy_{i_1} \wedge \cdots \wedge dy_{i_r},$$

we have

5.2.8.3 $$(f^* \omega)|_{f^{-1}(U)} = \sum_{i_1 < \cdots < i_r} (\omega_{i_1 \ldots i_r}^U \circ f) \, d(y_{i_1} \circ f) \wedge \cdots \wedge d(y_{i_r} \circ f).$$

This is the counterpart of 0.3.8, which we now declare outdated: from now on we stick to 5.2.8.3, even for vector spaces.

5.2.8.4. Example. Let $f : \mathbf{R}^2$ be defined by $f(t) = (\cos t, \sin t)$ and $\beta = x\,dy - y\,dx \in \Omega^1_\infty(\mathbf{R}^2)$. The pullback is $f^*\beta = \cos t\,d(\sin t) - \sin t\,d(\cos t) = dt$.

5.2.9. The exterior derivative

5.2.9.1. Proposition. *Let X be a d-dimensional C^q manifold and p an integer such that $1 \leq p \leq q - 1$. There exists a unique operator d sending $\Omega^r_p(X)$ into $\Omega^{r+1}_{p-1}(X)$ for all $0 \leq r \leq n - 1$ and satisfying the following conditions:*

(i)   *The restriction of d to $\Omega^r_p(X)$ is linear;*
(ii) $d(\alpha \wedge \beta) = d\alpha \wedge \beta + (-1)^{\deg \alpha}\alpha \wedge d\beta,$
(iii) $d \circ d = 0;$
(iv) *for $f \in C^p(X) = \Omega^0_q(X)$, df is the derivative in the sense of 5.2.6.*

This operation is called *exterior differentiation*, and $d\alpha$ is called the *(exterior) derivative* of α.

Proof. Notice first that if $U \in O(X)$ and $\alpha|_U = 0$, any operation d satisfying the four conditions gives $(d\alpha)|_U = 0$. In fact, for every $x \in U$ we can find by 3.1.2 a function f with support in U and equal to 1 on a neighborhood of x; thus $f\alpha = 0$, whence $d(f\alpha) = 0$. But $d(f\alpha) = df\,\alpha + f\,d\alpha$; since $df(x) = 0$ (recall that f is locally constant) and $f(x) = 1$, we get $d\alpha(x) = 0$, as asserted.

To show that d is unique, let $x \in X$ and $\omega \in \Omega^r_p(X)$ be given, as well as a chart (U, ϕ) at x. By 5.2.8.2, $\omega|_U$ can be uniquely written in the form

$$\sum_{i_1 < \cdots < i_r} \omega_{i_1 \ldots i_r}\, dx_{i_1} \wedge \cdots \wedge dx_{i_r}.$$

Let f be a function supported in U and equal to 1 in a neighborhood of x. The form

$$\tilde{\omega} = \sum_{i_1 < \cdots < i_r} f\omega_{i_1 \ldots i_r}\, d(fx_{i_1}) \wedge \cdots \wedge d(fx_{i_r})$$

clearly belongs to $\Omega^r_p(X)$. By construction, $\omega - \tilde{\omega}$ is zero on a neighborhood of x, so $(d\omega)(x) = (d\tilde{\omega})(x) = 0$ by the first paragraph. In addition, the four conditions on d show that

$$d\tilde{\omega} = \sum_{i_1 < \cdots < i_r} d(f\omega_{i_1 \ldots i_r}) \wedge d(fx_{i_1}) \wedge \cdots \wedge d(fx_{i_r}).$$

At x this boils down to

$$(d\omega)(x) = (d\tilde{\omega})(x) = \sum_{i_1 < \cdots < i_r} (d\omega_{i_1 \ldots i_r} \wedge dx_{i_1} \wedge \cdots \wedge dx_{i_r})(x),$$

which only depends on ω, proving the uniqueness of d. In addition, this shows that if d exists on X and on the open submanifold $U \subset X$, we have

$$(d\omega)|_U = d(\omega|_U).$$

Uniqueness and the preceding formula show that it is enough to define d on the domain of any chart (U, ϕ). Now by 5.2.6.5 the algebra homomorphism $\boxed{\phi}$ defined on 5.2.2.2 translates an operator d, acting on r-forms on U and satisfying conditions (i)–(iv) of the proposition, into an operator $\boxed{\phi} \circ d \circ \boxed{\phi}^{-1}$, acting on r-forms (old definition) on $\phi(U)$, and satisfying conditions (i)–(iv) of theorem 0.3.12. Since 0.3.12 guarantees the existence of such an operator, we conclude the existence of d as well.

There remains to see that $d\omega$ is of class C^{p-1} if ω is of class C^p. This follows from 5.2.2.3, 0.3.12 and the fact that (leaving U out of the notation)

$$\boxed{\phi}(d\omega) = d\left(\boxed{\phi}\omega\right). \qquad \square$$

5.2.9.2. Proposition. *If X and Y are manifolds and $f : X \to Y$ a morphism, we have*

5.2.9.3 $$d \circ f^* = f^* \circ d,$$

that is, the following diagram commutes:

5.2.9.4
$$
\begin{array}{ccc}
\Omega^r_p(X) & \xleftarrow{\ f^*\ } & \Omega^r_p(Y) \\
\downarrow{\scriptstyle d} & & \downarrow{\scriptstyle d} \\
\Omega^{r+1}_{p-1}(X) & \xleftarrow{\ f^*\ } & \Omega^{r+1}_{p-1}(Y)
\end{array}
$$

Proof. This follows from 5.2.9.1, 5.2.4.3 and 0.3.13. $\qquad \square$

5.2.9.5. Example. If Y is a submanifold of X and $\omega \in \Omega^r_1(X)$, the restrictions 5.2.5 satisfy

5.2.9.6 $$d(\omega|_Y) = (d\omega)|_Y.$$

5.2.9.7. Local expression. If (U, ϕ) is a chart on X and $\omega \in \Omega^r_1(X)$, it follows from conditions (i)–(iv) that

5.2.9.8 $$d\left(\sum_I \omega_I \, dx_I\right) = \sum_I d\omega_I \wedge dx_I.$$

5.2.10. Continuous families of differential forms. We now extend 0.3.15 to the case of manifolds.

5.2.10.1. Definition. Let X be a manifold of class C^q and p an integer such tha $0 \le p \le q-1$. A *continuous, one-parameter family of C^p forms of degree r on X* is a pair (I, α), where $I \subset \mathbf{R}$ is an interval and $\alpha : I \times X \to \Lambda^r T^* X$ satisfies the following conditions:

(i) $p \circ \alpha : I \times X \to X$ is the canonical projection;

(ii) for every chart (U, ϕ) on X, the map $\boxed{\phi}\alpha : I \times \phi(U) \to \Lambda^r(\mathbf{R}^n)^*$ (5.2.2.2) satisfies definition 0.3.15.1.

5.2.10.2. Of course it's enough to check (ii) for the charts in one atlas.

5.2.10.3. If I is open and $\alpha \in C^p(I \times X; T^*X)$, condition (ii) is automatically satisfied.

5.2.10.4. If $I = [a, b]$, we define $\int_a^b \alpha_t \, dt$ by

$$\left(\int_a^b \alpha_t \, dt \right)(x) = \int_a^b \alpha(t, x) \, dt \in \Lambda^r T_x^* X,$$

in the sense of 0.4.7. It follows from 0.3.15.5 and 5.2.10.1(ii) that $\int_a^b \alpha_t \, dt \in \Omega_p^r(X)$.

5.2.10.5. Proposition. *If α is a continuous family of differential forms of class C^1 we have*

$$d\left(\int_a^b \alpha_t \, dt \right) = \int_a^b d\alpha_t \, dt.$$

Proof. This follows from 0.3.15.6 by taking charts (cf. the proof of 5.2.9.1). $\qquad \square$

5.2.10.6. Lemma. *Let X and Y be C^p manifolds, and $F : [0, 1] \times X \to Y$ a map. Denote by F_t the map $X \ni x \mapsto F(t, x) \in Y$, and by $TF : [0, 1] \times TX \to TY$ the map such that $TF|_{\{t\} \times TX} = T(F_t)$. Assume that $F_t \in C^p(X; Y)$ for every $t \in [0, 1]$, and that $TF \in C^0([0, 1] \times TX; TY)$.*

Then, for every $\beta \in \Omega_{p-1}^k(Y)$, the pair $([0, 1], t \mapsto F_t^*\beta)$ is a continuous family of C^{p-1} differential forms on X.

Proof. Continuity being a local property, we start by fixing $(t_0, x) \in [0, 1] \times X$ and finding a chart (U, ϕ) at x, a chart (V, ψ) at $f(x)$ and an $\varepsilon > 0$ such that $F_t(U) \subset V$ for every $t \in [t_0 - \varepsilon, t_0 + \varepsilon]$. This is possible because TF, and consequently F, are continuous, so $F^{-1}(V)$ is open in $[0, 1] \times X$ and, by the definition of the product topology, we can find ε such that $[t_0 - \varepsilon, t_0 + \varepsilon] \times U \subset f^{-1}(V)$.

Now transfer $F|_{[t_0 - \varepsilon, t_0 + \varepsilon] \times U}$ and $\beta|_V$ to open sets $U' \in O(\mathbf{R}^m)$ and $V' \in O(\mathbf{R}^n)$ by means of the charts (U, ϕ) and (V, ψ), respectively. By 5.2.4.3, we'll be done if we prove the lemma for the particular case of open submanifolds of $\mathbf{R}^m$ and $\mathbf{R}^n$, a map $G : [0, 1] \times U' \to V'$, a form $\gamma \to \Omega_{p-1}^r(V')$, and a family $t \mapsto G_t^*\gamma$ in the sense of 0.3.7.

By 2.5.26.1, the map TG is just the pair (G_t, G_t'). By condition (ii) we have

$$G_t' \in C^0([0, 1] \times U'; L(\mathbf{R}^m; \mathbf{R}^n)),$$

which shows that $t \mapsto G_t^*\gamma$ is a continuous family of C^{p-1} forms (0.3.15.1), since $G_t^*\beta = (G_t')^*(\beta \circ G_t)$ (apply 0.3.7.1).

5.3. Volume Forms and Orientation

5.3.1. Our purpose is to define orientation on abstract manifolds. As in 0.1.13, we want to orient each $T_x X$, for $x \in X$, but in such a way that the choice is continuous. The definition in 0.1.13 indicates that we should use differential forms.

5.3.2. Definition. Let X be a d-dimensional C^p manifold, with $p \geq 1$. A *volume form* on X is a nowhere vanishing d-form $\omega \in \Omega_0^d(X)$ on X.

5.3.3. We have $\omega(x) \in \Lambda^d T_x^* X \setminus \{0\}$, so $\omega(x)$ determines an orientation for $T_x X$, and the map $x \mapsto \omega(x)$ is continuous. We can associate to $\omega(x)$ an element of $\mathcal{O}(T_x X)$, but for now we cannot talk about the continuity of the map thus defined. This will be possible when we give

$$\bigcup_{x \in X} \left(\{x\} \times \mathcal{O}(T_x X) \right) = \mathcal{O}(X)$$

a differentiable structure (5.3.28). For now we work using volume forms only.

5.3.4. Definition. Two volume forms ω and ω' are called *equivalent* if $\omega'(x) = f(x)\omega(x)$, with $f(x) > 0$ for every $x \in X$.

This is the same as saying that $\omega(x)$ and $\omega'(x)$ induce the same orientation on $T_x X$ (0.1.13).

5.3.5. Definition. An *orientation* for X is an equivalence class of volume forms; the choice of such an equivalence class makes X *oriented*. A manifold X is called *orientable* if it can be oriented, and *non-orientable* otherwise.

5.3.6. There exist non-orientable manifolds, for example, the Möbius strip (the reader should try to show this fact heuristically, and perhaps also rigorously). For proofs of non-orientability, see 5.3.18 and exercises 5.9.10 and 5.9.11.

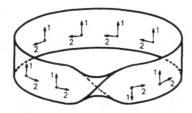

Figure 5.3.6

5.3.7. Theorem. *Let X be an orientable manifold. If X has k connected components, there are 2^k possible orientations for X.*

Proof. Let ω and ω' be orientations (orientations exist by assumption). We have $\omega(x) = f(x)\omega'(x)$ for every x, where f is continuous because $\omega, \omega' \in \Omega_0^d(X)$ (use 5.2.2.3). Thus the map

$$x \mapsto \operatorname{sgn}(f(x))$$

from X into the set $\{-1, +1\}$ (with the discrete topology) is continuous, and consequently constant on each connected component. Since there are k choices of sign, the total number of orientations is 2^k.   □

5.3.8. Examples of orientable manifolds

5.3.8.1. $\mathbf{R}^d$ is canonically oriented by the choice of the (constant) volume form

$$\omega_0 = dx_1 \wedge \cdots \wedge dx_d.$$

5.3.8.2. If X be an orientable manifold, every open submanifold U of X is orientable. In fact, the restriction of a nowhere vanishing volume form $\omega \in \Omega_0^d(X)$ to U is nowhere vanishing and lies in $\Omega_0^d(U)$ (5.2.5.1).

5.3.8.3. If X and Y are manifolds and $p : X \to Y$ a covering map, the pullback $p^*\omega$ of a volume form ω on Y is a volume form on X. For 5.2.4 gives

$$(p^*\omega)(x) = (T_x p)^*(\omega(p(x)));$$

since p is regular we conclude that $T_x p \in \operatorname{Isom}(T_x X; T_{p(x)} Y)$, so $(p^*\omega)(x) \neq 0$.

In particular, an orientation on Y gives rise to a canonically associated orientation on X.

5.3.8.4. A product manifold $X \times Y$ is orientable if and only if the factors X and Y are (see exercise 5.9.8).

For other examples one can use the next lemma, which will be necessary for other purposes as well:

5.3.9. Lemma. forms invariant under group of diffeos define form on quotient *Let X be a manifold of class C^q and $G \subset \operatorname{Diff}(X)$ a group of C^p diffeomorphisms acting properly discontinuously without fixed points on X. If $\alpha \in \Omega_p^r(X)$ is invariant under G, that is, if $g^*(\alpha) = \alpha$ for every $g \in G$, there exists a unique $\beta \in \Omega_p^r(X/G)$ such that $p^*\beta = \alpha$, where $p : X \to X/G$ is the canonical map.*

The condition that α is invariant under G is also necessary: if $p^*\beta = \alpha$, it follows from 5.2.4.4 and the identity $p \circ g = p$ that

$$(g^* \circ p^*)\beta = g^*\alpha = (p \circ g)^*\beta = p^*\beta = \alpha.$$

Proof. Take $y \in Y = X/G$ and $x \in p^{-1}(y)$, and set

5.3.9.1 $\beta(y) = ((T_x p)^{-1})^* \alpha(x).$

This makes sense because p is regular (2.4.9), so $T_x p \in \text{Isom}(T_x X; T_y Y)$ and $(T_x p)^{-1} \in \text{Isom}(T_y Y; T_x X)$. We must show that $\beta(y)$ does not depend on the choice of x. Choose another $x' \in p^{-1}(y)$, and take $g \in G$ such that $g(x) = x'$. Since $p(x) = (p \circ g)(x)$, we have

$$T_x(p \circ g) = T_x p = T_{g(x)} p \circ T_x g,$$

so $T_{g(x)} p = T_x p \circ (T_x g)^{-1}$ and $(T_{g(x)} p)^{-1} = T_x g \circ (T_x p)^{-1}$, since $T_x g$ is an isomorphism. Since $x' = g(x)$, we can write

$$
\begin{aligned}
\left((T_{x'} p)^{-1}\right)^* \alpha(x') &= \left((T_{g(x)} p)^{-1}\right)^* \left(\alpha(g(x))\right) \\
&= \left(T_x g \circ (T_x p)^{-1}\right)^* \left(\alpha(g(x))\right) \\
&= \left((T_x p)^{-1}\right)^* \left(g^*(\alpha)(x)\right) = \left((T_x p)^{-1}\right)^* \alpha(x),
\end{aligned}
$$

since α is invariant under G. This shows that $\beta(y)$ is well-defined.

We have $\alpha = p^* \beta$ by construction. There remains to check that β is of class C^p if α is. This is done by applying 5.2.2.3 to charts on X/G of the form $\psi = \phi \circ p^{-1}$ (cf. 2.4.9). $\qquad\square$

5.3.10. Examples (of forms on quotient)

5.3.10.1. The torus $T^d = \mathbf{R}^d / \mathbf{Z}^d$ (cf. 2.4.12.1). By 2.4.7.1 $\mathbf{Z}^d$ (as a group of translations of $\mathbf{R}^d$) acts properly discontinuously without fixed points. The forms dx_i defined in 5.2.7 are in $\Omega^1_\infty(\mathbf{R}^d)$ for every $i = 1, \dots, d$, and they are clearly invariant under $\mathbf{Z}^d$. Thus there exist on the torus d canonical one-forms, denoted by ω_i $(i = 1, \dots, d)$, such that $p^*(\omega_i) = dx_i$.

By 5.3.9 every form $\alpha \in \Omega^r_p(T^d)$ can be written in the form $\alpha = \sum_I \alpha_I \omega_I$, where $I = (i_1, \dots, i_r)$, $\omega_I = \omega_{i_1} \wedge \cdots \wedge \omega_{i_r}$, and $\alpha_I \in C^p(T^d)$.

Notice, however, that there are no functions $y_i : T^d \to \mathbf{R}$ such that $dy_i = \omega_i$ (exercice 5.9.13).

5.3.10.2. If α is a volume form on X, invariant under a group G that acts properly discontinuously without fixed points, there exists a unique volume form β on X/G such that $p^*(\beta) = \alpha$. This follows immediately from 5.3.9: since $\alpha(x) \neq 0$ and $T_x p$ is an isomorphism, we have $\beta(y) \neq 0$ by 5.3.9.1.

In particular, if X is orientable and admits an orientation invariant under G, the quotient X/G is orientable. This is the case of the torus T^d, which is canonically oriented by $\omega = \omega_1 \wedge \cdots \wedge \omega_d$, in the notation of 5.3.10.1.

5.3.11. Contractions. We will use contractions (0.1.18) to orient spheres and some projective spaces. Let ξ be a vector field on a manifold X (3.5.1); if X is of class C^p, the class of ξ can be up to C^{p-1}.

5.3.12. Definition and proposition. For every $\alpha \in \Omega^r_{p-1}(X)$, define the *contraction* of α by ξ as the form $\text{cont}(\xi)\alpha$ such that

$$\left(\text{cont}(\xi)\alpha\right)(x) = \text{cont}\left(\xi(x)\right)\left(\alpha(x)\right)$$

for every $x \in X$. Then $\mathrm{cont}(\xi)$ maps $\Omega^r_{p-1}(X)$ into $\Omega^{r-1}_{p-1}(X)$. In addition

$$\mathrm{cont}(\xi)(\alpha \wedge \beta) = \left(\mathrm{cont}(\xi)\alpha\right) \wedge \beta + (-1)^{\deg \alpha}\alpha \wedge \left(\mathrm{cont}(\xi)\beta\right)$$

for every $\alpha, \beta \in \Omega^r_{p-1}(X)$.

Proof. The last assertion follows from 0.1.20. There remains to show that $\mathrm{cont}(\xi)\alpha$ is of class C^{p-1}. Using charts we reduce to the case $X = U \in O(\mathbf{R}^n)$, $\alpha \in \underline{\Omega}^r_{p-1}(U)$ and $\xi \in C^{p-1}(U; \mathbf{R}^n)$. Then formula 0.1.19 shows that the map

$$\mathbf{R}^n \times \Lambda^r(\mathbf{R}^n)^* \ni (\xi, \alpha) \mapsto \mathrm{cont}(\xi)\alpha \in \Lambda^{r-1}(\mathbf{R}^n)^*$$

is bilinear, and we conclude by applying 0.2.8.3, 0.2.15.1 and the usual techniques of differential calculus. $\square$

5.3.13. Here's an explicit way to calculate $\mathrm{cont}(\xi)\alpha$ in the case $r = d = \dim X$. Denote by $\{x_i\}$ the local coordinates on U (5.2.8), and by

$$\frac{\partial}{\partial x_i} : U \to TX$$

the vector fields on U such that

$$\left\{\frac{\partial}{\partial x_i}\right\}_{i=1,\dots,d}$$

is the basis of $T_x X$ dual to the basis $\{dx_i(x)\}_{i=1,\dots,d}$ of $T^*_x X$, for every $x \in U$. Each $\partial/\partial x_i$ is of class C^{p-1} because its pullback by ϕ^{-1} is the constant vector field $e_i \in \mathbf{R}^d$ on $\phi(U)$. Similarly, if ξ is of class C^{p-1}, we have

5.3.14
$$\xi = \sum_{i=1}^d \xi_i \frac{\partial}{\partial x_i},$$

where the ξ are C^{p-1} vector fields on U.

Now write $\alpha|_U = a\,dx_1 \wedge \cdots \wedge dx_d$, using 5.2.8.2. By 0.1.22 and 5.2.7 we find

5.3.15 $$\mathrm{cont}(\xi)\alpha|_U = \sum_i (-1)^{i-1}a\xi_i\,dx_1 \wedge \cdots \wedge \widehat{dx_i} \wedge \cdots \wedge dx_d.$$

If α is a volume form we have $a(x) \neq 0$ for every $x \in U$, and formula 5.3.15 shows that:

5.3.16. Corollary. *A C^{p-1} volume form α on X determines an isomorphism between Ω^{d-1}_{p-1} and the space of C^{p-1} vector fields on X.* $\square$

5.3.17. Examples (of contractions)

5.3.17.1. Let X be a d-dimensional C^p submanifold of $\mathbf{R}^{d+1}$, contained in an open subset $U \subset \mathbf{R}^{d+1}$. Let $\xi \in C^{p-1}(U; TU)$ be a vector field on U (in the sense of 3.5.1) such that $\xi(x) \notin T_x X$ for every $x \in X$. (Such a vector field may not exist; it doesn't exist on a neighborhood U of a Möbius strip, for instance.)

If ω_0 is the canonical volume form on $\mathbf{R}^{d+1}$, we have $\mathrm{cont}(\xi)\omega_0 \in \Omega^d_{p-1}(U)$. We claim that $\mathrm{cont}(\xi)\omega_0|_X$ is a volume form on X. Clearly $\mathrm{cont}(\xi)\omega_0|_X$ is continuous on X, since it is continuous on U. To show that it does not vanish on X, take $x \in X$ and a basis $(\xi_1, \ldots, \xi_d)$ of $T_x X$. Then $\big(\xi(x), \xi_1, \ldots, \xi_d\big)$ is a basis for $T_x \mathbf{R}^{d+1}$, and we have

$$\big(\mathrm{cont}(\xi)\omega_0\big)(\xi_1, \ldots, \xi_d) = \omega_0\big(\xi(x), \xi_1, \ldots, \xi_d\big) \neq 0$$

because ω_0 is a volume form.

In section 6.4 we shall see that by taking $\xi(x)$ to be a unit vector normal to X we can define a canonical volume form on oriented submanifolds of $\mathbf{R}^{d+1}$.

5.3.17.2. The sphere $S^d \subset \mathbf{R}^{d+1}$. Take $U = \mathbf{R}^{d+1}$ and let ξ be the vector field $x \mapsto \theta_x^{-1}(x) \in T_x \mathbf{R}^{d+1}$, where $\theta_x : T_x \mathbf{R}^{d+1} \to \mathbf{R}^{d+1}$ is the canonical isomorphism defined in 2.5.12.3. By 5.3.17.1, $\sigma = \mathrm{cont}(\xi)\omega|_{S^d}$ defines a canonical orientation for S^d.

The explicit expression for σ is easily calculated using 5.3.15:

$$\sigma(x_1, \ldots, x_{d+1}) = \mathrm{cont}\left(\sum_{i=1}^{d+1} x_i e_i\right)(\omega_0)$$

$$= \sum_{i=1}^{d+1} (-1)^{i-1} x_i\, dx_i \wedge \cdots \wedge \widehat{dx_i} \wedge \cdots \wedge dx_{d+1}.$$

For S^1 the expression is

$$\sigma(x, y) = x\, dy - y\, dx.$$

For S^2:

$$\sigma(x, y, z) = x\, dy \wedge dz + y\, dz \wedge dx + z\, dx \wedge dy.$$

5.3.17.3. Remark. The canonical form σ is invariant under the action of $SO(d+1)$, the group of rotations of $\mathbf{R}^{d+1}$, on S^d. This is clear because both ξ and ω_0 are invariant under rotations (if f is a rotation, we have $f^*(\omega_0) = (\det f)\omega_0 = \omega_0$ by 0.1.12).

5.3.17.4. Every curve is orientable. By 3.4.1, any curve is diffeomorphic to $\mathbf{R}$ or S^1, and both of these are orientable (by 5.3.8.1 and 5.3.10.2, and using 2.6.13.1).

We now use lemma 5.3.9 to pass from spheres to projective spaces. We have the following result:

5.3.18. Theorem. *The projective space $P^d(\mathbf{R})$ is orientable if and only if d is odd.*

Proof. Set $s = -\mathrm{Id}_{\mathbf{R}^{d+1}}|_{S^d}$ and consider the subgroup $G = \{\mathrm{Id}_{S^d}, s\}$ of $\mathrm{Diff}(S^d)$. The action of G on S^d is properly discontinuous without fixed points, and the quotient S^d/G is $P^d(\mathbf{R})$ (2.4.12.2). In addition,

$$s^*\sigma = s^*\big(\mathrm{cont}(\xi)\omega_0\big) = \mathrm{cont}(\xi \circ s)(s^*\omega_0) = \mathrm{cont}(s)(s^*\omega_0)$$
$$= \mathrm{cont}(s)(\det s)\omega_0 = \mathrm{cont}(s)(-1)^{d+1}\omega_0$$
$$= (-1)^{d+1}\sigma.$$

Thus σ is invariant under G if d is odd; it follows from 5.3.9 that $P^d(\mathbf{R})$ is orientable.

Now let d be even, and assume that $P^d(\mathbf{R})$ is orientable. Let θ be a volume form on $P^d(\mathbf{R})$; the canonical map $p : S^d \to P^d(\mathbf{R})$ gives a volume from $\eta = p^*(\theta)$ on S^d (5.3.8.3), which we can write as $\eta = f\sigma$, where σ is the canonical form on S^d and $f : S^d \to \mathbf{R}^*$ is continuous. On the one hand we have $s^*\eta = s^*(p^*\theta) = (p \circ s)^*\theta = p^*\theta = \eta$ because $p \circ s = p$, and on the other $s^*\sigma = (-1)^{d+1}\sigma = -\sigma$ by the equation above; this implies $f \circ s = -f$, which is a contradiction because S^d is connected and $f \neq 0$ everywhere.

5.3.19. The Klein bottle (2.4.12.4) is non-orientable: exercise 5.9.10.

So far our only instrument to find out whether a manifold is orientable has been the definition, 5.3.5. We will now state a criterion involving charts.

5.3.20. Definition. Let X and Y be d-dimensional manifolds, oriented by the volume forms α and β, respectively. A diffeomorphism $f : X \to Y$ is said to *preserve orientation* if, for every $x \in X$, we can write $f^*\beta(x) = \lambda(x)\alpha(x)$ with $\lambda(x) > 0$ (notice that we already know that $\lambda(x) \neq 0$).

For example, $-\mathrm{Id}_{\mathbf{R}^{d+1}}|_{S^d} : S^d \to S^d$ preserves orientation if and only if d is odd.

5.3.21. Proposition. *Let X, Y, Z be oriented manifolds and $f : X \to Y$, $g : Y \to Z$ orientation-preserving diffeomorphisms. The composition $g \circ f$ is orientation-preserving.*

Proof. This follows from the identity $(g \circ f)^* = f^* \circ g^*$ (5.2.4.4). □

5.3.22. Definition. Let X be an oriented manifold. A chart (U, ϕ) is said to be *positively oriented* if $\phi \in \mathrm{Diff}(U; \phi(U))$ preserves orientation (that is, the induced orientations on U and $\phi(U)$, cf. 5.3.8.2 and 5.3.8.1).

5.3.23. Remark. Let (U, ϕ) and (V, ψ) be positively oriented charts on an oriented manifold X. The coordinate change

$$\psi \circ \phi^{-1} \in \text{Diff}\big(\phi(U \cap V); \psi(U \cap V)\big)$$

preserves the canonical orientation of $\mathbf{R}^d$. By 0.3.10.4 we have, for any $u \in \phi(U \cap V)$:

5.3.23.1 $$J(\psi \circ \phi^{-1})(u) > 0.$$

Now let X be an oriented manifold and $\big\{(U_i, \phi_i)\big\}_{i \in I}$ an atlas on X such that each U_i is connected. Let $s : \mathbf{R}^d \to \mathbf{R}^d$ be the diffeomorphism defined by

$$s(x_1, \ldots, x_d) = (-x_1, x_2, \ldots, x_d).$$

Replace each chart (U_i, ϕ_i) that is *not* positively oriented by the chart $(U, s \circ \phi)$, which is. This gives a new atlas on X, consisting solely of positively oriented charts. By 5.3.23 the coordinate changes $\phi_i \circ \phi_j^{-1}$ now preserve the canonical orientation of $\mathbf{R}^d$. Thus a necessary condition for a manifold to be orientable is that it have an atlas for which all coordinate changes preserve orientation. This condition is also sufficient:

5.3.24. Theorem. *A manifold X is orientable if and only if it has an atlas all of whose coordinate changes preserve orientation.*

Proof. Let $\big\{(U_i, \phi_i)\big\}_{i \in I}$ be such an atlas and $\big\{(U_i, \phi_i, \psi_i)\big\}_{i \in I}$ an associated partition of unity. Let ω_0 be the canonical volume form on $\mathbf{R}^d$. We claim that

$$\omega = \sum_{i \in I} \psi_i \phi_i^* \omega_0$$

defines an orientation on X. Since each ϕ_i and ψ_i is continuous, so is ω, and there remains to show that $\omega(x) \neq 0$.

Take $x \in X$, and choose i_0 such that $\psi_{i_0}(x) \neq 0$. For $i \neq i_0$ we have

$$\psi_i \phi_i^* \omega_0(x) = \psi_i(x)(\phi_i \circ \phi_{i_0}^{-1} \circ \phi_{i_0})^*(\omega_0)(x)$$
$$= \psi_i(x)\phi_{i_0}^*\big((\phi_i \circ \phi_{i_0}^{-1})^*(\omega_0)\big)(x).$$

But $(\phi_i \circ \phi_{i_0}^{-1})^*(\omega_0) = J(\phi_i \circ \phi_{i_0}^{-1})\omega_0$ by 0.3.10.4, so this becomes

$$\omega(x) = \bigg(\psi_{i_0}(x) + \sum_{i \neq i_0} \psi_i(x) J(\phi_i \circ \phi_{i_0}^{-1}) \circ \phi_{i_0}(x)\bigg)\phi_{i_0}^*(\omega)(x).$$

The factor in parentheses is a finite sum of non-negative terms, because $J(\phi_i \circ \phi_{i_0}^{-1}) > 0$ by 5.3.23.1, and one of the terms is strictly positive; thus $\omega(x) \neq 0$. $\square$

5.3.25. Remark. Rigorously speaking, we should have written

$$\omega = \sum_i \psi_i \phi_i^* \left(\omega|_{\phi_i(U_i)} \right)$$

and, more importantly, established an analogue of 0.3.10.4 for the new definition of differential forms (section 5.2). This concern is left to the reader.

We now give yet another characterization for orientable manifolds. If X is a manifold, set

5.3.26
$$\tilde{X} = \mathcal{O}(X) = \bigcup_{x \in X} \{x\} \times \mathcal{O}(T_x X)$$

(disjoint union). Let $p : \tilde{X} \to X$ be the map that takes $\alpha \in \mathcal{O}(T_x X)$ into $p(\alpha) = x$.

5.3.27. Theorem. *Let X be a manifold of class C^p. The set $\mathcal{O}(X) = \tilde{X}$ has a canonical C^p manifold structure such that $\tilde{X}$ is orientable and $p : \tilde{X} \to X$ is a double (i.e., two-fold) covering.*

Proof. We construct an atlas as follows. If (U, ϕ) is a chart on X, let $(\tilde{U}, \tilde{\phi})$ be defined by

$$\tilde{U} = \left\{ (x, \alpha(x)) : x \in U, \, \alpha(x) \in \mathcal{O}(T_x X), \, (T_x \phi)^* \omega_0 \in \alpha(x) \right\}$$

and $\tilde{\phi} = \phi \circ p$. Here ω_0 is the canonical volume form and $(T_x \phi)^* \omega_0 \in \alpha(x)$ means that the orientation induced on $T_x X$ by $(T_x \phi)^* \omega_0$ is exactly $\alpha(x)$.

Let's check that the pairs $(\tilde{U}, \tilde{\phi})$ form an atlas. The union of the $\tilde{U}$ is $\tilde{X}$: for $\omega = (x, \xi) \in \tilde{X}$, take a chart (U, ϕ) of X with U connected and $x \in U$. If $(T_x \phi)^* \omega_0 \in \xi$ we have $\omega \in (\tilde{U}, \tilde{\phi})$; otherwise $\omega \in (\tilde{U}, \tilde{\psi})$, where $\psi = s \circ \phi$ is constructed as in the last paragraph of 5.3.23.

Next, $\tilde{\phi}(\tilde{U}) = \phi(p(\tilde{U})) = \phi(U)$ is open in $\mathbf{R}^d$. Also, $\tilde{\phi} : \tilde{U} \to \phi(U)$ is bijective: it is surjective by construction and, if $\omega_1 = (x_1, \xi_1)$, $\omega_2 = (x_2, \xi_2) \in \tilde{U}$ are such that $\tilde{\phi}(\omega_1) = \tilde{\phi}(\omega_2)$, we get $\phi(p(\omega_1)) = \phi(p(\omega_2))$, hence $p(\omega_1) = p(\omega_2)$ because ϕ is bijective, and again $x_1 = x_2$. Finally, since $(T_x \phi)^* \omega_0$ is in both ξ_1 and ξ_2, the two equivalence classes are identical.

There remains to show that if $(\tilde{U}, \tilde{\phi})$ and $(\tilde{V}, \tilde{\psi})$ are charts on $\tilde{X}$, the map $\tilde{\psi} \circ \tilde{\phi}^{-1}$ is a C^p diffeomorphism between $\tilde{\phi}(\tilde{U} \cap \tilde{V})$ and $\tilde{\psi}(\tilde{U} \cap \tilde{V})$; in particular, this requires checking that $\tilde{\phi}(\tilde{U} \cap \tilde{V})$ is open in $\mathbf{R}^d$. Take $\omega \in \tilde{U} \cap \tilde{V}$, and $x = p(\omega)$. If ω_0 is the canonical volume form of $\phi(U \cap V)$, we see that $(T_x \phi)^* \omega_0$ and $(T_x \psi)^* \omega_0$ lie in the same element of $\mathcal{O}(T_x X)$, so $J(\psi \circ \phi^{-1})(\phi(x)) > 0$.

By continuity, there exists an open neighborhood W of x in $U \cap V$ such that $J(\psi \circ \phi^{-1}) \circ \phi$ is positive throughout W. Since $p^{-1}(W) \cap \tilde{U} \subset \tilde{U} \cap \tilde{V}$, we have

$$\tilde{\phi}(p^{-1}(W) \cap U) = \tilde{\phi}(p^{-1}(W)) \cap \tilde{\phi}(\tilde{U}) \subset \tilde{\phi}(\tilde{U} \cap \tilde{V}),$$

and, since $\tilde{\phi} = \phi \circ p$, we obtain

$$\phi(W) \cap \phi(U) \subset \tilde{\phi}(\tilde{U} \cap \tilde{V}).$$

But W is open in X, so $\phi(W \cap U)$ is open in $\mathbf{R}^d$. Thus, for every element $\tilde{\phi}(\omega) \in \tilde{\phi}(\tilde{U} \cap \tilde{V})$, there exists an open set $\phi(W \cap U)$ such that

$$\tilde{\phi}(\omega) \in \phi(W \cap U) \subset \tilde{\phi}(\tilde{U} \cap \tilde{V}),$$

which shows that $\tilde{\phi}(\tilde{U} \cap \tilde{V})$ is open in $\mathbf{R}^d$.

Finally, we can write

$$\tilde{\psi} \circ \tilde{\phi}^{-1} = (\psi \circ p) \circ (\phi \circ p)^{-1},$$

whence

$$\tilde{\psi} \circ \tilde{\phi}^{-1}|_{\tilde{\phi}(\tilde{U} \cap \tilde{V})} = \psi \circ \phi^{-1}|_{\phi(U \cap V)},$$

which shows that the coordinate change is of class C^p, as desired.

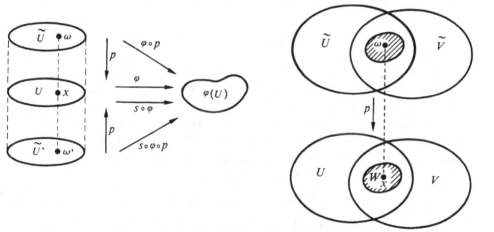

Figure 5.3.27

Before checking that $\tilde{X}$ is separable and Hausdorff, let's show that $\tilde{X}$ is orientable and a covering of X. Since

$$J(\tilde{\psi} \circ \tilde{\phi}^{-1}) = J(\psi \circ \phi^{-1}) > 0$$

by the preceding calculation, 5.3.24 guarantees that $\tilde{X}$ is oriented. Now consider the map $p : \tilde{X} \to X$. If $x \in X$ and (U, ϕ) is a chart at x, we have $p^{-1}(U) = \tilde{U} \cup \tilde{U}'$, where $\tilde{U}$ is the chart associated with (U, ϕ) and $\tilde{U}'$ is the chart associated with $(U, s \circ \phi)$, where s is as in the last paragraph of 5.3.23. Also, $\tilde{U}$ and $\tilde{U}'$ are diffeomorphic to U by construction, and they are disjoint because their elements have different orientations. This shows that p is a double cover.

To show that $\tilde{X}$ is a Hausdorff manifold, take $\omega \neq \omega'$. If $p(\omega) \neq p(\omega')$, we can take disjoint domains of charts U and U' on X, containing $p(\omega)$ and $p(\omega')$, respectively; thus $\omega \in \tilde{U}$, $\omega' \in \tilde{U}'$ and $\tilde{U} \cap \tilde{U}' = \emptyset$. If $p(\omega) = p(\omega')$, we have already seen that ω and ω' have disjoint neighborhoods, say $\tilde{U}$, the domain of the chart associated with some chart (U, ϕ) on X, and $\tilde{U}'$, the domain of the chart associated with $(U, s \circ \phi)$.

Finally, $\tilde{X}$ is separable because X is and $p : \tilde{X} \rightarrow X$ is a double cover (3.1.7.1). $\qquad\square$

5.3.28. Corollary. *Giving X an orientation is equivalent to choosing a map $f \in C^0(X; \tilde{X})$ such that $p \circ f = \mathrm{Id}_X$.*

Proof. Let ω be a volume form on X. For every $x \in X$, set $f(x) = \big(x, \alpha(x)\big)$, where $\alpha(x)$ is the orientation on $T_x X$ determined by $\omega(x)$. We have $p \circ f = \mathrm{Id}_X$ and $f : X \rightarrow \tilde{X}$; there remains to show that f is continuous. Let (U, ϕ) be a chart on X, positively oriented with respect to the orientation of X determined by ω, and $(\tilde{U}, \tilde{\phi})$ the associated chart on $\tilde{X}$ (see the proof of 5.3.27). Theorem 2.3.2(iii) says that f is continuous if $\tilde{\phi} \circ f \circ \phi^{-1}$ is. But certainly

$$\tilde{\phi} \circ f \circ^{-1} \circ \phi = \phi \circ p \circ f \circ \phi^{-1} = \phi \circ \phi^{-1} = \mathrm{Id}_{\phi(U)}$$

is continuous!

Conversely, take $f \in C^0(X; \tilde{X})$ such that $p \circ f = \mathrm{Id}_X$. We will show that X is orientable by using theorem 5.3.24. Let $x \in X$ and $f(x) = \big(x, \alpha(x)\big) \in \tilde{X}$. Then $\alpha(x)$ is an orientation on $T_x X$, so we can find a chart (U, ϕ) at $x \in X$ such that $T_x \phi$ preserves orientation (that is, takes the orientation $\alpha(x)$ of $T_x X$ into the canonical orientation of $T_{\phi(x)} \mathbf{R}^d$). Let $(\tilde{U}, \tilde{\phi})$ be the associated chart on $\tilde{X}$. Since f is continuous, $f^{-1}(U) \circ \tilde{U}$ is open in X and contains x. Thus the set of pairs of the form

$$\big(U \cap f^{-1}(\tilde{U}), \phi|_{U \cap f^{-1}(\tilde{U})}\big)$$

is an atlas of X with the following property: if $\big(U \cap f^{-1}(\tilde{U}), \phi|_{U \cap f^{-1}(\tilde{U})}\big)$ and $\big(V \cap f^{-1}(\tilde{V}), \psi|_{V \cap f^{-1}(\tilde{V})}\big)$ are two such charts and $x \in U \cap V \cap f^{-1}(\tilde{U}) \cap f^{-1}(\tilde{V})$, we have

$$J(\psi \circ \phi^{-1})\big(\phi(x)\big) > 0$$

by 5.3.23, since $T_x \phi$ and $T_x \psi$ preserve orientation. Thus all coordinate changes in this atlas are orientation-preserving, and X is oriented. $\qquad\square$

5.3.29. Theorem. *A connected manifold X is non-orientable if and only if $\tilde{X}$ is connected.*

Proof. We will use the following lemma, whose easy proof is left as an exercise (5.9.14):

5.3.30. Lemma. *Let* $p : \tilde{X} \to X$ *be a covering map and* $\tilde{X}'$ *a non-empty connected component of* $\tilde{X}$. *The restriction* $p|_{\tilde{X}'} : \tilde{X}' \to p(\tilde{X}')$ *is a covering map. In addition, if* X *is connected, so is* $p(\tilde{X}')$. $\qquad \square$

We want to show that X is orientable if and only if $\tilde{X}$ is disconnected. Assume first that $\tilde{X}$ is disconnected, and let $\tilde{X}'$ be a connected component of $\tilde{X}$. By the lemma, $p : \tilde{X}' \to p(\tilde{X}')$ is still a covering map, and $p(\tilde{X}')$, being a non-empty connected component of X, must be the whole of X. The multiplicity of the cover $\tilde{X}' \to X$ is constant by theorem 2.4.4, non-zero because $\tilde{X}'$ is non-empty, and < 2 because $\tilde{X}' \neq \tilde{X}$; thus $p|_{\tilde{X}'}$ is actually a diffeomorphism, and we see that X is orientable by applying 5.3.28 with $f = (p|_{\tilde{X}'})^{-1}$.

Conversely, let X be orientable. By 5.3.28 and its proof the map $f : X \to \tilde{X}$ given by $f(x) = (x, \alpha(x))$ is continuous, so $f(X)$ is connected. But we cannot have $f(X) = \tilde{X}$, since p is a double cover; this shows that $\tilde{X}$ has a non-trivial connected component, and so is disconnected. $\qquad \square$

5.3.31. Remarks

5.3.31.1. The antipodal map. There exists on $\tilde{X}$ a canonical involution s, which associates to $z = (x, \alpha(x))$ the pair $s(z) = (x, \beta(x))$, where $\beta(x)$ is the orientation opposite to $\alpha(x)$. The map s is an orientation-reversing diffeomorphism (exercise 5.9.14).

5.3.31.2. Let δ be a density on a manifold X. There exists on $\tilde{X}$ a canonical volume form α, defined by $|\alpha| = p^*\delta$ (exercise 5.9.14).

5.3.32. Manifolds-with-boundary

5.3.33. Definition. Let X be a d-dimensional manifold. A submanifold-with-boundary D of X is a closed subset $D \subset X$ such that, for every $x \in D$, either

(i) there exists an open subset U of X such that $x \in U \subset D$, or
(ii) X has a chart (U, ϕ) at x, with components $\phi(y) = (\eta_1(y), \ldots, \eta_d(y))$, such that
$$U \cap D = \{y : y \in U \text{ and } \eta_1(y) \leq 0\}.$$

In the first case there is a chart (U, ϕ) at x with $U \subset D$. In the second case, we will use the shorter notation
$$U \cap D = \{(x_1, \ldots, x_d) \in U : x_1 \leq 0\},$$

where $x_1 = \eta_1(x), \ldots, x_d = \eta_d(x)$ are the local coordinates.

5.3.34. Consequences

(i) In the first case, $x \in \mathring{D}$; here $\mathring{D}$ is the interior of D, an open subset of X, of dimension $d = \dim X$.

(ii) In the second case, x is in the frontier of D in X, since charts are homeomorphisms and points of the form $(0, x_2, \ldots, x_d)$ lie on the frontier of $U \cap D$. We call this frontier the *boundary* of D, and denote it by ∂D.

Finally, since D is closed, we see that the first case corresponds to charts associated with $x \in \overset{\circ}{D}$, and the second case with those associated with $x \in \partial D$.

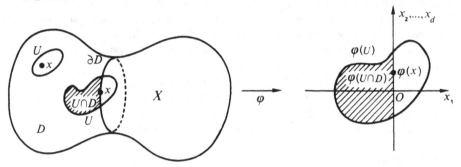

Figure 5.3.34

5.3.35. Theorem. boundary of submanifold-with-boundary is a subman *If D is a submanifold-with-boundary of a d-dimensional manifold X, the boundary ∂D is a $(d-1)$-dimensional submanifold of X.*

Proof. We will use the characterization of submanifolds stated in 2.6.15. For every $x \in \partial D$ there exists a chart (U, ϕ) centered at x and such that $U \cap D = \{y : y_1 \leq 0\}$, where y_1 denotes the first component of $\phi(y)$, for $y \in U$. By 5.3.34 we have $\partial D \cap U = y_1^{-1}(0)$. Now the function y_1 on U has nowhere vanishing derivative, being the composition of a diffeormorphism ϕ and the projection onto the first coordinate (5.2.6.3). Thus the conditions in 2.6.15 are satisfied.   □

5.3.36. Theorem. boundary has canon orienta if X is orienta *If X is an oriented manifold and D is a submanifold-with-boundary of X, the boundary ∂D has a canonical orientation.*

Proof. By 5.3.24 it is enough to start from a positively oriented atlas on X and construct an atlas on ∂D whose coordinate changes preserve orientation. We let this atlas be $\{(U \cap \partial D, \phi|_{\partial D})\}$, where the (U, ϕ) are positively oriented charts satisfying 5.3.33(ii). That these charts form an atlas follows from 5.3.35; we have to show that the coordinate changes have positive determinant.

Set $\psi \circ \phi^{-1} : (x_1, \ldots, x_d) \mapsto (f_1, \ldots, f_d)$. Since

$$\phi(U \cap V \cap \partial D) \subset \{0\} \times \mathbf{R}^{d-1} \quad \text{and} \quad \psi(U \cap V \cap \partial D) \subset \{0\} \times \mathbf{R}^{d-1},$$

we conclude that, for every point $(0, x_2, \ldots, x_d) \in \phi(U \cap V)$ the image $\psi \circ \phi^{-1}$ is also of the form $(0, f_2, \ldots, f_d)$. Thus $f_1(0, x_2, \ldots, x_d) = 0$, which implies $\partial f_1 / \partial x_i = 0$ for $i = 2, \ldots, d$.

We conclude that, for $V_1 = \phi(U) \cap (\{0\} \times \mathbf{R}^{d-1})$,

$$
J(\psi \circ \phi^{-1})|_{V_1} = \begin{vmatrix} \dfrac{\partial f_1}{\partial x_1} & 0 & \cdots & 0 \\ \dfrac{\partial f_2}{\partial x_1} & & & \\ \vdots & & J\big((\psi \circ \phi^{-1})|_{V_1}\big) & \\ \dfrac{\partial f_d}{\partial x_1} & & & \end{vmatrix} (0, \ldots, 0)
$$

$$
= \frac{\partial f_1}{\partial x_1}(0, \ldots, 0) J\big((\psi \circ \phi^{-1})|_{V_1}\big).
$$

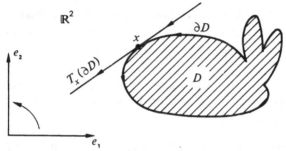

Figure 5.3.36

But $J(\psi \circ \phi^{-1})|_{V_1}$ is strictly positive by our choice of an atlas on X, and $\partial f_1 / \partial x_1(0, \ldots, 0)$ is non-negative because $f_1(x_1, \ldots, x_d)$ is negative for $x_1 \leq 0$ and zero for $x_1 = 0$. This shows that $J\big((\psi \circ \phi^{-1})|_{V_1}\big) > 0$, and this jacobian is equal to

$$
J\big(\psi|_{\partial D} \circ (\phi|_{\partial D})^{-1}\big),
$$

showing that the coordinate changes of our atlas are orientation-preserving. $\square$

5.3.37. Examples

5.3.37.1. The closed ball $\overline{B}(0, 1)$ is a submanifold-with-boundary of $\mathbf{R}^{d+1}$ and its boundary is S^d. The canonical orientation that S^d has as the boundary of $\overline{B}(0, 1)$ coincides with the orientation it was given in 5.3.17.2 (exercise 5.9.14).

5.3.37.2. Recall the notions of normal and unitary normal bundles introduced in section 2.7. We have $NUX = \partial(\overline{N^1 X})$ and $NU^\epsilon X = \partial(\overline{N^\epsilon X})$ (exercise 5.9.14).

5.4. De Rham Groups

For simplicity we assume, from here till the end of the chapter, that all manifolds are C^∞. We set

5.4.1 $\Omega^r(X) = \Omega^r_\infty(X)$.

5.4.2. Definition. An r-from α is said to be *closed* if $d\alpha = 0$, and *exact* if there exists an $(r-1)$-form β such that $d\beta = \alpha$, where d is the exterior derivative (5.2.9.1). We denote the sets of closed and exact r-forms by $F^r(X)$ and $B^r(X)$, respectively.

This definition can be written

5.4.3 $F^r(X) = d^{-1}(0) \cap \Omega^r(X)$, $B^r(X) = d(\Omega^{r-1}))$.

Since $d^2 = 0$, it follows that $B^r(X)$ is a vector subspace of $F^r(X)$.

5.4.4. Definition. Let X be an n-dimensional (C^∞) manifold. For $r = 0, \ldots, n$ we defined the *r-th de Rham group* of X to be the quotient

$$R^r(X) = F^r(X)/B^r(X).$$

5.4.5. Definition. Two closed forms $\alpha, \beta \in F^r(X)$ are said to be *homologous* if their difference is exact, that is, if they have the same image in the quotient $R^r(X)$.

5.4.6. Proposition. *For every r, R^r is a contravariant functor from the category of C^∞ manifolds into the category of real vector spaces. This means that every morphism $f \in C^\infty(X;Y)$ gives rise to an associated linear map $f^* : R^r(Y) \to R^r(X)$, in such a way that $(g \circ f)^* = f^* \circ g^*$ and $(\mathrm{Id}_X)^* = \mathrm{Id}_{R^r(X)}$.*

Proof. Take $\alpha \in F^r(Y)$, which is equivalent to $d\alpha = 0$. By 5.2.9.3 we have $d \circ f^* = f^* \circ d$, so $d(f^*\alpha) = 0$ implies $f^*\alpha \in F^r(X)$. Similarly, if $\beta \in B^r(Y)$, there exists γ such that $\beta = d\gamma$, so $f^*\beta = f^*(d\gamma) = d(f^*\gamma)$ and $f^*\beta \in B^r(X)$. This shows that, if α_1 and α_2 are homologous closed forms on Y, the forms $f^*\alpha_1$ and $f^*\alpha_2$ are also homologous. In other words, f^* is defined on the quotient $F^r(Y)/B^r(Y) = R^r(Y)$.

It is clear that $f^* : R^r(Y) \to R^r(X)$ is linear and that the relations $(g \circ f)^* = f^* \circ g^*$ and $(\mathrm{Id}_X)^* = \mathrm{Id}_{R^r(X)}$ hold (5.2.4.4). $\square$

5.4.7. Corollary. *If $f : X \to Y$ is a diffeomorphism, $f^* : R^r(Y) \to R^r(X)$ is an isomorphism.* $\square$

This gives a necessary condition for two manifolds to be diffeomorphic, but this condition is by no means sufficient. For instance, we will see that S^{2d+1} and $P^{2d+1}(\mathbf{R})$ have the same de Rham groups for every r, but the two manifolds are not diffeomorphic.

5.4.8. Remark. It is not true that if f is injective f^* is surjective, as is the case when f^* denotes the transpose of a linear map. Similarly, f surjective does not imply f^* injective. See exercises 5.9.16 and 5.9.19.

5.4.9. Examples

5.4.9.1. Proposition. *If X is a manifold with k connected components, $R^0(X) \simeq \mathbf{R}^k$.*

Proof. $F^0(X)$ is the set of closed 0-forms on X, or, by 5.2.9.1(iv), of functions $\alpha : X \to \mathbf{R}$ such that $d\alpha = 0$. Such functions are constant on each connected component, so $F^0(X) \simeq \mathbf{R}^k$ by 5.2.6.6.
    On the other hand, $B^0(X) = \{0\}$, so

$$R^0(X) = F^0(X)/\{0\} \simeq F^0(X) \simeq \mathbf{R}^k.$$

5.4.9.2. Proposition. *The group $R^1(S^1)$ is canonically isomorphic to $\mathbf{R}$.*

Proof. S^1 is diffeomorphic to $\mathbf{R}/\mathbf{Z}$ by 2.6.13.1; we must calculate $R^1(\mathbf{R}/\mathbf{Z})$. Let $p : \mathbf{R} \to \mathbf{R}/\mathbf{Z}$ be the canonical projection. By lemma 5.3.9, applied to $X = \mathbf{R}$ and $G = \mathbf{Z}$ acting on $\mathbf{R}$ by translations, we see that if $\alpha \in \Omega^1(\mathbf{R}/\mathbf{Z}) = F^1(\mathbf{R}/\mathbf{Z})$ the pullback $p^*\alpha \in \Omega^1(\mathbf{R})$ is periodic of period 1, so $p^*\alpha = f(t)\,dt$, where $f \in C^\infty(\mathbf{R};\mathbf{R})$ has period 1.
    We construct a map $\theta : F^1(\mathbf{R}/\mathbf{Z}) \to \mathbf{R}$ by setting

$$\theta(\alpha) = \int_0^1 f(t)\,dt.$$

If $\alpha \in B^1(\mathbf{R}/\mathbf{Z})$, we have $\alpha = dg$, whence $p^*\alpha = p^*(dg) = d(g \circ h) = h't\,dt$, where $g \circ p = h$ and h has period 1. But

$$\int_0^1 h'(t)\,dt = h(1) - h(0) = 0.$$

Conversely, if $\int_0^1 f(t)\,dt = 0$, there exists a function h such that $f = h'$ and h has period 1: to wit, the function

$$h(t) = \int_0^t f(s)\,ds,$$

since $h(1) = h(0) = \int_0^1 f(t)\,dt = 0$. In terms of $\mathbf{R}/\mathbf{Z}$ (using lemma 5.3.9 to carry out the transfer), this shows exactly that $B^1(\mathbf{R}/\mathbf{Z}) = \theta^{-1}(0)$, that is, the image of θ is isomorphic to $F^1(\mathbf{R}/\mathbf{Z})/B^1(\mathbf{R}/\mathbf{Z}) = R^1(\mathbf{R}/\mathbf{Z})$.
    There remains to show that θ is surjective. Since $\beta = dt \in F^1(\mathbf{R})$ has period 1, we see again by lemma 5.3.9 that there exists $\alpha \in F^1(\mathbf{R}/\mathbf{Z})$ such that $p^*\alpha = \beta$ and

$$\theta(\alpha) = \int_0^1 dt = 1 \neq 0. \qquad \square$$

A completely different proof of 5.4.9.2 will be given in section 5.8, and yet another in 7.2.1.

Recall that we mentioned in 4.2.24.2 the real cohomology groups $H^r(X)$ $(r = 0, 1, \ldots, \dim X)$ of a compact manifold X. We have the following result, which relates the $H^r(X)$ with the $R^r(X)$:

5.4.10. Theorem (de Rham). *Let X be a compact manifold. For every $r \leq \dim X$ we have $R^r(X) \simeq H^r(X)$. This isomorphism is functorial.*

Proof. See [ST67, chapter 6], [Hu69, chapter 4] or [War71, p. 206].   □

5.4.11. Corollary. *If X is a compact manifold, $\dim R^r(X) = b_r(X)$, the r-the Betti number of X (4.2.24.2). In particular, $\dim R^r(X) < \infty$.*   □

In the remainder of this chapter we will determine the de Rham groups of spheres, projective spaces and tori. Before laying out our plan of study, let's indicate in what way the notion of de Rham groups is connected with a classical problem in the differential calculus.

If U is an open subset of $\mathbf{R}^2$, the vector space $B^1(U) = \{df : f \in C^\infty(U)\}$ is called the set of *total differentials* on U. Let $\alpha \in \Omega^1(U)$ be given by

$$\alpha(x, y) = a(x, y)\, dx + b(x, y)\, dy.$$

We want to find a necessary and sufficient condition for α to be a total differential, that is, for there to be a function f such that

$$a(x, y) = \frac{\partial f}{\partial x} \quad \text{and} \quad b(x, y) = \frac{\partial f}{\partial y}.$$

Since $\dfrac{\partial^2 f}{\partial x \partial y} = \dfrac{\partial^2 f}{\partial y \partial x}$ (which is another way of saying that $d^2 f = 0$), a necessary condition is that $\dfrac{\partial a}{\partial y} = \dfrac{\partial b}{\partial x}$, that is, that

$$d\alpha = \left(-\frac{\partial a}{\partial y} + \frac{\partial b}{\partial x}\right) dx \wedge dy = 0,$$

or again that $\alpha \in F^1(U)$ be a closed form.

In general, this condition is not sufficient. For example, take $\mathbf{R}^2 \setminus \{0\}$ and

$$\alpha = \frac{x}{x^2 + y^2} dy - \frac{y}{x^2 + y^2} dx.$$

Then

$$d\alpha = \left(\frac{\partial}{\partial x}\left(\frac{x}{x^2 + y^2}\right) + \frac{\partial}{\partial y}\left(\frac{y}{x^2 + y^2}\right)\right) dx \wedge dy$$

$$= \left(\frac{x^2 + y^2 - 2x^2}{(x^2 + y^2)^2} + \frac{x^2 + y^2 - 2y^2}{(x^2 + y^2)^2}\right) dx \wedge dy = 0;$$

but if there existed $f \in C^\infty(U)$ such that $\alpha = df$, we'd have

$$\frac{\partial f}{\partial y} = \frac{x}{x^2 + y^2},$$

whence $f(x, y) = \arctan(y/x) + h(x)$ and

$$\frac{\partial f}{\partial x} = \frac{-y}{x^2 + y^2} + h'(x) = \frac{-y}{x^2 + y^2},$$

so that $h'(x) = 0$ and $h(x) =$ constant. But there is no way to define $\arctan(y/x)$ on $U = \mathbf{R}^2 \setminus \{0\}$: such a function would increase by 2π every time we go around the origin once. Thus α is not an exact differential, and $B^1(U) \neq F^1(U)$, that is, $R^1(U) \neq \{0\}$. (For a more rigorous proof, see 5.6.4.)

On the other hand, if $U =]0, 1[^2$, for example, the function f works, and we have $R^1(U) = \{0\}$. More generally, we will show that if U is a star-shaped set in $\mathbf{R}^d$, all the de Rham groups $R^r(U)$ are trivial for $r > 0$ (5.6.1). This will be done by using the notion of the Lie derivative (section 5.5).

5.4.12. More about de Rham groups. Here is a guide to our further study of de Rham groups:

(1) The fundamental formula $L_\xi = d \circ \mathrm{cont}(\xi) + \mathrm{cont}(\xi) \circ d$ (theorem 5.5.8);
(2) Poincaré's lemma (5.6.1);
(3) Calculation of $R^r(S^d)$ and $R^r(P^d(\mathbf{R}))$ (section 5.7);
(4) Calculation of $R^r(T^d)$ (section 5.8);
(5) The isomorphism $R^d(X) \simeq \mathbf{R}$, for X an orientable, compact, connected, d-dimensional manifold (7.2.1).

5.4.13. Remarks. Poincaré's lemma says that locally all de Rham groups (except the 0-th) are trivial. This no longer holds globally; thus (3), (4) and (5) above are reached by a process of globalization.

The de Rham groups of the sphere are calculated by considering it as a union of two open sets diffeomorphic to $\mathbf{R}^d$ (and thus having trivial de Rham groups), and glued along a sphere of dimension one less. One then uses recurrence, starting with $R^1(S^1) \simeq \mathbf{R}$ (5.4.9.2). The de Rham groups of real projective spaces can then be computed because the sphere is a double cover of projective space.

The groups of the torus are found by a completely different, algebraic method: averaging a form $\alpha \in F^r(T^d)$ under the transitive action of $\mathbf{R}^d$ on $T^d = \mathbf{R}^d/\mathbf{Z}^d$. This leads to forms that are invariant under $\mathbf{R}^d$, hence constant.

Fact (5) above is the fundamental result in degree theory, the topic of chapter 7.

From (3) and (4) it follows, by corollary 5.4.7, that S^d and T^d are not diffeomorphic. This result could be obtained more simply via the fundamental groups: S^d is simply connected and T^d isn't. But de Rham groups differentiate between S^4 and $S^2 \times S^2$, for example, whereas fundamental groups don't. To see this, apply the formula

5.4.14 $$R^r(X \times Y) \simeq \sum_{p+q=r} R^p(X) \otimes R^q(Y),$$

which holds for compact manifolds (for a proof involving the essentially equivalent cohomology groups $H^r(X \times Y)$, see [Gre67, p. 198]). Then (3) gives $R^2(S^2 \times S^2) \simeq \mathbf{R}^2$, while $R^2(S^4)$ is trivial.

Actually we will give in 6.3.3 another proof for the fact that $S^2 \times S^2$ and S^4 are not diffeomorphic, as an application of Stokes' theorem. Although purely negative, this result is certainly not trivial: the skeptical reader is welcome to try his hand at it.

5.5. Lie Derivatives

5.5.1. Let X be a manifold (of class C^∞) and I an open interval in $\mathbf{R}$, considered as a C^∞ manifold. Let $F \in C^\infty(I \times X; X)$ and $\alpha \in \Omega^r(X)$ (recall the convention in 5.4.1). We associate to F the maps $F_t : X \to X$ defined by

5.5.2 $$F_t(x) = F(t, x);$$

since $F_t \in C^\infty(X; X)$, we have $F_t^* \alpha \in \Omega^r(X)$, whence a map $t \mapsto F_t^* \alpha$ from I into $\Omega^r(X)$.

5.5.3. Lemma. *For every point $x \in X$, the map $t \mapsto (F_t^* \alpha)(x)$ belongs to $C^\infty(I; \Lambda^r(T_x X)^*)$. Moreover, if we set*

$$D_F \alpha = \frac{\partial(F_t^* \alpha)}{\partial t} : x \mapsto \frac{\partial((F_t^* \alpha)(x))}{\partial t},$$

we have $(D_F \alpha)(t) \in \Omega^r(X)$ for every $t \in I$.

Proof. We just emulate the proof of lemma 5.2.10.6. With the same notation, but substituting I for $[0, 1]$, we get $G_t' \in C^\infty(I \times U'; L(\mathbf{R}^m; \mathbf{R}^n))$. This means that $t \mapsto G_t^* \gamma$ is differentiable, proving 5.5.3.   □

This actually shows that $t \mapsto G_t^* \gamma$ is continuously differentiable. Thus we can apply 5.2.10.5 to the family $t \mapsto (D_F \alpha)(t)$; since differentiation and integration are inverse to each other, we have proved the following fact:

5.5.4. Lemma. $d(D_F \alpha) = D_F(d\alpha)$.   □

5.5.5. Assume, in addition, that for every $t \in I$ we have $F_t \in \mathrm{Diff}(X)$, and that $F_0 = \mathrm{Id}_X$. The picture then is the same as in 3.5.14, and we obtain a C^∞ vector field ξ on X.

If ξ does not depend on time, that is, if F is a one-parameter group of diffeomorphisms, ξ determines F, so $D_F\alpha$ only depends on α and ξ. More generally, we will show that $(D_F\alpha)(0)$ only depends on α and $\xi(0)$. Intuitively, this follows from the fact that $\xi(t, x)$ is the "first derivative of F at t", and from the assumption $F_0 = \mathrm{Id}_X$. For the proof, we will establish an explicit formula (theorem 5.5.8), where $D_F\alpha$ is expressed in terms of ξ and exterior derivatives only.

5.5.6. Definition. Let $\Omega = \bigoplus \Omega^r$ be a graded algebra. A *derivation* u of degree k is a sequence of linear maps $u_r : \Omega^r \to \Omega^{r+k}$, for every r such that Ω^{r+k} exists, such that, for every $\alpha \in \Omega^r$ and $\beta \in \Omega^s$, we have

$$u_{r+s}(\alpha \wedge \beta) = u_r(\alpha) \wedge \beta + \alpha \wedge u_s(\beta).$$

An *antiderivation* of degree k is a similar sequence such that

$$u_{r+s}(\alpha \wedge \beta) = u_r(\alpha) \wedge \beta + (-1)^r \alpha \wedge u_s(\beta).$$

5.5.7. Examples

5.5.7.1. If X is a manifold, exterior differentiation d is an antiderivation of degree 1 of the graded algebra $\Omega^*(X)$, according to 5.2.9.1(ii).

5.5.7.2. Formula 5.3.12 shows that for every C^∞ vector field ξ on X, contraction by ξ is an antiderivation of degree -1 of $\Omega^*(X)$.

5.5.7.3. Let u and v be antiderivations of degree 1 and -1, respectively. Then $u \circ v + v \circ u$ is a derivation of degree 0 (this is immediate). In particular, if ξ is a C^∞ vector field on a C^∞ manifold X, the operator

$$L_\xi = d \circ \mathrm{cont}(\xi) + \mathrm{cont}(\xi) \circ d$$

is a derivation of degree 0.

5.5.7.4. Consider again a map $F \in C^\infty(I \times X; X)$ such that $F_t \in \mathrm{Diff}(X)$ for every t, and $F_0 = \mathrm{Id}_X$. Recall the map $D_F(t_0) : \Omega^*(X) \to \Omega^*(X)$, defined in 5.5.3 by

$$D_F(\alpha)(t_0) = \frac{\partial}{\partial t}\big(F_t^*(\alpha)\big)_{t=t_0}.$$

We have

$$D_F(\alpha \wedge \beta)(t_0) = \left(\frac{\partial\big(F_t^*(\alpha \wedge \beta)\big)}{\partial t}\right)_{t=t_0} = \left(\frac{\partial\big(F_t^*(\alpha) \wedge F_t^*(\beta)\big)}{\partial t}\right)_{t=t_0}$$

by 5.2.4.2. Since the exterior product is bilinear, and we're differentiating with respect to t, we obtain, by 0.2.8.3 and 0.2.15.1:

$$D_F(\alpha \wedge \beta)(t_0) = \left(\frac{\partial\big(F_t^*(\alpha)\big)}{\partial t}\right)_{t=t_0} \wedge F_{t_0}^*(\beta) + F_{t_0}^*(\alpha) \wedge \left(\frac{\partial\big(F_t^*(\beta)\big)}{\partial t}\right)_{t=t_0},$$

or finally

$$D_F(\alpha \wedge \beta)(t_0) = D_F(\alpha)(t_0) \wedge F_{t_0}^*(\beta) + F_{t_0}^*(\alpha) \wedge D_F(\beta)(t_0).$$

In particular, since $F(0) = \mathrm{Id}_X$, we get $F_0 = \mathrm{Id}_{\Omega^*(X)}$, so that

5.5.7.5 $D_F(\alpha \wedge \beta)(0) = D_F(\alpha)(0) \wedge \beta + \alpha \wedge D_F(\beta)(0),$

which shows that $D_F(0)$ is a derivation of degree 0.

5.5.8. Theorem. *We have*

$$D_F(0) = d \circ \mathrm{cont}\big(\xi(0)\big) + \mathrm{cont}\big(\xi(0)\big) \circ d = L_{\xi(0)}.$$

5.5.9. Definition. The form $L_{\xi(0)}\alpha$ defined in 5.5.8 is called the *Lie derivative* of α with respect to the vector field $\xi(0)$.

Proof of 5.5.8. The proof is analogous to the one used in 5.2.9.1 to show the existence of d. The idea is that the algebra $\Omega^*(X)$ is generated (as an algebra) by $\Omega^0(X) = C^\infty(X)$, and that $B^1(X) = d\big(\Omega^0(X)\big)$. Examples 5.5.7.2 and 5.5.7.5 show that $D_F(0)$ and $L_{\xi(0)} = d\circ\mathrm{cont}\big(\xi(0)\big)+\mathrm{cont}\big(\xi(0)\big)\circ d$ are derivations of degree 0 on $\Omega^*(X)$; if they coincide on $\Omega^0(X)$ and on $d\big(\Omega^0(X)\big)$, they will coincide on $\Omega^*(X)$. So this is what we must show first:

5.5.9.1. Lemma. *The derivations $D_F(0)$ and $L_{\xi(0)}$ coincide on $\Omega^0(X)$ and on $d\big(\Omega^0(X)\big)$.*

Proof. This amounts to saying that $D_F(f)(0) = L_{\xi(0)}(f)$ and $D_F(df)(0) = L_{\xi(0)}(df)$ for every $f \in C^\infty(X)$. We first show that $D_F(0)$ and $L_{\xi(0)}$ commute with d, which makes equality for df a consequence of equality for f.

We have seen (lemma 5.5.4) that $d(D_F\alpha) = D_F(d\alpha)$, that is, $d \circ D_F = D_F \circ d$. As for $L_{\xi(0)}$, notice that, since $d^2 = 0$, we have

$$L_{\xi(0)} \circ d = \big(d \circ \mathrm{cont}(\xi(0)) + \mathrm{cont}(\xi(0)) \circ d\big) = d \circ \mathrm{cont}\big(\xi(0)\big) \circ d,$$

and $d \circ L_{\xi(0)}$ expands to the same result.

Now we show that $L_{\xi(0)}$ and $D_F(0)$ coincide on $C^\infty(X)$. Take $f \in C^\infty(X)$ and $x \in X$. We have

$$D_F(f)(0)(x) = \left(\frac{\partial (F_t^* f)(x)}{\partial t}\right)_{t=0}$$

by 5.5.3. Since f is a function, $F_t^* f = f \circ F_t$. If β denotes the curve $t \mapsto F_t(x) = F(t, x)$ in X, we have

$$D_F(f)(0)(x) = \frac{\partial}{\partial t}\big(f \circ F_t(x)\big)(0) = \frac{\partial (f \circ \beta)}{\partial t}(0) = df\big(\beta(0)\big)\big(\beta'(0)\big).$$

Now $\beta(0) = x$ because $F_0 = \mathrm{Id}_X$, and $\beta'(0) = \xi(0)$ by the definition of ξ (3.5.14). Thus $D_F(f)(0)(x) = df(x)\big(\xi(0)\big)$. On the other hand,

$$\big(L_{\xi(0)} f\big)(x) = \big(d \circ \mathrm{cont}\,\xi(0) + (\mathrm{cont}\,\xi(0)) \circ d\big)(f)(x).$$

Since f is a zero-form, we have $\mathrm{cont}(\xi(0))f = 0$, and

$$(L_{\xi(0)}f)(x) = (\mathrm{cont}\,\xi(0) \circ d)(f)(x) = df(x)(\xi(0))$$

by the definition of contraction. This completes the proof of the lemma. □

Now we must show that

$$(D_F\alpha)(0)(x) = (L_{\xi(0)}\alpha)(x)$$

for every α and every $x \in X$. Fix x and choose a chart (U, ϕ) at x, as well as functions $f, g \in \Omega^0(X)$ such that $f = 1$ on a neighborhood of x, both f and g are supported in U and $g = 1$ on $\mathrm{supp}\,f$. Notice first that

$$\big(D_F(f\alpha)\big)(0)(x) = (D_F\alpha)(0)(x) \text{ and } \big(L_{\xi(0)}(f\alpha)\big)(0)(x) = (L_{\xi(0)}\alpha)(0)(x);$$

this is because $D_F(0)$ and $L_{\xi(0)}$ are derivations, $(D_F)(0)(x) = 0$ (since f is constant in a neighborhood of x) and $(L_{\xi(0)}f)(x) = 0$ (since $L_{\xi(0)}f = \mathrm{cont}(\xi(0))(df)$ and df vanishes in an neiborhood of x). Thus it is enough to prove our equality for $\beta = f\alpha$.

The advantage of this is that β can be written in the form

$$\beta|_U = \sum_{i_1 < \cdots < i_r} \beta_{i_1\ldots i_r}\, dx_{i_1} \wedge \cdots \wedge dx_{i_r},$$

with $\mathrm{supp}\,\beta_{i_1\ldots i_r} \subset \mathrm{supp}\,f$ (5.2.8.1). Unfortunately the functions x_i are not defined on X; but the gx_i are, and, since $d(gx_i) = x_i\,dg + g\,dx_i$ and

$$\mathrm{supp}(dg) \cap \mathrm{supp}\,\beta_{i_1\ldots i_r} \subset \mathrm{supp}(dg) \cap \mathrm{supp}\,f = \emptyset,$$

we get

$$\beta = \sum_{i_1 < \cdots < i_r} \beta_{i_1\ldots i_r}\, d(gx_{i_1}) \wedge \cdots \wedge d(gx_{i_r}),$$

which proves the theorem. □

Now we can calculate $(D_F\alpha)(t)$ for every t. We introduce the maps F_h^t as in 3.5.12, and the family $h \mapsto F_{t+h}^*\alpha = F_t^*(F_h^{t*}\alpha)$. Since F_t^*, again for fixed t, commutes with differentiation with respect to h, we get

$$(D_F\alpha)(t) = F_t^*\big(D_{F_0^t}\alpha)(0)\big) = F_t^*(L_{\xi(t)}\alpha),$$

by 5.5.9 and the definition of $\xi(t)$. Whence the formula

5.5.10 $\qquad (D_F\alpha)(t) = F_t^*\big(d(\mathrm{cont}(\xi(t))\alpha) + \mathrm{cont}(\xi(t))(d\alpha)\big).$

5.5.11. Theorem. *Let $F \in C^\infty(I \times X; X)$, where $I \subset \mathbf{R}$ is an open interval, be a one-parameter family of diffeomorphisms. If $\alpha \in F^r(X)$ and $[a, b] \subset I$, the pullbacks $F_a^*\alpha$ and $F_b^*\alpha$ are homologous. In other words, F_a and F_b induce the same homomorphisms $R^r(X) \to R^r(X)$ on de Rham groups (5.4.6).*

Proof. We have

$$F_b^* \alpha - F_a^* \alpha = \int_a^b \frac{\partial(F_t^* \alpha)}{\partial t} \, dt.$$

But

$$\frac{\partial(F_t^* \alpha)}{\partial t} = (D_F \alpha)(t) = F_t^* \big(d(\mathrm{cont}(\xi(t))\alpha) + \mathrm{cont}(\xi(t))(d\alpha) \big)$$

by 5.5.10. Now $d\alpha = 0$ because $\alpha \in F^r(X)$, and since d commutes with F_t^*, we have

$$F_b^* \alpha - F_a^* \alpha = \int_a^b d\big(F_t^* (\mathrm{cont}(\xi(t))\alpha) \big) \, dt.$$

An application of 5.2.10.5 gives

5.5.12 $$F_b^* \alpha - F_a^* \alpha = d\left(\int_a^b F_t^* \big(\mathrm{cont}(\xi(t))\alpha\big) \, dt \right),$$

which shows that $F_b^* \alpha - F_a^* \alpha \in B^r(X)$, that is, $F_b^* \alpha$ and $F_a^* \alpha$ are homologous.   □

5.6. Star-shaped Sets and Poincaré's Lemma

5.6.1. Poincaré's lemma. *If U is a star-shaped open subset of a finite-dimensional vector space E, we have $R^r(U) = 0$ for every $r > 0$.*

Proof. Let U be star-shaped at 0, that is, $tx \in U$ for every $x \in U$ and $t \in [0,1]$. We consider the C^∞ family of diffeomorphisms F defined by $F(t,x) \mapsto tx$ for every $t \in [0,1]$. If $\alpha \in F^r(U)$, 5.5.11 shows that $F_b^* \alpha$ and $F_a^* \alpha$ are homologous for any $a, b \in {]}0,1{[}$.

Figure 5.6.1

If we could show that this is still true for $a = 0$ and $b = 1$, we'd have proved that α is homologous to zero: for $F_1^*\alpha = \alpha$ because $F_1 = \mathrm{Id}_U$, and $F_0^*\alpha = 0$ because

$$(F_0^*\alpha)(x)(\xi_1, \ldots, \xi_r) = \alpha(0)(0 \cdot \xi_1, \ldots, 0 \cdot \xi_r) = 0$$

for every $r > 0$. But we can't just extend to $a = 0$ and $b = 1$ for two reasons: F_0 is not a diffeomorphism, and there is no $\varepsilon > 0$ such that $tx \in U$ for all $t \in {]}1, 1 + \varepsilon[$ and $x \in U$.

What we'll do is simply show that formula 5.5.12 still holds as $a \to 0$ and $b \to 1$, by explicitly computing the right-hand side. First we must find $\xi(t)$. Apply definition 3.5.14: the curve β going through x is

$$x \mapsto \frac{t + s}{t} x,$$

so that $\xi(t) = \frac{1}{t}x$. Also $\big(T_x(F_t)\big)(\xi) = t\xi$ for every ξ, so

$$F_t^*\big(\mathrm{cont}(\xi(t))\alpha\big)(x)(\xi_1, \ldots, \xi_r) = \alpha(tx)\left(\frac{1}{t}x, t\xi_1, \ldots, t\xi_r\right)$$

$$= t^{r-1}\alpha(tx)(x, \xi_1, \ldots, \xi_r)$$

(notice that this is the form that comes out of the blue in [Car70, I.2.13.2]).

Thus, if $r > 0$, we have

$$(F_b^*\alpha - F_a^*\alpha)(x)(\xi_1, \ldots, \xi_r) = d\left(\int_a^b t^{r-1}\alpha(tx)(x, \xi_1, \ldots, \xi_r)\, dt\right)$$

for every closed form α, every $x, \xi_1, \ldots, \xi_r$ and every $[a, b] \subset {]}0, 1[$. The left-hand side approaches α as $a \to 0$ and $b \to 1$, and the right-hand side approaches

$$d\left(\int_0^1 t^{r-1}\alpha(tx)(x, \xi_1, \ldots, \xi_r)\, dt\right).$$

This shows that α is homologous to 0. $\square$

5.6.2. Remark. Even if α has compact support in U, the construction above for a form β such that $\alpha = d\beta$ does not necessarily yield a form with compact support. In fact, we will see in 7.1.2 that such a form β with compact support may not exist; we will find necessary and sufficient conditions for its existence in the case $\deg \alpha = \dim E$.

In section 5.7 we will need the following generalization of Poincaré's lemma:

5.6.3. Theorem. *If U is a star-shaped open set in $\mathbf{R}^d$ and X is an arbitrary manifold, we have $R^r(U \times X) \simeq R^r(X)$ for all r.*

Proof. We could just combine 5.4.14, 5.4.9.1 and 5.6.1, but we will instead give a proof that does not involve 5.4.14. For $r = 0$, just apply 5.4.9.1. So let $r > 0$, and assume that U is star-shaped at 0. Let $p : U \times X \to U$

and $q : U \times X \to X$ be the canonical projections and i the zero section, that is, $i : X \ni x \mapsto (x,0) \in U \times X$. As in the proof of 5.6.1, introduce $G : [0,1] \times U \times X \to U \times X$, by setting $G(t,u,x) = (tu,x)$. The same argument used there shows that, for every closed form α on $U \times X$, the forms $\alpha = G_1^* \alpha$ and $G_0^* \alpha$ are homologous. But G_0 is just $(u,x) \mapsto (0,x)$, that is, $i \circ q$. Saying that every closed form α is homologous to $(i \circ q)^* \alpha$ is the same as saying that the induced homomorphism $(i \circ q)^* : R^r(U \times X) \to R^r(U \times X)$ is the identity. But $(i \circ q)^* = q^* \circ i^*$, where $q^* : R^r(X) \to R^r(U \times X)$ and $i^* : R^r(U \times X) \to R^r(X)$, so $q^* \circ i^* = \mathrm{Id}_{R^r(U \times X)}$. On the other hand, $i^* \circ q^* = \mathrm{Id}_{R^r(X)}$ because $q \circ i = \mathrm{Id}_X$; this shows that $R^r(X)$ and $R^r(U \times X)$ are isomorphic. $\qquad\square$

5.6.4. Corollary. *We have*

$$R^1\big(\mathbf{R}^2 \setminus \{0\}\big) \simeq R^1(\mathbf{R}_+^* \times S^1) \simeq R^1(S^1) \simeq \mathbf{Z}.$$

Proof. Using polar coordinates one shows that $\mathbf{R}^2 \setminus \{0\}$ and $\mathbf{R}_+^* \times S^1$ are diffeomorphic (see 6.5.8 for details). One concludes by applying 5.6.3, 5.4.7 and 5.4.9.2. $\qquad\square$

5.7. De Rham Groups of Spheres and Projective Spaces

5.7.1. Theorem. *The de Rham groups $R^r(S^d)$ of the sphere are zero, except for $R^0(S^d)$ and $R^d(S^d)$, which are isomorphic to $\mathbf{R}$.*

Proof. Example 5.4.9.1 shows that $R^0(S^d) \simeq \mathbf{R}$ because S^d is connected. Example 5.4.9.2 shows that $R^1(S^1) \simeq \mathbf{R}$.

The idea of the proof is the following: let N and S be the north and south poles of S^d. The stereographic projections i_N and i_S, from the north and south poles, respectively, are diffeomorphisms between $\mathbf{R}^d$ and their images (exercise 2.8.7):

5.7.1.1
$$\begin{cases} i_N \in \mathrm{Diff}\big(\mathbf{R}^d; U = S^d \setminus N\big), \\ i_S \in \mathrm{Diff}\big(\mathbf{R}^d; V = S^d \setminus S\big). \end{cases}$$

Since $\mathbf{R}^d$ is star-shaped, its de Rham groups are trivial (5.6.1). From 5.4.7 we get:

5.7.1.2. Proposition. *The de Rham groups of U and V, open subsets of S^d, are trivial.* $\qquad\square$

We next consider the open set $U \cap V \subset S^d$. Take $(y_1, \ldots, y_d, u) \in U \cap V$; then $u \in {]-1,1[}$. Set $y = (y_1, \ldots, y_d)$; we have $y/\|y\| \in S^{d-1}$, and the map

5.7.1.3
$$U \cap V \ni (y,u) \mapsto \left(u, \frac{y}{\|y\|}\right) \in {]-1,1[} \times S^{d-1}$$

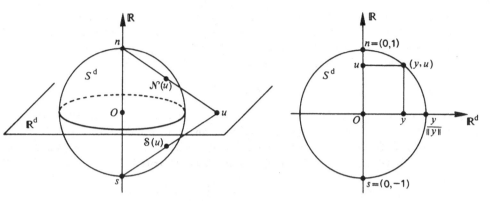

Figure 5.7.1

is a diffeomorphism between $U \cap V$ and $]-1,1[\times S^{d-1}$. Since $]-1,1[\times S^{d-1}$ has same de Rham groups as S^{d-1} (theorem 5.6.3), we can work by induction on d, distinguishing between the cases $0 < r < d$ and $r = d$ (since $R^0(S^d)$ is already known).

5.7.1.4. First case: $0 < r < d$ (this implies $d > 1$). Consider first the case $r = 1$, $d > 1$. Take $\alpha \in F^1(S^d)$. By Poincaré's lemma (5.6.1) and corollary 5.4.7, the restrictions $\alpha|_U$ and $\alpha|_V$ are exact, so there exist functions $f \in C^\infty(U)$ and $g \in C^\infty(V)$ such that $\alpha|_U = df$ and $\alpha|_V = dg$.

On the open set $U \cap V$, these functions satisfy

$$d\big((f - g)|_{U\cap V}\big) = (df - dg)|_{U\cap V} = (\alpha - \alpha)|_{U\cap V} = 0,$$

so $f - g$ is constant on $U \cap V$ (which is connected, being diffeomorphic to $]-1,1[\times S^{d-1}$ when $d > 2$). Thus we can match f and g by adding to g the constant k that gives the difference $f - g$ on $U \cap V$; that is, we can set $\hat{f} = f$ on U and $\hat{f} = g + k$ on V. Clearly $\hat{f}$ is C^∞ on S^d, and its derivative is α; this shows that α is exact, and consequently that $R^1(S^d) = 0$ for $d > 1$.

Now assume known that $R^k(S^d)$ is zero for $1 \le k \le r - 1 < d$, and consider $R^r(S^d)$ with $r > 1$ (which implies $d > 2$).

If $\alpha \in F^r(S^d)$, proposition 5.7.1.2 guarantees the existence of forms $\beta \in \Omega^{r-1}(U)$ and $\gamma \in \Omega^{r-1}(V)$ such that $\alpha|_U = d\beta$ and $\alpha|_V = d\gamma$. Thus $d\big((\beta - \gamma)|_{U\cap V}\big) = 0$, whence

$$(\beta - \gamma)|_{U\cap V} \in F^{r-1}(U \cap V).$$

Since $R^{r-1}(U \cap V) \simeq R^{r-1}\big(]-1,1[\times S^{d-1}\big)$ by 5.7.1.3, an application of 5.6.3 gives $R^{r-1}(U \cap V) \simeq R^{r-1}(S^{d-1})$. By the induction assumption, $R^{r-1}(S^{d-1}) = 0$, and we conclude that $(\beta - \gamma)|_{U\cap V}$ is exact: there exists $\omega \in \Omega^{r-2}(U \cap V)$ such that $(\beta - \gamma)|_{U\cap V} = d\omega$.

We now have information on the three open sets U, V and $U \cap V$. To match this information, we use a partition of unity subordinated to the cover $\{U, V\}$ of S^d. More precisely, pick a bump function f around s and supported in U (3.1.2), and form $g = 1 - f$. Clearly $\{(U, f), (V, g)\}$ is a partition of unity subordinated to $\{U, V\}$, and f (resp. g) is equal to 1 on a neighborhood of S (resp. N). Now consider

5.7.1.5 $\hat{\beta} = f\beta + g\gamma - df \wedge \omega.$

We have $f\beta \in \Omega^{r-1}(U)$ and $g\gamma \in \Omega^{r-1}(V)$. In addition, df and dg are supported in $U \cap V$, because f and g are constant on neighborhoods of s and n; thus $df \wedge \omega$ makes sense, and belongs to $\Omega^{r-1}(U \cap V)$. So in fact $\hat{\beta}$ is an $(r - 1)$-form on the whole of S^d, and we can write

$$d\hat{\beta} = df \wedge \beta + f\, d\beta + dg \wedge \gamma + g\, d\gamma - df \wedge d\omega.$$

Now $df = -dg$ because $f + g = 1$, so

$$d\hat{\beta} = df \wedge (\beta - \gamma - d\omega) + f\, d\beta + g\, d\gamma.$$

On the other hand, $(\beta - \gamma)|_{U \cap V} = d\omega$ and $d\beta = \alpha|_U$, $d\gamma = \alpha|_V$. Since f has support in U and g has support in V, we finally get

$$d\hat{\beta} = (f + g)\alpha = \alpha,$$

which proves that α is exact, completing the induction step.

5.7.1.6. Second case: $R^d(S^d)$. We have seen (5.4.9.2) that $R^1(S^1) \simeq \mathbf{R}$. Suppose that $R^{d-1}(S^{d-1}) \simeq \mathbf{R}$ and consider $R^d(S^d)$. Taking $\alpha \in F^d(S^d)$, we can find as in 5.7.1.4 two forms $\beta \in \Omega^{d-1}(U)$ and $\gamma \in \Omega^{d-1}(V)$ such that $\alpha|_U = d\beta$ and $\alpha|_V = d\gamma$, or again $d\big((\beta - \gamma)|_{U \cap V}\big) = 0$.

Now we have assumed that $R^{d-1}(S^{d-1}) \simeq \mathbf{R}$. Since $R^{d-1}(U \cap V) \simeq R^{d-1}(S^{d-1})$ by 5.7.1.3 and 5.6.3, and $R^{d-1}(U \cap V)$ is a one-dimensional vector space, there exists $\omega_0 \in \Omega^{d-1}(U \cap V)$ such that

$$\beta - \gamma = k\omega_0 + d\delta,$$

where $k \in \mathbf{R}$ and $\delta \in \Omega^{d-2}(U \cap V)$. Using the same f and g as in 5.7.1.4, set

5.7.1.7 $\hat{\beta} = f\beta + g\gamma - df \wedge \delta;$

this makes sense because f and β are supported in U, g and γ in V and df and δ in $U \cap V$. Also $\hat{\beta} \in \Omega^{d-1}(S^d)$, and we have

$$\begin{aligned}
d\hat{\beta} &= f\, d\beta + g\, d\gamma + df \wedge \beta + dg \wedge \gamma - df \wedge d\delta \\
&= (f + g)\alpha + df \wedge \big(\beta - \gamma - (\beta - \gamma - k\omega_0)\big) \\
&= \alpha + k\, df \wedge \omega_0.
\end{aligned}$$

Thus α and $k\, df \wedge \omega_0$ are homologous for every α; this means that the class of $df \wedge \omega_0$ generates $R^d(S^d)$, which must have dimension at most 1. Since S^d is oriented, 6.3.2 implies that the dimension in 1. $\square$

5.7.2. Corollary. *The de Rham groups* $R^r\big(P^d(\mathbf{R})\big)$ *are zero, except for* $R^0\big(P^d(\mathbf{R})\big)$ *and* $R^{2d+1}\big(P^{2d+1}(\mathbf{R})\big)$, *which are isomorphic to* $\mathbf{R}$.

Proof. Let s be the antipodal map on S^d, that sends $(x_1, \ldots, x_{d+1})$ to $(-x_1, \ldots, -x_{d+1})$. Let $p : S^d \to P^d(\mathbf{R}) = S^d/\{\mathrm{Id}_{S^d}, s\}$ be the double cover of $P^d(\mathbf{R})$ by S^d (2.4.11.2). Take $\alpha \in F^r(P^d)$; the pullback $\beta = p^*\alpha$ is in $F^r(S^d)$ by the proof of 5.4.6, and $s^*\beta = \beta$ because $s^*\beta = s^* \circ p^*\alpha = (p \circ s)^*\alpha = p^*\alpha$.

For $r = 0$ we have $R^0\big(P^d(\mathbf{R})\big) \simeq \mathbf{R}$ by 5.4.9.1.

For $0 < r < d$ we have $R^r(S^d) = 0$, so for any $\beta \in F^r(S^d)$ there exists $\gamma \in \Omega^{r-1}(S^d)$ such that $\beta = d\gamma$. Since $s^* \circ d = d \circ s^*$ (5.2.13), we get $d(s^*\gamma) = s^* \circ d\gamma = s^*\beta = \beta$, and

$$\beta = d\left(\frac{\gamma + s^*\gamma}{2}\right).$$

Define $\delta \in \Omega^{r-1}(S^d)$ by $\delta = (\gamma + s^*\gamma)/2$. Clearly d is invariant under the action of $G = \{\mathrm{Id}_{S^d}, s\}$, so lemma 5.3.9 gives a unique $\varepsilon \in \Omega^{r-1}\big(P^d(\mathbf{R})\big)$ such that $p^*\varepsilon = \delta$. Then $\beta = p^*\alpha = d\delta$ becomes $p^*\alpha = p^*(d\varepsilon)$ (again because $p^* \circ d = d \circ p^*$), and this implies $\alpha = d\varepsilon$ because $T_x p$ is an isomorphism. This shows that $\alpha \in F^r\big(P^d(\mathbf{R})\big)$ is exact, so $R^r\big(P^d(\mathbf{R})\big) = 0$ for every $0 < r < d$.

There remains the case $r = d$. We have $\beta = p^*\alpha \in R^d(S^d) \simeq \mathbf{R}$; denoting by σ the canonical volume form on S^d, we can find $k \in \mathbf{R}$ and $\gamma \in \Omega^{d-1}(S^d)$ such that $\beta = k\sigma + d\gamma$. Now $s^*\beta = \beta$ and $s^*\sigma = (-1)^{d+1}\sigma$, so $\beta = (-1)^{d+1}k\sigma + d(s^*\gamma)$, and

$$\beta = k\frac{1 + (-1)^{d+1}}{2}\sigma + d\delta,$$

where $\delta = (\gamma + s^*\gamma)/2$ is invariant under the action of $\{\mathrm{Id}_{S^d}, s\}$.

If d is even, we have $\beta = d\delta$, and there exists $\varepsilon \in \Omega^{d-1}\big(P^d(\mathbf{R})\big)$ such that $\alpha = d\varepsilon$, which shows that $R^d\big(P^d(\mathbf{R})\big) = 0$.

If d is odd, $\beta = k\sigma + d\delta$ with $s^*\delta = \delta$; but also

$$s^*\sigma = (-1)^{d+1}\sigma = \sigma.$$

Thus an application of 5.3.9 gives $\varepsilon \in \Omega^{d-1}\big(P^d(\mathbf{R})\big)$ and $\omega \in \Omega^d\big(P^d(\mathbf{R})\big)$ such that $\alpha = k\omega + d\varepsilon$, whence $\dim R^d\big(P^d(\mathbf{R})\big) \leq 1$. By theorem 5.3.18, $P^d(\mathbf{R})$ is orientable for odd d, so $\dim R^d\big(P^d(\mathbf{R})\big) = 1$ by 6.3.2. $\qquad\square$

5.8. De Rham Groups of Tori

5.8.1. Theorem. *For every d and r, the de Rham groups of the torus are given by $R^r(T^d) \simeq \mathbf{R}^{\binom{d}{r}}$.*

Here's an idea of the proof. Given $\alpha \in F^r(T^d)$, we define the average $\overline{\alpha}$ of α under the group of translations of T^d. This average is invariant under translations and so must have constant coefficients. Invariant r-forms form a vector subspace Inv^r of $F^r(X)$, which is isomorphic to $\Lambda^r(\mathbf{R}^d)^*$ and consequently has dimension $\binom{d}{r}$. Since α and $\overline{\alpha}$ are homologous, Inv^r and $B^r(T^d)$ span $F^r(T^d)$. Finally we show that $\mathrm{Inv}^r \cap B^r(T^d) = \{0\}$, so Inv^r is isomorphic to $F^r(T^d)/B^r(T^d) = R^r(T^d)$.

Proof. Consider the C^∞ one-parameter group of diffeomorphisms of $\mathbf{R}^d$ of the form $G_t : (x_1, \ldots, x_d) \mapsto (x_1 + t, x_2, \ldots, x_d)$. By passing to the quotient we obtain a one-parameter group of diffeomorphisms of T^d, which we still denote by G_t. For every $t, s \in \mathbf{R}$ we have

5.8.2 $G_{t+1} = G_t$ and $G_{t+s} = G_t \circ G_s.$

Now take $\alpha \in F^r(T^d)$. Since $G \in C^\infty(\mathbf{R} \times T^d; T^d)$ we are in the situation of section 5.5, and we can introduce the average of α under the action of G_t:

5.8.3 $\mu_1(\alpha) = \displaystyle\int_0^1 G_\theta^* \alpha \, d\theta.$

This average satisfies $\mu_1(\alpha) \in F^r(T^d)$ by 5.2.10.5 and is G_t-invariant for every t, that is,

5.8.4 $G_t^*\big(\mu_1(\alpha)\big) = \mu_1(\alpha).$

This is because G_t^* is linear and commutes with the integral sign, so we get by 5.8.2:

$$G_t^*\big(\mu_1(\alpha)\big) = G_t^*\left(\int_0^1 G_\theta^*\alpha \, d\theta\right) = \int_0^1 G_t^*(G_\theta^*\alpha) \, d\theta = \int_0^1 G_{t+\theta}^*\alpha \, d\theta$$

$$= \int_t^{t+1} G_\theta^*\alpha \, d\theta = \int_0^1 G_\theta^*\alpha \, d\theta = \mu_1(\alpha).$$

5.8.5. Next we show that $\mu_1(\alpha)$ and α are homologous. By formula 5.5.12, we have

$$G_t^*\alpha - G_0^*\alpha = G_t^*\alpha - \alpha = d\left(\int_0^\infty G_t^*\big(\mathrm{cont}(\xi)\alpha\big) \, dt\right) = d\beta_t.$$

Since the β_t also form a C^∞ family, it follows from 5.2.10.5 that

$$\mu_1(\alpha) = \int_0^1 G_t^*\alpha \, dt = \int_0^1 (\alpha + d\beta_t) \, dt = \alpha + \int_0^1 d\beta_t \, dt = \alpha + d\left(\int_0^1 \beta_t \, dt\right),$$

proving our claim.

Now consider the C^∞ one-parameter group of diffeomorphisms H_t : $(x_1, x_2, \ldots, x_d) \mapsto (x_1, x_2 + t, \ldots, x_d)$ on $\mathbf{R}^d$. By the same process used above we obtain a one-parameter group of diffeomorphisms H_t of T^d and an average

5.8.6
$$\mu_2(\alpha) = \int_0^1 H_t^* \alpha \, dt$$

such that $H_t^*\big(\mu_2(\alpha)\big) = \mu_2(\alpha)$ for every t and $\mu_2(\alpha)$ is homologous to α.

5.8.7. In particular, $\mu_2\big(\mu_1(\alpha)\big)$ is homologous to α because $\mu_1(\alpha)$ is. We claim that $\mu_2\big(\mu_1(\alpha)\big)$ is invariant under both the H_t^* and the G_t^*.

Invariance under H_t^* is proved as in 5.8.4. For G_t^*, notice that

$$G_t^*\big(\mu_2(\mu_1(\alpha))\big) = G_t^*\left(\int_0^1 H_s^*(\mu_1(\alpha)) \, ds\right) = \int_0^1 G_t^* H_s^*\big(\mu_1(\alpha)\big) \, ds$$

because the G_t^* are linear. But it is clear from the definitions that the H_s and the G_t commute, so $G_t^* \circ H_s^* = H_s^* \circ G_t^*$, and, by 5.8.4 and 5.8.6,

$$G_t^*\big(\mu_2(\mu_1(\alpha))\big) = \int_0^1 H_s^* G_t^*(\mu_1(\alpha)) \, ds = \int_0^1 H_s^*(\mu_1(\alpha)) \, ds = \mu_2\big(\mu_1(\alpha)\big).$$

We continue in this way, introducing the diffeomorphisms of T^d derived from the translations $(x_1, \ldots, x_i, \ldots, x_d) \mapsto (x_1, \ldots, x_i + t, \ldots, x_d)$, and the corresponding averages $\mu_i(\alpha)$. Then we consider

5.8.8
$$\overline{\alpha} = \mu_d\big(\mu_{d-1}(\cdots(\mu_2(\mu_1(\alpha)))\cdots)\big).$$

It follows as in 5.8.7 that $\overline{\alpha}$ is homologous to α and invariant under all translations. Now recall the notation $\alpha = \sum_I \alpha_I \omega_I$ introduced in 5.3.10.1. Since the forms ω_I are invariant under the action of G_t^*, H_t^*, etc., we conclude that the functions $\overline{\alpha}_I$ such that

$$\overline{\alpha} = \sum_I \overline{\alpha}_I \omega_I$$

are invariant under translation, that is, they are constants $k_I \in \mathbf{R}$.

The forms $\sum_I k_I \omega_I$, for $k_I \in \mathbf{R}$, form a real vector space Inv^r having $\{\omega_I : I = (i_1, \ldots, i_r)\}$ as a basis. Thus Inv^r is isomorphic to $\Lambda^r(\mathbf{R}^d)^*$, and its dimension is $\binom{d}{r}$. Also, $\mathrm{Inv}^r \subset F^r(T^d)$, because $d\omega_I = 0$ by example 5.3.10.1.

Saying that an arbitrary $\alpha \in F^r(T^d)$ is homologous to $\overline{\alpha} \in \mathrm{Inv}^r$ is saying that $B^r(T^d)$ and Inv^r together span $F^r(T^d)$. There remains to show that $B^r(T^d) \cap \mathrm{Inv}^r = \emptyset$. Assume $\alpha = d\beta$ with $\beta \in \Omega^{r-1}(T^d)$ and $\alpha \in \mathrm{Inv}^r$. The averaging operator $^-$ makes sense whether or not β is closed, and it commutes with d by 5.2.10.5; thus $\overline{\alpha} = \overline{d\beta} = d\overline{\beta}$. But $\overline{\beta} \in \mathrm{Inv}^{r-1}$, so $\overline{\beta} \in F^{r-1}(T^d)$ as shown above. This implies $d\overline{\beta} = 0$ and $\overline{\alpha} = \alpha = 0$. $\quad\square$

5.8.9. Remark. The same idea can be used to compute de Rham groups of several other manifolds operated on by compact Lie groups G. One needs to know the Haar measure $d\delta$ of G (see exercise 3.6.5), and use it to average forms $\alpha \in F^r(X)$:

$$\overline{\alpha} = \int_{g \in G} g^* \alpha \, d\delta.$$

As above, $\overline{\alpha}$ is homologous to α and invariant under the action of G.

For $X = S^d$ and $G = SO(d+1)$, for example, it is easily checked that the only non-zero forms $\alpha \in \Omega^r(S^d)$ invariant under G are constant functions $(r = 0)$ or multiples of the canonical volume form $(r = d)$. This gives another proof for 5.7.1, avoiding the reference to chapter 6.

5.9. Exercises

5.9.1. Let ξ be a one-form on S^2. Assume that ξ is invariant under rotation, that is, $s^*\xi = \xi$ for any $s \in SO(3)$. Prove that $\xi = 0$.

5.9.2. What are the d-forms ω on S^d such that $s^*\omega = \omega$ for every $s \in SO(d+1; \mathbf{R})$? (Hint: treat the case $d = 3$ first).

5.9.3. What are the $(d-1)$-forms on $\mathbf{R}^d$ such that $s^*\omega = \omega$ for every $s \in SL(d)$? (Hint: treat the case $d = 3$ first. Recall that $SL(d) = \{f \in \text{Isom}(\mathbf{R}^d; \mathbf{R}^d) : \det f = 1\}$.)

5.9.4. Let X be a d-dimensional manifold. Denote by $s : \Lambda^d T^* X \to \Lambda^d T^* X$ the map that takes α into $-\alpha$, and by N the zero section of $\Lambda^d T^* X$, that is, the set $\{0 \in \Lambda^d T^*_x X : x \in X\}$.

(a) Show that N is a closed submanifold of $\Lambda^d T^* X$.
(b) Show that the group $G = \{\text{Id}_{\Lambda^d T^* X}, s\}$ acts properly discontinuously without fixed points on $\Lambda^d T^* X \setminus N$.
(c) Denote by $p : (\Lambda^d T^* X \setminus N)/G \to X$ the map obtained from the canonical projection $\Lambda^d T^* X \to X$ by restricting and passing to the quotient. Show that a density on X determines a map $\delta : X \to (\Lambda^d T^* X \setminus N)/G$ such that $p \circ \delta = \text{Id}_X$, and conversely.

5.9.5. Consider $P^n(\mathbf{R})$, the n-dimensional projective space, with its canonical C^∞ structure, and let $\pi : \mathbf{R}^{n+1} \to P^n(\mathbf{R})$ be the canonical projection. Let α be a p-form $(p \leq n)$ on $\mathbf{R}^{n+1} \setminus 0$.

5.9.5.1. Show that a necessary and sufficient condition for the existence of a form β on $P^n(\mathbf{R})$ such that $\pi^*\beta = \alpha$ is that α can be written in the form

$$\alpha(x) = \sum_{0 \leq i_0 < \cdots < i_p \leq n} a_{i_0 \ldots i_p}(x) \eta_{i_0 \ldots i_p}(x),$$

where $\eta_{i_0\ldots i_p}(x)$ stands for the differential form

$$\sum_{j=0}^{p}(-1)^j x_{i_j}\,dx_{i_0}\wedge\cdots\wedge\widehat{dx_{i_j}}\wedge\cdots\wedge dx_{i_p},$$

and each $a_{i_0\ldots i_p}$ is a function on $\mathbf{R}^{n+1}\setminus 0$, homogeneous of degree $(-p-1)$. (Hint: use charts, cf. exercise 2.8.30. To show sufficiency, do 5.9.5.2(c) first.)

5.9.5.2. Determine all forms α having the property above, in the following cases:

(a) n arbitrary and $p=0$;
(b) $n=1$ and $p=1$;
(c) $n=2$ and $p=1$;
(d) n arbitrary and $p=n$.

5.9.5.3. Show that if n is even every n-form on $P^n(\mathbf{R})$ vanishes at one point at least. Show that if n is odd there exist nowhere vanishing n-forms on $P^n(\mathbf{R})$.

5.9.5.4. Assume that $p=2q-1$, with $2q\le n$. Show that the $2q$-form on $\mathbf{R}^{n+1}\setminus 0$ given by

$$\alpha_1(x)=(2q-1)r^{-2q}\sum_{i_0<i_1<\cdots<i_p}c_{i_0i_1\ldots i_p}\,dx_{i_0}\wedge dx_{i_1}\wedge\cdots\wedge dx_{i_p},$$

where $r^2=x_0^2+\cdots+x_n^2$ and $c_{i_0i_1\ldots i_p}\in\mathbf{R}$, can be written in the form $\alpha_1=\pi^*\beta_1$, where β_1 is a form on $P^n(\mathbf{R})$.

5.9.5.5. Assume that $n\ge 3$. Let $(a_{ij})_{0\le i,j\le n}$ be a skew-symmetric matrix whose entries are C^∞ functions on $\mathbf{R}^{n+1}\setminus 0$, homogeneous of degree -2. Show that the differential form on $\mathbf{R}^{n+1}\setminus 0$ given by

$$\alpha_2=\sum_{i,j,k,l=0}^{n}\left(a_{ij}a_{kl}+2\sum_{m=0}^{n}a_{ij}x_m\frac{\partial a_{ml}}{\partial x_k}+\sum_{m,p=0}^{n}x_m x_p\frac{\partial a_{mj}}{\partial x_i}\frac{\partial a_{pl}}{\partial x_k}\right)$$
$$\times dx_i\wedge dx_j\wedge dx_k\wedge dx_l$$

can be written in the form $\alpha_2=\pi^*\beta_2$, where β_2 is a form on $P^n(\mathbf{R})$. Express β_2 in terms of the usual charts on $P^n(\mathbf{R})$ (exercise 2.8.30).

5.9.5.6. Still $n\ge 3$. Let α_3 be the form

$$\alpha_3(x)=r^{-4}\,dx_0\wedge dx_1\wedge dx_2\wedge dx_3$$

on $\mathbf{R}^{n+1}\setminus 0$. Is there a form β_1 on $P^n(\mathbf{R})$ such that $\pi^*\beta_1=\alpha_3$?

5.9.6. Let s be the antipodal map on S^2. Show that the map $(x,y)\mapsto\big(s(x),s(y)\big)$ from $S^2\times S^2$ into itself generates a group G of order two that acts properly discontinuously without fixed points. Show that the quotient manifold $(S^2\times S^2)/G$ is orientable. Is the product $P^2(\mathbf{R})\times P^2(\mathbf{R})$ orientable?

5.9.7. Let $f : \mathbf{R}^d \to \mathbf{R}$ be a C^∞ submersion. Show that the manifold $f^{-1}(0)$ is orientable.

5.9.8. Show that the product of two C^∞ manifolds is orientable if and only if each of the factors is.

5.9.9. Show that the tangent bundle TX of *any* manifold X is orientable.

5.9.10. Show that the Klein bottle is non-orientable.

5.9.11. The Möbius strip

(a) Show that the image of $[0, 4\pi] \times \left[0, \frac{3}{4}\right[$ in $\mathbf{R}^3$ under the map

$$(\theta, r) \mapsto \left(\cos\theta\left(1 + r\cos\frac{\theta}{2}\right), \sin\theta\left(1 + r\cos\frac{\theta}{2}\right), r\sin\frac{\theta}{2}\right)$$

is a submanifold of $\mathbf{R}^3$.

(b) Show that this manifold is not orientable.

5.9.12. Let X and Y be manifolds and $D \subset X$ a submanifold-with-boundary of X. Show that $D \times Y \subset X \times Y$ is a submanifold-with-boundary of $X \times Y$ and that $\partial(D \times Y) = \partial D \times Y$. If $E \subset Y$ is a submanifold-with-boundary of Y, is $D \times E \subset X \times Y$ a submanifold-with-boundary of Y? (Write down the frontier $\partial(D \times E)$ anyway.)

5.9.13. Show that there exists no function f on the torus T^d satisfying $df = \omega_1$, where ω_1 is defined in 5.3.10.1.

5.9.14. Prove in detail the results in 5.3.30, 5.3.31.1, 5.3.31.2, 5.3.37.1 and 5.3.37.2.

5.9.15. Show that the direct sum $R^*(X) = \sum_{k=0}^d R^k(X)$ of the de Rham groups of a d-dimensional manifold X has a canonical algebra structure.

5.9.16. Let Y be a submanifold of a manifold X, and $i : Y \to X$ the canonical injection. Assume that there exists a C^∞ map $r : X \to Y$ such that $r|_Y = \mathrm{Id}\,|_Y$. Show that $i^* : R^k(X) \to R^k(Y)$ is surjective for every k. Can there be such an r for the canonical inclusion $S^{d-1} \subset S^d$?

5.9.17. Identify $\mathbf{R}^4$ with $\mathbf{C}^2$ by $(x, y, z, t) = (u = x + iy, v = z + it)$, and let G be the group generated by

$$(u, v) \mapsto \left(\exp\left(\frac{2i\pi}{n}\right)u, \exp\left(\frac{2i\pi}{n}\right)v\right),$$

for a fixed integer $n > 0$. Show that the action of G on $S^3 \subset \mathbf{R}^4$ is properly discontinuous without fixed points. Set $Y = S^3/G$; is Y orientable? Find Y for $n = 1$ and $n = 2$? Calculate $R^k(Y)$ for $k = 0, 1, 2, 3$ and arbitrary n.

5.9.18. Compute the de Rham groups of the Klein bottle.

5.9.19. Find counterexamples to illustrate 5.4.8.

CHAPTER 6

Integration of Differential Forms

We have seen that manifolds do not have a canonical measure, but it can be shown (6.1.3) that there is a canonical way of integrating a d-form over an oriented d-dimensional manifold. This is a fundamental fact. It provides the framework for Stokes' theorem, an essential tool that relates submanifolds-with-boundary with their boundaries.

The power of Stokes' theorem is illustrated by several applications, for example: a sphere cannot be diffeomorphic to a product of manifolds (6.3.3); an automorphism of the disk must have at least one fixed point (6.3.5).

Next we define the volume of a submanifold of Euclidean space (section 6.5 and definition 6.6.3), thus generalizing the usual idea of arclength and surface area. We find explicit formulas for the volume of spheres and balls in any dimension (6.5.7). As an application to physics, we prove Archimedes's theorem (6.5.15). In addition, we use Stokes' formula in a proof of the isoperimetric inequality in arbitrary dimension.

The last three sections are devoted to the volume of tubular neighborhoods of manifolds embedded in Euclidean space (cf. chapter 2). This calculation is long and involved, and yields a formula (6.9.9) that expresses this volume in terms of integrals over the manifold of certain invariants, called the Weyl curvatures. In the case of curves the formula has only one term, and we get an explicit value for the volume.

6.1. Integrating Forms of Maximal Degree

Let X be an oriented, d-dimensional manifold. If $\omega \in \Omega_0^d(X)$ is a volume form (5.3.3) compatible with the orientation of X, there is a Lebesgue measure on X canonically associated to ω, arising from the density $|\omega|$ (cf. 3.3.5 and 3.3.11). This Lebesgue measure is still denoted by $|\omega|$.

Now let α be a d-form. Since $\omega(x)$ is non-zero for every x, there exists $f : X \to \mathbf{R}$ such that $\alpha = f\omega$ (cf. the proof of 3.3.8).

We want to say that α is *integrable* over X, and write $\alpha \in \Omega_{\text{int}}^d(X)$, if f lies in the space $C_{|\omega|}^{\text{int}}(X)$ of functions on X integrable with respect to $|\omega|$:

6.1.1
$$\alpha \in \Omega_{\text{int}}^d(X) \Leftrightarrow f \in C_{|\omega|}^{\text{int}}(X).$$

In that case we want to write

6.1.2
$$\int_X \alpha = \int_X f\,|\omega|.$$

All of this will only make sense if it is independent of the volume form ω. If $\omega' \in \Omega_0^d(X)$ is another volume form, also compatible with the orientation, we have $\omega' = h\omega$, with $h \in C^0(X; \mathbf{R}_+^*)$ (5.3.4). In particular, $|\omega'| = h|\omega|$, and 0.4.3.2 says exactly that 6.1.1 and 6.1.2 do not depend on the choice of ω.

6.1.3. Theorem and definition. *Let X be an oriented, d-dimensional manifold. The vector space $\Omega_{\text{int}}^d(X)$ is well-defined by equations 6.1.1 and 6.1.2, and the map*

$$\alpha \mapsto \int_X \alpha.$$

is a real linear functional on this vector space. The scalar $\int_X \alpha$ is called the integral of α over X. $\qquad\square$

6.1.4. Examples

6.1.4.1. If $\alpha \in \Omega_0^d(X)$ has compact support, the function f such that $\alpha = f\omega$ is continuous with compact support, hence integrable. Thus $\alpha \in \Omega_{\text{int}}^d(X)$.

6.1.4.2. In particular, if $X = \mathbf{R}^d$, the support of $\alpha \in \Omega_0^d(\mathbf{R}^d)$ is compact and $\omega = dx_1 \wedge \cdots \wedge dx_d$ (5.3.8.1), we have

$$\int_{\mathbf{R}^d} \alpha = \int_{\mathbf{R}^d} f\,\delta_0,$$

where $\alpha = f\omega$ and δ_0 is the Lebesgue measure on $\mathbf{R}^d$.

6.1.4.3. Let the orientation of X be determined by the choice of a volume form ω. The form $-\omega$ determines the opposite orientation, and we denote by $-X$ the same manifold with that orientation. Then $\alpha \in \Omega_{int}^d(X)$ if and only if $\alpha \in \Omega_{int}^d(-X)$, and

$$\int_{-X} \alpha = -\int_X \alpha.$$

In fact, if $f \in C_\mu^{int}|\omega|$ is such that $\alpha = f\omega$, we have $\alpha = (-f)(-\omega)$, whence

$$\int_{-X} \alpha = \int_X (-f)\,|-\omega| = -\int_X f\,|\omega|.$$

6.1.4.4. Proposition. *If X and Y are oriented manifolds and $f : X \to Y$ is an orientation-preserving diffeomorphism, $\beta \in \Omega_{int}^d(Y)$ implies $f^*\beta \in \Omega_{int}^d(X)$, and*

$$\int_X f^*\beta = \int_Y \beta.$$

Proof. Let ω' be a volume form on Y compatible with the orientation of Y, and $g : Y \to \mathbf{R}$ the function such that $\beta = g\omega'$. Set $\omega = f^*\omega'$. Since f preserves orientation, ω is a volume form on X compatible with the orientation of X, and the associated densities satisfy

$$f^*|\omega'| = |\omega|.$$

We also have $f^*\beta = f^*(g\omega') = (g \circ f)(f^*\omega') = (g \circ f)\omega$. By definition,

$$\int_X f^*\beta = \int_X (g \circ f)|\omega| \qquad \text{and} \qquad \int_Y \beta = \int_Y g\,|\omega'|;$$

now the proposition follows from 3.3.16.   □

6.1.4.5. Proposition. *If X and Y are oriented manifolds and $f : X \to Y$ is an orientation-reversing diffeomorphism, $\beta \in \Omega_{int}^d(Y)$ implies $f^*\beta \in \Omega_{int}^d(X)$, and*

$$\int_X f^*\beta = -\int_Y \beta.$$

Proof. This follows from 6.1.4.3 and 6.1.4.4.   □

6.1.4.6. Let X be an oriented manifold and (U_i, ϕ_i, ψ_i) a partition of unity subordinate to positively oriented charts (U_i, ϕ_i). Let $\alpha \in \Omega_0^d(X)$ have compact support. We can write $\alpha = \sum_i \psi_i\alpha$, where only a finite number of $\psi_i\alpha$ is non-zero because α has compact support. We have $\alpha \in \Omega_{int}^d(X)$ by 6.1.4.1, and, by linearity,

$$\int_X \alpha = \sum_i \int_X \psi_i\alpha.$$

Now $\psi_i\alpha$ is a form with compact support on the domain U_i of the chart (U_i, ϕ_i), and $\phi_i : U_i \to \phi_i(U_i)$ is a diffeomorphism. Since the chart is positively oriented, 6.1.4.4 and 6.1.4.9 imply that

$$\int_X \psi_i \alpha = \int_{U_i} \psi_i \alpha = \int_{\phi_i(U_i)} (\phi_i^{-1})^* (\psi_i \alpha) = \int_{\mathbf{R}^d} (\phi_i^{-1})^* (\psi_i \alpha).$$

Thus

$$\int_X \alpha = \sum_i \int_{\mathbf{R}^d} (\phi_i^{-1})^* (\psi_i \alpha).$$

Notice that if we knew nothing about densities this formula could serve as a starting point to define $\int_X \alpha$.

6.1.4.7. Proposition. *Let X and Y be oriented manifolds, of dimension d and e, respectively. Give $X \times Y$ its canonical orientation (5.3.8.4), and let $p : X \times Y \to X$ and $q : X \times Y \to Y$ be the canonical projections. If $\alpha \in \Omega^d_{\text{int}}(X)$, $\beta \in \Omega^e_{\text{int}}(X)$ and $p^* \alpha \wedge q^* \beta \in \Omega^{d+e}_{\text{int}}(X \times Y)$, we have*

$$\int_{X \times Y} p^* \alpha \wedge q^* \beta = \left(\int_X \alpha \right) \left(\int_Y \beta \right).$$

Proof. Let ω and ω' be volume forms on X and Y, respectively, and f and g functions such that $\alpha = f\omega$ and $\beta = g\omega'$. We have

$$p^* \alpha = (f \circ p) p^* \omega, \qquad q^* \beta = (g \circ q) q^* \omega'$$

and

$$p^* \alpha \wedge q^* \beta = (f \circ p)(g \circ q) p^* \omega \wedge q^* \omega'.$$

Applying the results from 3.3.18 to the densities $|\omega|$ and $|\omega'|$, we conclude that

$$|p^* \omega \wedge q^* \omega'| = |\omega| \otimes |\omega'|,$$

the product measure of $|\omega|$ and $|\omega'|$.

By assumption, $p^* \omega \wedge q^* \omega'$ is compatible with the orientation of $X \times Y$. From 3.3.18.7 we get

$$\begin{aligned}
\int_{X \times Y} p^* \alpha \wedge q^* \beta &= \int_{X \times Y} (f \circ p)(g \circ q) |p^* \omega \wedge q^* \omega'| \\
&= \int_{X \times Y} (f \circ p)(g \circ q) |\omega| \otimes |\omega'| \\
&= \int_X f \left(\int_{\{x\} \times Y} (g \circ q) \, d|\omega'| \right) d|\omega| \\
&= \int_X f\omega \int_Y g\omega' = \int_X \alpha \int_Y \beta. \qquad \square
\end{aligned}$$

6.1.4.8. Proposition. *Let X and Y be d-dimensional manifolds and $p : X \to Y$ a k-fold covering. Assume Y is oriented and give X the orientation induced by p (5.3.8.3). If $\alpha \in \Omega^d_{\text{int}}(Y)$ we have $p^* \alpha \in \Omega^d_{\text{int}}(X)$ and*

$$\int_X p^* \alpha = k \int_Y \alpha.$$

Proof. Since continuous functions with compact support are dense in $C^{int}_{|\omega|}$, we can restrict ourselves to the case that $\alpha \in \Omega^d_0(Y)$ has compact support. Consider on Y a partition of unity (U_i, ϕ_i, ψ_i) such that each (U_i, ϕ_i) is a chart on Y and each inverse image $p^{-1}(U_i)$ is a union of pairwise disjoint open sets U^j_i such that $p : U^j_i \to U_i$ is a diffeomorphism (cf. 2.4.1, 2.4.4 and 3.2.4). By construction, the restriction of p to U^j_i preserves orientation.

By 6.1.4.6,

$$\int_Y \alpha = \sum_i \int_Y \psi_i \alpha = \sum_i \int_{U_i} \psi_i \alpha.$$

Since each $p : U^j_i \to U_i$ is an orientation-preserving diffeomorphism, we have

$$\int_{U_i} \psi_i \alpha = \int_{U^j_i} p^*(\psi_i \alpha)$$

by 6.1.4.4, so that

$$\sum_{j=1}^k \left(\int_{U^j_i} p^*(\psi_i \alpha) \right) = k \int_{U_i} \psi_i \alpha.$$

On the other hand,

$$\int_{p^{-1}(U_i)} p^*(\psi_i \alpha) = \int_{p^{-1}(U_i)} (\psi_i \circ p)(p^* \alpha),$$

whence

$$k \int_Y \alpha = \sum_i k \int_{U_i} \psi_i \alpha = \sum_i \int_{p^{-1}(U_i)} (\psi_i \circ p)(p^* \alpha) = \sum_i \int_X (\psi_i \circ p)(p^* \alpha),$$

since $\psi_i \circ p$ is actually supported in $p^{-1}(U_i)$. The integral and the sum commute because the sum is locally finite and α has compact support, so

$$k \int_Y \alpha = \int_X p^* \alpha$$

because $\sum \psi_i = 1$ implies $\sum \psi_i \circ p = 1$.   □

6.1.4.9. Proposition. *Let X be an oriented manifold and $U \subset X$ an open submanifold with the induced orientation. If $\alpha \in \Omega^d_{int}(X)$ we have $\alpha|_U \in \Omega^d_{int}(U)$ and*

$$\int_U (\alpha|_U) = \int_X \chi_U \alpha,$$

where χ_U is the characteristic function of U.

Proof. If ω is a volume form on X and $\alpha = f\omega$, we have $f \in C^{int}_{|\omega|}(X)$. Since χ_U is measurable and bounded, $\chi_U f$ is also integrable with respect to $|\omega|$, hence also with respect to $|\omega||_U$ [Gui69, p. 15 and 5]. The proposition follows from the equation $\alpha|_U = (\chi_U f)(\omega|_U)$.   □

6.1.4.10. Notation. Let X be an oriented manifold and $D \subset X$ a compact set. If $\alpha \in \Omega_0^d(X)$, we have $\chi_D \alpha \in \Omega_{\text{int}}^d(X)$, where χ_D is the characteristic function of D. We set

$$\int_D \alpha = \int_X \chi_D \alpha.$$

6.1.4.11. Proposition (continuity of the integral). *If $t \mapsto \alpha(t)$ is a continuous family (5.2.10) of continuous d-forms on a compact, oriented, d-dimensional manifold X, the map $t \mapsto \int_X \alpha(t)$ is continuous.*

Proof. Let (U_i, ϕ_i, ψ_i) be a partition of unity of X, used in applying definition 5.2.10. By the last formula in 6.1.4.6 we have

$$\int_X \alpha(t) = \sum_i \int_{U_i} \psi_i \alpha(t) = \sum_i \int_{\phi_i(U_i)} \phi_i^{-1*}(\psi_i \alpha(t)),$$

and it is enough to show that each term is continuous. This follows from 5.2.10.1 and 6.1.4.2. $\square$

The formulas in 6.1.4.6 are interesting from the theoretical point of view, but for practical calculations it is preferable to use the following result:

6.1.5. Theorem. *Let X be a compact, oriented, d-dimensional manifold, and α a continuous d-form on X. Assume there exists a finite family $\{(U_i, V_i, \tau_i)\}$ of triples, consisting of open sets $U_i \subset \mathbf{R}^d$, compact sets $V_i \subset U_i$ and C^p maps $\tau_i : U_i \to X$ such that the following conditions are satisfied:*

(a) *$\bigcup_i \tau_i(V_i) = X$;*

(b) *for each i, the difference $V_i \setminus \mathring{V}_i$ has measure zero, $\tau_i(\mathring{V}_i)$ is open in X and $\tau_i|_{\mathring{V}_i}$ is an orientation-preserving diffeomorphism onto its image;*

(c) *if i and j are distinct, $\tau_i(\mathring{V}_i) \cap \tau_j(\mathring{V}_j) = \emptyset$.*

Then

$$\int_X \alpha = \sum_i \int_{\tau_i(\mathring{V}_i)} \alpha = \sum_i \int_{V_i} \tau_i^* \alpha.$$

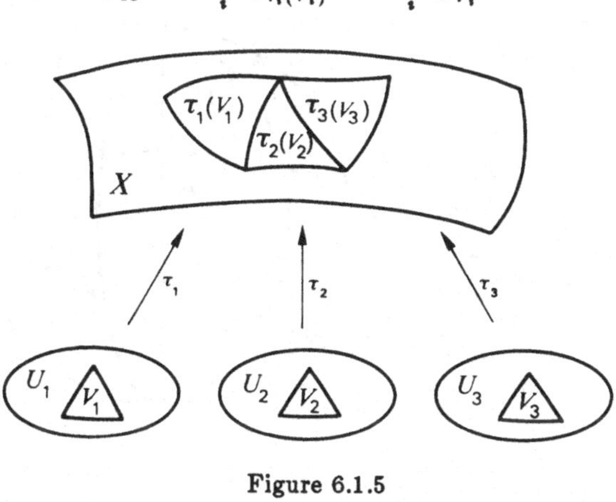

Figure 6.1.5

This is a long statement, but it expresses a simple result: one can compute the integral piecemeal, the pieces being open sets (generally domains of charts) that cover the whole manifold up to a set of measure zero.

Proof. Since $V_i \setminus \mathring{V}_i$ has measure zero in $\mathbf{R}^d$ and τ_i is of class C^p, the image $\tau_i(V_i \setminus \mathring{V}_i)$ also has measure zero by 3.3.17.3. Thus $\bigcup_i \tau_i(V_i \setminus \mathring{V}_i)$ has measure zero and

$$\int_X \alpha = \int_{X \setminus \bigcup_i \tau_i(V_i \setminus \mathring{V}_i)} \alpha = \int_{\bigcup_i \tau_i(V_i) \setminus \bigcup_i \tau_i(V_i \setminus \mathring{V}_i)} \alpha.$$

But V_i is compact and τ_i is continuous, so the image of the frontier is the frontier of the image:

$$\tau_i(V_i \setminus \mathring{V}_i) = \tau_i(V_i) \setminus \tau_i(\mathring{V}_i).$$

Thus we can remove the frontiers:

$$\bigcup_i \tau_i(V_i) \setminus \bigcup_i \tau_i(V_i \setminus \mathring{V}_i) = \bigcup_i \tau_i(\mathring{V}_i).$$

Since the $\tau_i(\mathring{V}_i)$ are disjoint, we have

$$\int_X = \sum_i \int_{\tau_i(\mathring{V}_i)} \alpha = \sum_i \int_{\mathring{V}_i} \tau_i^* \alpha. \qquad \cdot \ \Box$$

6.1.6. Example: the latitude-longitude chart on the sphere (other examples of this nature can be found in physics books). Here $X = S^2$, and we apply 6.1.5 to the single triple (U, V, τ), where $U = \mathbf{R}^2$, $V = [-\pi, \pi] \times [-\pi/2, \pi/2]$ and

$$\tau : (\phi, \theta) \mapsto (\cos \phi \cos \theta, \sin \phi \cos \theta, \sin \theta).$$

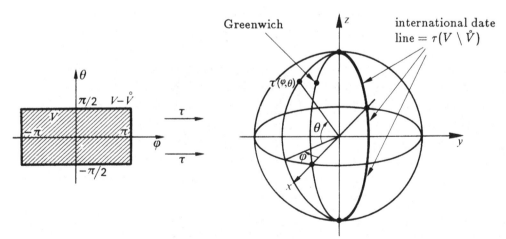

Figure 6.1.6

We have $\tau \in C^\infty(\mathbf{R}^2; S^2)$ by 2.6.7 and $V \setminus \mathring{V}$ has measure zero; also, $\tau|_{\mathring{V}}$ is a diffeomorphism onto its image. (If ϕ represents longitude with respect to the Greenwich meridian, $\tau(V \setminus \mathring{V})$ is the international date line, where the date changes if you move one inch.) If we consider S^2 with its canonical orientation, τ preserves orientation (5.3.20): the canonical volume form of S^2 is

$$\sigma = x\,dy \wedge dz + y\,dz \wedge dx + z\,dx \wedge dy$$

(5.3.17.2), so $\tau^*\sigma = \cos\theta\,d\phi \wedge d\theta$ by 5.2.8.3, and $\cos\theta$ is positive for $\theta \in {]{-\pi/2}, \pi/2[}$.

By 6.1.5 we have, for every continuous function f on S^2:

$$\int_{S^2} f\sigma = \int_{-\pi}^{\pi} \int_{-\pi/2}^{\pi/2} f(\cos\phi\cos\theta, \sin\phi\cos\theta, \sin\theta) \cdot \cos\theta \cdot |d\phi \wedge d\theta|,$$

the Lebesgue measure on $\mathbf{R}^2$ being given by $\delta_0 = |d\phi \wedge d\theta|$.

For example, setting $f = 1$ we obtain the surface area of S^2, which is $\int_{S^2} \sigma = 4\pi$. See also 6.5.7.

6.2. Stokes' Theorem

6.2.1. Theorem. *Let X be an oriented, d-dimensional manifold, $\omega \in \Omega_1^{d-1}(X)$ a form on X and D a compact submanifold-with-boundary of X. If $i : \partial D \to X$ denotes the canonical injection, we have*

$$\int_D d\omega = \int_{\partial D} i^*\omega.$$

In particular, $\int_D d\omega = 0$ if $\partial D = \emptyset$.

Recall the definition of $\int_D d\omega$ from 6.1.4.10, and that ∂D has a canonical orientation (5.3.36). In the notation of 5.2.5.1 Stokes' theorem can also be written

$$\int_D d\omega = \int_{\partial D} i^*\omega.$$

Stokes' theorem generalizes the classical formula $\int_a^b f'(t)\,dt = f(b) - f(a)$: with $D = [a, b]$ we get $\partial D = \{a\} \cup \{b\}$, where point a gets a minus sign and point b a plus sign.

Proof. We do this in several steps.

6.2.1.1. First step. We have

$$\int_D d\omega = \int_X \chi_D\,d\omega.$$

Let $\{(U_i, \phi_i, \theta_i)\}$ be a partition of unity associated with charts (U_i, ϕ_i) which are positively oriented and (if they intersect D) of one of the two types described in 5.3.34. The orientation of ∂D is such that the charts $(U_i \cap \partial D, \phi_i|_{\partial D})$ are positively oriented (5.3.36).

Since D is compact, $D \cap U_i \neq \emptyset$ for only finitely many indices i. Thus

$$\int_X \chi_D \, d\omega = \int_X \chi_D \, d\left(\sum_i \psi_i \omega\right),$$

and the sum is finite. Since the operator d, integration over D, the restriction i^* and integration over ∂D are all additive, it's enough to prove the theorem for $\omega \in \Omega_1^{d-1}(X)$ with compact support contained in U, where (U, ϕ) is a chart of one of the types in 5.3.34.

6.2.1.2. Second step. Assume that ω has compact support contained in U. Since $\phi : U \to \phi(U)$ is a diffeomorphism onto an open subset of $\mathbf{R}^d$, we can apply 6.1.4.4 to get

$$\int_D d\omega = \int_X \chi_D \, d\omega = \int_U \chi_D \, d\omega = \int_{\phi(U)} (\phi^{-1})^*(\chi_D \, d\omega)$$

$$= \int_{\phi(U)} (\chi_D \circ \phi^{-1})(\phi^{-1})^*(d\omega).$$

Notice that $\chi_D \circ \phi^{-1} = \chi_{\phi(D)}$, that $(\phi^{-1})^*(d\omega)$ is supported in U and that d commutes with $(\phi^{-1})^*$. Thus we can write

$$\int_D d\omega = \int_{\mathbf{R}^d} \chi_{\phi(D)} d\big((\phi^{-1})^*\omega\big),$$

and similarly

$$\int_{\partial D} \omega|_{\partial D} = \int_{\mathbf{R}^{d-1}} (\phi^{-1}|_{\partial D})^*(i^*\omega).$$

Now $\phi(\partial D)$ is contained in $\{0\} \times \mathbf{R}^{d-1}$ (5.3.33), so that

$$\int_{\partial D} \omega|_{\partial D} = \int_{\{0\} \times \mathbf{R}^{d-1}} \big((\phi^{-1})^*\omega\big)\big|_{\{0\} \times \mathbf{R}^{d-1}}.$$

Setting $\alpha = (\phi^{-1})^*\omega$, we see that it is enough to prove the theorem for $X = \mathbf{R}^d$, $\alpha \in \Omega_0^d(\mathbf{R}^d)$ with compact support, and either $D = \mathbf{R}^d$ or $D =]-\infty, 0[\times \mathbf{R}^{d-1}$. In the first case $\partial D = \emptyset$, in the second $\partial D = \{0\} \times \mathbf{R}^{d-1}$.

6.2.1.3. Third step. Assume that $D = \mathbf{R}^d$ and $\partial D = \emptyset$; we must show that $\int_{\mathbf{R}^d} d\alpha = 0$. Write

$$\alpha = \sum_{i=1}^d \alpha_i \, dx_i \wedge \cdots \wedge \widehat{dx_i} \wedge \cdots \wedge dx_d$$

(5.2.7.2). Since α has compact support, we can find a closed cube Q such that $\operatorname{supp} \alpha \subset \mathring{Q}$; set $Q = Q_1 \times Q_{d-1}$, where $Q_1 = [a, b] \subset \mathbf{R}$ and Q_{d-1} is a cube in $\mathbf{R}^{d-1}$.

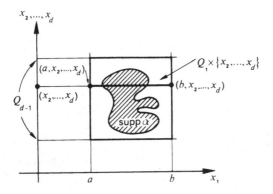

Figure 6.2.1.3

We have

$$d\alpha = \left(\sum_{i=1}^{d} (-1)^{i+1} \frac{\partial \alpha_i}{\partial x_i} \right) \omega_0,$$

with $\omega_0 = dx_1 \wedge \cdots \wedge dx_d$. By Fubini's theorem (0.4.5.1) we have

$$\int_{\mathbf{R}^d} \frac{\partial \alpha_1}{\partial x_1} |\omega_0| = \int_Q \frac{\partial \alpha_1}{\partial x_1} |\omega_0| = \int_{Q_1 \times Q_{d-1}} \frac{\partial \alpha_1}{\partial x_1} |\omega_0|$$

$$= \int_{Q_{d-1}} \left(\int_{Q_1 \times \{(x_2,\ldots,x_d)\}} \frac{\partial \alpha_1}{\partial x_1} \right) |dx_2 \wedge \cdots \wedge dx_d|.$$

On the other hand,

$$\int_{Q_1 \times \{(x_2,\ldots,x_d)\}} \frac{\partial \alpha_1}{\partial x_1} = \alpha_1(b, x_2, \ldots, x_d) - \alpha_1(a, x_2, \ldots, x_d) = 0$$

because the support of α does not intersect ∂Q; thus

$$\int_{\mathbf{R}^d} \frac{\partial \alpha_1}{\partial x_1} |\omega_0| = \int_Q \frac{\partial \alpha_1}{\partial x_1} |\omega_0| = 0.$$

A similar argument shows that $\int_Q \frac{\partial \alpha_i}{\partial x_i} |\omega_0|$ is zero for every i, as we wished to prove.

6.2.1.4. Fourth step. Now assume that $D =]-\infty, 0[\times \mathbf{R}^{d-1}$ and $\partial D = \{0\} \times \mathbf{R}^{d-1}$. Again we consider a cube Q' such that $\operatorname{supp} \alpha \subset \mathring{Q}'$, and we set

$$Q = Q' \cap (\mathbf{R}_- \times \mathbf{R}^{d-1}) = [a, 0] \times Q_{d-1}.$$

Expressing α as before, we will show that $\int_Q \frac{\partial \alpha_i}{\partial x_i} |\omega_0| = 0$ for $i = 2, \ldots, d$.

Indeed, if $Q_{d-1} = [a_2, b_2] \times \cdots \times [a_d, b_d]$, we find, as above:

$$\int_{a_i}^{b_i} \frac{\partial \alpha_i}{\partial x_i} (x_1, \ldots, x_{i-1}, t, x_{i+1}, \ldots, x_d) \, dt = \alpha_i(x_1, \ldots, x_{i-1}, b_i, x_{i+1}, \ldots, x_d)$$

$$- \beta_i(x_1, \ldots, x_{i-1}, a_i, x_{i+1}, \ldots, x_d).$$

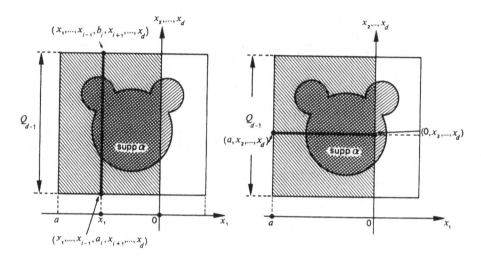

Figure 6.2.1.4

But $(x_1, \ldots, x_{i-1}, b_i, x_{i+1}, \ldots, x_d)$ and $(x_1, \ldots, x_{i-1}, a_i, x_{i+1}, \ldots, x_d)$ lie on $\partial Q'$ for every

$$(x_1, \ldots, \widehat{x_i}, \ldots, x_d) \in [a, 0] \times [a_2, b_2] \times \cdots \times [\widehat{a_i, b_i}] \times \cdots \times [a_d, b_d];$$

this implies

$$\int_{a_i}^{b_i} \frac{\partial \alpha_i}{\partial x_i}(x_1, \ldots, x_{i-1}, t, x_{i+1}, \ldots, x_d)\, dt = 0,$$

showing that $\int_Q \frac{\partial \alpha_i}{\partial x_i}|\omega_0| = 0$.

For $\int_Q \frac{\partial \alpha_1}{\partial x_1}|\omega_0|$ the calculation is different, because in general a non-zero term is left over:

$$\int_{Q_1}\left(\int_a^0 \frac{\partial \alpha_1}{\partial x_1}|dx_1|\right)|dx_2 \wedge \cdots \wedge dx_d|$$
$$= \int_{Q_1}(\alpha_1(0, x_2, \ldots, x_d) - \alpha_1(a, x_2, \ldots, x_d))|dx_2, \ldots, dx_d|.$$

We still have $\alpha_1(a, x_2, \ldots, x_d) = 0$, so that

$$\int_{\mathbf{R}^d} \chi_D\, d\alpha = \int_{\{0\} \times Q_{d-1}} \alpha_1(0, x_2, \ldots, x_d)|dx_2 \wedge \cdots \wedge dx_d|.$$

Since $\alpha = \sum_{i=1}^d \alpha_i\, dx_1 \wedge \cdots \wedge \widehat{dx_i} \wedge \cdots \wedge dx_d$ and $dx_1|_{\partial D} = 0$, we have

$$\alpha|_{\partial D} = \alpha_1(0, x_2, \ldots, x_d)\, dx_2 \wedge \cdots \wedge dx_d,$$

whence

$$\int_{\mathbf{R}^d} \chi_D \, d\alpha = \int_{\{0\} \times Q_{d-1}} \alpha|_{\partial D} = \int_{\mathbf{R}^{d-1}} \alpha|_{\partial D},$$

concluding the proof. □

6.2.2. Remarks. One can define abstract manifolds-with-boundary and differential forms on them, and prove an appropriate version of Stokes' theorem; see [Lan69, pages 437 and ii]. One can also prove a similar theorem about densities. Finally, there is a version of Stokes' theorem for "submanifolds-with-boundary" whose frontier is less regular than a codimension-one submanifold; see [Lan69, p. 459].

A theory parallel to the one developed above is built on the notions of *chains* on manifolds, their *boundaries* and *integrals* over them; Stokes' theorem can be translated into this language as well. See [Spi79, vol. I, chapter 8].

6.3. First Applications of Stokes' Theorem

6.3.1. Proposition. *Let X be a compact, oriented, d-dimensional manifold. For every form $\omega \in \Omega_1^{d-1}(X)$ we have*

$$\int_d \omega = 0.$$

Proof. X is submanifold-with-boundary of itself, with empty boundary. □

6.3.2. Corollary. *If X is a compact, oriented, d-dimensional manifold, $R^d(X)$ has dimension at least one.*

Proof. Let ω be a volume form; since $\int_X \omega > 0$, we cannot have $\omega = d\alpha$. □

6.3.3. Corollary. *Let p and q be integers ≥ 1. The sphere S^{p+q} cannot be homeomorphic to $X \times Y$, where X and Y are orientable manifolds of dimension p and q, respectively.*

Proof. Since $p \geq 1$ and $q \geq 1$, we know that the de Rham group $R^p(S^{p+q})$ is zero (5.7.1). If S^{p+q} were homeomorphic to $X \times Y$, we'd have $R^p(X \times Y) = 0$ (5.4.7). We will show that this leads to a contradiction. Let $p : X \times Y \to X$ be the canonical projection, and ω a volume form on X. For fixed $y \in Y$, the map $p : X \times \{y\} \to X$ is an orientation-preserving diffeomorphism (assuming we have oriented $X \times Y$ accordingly). Thus

$$0 < \int_X \omega = \int_{X \times \{y\}} p^*\omega|_{X \times \{y\}}.$$

But ω is a form of maximal degree on X, so $d\omega = 0$ and

$$d(p^*\omega) = p^*(d\omega) = 0,$$

and the hypothesis that $R^p(X \times Y) = 0$ would imply the existence of $\alpha \in \Omega^{p-1}(X \times Y)$ such that $p^*\omega = d\alpha$. This would entail

$$\int_{X \times \{y\}} p^*\omega|_{X \times \{y\}} = \int_{X \times \{y\}} d\alpha|_{X \times \{y\}} = \int_{X \times \{y\}} d(\alpha|_{X \times \{y\}}) = 0$$

by proposition 6.3.1.   □

6.3.4. Proposition. *Let D be a compact, non-empty, C^1 submanifold-with-boundary of $\mathbf{R}^d$ and $V \subset \mathbf{R}^d$ an open neighborhood of D. If f is a C^1 map from V into ∂D, the restriction $f|_{\partial D}$ cannot be the identity.*

Proof. Write $f = (f_1, \ldots, f_d)$, let $j : \partial D \to \mathbf{R}^d$ be the canonical injection and $x_1, \ldots, x_d : \mathbf{R}^d \to \mathbf{R}$ the canonical projections. Consider the integrals

$$\int_{\partial D} j^*(x_1 \, dx_2 \wedge \cdots \wedge dx_d) \quad \text{and} \quad \int_{\partial D} j^*(f_1 \, df_2 \wedge \cdots \wedge df_d).$$

If f were such that $f|_{\partial D} = \mathrm{Id}_{\partial D}$, these two integrals would be the same. By Stokes' theorem we'd have

$$\int_D dx_1 \wedge \cdots \wedge dx_d = \int_D df_1 \wedge \cdots \wedge df_d.$$

The integrand on the right-hand side is identically zero, because for any $m \in V$ the d forms $df_i(m) = f^*(dx_i)(m)$ live in the $(d-1)$-dimensional space $T^*_{f(m)}(\partial D)$, and so must be linearly dependent. But the left-hand side is the volume of D, hence strictly positive: contradiction.   □

Intuitively, this says that a membrane cannot retract into its boundary without being punctured somewhere.

6.3.5. Corollary (the Brouwer fixed point theorem). *Let U be an open neighborhood of $\overline{B}(0,1)$ in $\mathbf{R}^d$. Any map $g \in C^1(U; \mathbf{R}^d)$ taking $\overline{B}(0,1)$ inside itself has a fixed point in $\overline{B}(0,1)$, that is, there exists $x \in \overline{B}(0,1)$ such that $g(x) = x$.*

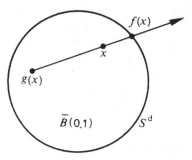

Figure 6.3.5

Proof. By contradiction: assume that $g(x) \neq x$ for every $x \in \overline{B}(0,1)$. We can define a new map f on $\overline{B}(0,1)$ by the following rule: $f(x)$ is the unique point where the half-line originating at $g(x)$ and containing x intersects S^d. More explicitly, $f(x) = x + tu$, where

$$u = \frac{x - g(x)}{\|x - g(x)\|}, \qquad t = -(x \mid u) + \sqrt{1 - \|x\|^2 + (x \mid u)^2}.$$

It is easy to see that f is actually defined and C^1 on a neighborhood of $\overline{B}(0,1)$, and by construction its image is contained in $\partial \overline{B}(0,1) = S^d$. This contradicts the proposition. $\qquad\square$

6.3.6. Remark. Brouwer's theorem still holds for $g \in C^0\big(\overline{B}(0,1); \overline{B}(0,1)\big)$: see exercise 6.10.4.

6.3.7. Proposition. *Let $D = \overline{B}(0,1) \subset \mathbf{R}^2$. If ξ is a C^1 vector field on $\mathbf{R}^2$ such that, for every $x \in S^1 = \partial D$, we have $\xi(x) = \lambda x$ with $\lambda \in \mathbf{R}^*_-$, there exists $y \in D$ such that $\xi(y) = 0$.*

Actually, ξ just has to be defined on an open neighborhood of $\overline{B}(0,1)$. The existence of $\lambda < 0$ such that $\xi(x) = \lambda x$ means that all around S^1 the field is radial, pointing inward.

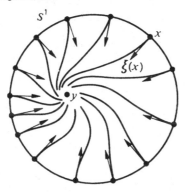

Figure 6.3.7

Proof. By contradiction. If $\xi(x) \neq 0$ for all $x \in \overline{B}(0,1)$, we can define

$$\eta(x) = \frac{\xi(x)}{\|\xi(x)\|}$$

on an open neighborhood U of $\overline{B}(0,1)$. We have $\eta \in C^1(U; S^1)$. Let σ be the canonical form on S^1 (5.3.17.2). Since $\eta^*\sigma$ is a one-form on $\mathbf{R}^d$, we can apply Stokes' theorem to $\mathbf{R}^2$ to get

$$\int_{\partial D} (\eta^*\sigma)|_{\partial D} = \int_D d(\eta^*\sigma|_{\partial D}) = \int_D \eta^*(d\sigma|_{\partial D}).$$

But $d\sigma$ is a two-form and ∂D has dimension one, so $d\sigma|_{\partial D} = 0$ (cf. 5.2.5.2). On the other hand, the restriction of η to $S^1 = \partial D$ is the map $x \mapsto -x$, so $\eta|_{S^1}$ is an orientation-preserving diffeomorphism. By 6.1.4.4 we get the contradiction

$$\int_{S^1} \eta^*\sigma = \int_{S^1} \sigma > 0. \qquad \square$$

Intuitively, this says that the integral curves of ξ (1.2.2) must converge to a point in D. Actually, this is not quite true; there may be periodic integral curves going around the fixed point, and the other integral curves will either spiral into a periodic curve or converge to the fixed point. What we can say is that the integral curves define a strict retraction of D into itself and that they must stop somewhere.

6.3.8. Proposition. *Let $D = \overline{B}(0,1) \subset \mathbf{R}^2$ and let E be the submanifold-with-boundary obtained by removing from D a certain number $q > 1$ of pairwise disjoint open disks D_i. If ξ is a C^1 vector field on $\mathbf{R}^2$ normal to ∂E and pointing inward, there exists $y \in D$ such that $\xi(y) = 0$.*

Saying that ξ is normal to ∂E and points inward means that for every $x \in S^1 = \partial D$ there exists $\lambda(x) \in \mathbf{R}^*_-$ such that $\xi(x) = \lambda(x)x$, and for every $i = 1, \ldots, q$ and $x \in \partial D_i$ there exists $\lambda(x) \in \mathbf{R}^*_+$ such that $\xi(x) = \lambda(x)(x - x_i)$, where x_i is the center of D_i.

Proof. We work as in the proof of 6.3.7 and get

$$\int_{\partial E} (\eta^*\sigma)|_{\partial E} = \int_E d(\eta^*\sigma) = 0.$$

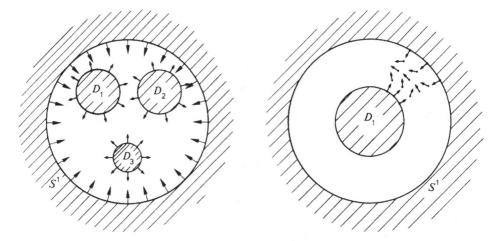

Figure 6.3.8

We can partition ∂E as $S^1 \cup \bigcup_{i=1}^q \partial D_i$, so that

$$\int_{\partial E} (\eta^* \sigma)|_{\partial E} = \int_{S^1} \eta^* \sigma + \sum_{i=1}^q \int_{\partial D_i} \eta^* \sigma.$$

Since the map $\eta : \partial D_i \to S^1$ is an orientation-reversing diffeomorphism, we have $\int_{\partial D_i} \eta^* \sigma = - \int_{S^1} \sigma$, while on the other hand $\int_{S_i} \eta^* \sigma = \int_{S^1} \sigma > 0$. Thus we obtain $0 = (1 - q) \int_{S^1} \sigma$, which is absurd if $q \neq 1$. $\square$

6.3.9. Remarks. For $q = 1$ the result is false (figure 6.3.8).

In 6.3.7 and 6.3.8 it is enough to require that $\xi \in C^0(D; \mathbf{R}^2)$ and $\xi \in C^0(E; \mathbf{R}^2)$, respectively (exercise 6.10.4). It is also enough that ξ point inward (without being normal): see 7.4.19.

6.4. Canonical Volume Forms

Recall from 0.1.15.5 that an oriented Euclidean space F has a canonical volume form λ_F. If V is an oriented submanifold of E, its tangent space $T_x V$ at $x \in V$ admits a canonical volume form $\lambda_{T_x V}$, because $T_x V$ inherits from E a Euclidean structure and an orientation. Notice that here we are implicitly identifying $T_x V \subset E$ and $T_x V$, the tangent space to the abstract manifold V, by means of θ_x (2.5.22.1 and 2.5.12).

6.4.1. Proposition. *Let E be a Euclidean vector (or affine) space and V an oriented C^p submanifold of E. There is a canonical volume form $\omega \in \Omega_{p-1}^d(V)$ on V, defined by $\omega(x) = \lambda_{T_x V}$ for all $x \in V$.*

Proof. To show that ω is of class C^{p-1}, consider positively oriented parametrizations of V, that is, pairs (U, g) with U open in $\mathbf{R}^d$ and $g \in \mathrm{Diff}(U; g(U))$. If $\{e_1, \ldots, e_d\}$ is the canonical basis of $\mathbf{R}^d$, formula 5.2.4.1 gives, for every $u \in U$:

$$(g^* \omega)(u)(e_1, \ldots, e_d) = \omega(g(u))((T_u g)(e_1), \ldots, (T_u g)(e_d)).$$

Since U and $g(U)$ are open subsets of vector spaces, each $(T_u g)(e_i)$ is equal to $\partial g / \partial x_i$, up to isomorphisms identifying the tangent spaces with the ambient vector spaces. Thus we have, by 0.1.15.6:

$$(g^* \omega)(u)(e_1, \ldots, e_d) = \omega(g(u)) \left(\frac{\partial g}{\partial x_1}, \ldots, \frac{\partial g}{\partial x_d} \right) = \sqrt{\det \left(\frac{\partial g}{\partial x_i} \, \Big| \, \frac{\partial g}{\partial x_j} \right)}.$$

Now g is of class C^p because V is, so

6.4.1.1 $$g^* \omega = \sqrt{\det \left(\frac{\partial g}{\partial x_i} \, \Big| \, \frac{\partial g}{\partial x_j} \right)} \, dx_1 \wedge \cdots \wedge dx_d \in \Omega_{p-1}^d(U).$$

It immediately follows that $\omega \in \Omega_{p-1}^d(V)$. $\square$

6.4.2. Examples

6.4.2.1. If $V \in O(\mathbf{R}^d)$, we get $\omega = \omega_0 = (dx_1 \wedge \cdots \wedge dx_d)|_V$.

6.4.2.2. Let $V = \{(x, y, f(x, y)) : (x, y) \in A\} \subset \mathbf{R}^3$, oriented so that the parametrization $g : (x, y) \mapsto (x, y, f(x, y))$ preserves orientation. If $p = \partial f/\partial x$ and $q = \partial f/\partial y$, we have

$$\frac{\partial g}{\partial x} = (1, 0, p) \qquad \text{and} \qquad \frac{\partial g}{\partial y} = (0, 1, q),$$

so the determinant of $\left(\dfrac{\partial g}{\partial x_i} \,\Big|\, \dfrac{\partial g}{\partial x_j} \right)$ is equal to

$$\begin{vmatrix} 1 + p^2 & pq \\ pq & 1 + q^2 \end{vmatrix} = 1 + p^2 + q^2.$$

Thus, for this parametrization g and for the canonical volume form ω on V, we have

6.4.2.3 $$g^*\omega = \sqrt{1 + p^2 + q^2}\; dx \wedge dy.$$

For example, the integral of a function $G(x, y, z)$ on V with respect to the measure associated with the density $|\omega|$ is given by

$$\int_A G(x, y, f(x, y)) \sqrt{1 + p^2 + q^2}\; dx\, dy.$$

It makes sense to ask whether the canonical volume form just defined, applied to $V = S^d \subset \mathbf{R}^{d+1}$, coincides with the canonical volume form on the sphere introduced in 5.3.17.2. The answer is provided by the following, more general, result (cf. figure 6.4.3):

6.4.3. Proposition and definition. *Let V be an oriented, codimension-one, C^p submanifold of a Euclidean vector space E. The canonical normal vector field to V is the unique vector field $\nu \in C^{p-1}(V; E)$ defined by the following conditions:*

(i) $\nu(x) \in \theta_x\big((T_x V)^\perp\big)$ *for all $x \in V$;*
(ii) $\|\nu(x)\| = 1$;
(iii) *if $\{e_1, \ldots, e_d\}$ is a positively oriented basis for $T_x V$, then $\{\nu(x), \theta_x(e_1), \ldots, \theta_x(e_d)\}$ is a positively oriented basis for E.*

6.4.4. Remark. If $D \subset E$ is a submanifold-with-boundary of an oriented space E and $V = \partial D$, we call ν the *pointing outward* normal vector field (figure 6.4.3 and exercise 6.10.7).

Proof. Conditions (i) and (iii) determine $\nu(x)$ up to a positive scalar, so (ii) takes care of uniqueness. There remains to show that ν is of class C^{p-1}. If ω is the canonical volume form on E and σ the canonical volume form on V, we will show that

6.4.5 $$\sigma = \operatorname{cont}(\nu)\omega.$$

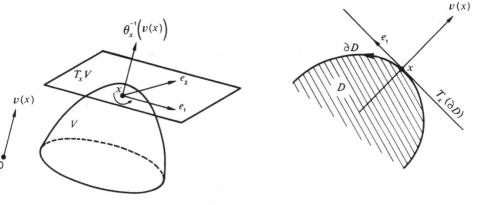

Figure 6.4.3

If this is true, ν is C^{p-1} by corollary 5.3.16, since σ is C^{p-1} and ω is C^{∞}.

In order to establish 6.4.5 it is enough to show that σ and $\mathrm{cont}(\nu)\omega$ give the same result on one basis of $T_x V$, since they are both volume forms on V (5.3.17.1). Let $\{e_2, \ldots, e_d\}$ be a positively oriented orthonormal basis for $T_x V$. We have $\sigma(e_2, \ldots, e_d) = 1$ by 0.1.15.1 and

$$\big(\mathrm{cont}(\nu)\omega\big)(e_2, \ldots, e_d) = \omega\big(\theta_x^{-1}(\nu(x)), e_2, \ldots, e_d\big),$$

where $\{\theta_x^{-1}(\nu(x)), e_2, \ldots, e_d\}$ is a positively oriented orthonormal basis for $\theta_x^{-1}(E)$. Thus

$$\sigma(e_2, \ldots, e_d) = \big(\mathrm{cont}(\nu)\omega\big)(e_2, \ldots, e_d) = 1,$$

as we wished to prove. $\qquad\square$

6.4.6. Example. In 5.3.17.2 we took $\sigma = \int(\nu)\omega_0$ on the sphere S^d; this is then the canonical form introduced in 6.4.3.

6.4.7. Computation of ν

6.4.7.1. If (U, g) is a positively oriented parametrization of a hypersurface V in a d-dimensional Euclidean space E, we have

$$\nu = \frac{\dfrac{\partial g}{\partial x_1} \wedge \cdots \wedge \dfrac{\partial g}{\partial x_{d-1}}}{\left\| \dfrac{\partial g}{\partial x_1} \wedge \cdots \wedge \dfrac{\partial g}{\partial x_{d-1}} \right\|}.$$

Here $\wedge$ denotes the exterior product in ΛE and $\|\cdot\|$ the associated norm (0.1.15.1). In dimension three $\wedge$ is just the cross product $\times$.

In particular, let $g : (x, y) \mapsto (x, y, f(x, y))$ be a parametrization for a surface V in $\mathbf{R}^3$, oriented so that g preserves orientation. Setting $p =$

$\partial f/\partial x$ and $q = \partial f/\partial y$, we obtain

$$\nu = \left(\frac{-p}{\sqrt{1 + p^2 + q^2}}, \frac{-q}{\sqrt{1 + p^2 + q^2}}, \frac{1}{\sqrt{1 + p^2 + q^2}} \right).$$

6.4.7.2. Lemma. *Let S be an oriented surface in $\mathbf{R}^3$, with canonical volume form σ and canonical normal vector field $\nu = (\nu_1, \nu_2, \nu_3)$. Denoting by x, y, z the coordinate functions on $\mathbf{R}^3$, we can write*

$$\nu_1 \sigma|_S = dy \wedge dz|_S; \quad \nu_2 \sigma|_S = dz \wedge dx|_S; \quad \nu_3 \sigma|_S = dx \wedge dy|_S.$$

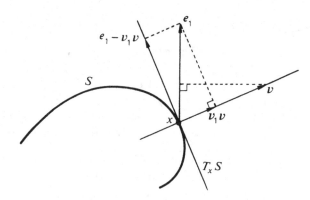

Figure 6.4.7.2

Proof. Let (e_1, e_2, e_3) be the canonical basis of $\mathbf{R}^3$ and ω_0 the canonical volume form. By 6.4.5 we have $\sigma = \text{cont}(\nu)(\omega_0)$. In addition, $\nu_1 = (e_2 \mid \nu)$, so that

$$(\nu_1 \nu - e_1 \mid \nu) = \nu_1 \|\nu\|^2 - (e_1 \mid \nu) = 0,$$

that is, $\nu_1 \nu - e_1$ is tangent to S. But $\nu_1 \sigma = \text{cont}(\nu_1 \nu)\omega_0$ and $dy \wedge dz = \text{cont}(e_1)\omega_0$, so

$$\nu_1 \sigma - dy \wedge dz|_S = \text{cont}(\nu_1 \nu - e_1)\omega_0|_S.$$

Thus for any two vectors $a, b \in TS$ we have

$$(\nu_1 \sigma - dy \wedge dz)|_S(a, b) = \omega_0(\nu_1 \nu - e_1, a, b) = 0$$

because all three vectors on the right-hand side lie in the two-dimensional space TS. $\qquad\qquad\square$

6.4.7.3. Remark. The relation $\nu_1 \sigma|_S = dy \wedge dz|_S$ is exactly the infinitesimal version of the formula that says that the orthogonal projection of a plane onto another multiplies areas by $\cos\alpha$, where α is the angle between the planes.

6.5. Volume of a Submanifold of Euclidean Space

6.5.1. Definition. Let E be a finite-dimensional Euclidean space, V an oriented, d-dimensional submanifold and σ its canonical volume form. The *volume* of V (or *area* if $d = 2$, or *length* if $d = 1$) is the (possibly infinite) integral $\int_V \sigma$, which we denote by $\mathrm{vol}(V)$ (or $\mathrm{area}(V)$, or $\mathrm{leng}(V)$).

This makes sense, because the integral is $\int_V |\sigma|$ and the function 1 is positive.

6.5.2. Remarks. If V is an open submanifold of E, $\mathrm{vol}(V)$ is finite if V is relatively compact. The converse doesn't hold.

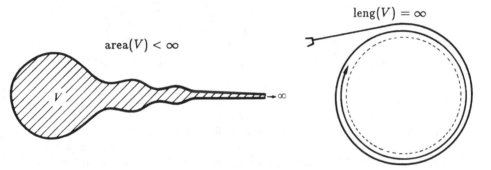

Figure 6.5.2

If $\dim V < \dim E$ we can have V relative compact and $\mathrm{vol}(V) = \infty$. Example: $E = \mathbf{R}^2$ and V asymptotically approaching a circle.

6.5.3. The case of curves. An elementary definition for the length of a curve $(]a, b[, \theta)$ is the upper bound of the lengths of inscribed polygonal curves. It can be shown [Dix68, chapter 53] that this number is equal to

$$\int_a^b \|\theta'(t)\| \, dt \in \mathbf{R}_+ \cup \{+\infty\}.$$

On the other hand, if θ is a diffeomorphism onto its image V, the length introduced in 6.5.1 is

$$\int_V \sigma = \int_{]a,b[} \theta^* \sigma,$$

where σ is the canonical volume form on V. But $\theta^* \sigma = \|\theta'(t)\|$ by 6.4.1.1, so the two notions coincide.

In fact, the elementary notion can be applied also to *arcs of curve*, which are defined on *closed* intervals $[a, b]$. (It no longer makes sense to talk about θ being a diffeomorphism.) It is clear that the length of such an arc is the same as the length of the curve obtained by omitting the endpoints.

6.5.4. Caution. For $d \geq 2$ the comments above no longer hold. Already for $d = 2$, the area of V is not the upper bound of the areas of polyhedra inscribed in V. For a counterxample, consider the Chinese lantern in the figure: given a fixed cylinder, we can inscribe in it polyhedra with arbitrarily large area. (The reader should work out the details: the trick is to increase the ratio between the number of layers and the number of sides of the polygons.)

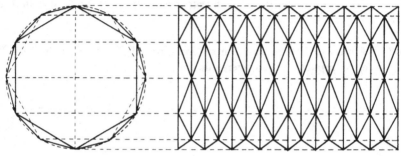

Figure 6.5.4

6.5.5. Volume of balls and spheres. It is easy to show (6.10.17) that $\mathrm{vol}\big(B_d(0, r)\big) = r^d \, \mathrm{vol}\big(B_d(0, 1)\big)$ and $\mathrm{vol}\big(S_d(r)\big) = r^{d-1} \, \mathrm{vol}\big(S_d(1)\big)$ (see 0.0.3 for notation). We also have the following result, which will follow from 6.9.13 or 6.5.9, but can be proved already using Stokes' theorem:

6.5.6. Lemma. $\mathrm{vol}\big(B_{d+1}(0, 1)\big) = \dfrac{1}{d+1} \, \mathrm{vol}(S^d)$.

Proof. Let

$$\sigma = \sum_{i=1}^{d+1} (-1)^{i-1} x_i \, dx_1 \wedge \cdots \wedge \widehat{dx_i} \wedge \cdots \wedge dx_{d+1}$$

(5.3.17.2) and $\omega_0 = dx_1 \wedge \cdots \wedge dx_{d+1}$ be the canonical volume forms on $\mathbf{R}^{d+1}$ and S^d, respectively. By Stokes' theorem, we have

$$\mathrm{vol}(S^d) = (d+1) \int_{B_{d+1}(0,1)} \omega_0 = (d+1) \, \mathrm{vol}\big(B_{d+1}(0,1)\big). \qquad \square$$

6.5.7. Theorem.

$$\mathrm{vol}(S^{2d}) = \frac{2^{d+1}\pi^d}{1 \times 3 \times \cdots \times (2d-1)} \qquad \text{and} \qquad \mathrm{vol}(S^{2d+1}) = \frac{(2\pi)^{d+1}}{d!}.$$

6.5.8. Lemma. $\mathrm{vol}(S^d) = 2^{d+1} \dfrac{\left(\displaystyle\int_0^\infty e^{-t^2} \, dt\right)^{d+1}}{\displaystyle\int_0^\infty t^d e^{-t^2} \, dt}.$

Proof. We introduce *spherical coordinates*, that is, the map $f : \mathbf{R}_+^* \times S^d \mapsto \mathbf{R}^{d+1} \setminus 0$ defined by $f(t, x) = tx$. This map is an orientation preserving diffeomorphism (immediate). We claim that it also satisfies

6.5.9 $$f^* \omega_0 = t^d \, dt \wedge \sigma,$$

where ω_0 and σ denote the canonical volume forms on $\mathbf{R}^{d+1}$ and S^d, respectively. (Actually $dt \wedge \sigma$ stands for $p^*(dt) \wedge q^*(\sigma)$, where $p : \mathbf{R}_+^* \times S^d \to \mathbf{R}_+^*$ and $q : \mathbf{R}_+^* \times S^d \to S^d$ are the canonical projections.)

To prove 6.5.9, embed S^d in $\mathbf{R}^{d+1}$ and consider the submanifold

$$\mathbf{R}_+^* \times S^d \subset \mathbf{R}_+^* \times \mathbf{R}^{d+1} \subset \mathbf{R} \times \mathbf{R}^{d+1}.$$

Thus f is the restriction to $\mathbf{R}_+^* \times S^d$ of the map $\hat{f} \in C^\infty(\mathbf{R} \times \mathbf{R}^{d+1}; \mathbf{R}^{d+1})$ given by $\hat{f}(t, x) = tx$. Since these are all vector spaces, we can work with formula 5.2.8.3. For a change, let's use coordinates $(t, x_1, \ldots, x_{d+1}) \in \mathbf{R} \times \mathbf{R}^{d+1}$. We have

$$\omega_0 = dx_1 \wedge \cdots \wedge dx_{d+1}$$

and $\hat{f}(t, x_1, \ldots, x_{d+1}) = (tx_1, \ldots, tx_{d+1})$. By 5.2.8.3 and 5.3.17.2,

$$\begin{aligned}
f^* \omega_0 &= d(tx_1) \wedge \cdots \wedge d(tx_{d+1}) \\
&= (x_1 \, dt + t \, dx_1) \wedge \cdots \wedge (x_{d+1} \, dt + t \, dx_{d+1}) \\
&= t^{d+1} \, dx_1 \wedge \cdots \wedge dx_{d+1} \\
&\quad + t^d \, dt \wedge \left(\sum_i (-1)^i x_i \, dx_1 \wedge \cdots \wedge \widehat{dx_i} \wedge \cdots \wedge dx_{d+1} \right) \\
&= t^{d+1} \omega_0 + t^d \, dt \wedge \sigma.
\end{aligned}$$

But $\omega_0|_{S^d} = 0$ by a dimension argument (5.2.5.2), which proves 6.5.9. For a proof without the use of coordinates, see 6.10.26.

Now consider the function $g : y \to e^{-\|y\|^2}$ on $\mathbf{R}^{d+1} \setminus 0$, which is C^∞. We have $(g \circ f)(t, x) = e^{-t^2}$; applying 6.1.4.4 and Fubini's theorem (0.4.5.1), we get

$$\begin{aligned}
\int_{\mathbf{R}^{d+1}\setminus 0} e^{-\|y\|^2} \omega_0 &= \int_{\mathbf{R}_+^* \times S^d} e^{-t^2} t^d \, dt \wedge \sigma \\
&= \left(\int_0^\infty e^{-t^2} t^d \, dt \right) \left(\int_{S^d} \sigma \right) = \mathrm{vol}(S^d) \int_0^\infty e^{-t^2} \, dt.
\end{aligned}$$

On the other hand, $\{0\}$ has measure zero in $\mathbf{R}^{d+1}$, so

$$\begin{aligned}
\int_{\mathbf{R}^{d+1}\setminus 0} e^{-\|y\|^2} \omega_0 &= \int_{\mathbf{R}^{d+1}} e^{-(y_1^2 + \cdots + y_{d+1}^2)} \, dy_1 \wedge \cdots \wedge dy_{d+1} \\
&= \prod_{i=1}^{d+1} \left(\int_{-\infty}^{+\infty} e^{-y_i^2} \, dy_i \right) = 2^{d+1} \left(\int_0^\infty e^{-t^2} \, dt \right)^{d+1}.
\end{aligned}$$

Setting $I_d = \int_0^\infty e^{-t^2} t^d \, dt$ for every integer $d \geq 0$, we get

$$\text{vol}(S^d) = \frac{(2I_0)^{d+1}}{I_d},$$

which proves lemma 6.5.8.   □

Proof of 6.5.7. We start by calculating I_0 and I_1. We have

$$I_1 = \int_0^\infty te^{-t^2} \, dt = \left[-\frac{1}{2} e^{-t^2} \right]_0^\infty = \frac{1}{2}.$$

On the other hand, the length of the circle is $\text{leng}(S^1) = (2I_0)^2/I_1 = 2\pi$ (exercise), so $I_0 = \sqrt{\pi}/2$.

For d arbitrary, we use recurrence. Integrating by parts,

$$I_d = \frac{1}{2} \int_0^\infty 2te^{-t^2} t^{d-1} \, dt = \frac{d-1}{2} \int_0^\infty t^{d-2} e^{-t^2} \, dt = \frac{d-1}{2} I_{d-2}.$$

This gives

$$I_{2d} = \frac{1}{2} \cdot \frac{3}{2} \cdots \frac{2d-3}{2} \cdot \frac{2d-1}{2} I_0 \quad \text{and} \quad I_{2d+1} = \frac{2}{2} \cdot \frac{4}{2} \cdots \frac{2d}{2} I_1,$$

or again

$$I_{2d} = 1 \times 3 \times \cdots \times (2d-1) \times \frac{\sqrt{\pi}}{2^{d+1}}$$

and

$$I_{2d+1} = \frac{d!}{2}.$$ □

The values of I_d are classic and often expressed in terms of the gamma function; thus the volumes of balls and spheres, too, can be expressed in terms of the gamma function.

6.5.10. Proposition. *Let V and W be oriented submanifolds of finite-dimensional Euclidean spaces E and F, respectively. We have*

$$\text{vol}(V \times W) = \text{vol}(V) \times \text{vol}(W).$$

Proof. In the notation of 6.4.1, the canonical volume form on $V \times W$ is defined by $\omega(v, w) = \lambda_{T_{(v,w)}(V \times W)}$ for every $v \in V$ and $w \in W$. But $T_{(v,w)}(V \times W)$ is canonically isomorphic to $T_v(V) \times T_w(W)$, and

$$\lambda_{T_{(v,w)}(V \times W)} = \lambda_{T_v(V)} \wedge \lambda_{T_w(W)}$$

(cf. 2.5.18 and 3.3.18). The result now follows from Fubini's theorem (0.4.5.1).   □

6.5.11. Example. The torus $T^d = \left(S^1\left(\frac{1}{\sqrt{d}}\right)\right)^d \subset \mathbf{R}^{2d}$ (2.1.6.3) has volume $\left(\frac{2\pi}{\sqrt{d}}\right)^d$.

This is the volume of one particular embedded torus. Let's look now at the abstract torus $T^d = \mathbf{R}^d/\mathbf{Z}^d$. The canonical volume form ω_0 on $\mathbf{R}^d$ is invariant under $\mathbf{Z}^d$ (5.3.10.2) and defines a canonical volume form ω on T^d. Let's define the volume of T^d as $\int_{T^d} \omega$. I claim that

$$\int_{T^d} \omega = 1.$$

In fact, let $p : \mathbf{R}^d \to \mathbf{R}^d/\mathbf{Z}^d$ be the canonical covering map. If $Q = \,]0,1[^d$ is the open unit cube, $p : Q \to p(Q)$ is an orientation-preserving diffeomorphism. Then $p^*\omega = \omega|_Q$ by 5.3.10.2, and

$$1 = \int_{\overline{Q}} \omega_0 = \int_Q \omega_0 = \int_Q p^*\omega = \int_{p(Q)} \omega$$

by 6.1.4.4. Since $p(\overline{Q}) = T^d$ and $\overline{Q} \setminus Q$ has measure zero, we get

$$\int_{p(Q)} \omega = \int_{p(\overline{Q})} \omega = \int_{T^d} \omega = \mathrm{vol}(T^d).$$

6.5.12. Centers of mass. Let D be a *compact* submanifold-with-boundary of an oriented, Euclidean, finite-dimensional vector space E. Let ω_0 be the canonical volume form on E, and denote by $|\omega_0|$ the associated density as well as the measure derived from it. If χ_D is the characteristic function of D and $f : E \to F$ is a continuous function into a finite-dimensional vector space F, we have $\chi_D f \in C^{\mathrm{int}}_{|\omega_0|}(E)$ in the sense of 0.4.7. So we set

6.5.13 $$\int_D f\omega_0 = \int_E \chi_D f|\omega_0|.$$

In particular, $\int_D \mathrm{Id}_E\, \omega_0$ makes sense because $\mathrm{Id}_E : E \to E$ is continuous.

6.5.14. Proposition and definition. *Let D be a compact submanifold-with-boundary of an affine space E. Fix an origin, an orientation and a Euclidean structure for E, and let ω_0 be the canonical volume form of the Euclidean vector space thus obtained. The point of E defined by*

$$\mathrm{cent}(D) = \frac{\int_D \mathrm{Id}_E\, \omega_0}{\int_D \omega_0}$$

does not depend on any of the choices above; we call it the center of mass of D.

Proof. If we change the Euclidean structure of the orientation of E, the canonical volume form becomes $\omega_0' = k\omega_0$, where k is a constant, and cent(D) doesn't change. If we move the origin of the vector space to $x_0 \in E$, thus obtaining a new vector space E', we have $\mathrm{Id}_E = \mathrm{Id}_{E'} + x_0$, so

$$\int_D \mathrm{Id}_E\, \omega_0 = \int_D \mathrm{Id}_{E'}\, \omega_0 + x_0 \int_D \omega_0,$$

whence

$$\frac{\int_D \mathrm{Id}_E\, \omega_0}{\int_D \omega_0} = \frac{\int_D \mathrm{Id}_{E'}\, \omega_0}{\int_D \omega_0} + x_0;$$

this says that, in the affine space, the vectors

$$\frac{\int_D \mathrm{Id}_E\, \omega_0}{\int_D \omega_0} \in E \qquad \text{and} \qquad \frac{\int_D \mathrm{Id}_{E'}\, \omega_0}{\int_D \omega_0} \in E'$$

correspond to the same point.   □

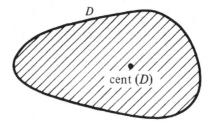

Figure 6.5.14

See 6.10.13 and 6.10.25 for examples.

6.5.14.1. Remark. This generalizes the notion of the barycenter, or average, of a finite number of points. More generally, if we have a *density* function $\lambda \in C^0(D)$, the mass of D will be $\int_D \lambda \omega_0$ and its center of mass will be the point

$$\frac{\int_D \mathrm{Id}_E\, \lambda \omega_0}{\int_D \lambda \omega_0}.$$

6.5.15. Theorem (Archimedes). *Let D be a compact submanifold-with-boundary of $\mathbf{R}^3$, ν the normal unitary vector field pointing out of D (6.4.4), $G = \mathrm{cent}(D)$ the center of mass of D and $z : m \mapsto (m \mid e_3)$ the height function (where $\{e_1, e_2, e_3\}$ is the canonical basis of $\mathbf{R}^3$). If σ denotes the canonical area form on the boundary $S = \partial D$ of D (5.3.36 and 6.4.1) and $\times$ denotes the cross product in $\mathbf{R}^3$ (0.1.17), we have*

$$\int_S z\nu\sigma = \mathrm{vol}(D)e_3 \tag{i}$$

and, for any $m \in \mathbf{R}^3$,

$$\int_{n \in S} (\overrightarrow{mv} \times z\nu)\sigma = \mathrm{vol}(D)(\overrightarrow{mG} \times e_3). \tag{ii}$$

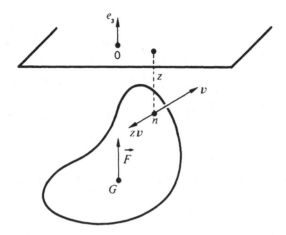

Figure 6.5.16

6.5.16. Physical interpretation. We consider D as a solid immersed in a liquid of density one whose surface coincides with the e_1e_2-plane. At each point $x \in S$ there is an element of force $zv\sigma$ acting on D, where σ is the element of area, v is the direction of the force and perpendicular to S, and z is the pressure, equal to density times depth and pointing toward the interior of D (assuming $z < 0$). Then 6.1.15 says that:

(i) the resultant force $\vec{F}$ acting on D is vertical, pointing up, and its intensity is equal to the volume of D;

(ii) the resultant torsor is equivalent to the sole force $\vec{F}$, applied at G.

Proof of 6.5.15. (i) Denoting the coordinate functions by x, y, z we have, by 6.4.7.2:

$$\int_S zv\sigma = \left(\int_S z\, dy \wedge dz,\ \int_S z\, dz \wedge dx,\ \int_S z\, dx \wedge dy \right).$$

An application of Stokes' theorem gives

$$\int_S zv\sigma = \left(\int_D 0,\ \int_D 0,\ \int_D dz \wedge dx \wedge dy \right) = (0, 0, \mathrm{vol}(D)).$$

(ii) Set $m = (a, b, c)$, $n = (x, y, z)$ and $v = (v_1, v_2, v_3)$. We have

$$\overrightarrow{mn} \times zv = z\big((y - b)v_3 - (z - c)v_2, (z - c)v_1 - (x - a)v_3, $$
$$(x - a)v_1 - (y - b)v_2 \big),$$

whence

$$\int_S (\overrightarrow{mn} \times z\nu)\sigma = \left(\int_S z\big((y-b)\, dx \wedge dy - (z-c)\, dz \wedge dx\big), \right.$$

$$\int_S z\big((z-c)\, dx \wedge dz - (x-a)\, dx \wedge dy\big),$$

$$\left. \int_S z\big((x-a)\, dz \wedge dx - (y-b)\, dy \wedge dz\big) \right).$$

By Stokes' theorem,

$$\int_S (\overrightarrow{mn} \times z\nu)\sigma = \left(\int_D (y-b)\, dx \wedge dy \wedge dz, -\int_D (x-a)\, dx \wedge dy \wedge dz, 0 \right).$$

Setting $G = (G_1, G_2, G_3)$, we get

$$\int_S (\overrightarrow{mn} \times z\nu)\sigma = \big((G_2 - b)\operatorname{vol}(D), (G_1 - a)\operatorname{vol}(D), 0\big).$$

On the other hand,

$$\overrightarrow{mG} \times \big(\operatorname{vol}(D)e_3\big) = (G_1 - a, G_2 - b, G_3 - c) \times \big(0, 0, \operatorname{vol}(D)\big)$$
$$= \big((G_2 - b)\operatorname{vol}(D), -(G_1 - a)\operatorname{vol}(D), 0\big). \qquad \square$$

6.6. Canonical Density on a Submanifold of Euclidean Space

Here we consider a possibly non-oriented submanifold V and want to define $\operatorname{vol}(V)$. Thanks to 0.1.26 (and using the μ introduced there) we have the following result:

6.6.1. Proposition. *A C^p submanifold V of a Euclidean vector space E has a canonical density δ, of class C^{p-1}, given by*

$$x \mapsto \delta(x) = \mu_{T_x V}.$$

Proof. This is like the proof of 6.4.1; we're setting $\delta(x) = |\omega(x)|$, where $\omega(x)$ is defined as in 6.4.1 by orienting $T_x V$ (and thus V, locally) either way. $\qquad \square$

6.6.2. Proposition. *If g is any parametrization of V, we have*

$$g^*\delta = \sqrt{\left| \det\left(\frac{\partial g}{\partial x_i} \,\middle|\, \frac{\partial g}{\partial x_j} \right) \right|}\, |dx_1 \wedge \cdots \wedge dx_d|.$$

Proof. This is formula 0.1.27. $\qquad \square$

6.6.3. Definition. Let V be a submanifold of a Euclidean space E. The *volume* of V, denoted by $\mathrm{vol}(V)$, is the (possibly infinite) integral $\int_V \delta$. (If $d = 1$ or 2, we talk about *length* or *area*, respectively.)

6.6.4. Remark. If V is oriented, we're back to definition 6.5.1.

6.6.5. For examples, both general and explicit, see exercises 6.10.9, 6.10.11, and 6.10.14 to 6.10.21.

6.6.6. Remark. As in 6.5.2, if V is relatively compact and has the same dimension as E, the volume of V is finite. But if $\dim V < \dim E$ this is no longer true in general. Similarly, $\mathrm{vol}(V) < \infty$ and $\dim(V) = \dim(E)$ do not imply that V is relatively compact.

6.6.7. Now let X be an abstract manifold and δ a density on X. The *volume* (or *area*, or *length*) of X, denoted by $\mathrm{vol}(X, \delta)$, is the (possibly infinite) integral $\int_X \delta$. This is the same as the total mass of the measure associated with δ.

6.6.8. Proposition. *Let X be a compact manifold and δ a density on X. Assume that a group G acts on X properly discontinuously without fixed points, and that δ is invariant under G. There exists a unique density $\underline{\delta}$ on X/G such that $p^*\underline{\delta} = \delta$, where $p : X \to X/G$ is the canonical covering map. In addition, if G is finite we have*

$$\mathrm{vol}(X, \delta) = \#G \cdot \mathrm{vol}(X/G; \underline{\delta}).$$

Proof. The existence of δ follows form the proof of lemma 5.3.9. The formula is proved like 6.1.4.8.   □

6.6.9. The isoperimetric inequality

6.6.9.1. Theorem. *Let $C \subset \mathbf{R}^n$ be a compact set whose frontier $H = \partial C$ is a codimension-one C^∞ submanifold of $\mathbf{R}^n$, and denote by $B = B_d(0, 1)$ the unit ball in $\mathbf{R}^n$. The volumes of C and ∂C satisfy the following inequality:*

$$\frac{\left(\mathrm{vol}(H)\right)^d}{\left(\mathrm{vol}(C)\right)^{d-1}} \geq \frac{\left(\mathrm{vol}(S^{d-1})\right)^d}{\left(\mathrm{vol}(B)\right)^{d-1}}.$$

In addition, equality holds if and only if C is itself a ball.

6.6.9.2. Remarks. Notice that this quotient is invariant under homothety.

In the case $d = 2$ this inequality is easy, and we include a very simple proof of it in section 9.3. The case $d \geq 3$ is much more subtle, and wasn't solved until long after the plane case (see [Ber87, section 12.11] for details). The proof given here is recent, and due to Gromov; it is the only known proof based on Stokes' theorem.

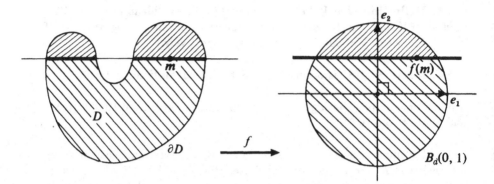

Figure 6.6.9.1

Proof. Let $\{e_i\}_{i=1,\ldots,d}$ be the canonical basis for $\mathbf{R}^d$. We can assume, without loss of generality, that $\mathrm{vol}(C)$ is the volume of the unit ball B of X. The fundamental idea is to define a map $f : C \to B$ as follows:

For $m \in C$, we let $H_1(m) = x_1^{-1}(x_1(m))$ be the hyperplane parallel to $x_1 = 0$ and containing m, and let $\hat{H}_1(m)$ be the hyperplane in the same direction that partitions B into two subsets with the same volumes as the subsets of C on each side of $H_1(m)$. In other words, $\hat{H}_1(m) = x_1^{-1}(\alpha_1)$, where α_1 is defined by the condition

$$\mathrm{vol}\big(C \cap x_1^{-1}([x_1(m), \infty[)\big) = \mathrm{vol}\big(B \cap x_1^{-1}([\alpha_1, \infty[)\big).$$

We next define two $(d-2)$-dimensional affine subspaces $H_2(m)$ and $\hat{H}_2(m)$, parallel to the intersection $x_1 = x_2 = 0$, by an analogous construction, with $C \cap H_1(m)$ instead of C and $B \cap \hat{H}_1(m)$ instead of B. We continue in this fashion until we obtain lines $H_{d-1}(m)$, $\hat{H}_{d-1}(m)$ and finally points $H_d(m) = \{m\}$ and $\hat{H}_d(m) = \{f(m)\}$; this is the definition of $f(m)$. By construction, the jacobian $Jf(m) = \left(\dfrac{\partial f_i}{\partial x_j}(m)\right)$ is of the form

$$\begin{pmatrix} \lambda_1(m) & ? & \cdots & ? \\ 0 & \lambda_2(m) & \cdots & ? \\ \vdots & \vdots & \ddots & \vdots \\ 0 & 0 & \cdots & \lambda_d(m) \end{pmatrix},$$

where entries below the diagonal are zero and entries above the diagonal don't matter to us. (The differentiability of f is problematic on the "inner folds" of the boundary, cf. proof of unicity below, but we shall ignore these problems.)

By construction and Fubini's theorem we easily verify that f preserves volume, that is,

$$\prod_{i=1}^{d} \lambda_i(m) = 1.$$

Now consider f as a vector field on C; its norm satisfies $\|f(m)\| \le 1$ since $f(m)$ belongs to the unit ball B. We apply Stokes' theorem to f, C and $H = \partial C$, as follows: we compute $\int_H \big(f(h) \mid \nu(h)\big)dh = \int_H \big(\sum_i f_i \nu_i\big)\,dh$, where dh is the canonical measure on the submanifold H of X and ν is the unit normal vector to H point outward. Lemma 6.4.7.2 extends to any dimension, yielding

$$\nu_i\, dh|_H = (-1)^{i-1}\, dx_1 \wedge \cdots \wedge \widehat{dx_i} \wedge \cdots \wedge dx_d.$$

Then Stokes' theorem gives

$$\int_H \left(\sum_i f_i \nu_i\right)\,dh = \int_C \sum_i \frac{\partial f_i}{\partial x_i},$$

where dm is the Lebesgue measure on X (and C). The scalar $\sum_i \dfrac{\partial f_i}{\partial x_i}$ is called the *divergence* of the vector field f on $\mathbf{R}^d$, and denoted by $\operatorname{div} f$. We have

$$\operatorname{div} f(m) = \sum_{i=1}^d \frac{\partial f_i}{\partial x_i}(m) = \sum_{i=1}^d \lambda_i(m) \ge d,$$

the last relation coming from the classical inequality between the arithmetic and geometric means and the fact that $\prod_{i=1}^d = 1$. Since $\|f\| \le 1$, we always have $\big|(f \mid \nu)\big| \le 1$. We finally get

$$\operatorname{vol}(B) = \operatorname{vol}(C) = \int_C dm \le d\int_C \operatorname{div} f\, dm = d\int_H \big(f(h), \nu(h)\big)\, dh$$

$$\le d\int_H dh = d\operatorname{vol}(H).$$

We conclude that $\operatorname{vol}(H) \ge d\operatorname{vol}(B) = \operatorname{vol}(S^{d-1})$, by 6.5.6. This is what we wished to show.

Suppose from now on that equality holds. This means, first, that for every point $m \in C$ the entries $\lambda_i(m)$ are equal and have the value 1. Next, $(f \mid \nu) = 1$ at every point of the boundary H of C. This precludes figures like 6.6.9.2, since the points of the thickened part of the frontier (the "inner folds") have their image in the interior of B. In particular, all lines $H_{d-1}(m)$ intersect H in only two points.

Since $\dfrac{\partial f_i}{\partial x_i} = 1$, the map f, after a translation of C if necessary, takes the form

$$f(x_1, \ldots, x_d) = \big(x_1, x_2 + a(x_1), x_3 + b(x_1, x_2), \ldots\big).$$

Using the fact that $(f|\nu) = 1$ on H, we get $f = \nu$. Now consider the section $K \cap C$ of C by an affine plane K in the direction $x_3 = x_4 = \cdots = x_d = 0$.

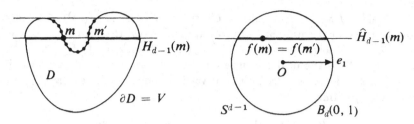

Figure 6.6.9.2

The relation above for H and the condition $f = \nu$ show that

$$\frac{2x_1}{x_1} = \frac{2\big(x_2 + a(x_1)\big)\big(1 + a'(x_1)\big)}{x_2 + a(x_1)},$$

so that $a'(x_1) = 0$. After translating to eliminate a, we see that the restriction of f to $C \cap K$ is the identity, so that the section $K \cap C$ is identical with the disc $K \cap B$. But the choice of orthonormal coordinates is arbitrary, so every section of C by an affine plane is a disc of radius ≤ 1.

Since f is surjective, there exists at least one such disc D of radius 1. Take two diametrically opposed points m_1 and m_2 on D, and consider an arbitrary affine plane containing m_1 and m_2. Since $P \cap C$ is a disc of radius ≤ 1, it must be a disc of radius 1 on which m_1 and m_2 are diametrically opposed. By varying P we conclude that C must be the ball of radius 1 centered at the midpoint of m_1 and m_2.   □

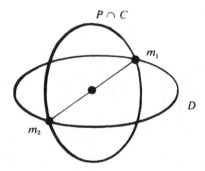

Figure 6.6.9.3

6.7. Volume of Tubes I

Let X be a d-dimensional submanifold of an n-dimensional Euclidean affine space E. To simplify matters, we assume till the end of the chapter that X is C^∞. We denote by $i : X \to E$ the canonical injection.

We recall the definitions and notation from section 2.7, applied to the case $f = i$. For $x \in X$, set

6.7.1
$$N_x X = \left(\theta(T_x X) \right)^\perp,$$

where the orthogonal complement is taken with respect to the Euclidean structure of E, and

6.7.2
$$N X = \left\{ (x, v) : x \in X, v \in N_x X \right\}.$$

NX is an n-dimensional, C^∞ submanifold of $X \times E$ (2.7.7). Set also

6.7.3
$$N_x^\varepsilon X = \left\{ v \in N_x X : \|v\| < \varepsilon \right\},$$

6.7.4
$$N^\varepsilon X = \left\{ (x, v) : x \in X, v \in N_x^\varepsilon(X) \right\} = \left\{ v \in NX : \|v\| < \varepsilon \right\}.$$

Let can $: NX \to E$ be the canonical map $(x, v) \mapsto x + v$, introduced in 2.7.5, and set

6.7.5
$$\mathrm{Tub}^\varepsilon(X) = \mathrm{can}(N^\varepsilon X).$$

We know from 2.7.12 that if X is compact there exists $\varepsilon > 0$ such that $\mathrm{can}\,|_{N^\varepsilon X}$ is a diffeomorphism onto $\mathrm{Tub}^\varepsilon X$.

Introduce also

6.7.6
$$NU_x X = \left\{ (x, v) \in NX; \|v\| = 1 \right\}, \quad NUX = \bigcup_{x \in X} NU_x X.$$

Recall that NUX is called the unitary normal fibre bundle (2.7.4) and is a C^∞ submanifold of NX (2.7.7).

Let $p : NX \to X$ be the canonical projection

6.7.7
$$p : NX \ni (x, v) \mapsto x \in X.$$

We finally set

6.7.8
$$T_v(N_x X) = \left(T_{(x,v)} p \right)^{-1}(0) \subset T_{(x,v)}(NX).$$

The equality is due to the fact that $N_x X = p^{-1}(x) \subset NX$ is a submanifold of NX, which in turn follows from the result, to be demonstrated in 6.7.12, that p is a submersion.

Since X is a submanifold of a Euclidean space E, we can consider the canonical density δ on X (6.6.1).

6.7.9. Lemma. *There exists on the normal bundle NX a C^∞ canonical density Δ, characterized by the condition that, for any orthonormal basis $\{\eta_{d+1}, \ldots, \eta_n\}$ of $T_v(N_x X)$ and any $\lambda_1, \ldots, \lambda_d \in T_{(x,v)} NX$, we have*

$$\Delta(\lambda_1, \ldots, \lambda_d, \eta_{d+1}, \ldots, \eta_n) = \delta\left(T_{(x,v)} p(\lambda_1), \ldots, T_{(x,v)} p(\lambda_d) \right).$$

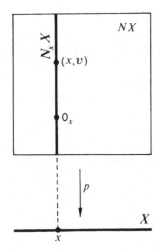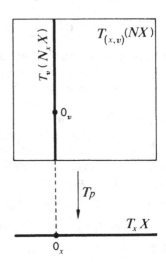

Figure 6.7.7

6.7.10. Caution. In general, Δ is not the canonical density on NX as a submanifold of $E \times E$.

Proof. We first must show that the definition of Δ makes sense algebraically, then that Δ is C^∞. To simplify the notation somewhat, we will often ignore the isomorphisms θ, θ^{-1}, etc.

To see that Δ is well-defined, let $(x, v) \in NX$ be a fixed point. Since $T_v(N_xX)$ is a subspace of $T_{(x,v)}NX$, we have an associated orthogonal projection $q : T_{(x,v)}NX \to T_v(N_xX)$. Choose arbitrary orientations for T_xX and N_xX (hence also for $T_v(N_xX)$). Next choose volume forms α and β compatible with these orientations: for α take the canonical volume form on T_xX (the one that assigns the value 1 to any positively oriented orthonormal basis) and for β the form corresponding to the canonical density δ on X (6.6.1).

Set $\Delta(x, v) = |p^*\beta \wedge q^*\alpha|$. We have

$$(p^*\beta \wedge q^*\alpha)(\lambda_1, \ldots, \lambda_d, \eta_{d+1}, \ldots, \eta_n) = p^*\beta(\lambda_1, \ldots, \lambda_d)q^*\alpha(\eta_{d+1}, \ldots, \eta_n)$$

because $\eta_i \in T_v(N_xX) = (T_{(x,v)}p^{-1})(0)$ (cf. 0.1.2), so that

$$p^*\beta(\lambda_1, \ldots \eta_i, \ldots, \lambda_d) = \beta(\ldots, T_{(x,v)}p(\eta_i), \ldots) = \beta(\ldots, 0, \ldots) = 0.$$

Now the basis $\{\eta_{d+1}, \ldots, \eta_n\}$ of $T_v(N_xX)$ is orthonormal, so

$$\alpha(T_{(x,v)}q(\eta_{d+1}), \ldots, T_{(x,v)}q(\eta_n)) = \pm 1,$$

and, since β corresponds to the density δ, we get

$$\Delta(x, v)(\lambda_1, \ldots, \lambda_d, \eta_{d+1}, \ldots, \eta_n) = \delta(T_{(x,v)}p(\lambda_1), \ldots, T_{(x,v)}p(\lambda_d)).$$

This formula characterizes Δ because, as we shall soon see, p is a submersion, so there are exists a basis $\{\lambda_1, \ldots, \lambda_d, \eta_{d+1}, \ldots, \eta_n\}$ of $T_{(x,v)}NX$ such that $\{T_{(x,v)}p(\lambda_1), \ldots, T_{(x,v)}p(\lambda_d)\}$ is a basis for $T_x X$.

We now have to show that Δ is C^∞. By 3.3.5 and 3.3.6, it is enough to prove that $(\phi^{-1})^* \Delta$ is C^∞ for all charts ϕ in one given atlas on NX. We consider parametrizations of the following type:

Let (U, h) be a parametrization of X at x (2.1.2), that is, $U \in O_0(\mathbf{R}^d)$, $h \in \mathrm{Diff}(U; h(U))$ and $h(0) = x$. The vectors $\partial h / \partial u_i(0)$ $(1 \le i \le d)$ are linearly independent, so we can choose a basis for E of the form

$$\left\{ \frac{\partial h}{\partial u_1}(0), \ldots, \frac{\partial h}{\partial u_d}(0), e_{d+1}, \ldots, e_n \right\}$$

(actually we have $\partial h / \partial u_i(0) \in T_x X$, but we're identifying these vectors with their images in E under θ).

Shrinking U if necessary, we can assume that

$$\left\{ \frac{\partial h}{\partial u_1}(u), \ldots, \frac{\partial h}{\partial u_d}(u), e_{d+1}, \ldots, e_n \right\}$$

is still a basis for E for every $u \in U$, thanks to the continuity of partial derivatives. Next we apply Gram-Schmidt to this basis, getting an orthonormal basis

$$\left\{ \xi_1(u), \ldots, \xi_d(u), \nu_{d+1}(u), \ldots, \nu_n(u) \right\}$$

for every $u \in U$. This basis varies C^∞ with u, because Gram-Schmidt only involves C^∞ operations; in addition, its first d elements $\xi_1(u), \ldots, \xi_d(u)$ form a basis for $\theta(T_{h(u)}X)$ (since they are linear combinations of the $\partial h / \partial i(u)$), so the last d elements $\nu_1(u), \ldots, \nu_d(u)$ form an orthonormal basis for $\big(\theta(T_{h(u)}X)\big)^\perp = N_{h(u)}X$.

Now to define the parametrization. Since NX has dimension n, we consider the open set $U \times \mathbf{R}^{n-d} \subset \mathbf{R}^n$ and define $H \in C^\infty(U \times \mathbf{R}^{n-d}; NX)$ by

6.7.11
$$H(u, t_{d+1}, \ldots, t_n) = \left(h(u), \sum_{j=d+1}^{n} t_j \nu_j(u) \right).$$

In the sequel we will write $t = (t_{d+1}, \ldots, t_n)$. We have $h(u) = x \in X$ and, by the discussion above, $\sum_{j=d+1}^{n} t_j \nu_j(u) \in N_{h(u)}X$. Thus $H(u, t) \in NX$. Since $\sum_{j=d+1}^{n} t_j \nu_j(u)$ ranges over $N_{h(u)}X$ when t ranges over $\mathbf{R}^{n-d}$, we have

$$H(U \times \mathbf{R}^{n-d}) = \bigcup_{u \in U} \big(\{h(u)\} \times N_{h(u)}X \big).$$

Let us set, as in 6.7.3,

$$N\big(h(U)\big) = \big\{ (x, v) : x \in h(U), v \in N_x X \big\}.$$

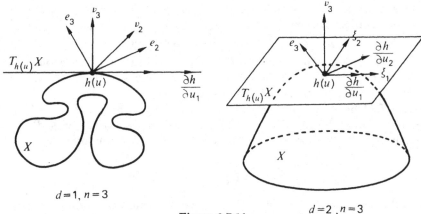

$d=1,\ n=3$

Figure 6.7.11

$d=2\ ,\ n=3$

Then $H(U \times \mathbf{R}^{n-d}) = N(h(U))$, an open subset of NX. In addition H is injective (because h is, as well as the map $t \mapsto \sum_{j=d+1}^{n} t_j \nu_j(u)$ for fixed u), so in fact $H \in C^{\infty}(U \times \mathbf{R}^{n-d}; N(h(U)))$ is a bijection. There remains to show that it is an immersion. For that, since $N(h(U)) \subset NX \subset E \times E$, we can consider H as a map from $U \times \mathbf{R}^{n-d}$ into $E \times E$, and calculate the rank of its jacobian H'.

First, $h : U \times X \subset E$ is a map with values in E; call its coordinate functions $h_1, \ldots, h_n$. Similarly, since $\nu_j(U) \in N_{h(u)}X \subset E$, we call its coordinates $\nu_j^{(k)}(u)$ $(k = 1, \ldots, n)$. Then

$$H(u,t) = \left(h_1(u), \ldots, h_n(u), \sum_{j=d+1}^{n} t_j \nu_j^{(1)}(u), \ldots, \sum_{j=d+1}^{n} t_j \nu_j^{(n)}(u) \right)$$

(considered as a point in $E \times E$), where we have written $u = (u_1, \ldots, u_d)$ and $t = (t_{d+1}, \ldots, t_n)$. The jacobian is then

$$\begin{pmatrix}
\dfrac{\partial h_1}{\partial u_1} & \cdots & \dfrac{\partial h_1}{\partial u_d} & 0 & \cdots & 0 \\[2mm]
\vdots & \ddots & \vdots & & & \\[2mm]
\dfrac{\partial h_n}{\partial u_1} & \cdots & \dfrac{\partial h_n}{\partial u_d} & 0 & \cdots & 0 \\[2mm]
\displaystyle\sum_{j=d+1}^{n} t_j \dfrac{\partial \nu_j^{(1)}(u)}{\partial u_1} & \cdots & \displaystyle\sum_{j=d+1}^{n} t_j \dfrac{\partial \nu_j^{(1)}(u)}{\partial u_d} & \nu_{d+1}^{(1)}(u) & \cdots & \nu_n^{(1)}(u) \\[2mm]
\vdots & \ddots & \vdots & & & \\[2mm]
\displaystyle\sum_{j=d+1}^{n} t_j \dfrac{\partial \nu_j^{(n)}(u)}{\partial u_1} & \cdots & \displaystyle\sum_{j=d+1}^{n} t_j \dfrac{\partial \nu_j^{(n)}(u)}{\partial u_d} & \nu_{d+1}^{(n)}(u) & \cdots & \nu_n^{(n)}(u)
\end{pmatrix},$$

where the first n columns are the derivatives with respect to $u_1, \ldots, u_d$ and the last $n - d$ columns are the derivatives with respect to $t_{d+1}, \ldots, t_n$.

This matrix can be more simply written as

$$H' = \begin{pmatrix} \dfrac{\partial h}{\partial u_1} & \cdots & \dfrac{\partial h}{\partial u_d} & 0 & \cdots & 0 \\[2mm] \sum_j t_j \dfrac{\partial \nu_j}{\partial u_1} & \cdots & \sum_j t_j \dfrac{\partial \nu_j}{\partial u_d} & \nu_{d+1} & \cdots & \nu_n \end{pmatrix}.$$

In addition we set $(h') = \left(\dfrac{\partial h}{\partial u_1} \cdots \dfrac{\partial h}{\partial u_d} \right)$ and $(\nu) = (\nu_{d+1} \cdots \nu_n)$.

Now (h') has rank d because h is a parametrization of X, and (ν) has rank $n - d$ because $\{\nu_{d+1}(u), \ldots, \nu_n(u)\}$ is a basis for $N_{h(u)}X$. Thus H' has maximal rank n, and we have shown that H is a C^∞ parametrization of NX.

6.7.12. Remark. From this we see that $p : NX \to X$ is a submersion, as follows: $h^{-1} \circ p \circ H$ takes (u, t) into u, so it has rank $d = \dim X$. On the other hand, H and h are regular, so rank $p \geq \mathrm{rank}(h^{-1} \circ p \circ H) = d$. Since p has values in X, which is d-dimensional, we conclude that p has maximal rank.

To conclude the proof of lemma 6.7.9, we still have to show that, for the charts inverse to the parametrizations defined in 6.7.11, condition 3.3.5 is satisfied. Let $\{e_1, \ldots, e_d\}$ and $\{f_{d+1}, \ldots, f_n\}$ be the canonical bases of $\mathbf{R}^d$ and $\mathbf{R}^{n-d}$, respectively, and $\delta_1 = |du_1 \wedge \cdots \wedge du_d|$ and $\Delta_1 = |du_1 \wedge \cdots \wedge du_d \wedge dt_{d+1} \wedge \cdots \wedge dt_n|$ the canonical densities on $\mathbf{R}^d$ and $\mathbf{R}^n$.

There exist scalars $z(u)$ and $Z(u, t)$ such that

6.7.13 $\qquad H^*\Delta = Z(u, t)\Delta_1 \qquad$ and $\qquad h^*\delta = z(u)\delta_1.$

To prove condition 3.3.5, we must show that $Z(u, t)$ is C^∞. We will show that

6.7.14 $\qquad Z(u, t) = z(u) = \delta \left(\dfrac{\partial h}{\partial u_1}(u), \ldots, \dfrac{\partial h}{\partial u_d}(u) \right)$

(here the $\partial h / \partial u_j(u)$ are regarded as elements of $T_{h(u)}X$, and δ is the canonical density on the submanifold X of the Euclidean space E). In fact, notice that

$$(H^*\Delta)\big(h(u), t\big)(e_1, \ldots, e_d, f_{d+1}, \ldots, f_n)$$
$$= \Delta\big(h(u), t\big)\big(T_{(u,t)}H(e_1), \ldots T_{(u,t)}H(e_d), T_{(u,t)}H(f_{d+1}), \ldots T_{(u,t)}H(f_n)\big).$$

In view of the form of H', we have

$$T_{(u,t)}H(f_j) = \theta^{-1}_{(h(u),t)}(\nu_j) = \eta_j \in T_v(N_{h(u)}X)$$

for $j = d+1, \ldots, n$, and $\{\eta_{d+1}, \ldots, \eta_n\}$ is an orthonormal basis for the space $T_v(N_{h(u)}X)$ by the choice of the $\nu_j(u)$. It follows from the construction of

Δ that

$$(H^*\Delta)(h(u), t)(e_1, \ldots, e_d, f_{d+1}, \ldots, f_n)$$
$$= \delta\big(T_u(p \circ H)(e_1), \ldots, T_u(p \circ H)(e_d)\big)$$
$$= \delta\left(\frac{\partial h}{\partial u_1}(u), \ldots, \frac{\partial h}{\partial u_d}(u)\right),$$

again because of the form of H'.

Now h is a C^∞ parametrization, so

$$Z(u, t) = \delta\left(\frac{\partial h}{\partial u_1}(u), \ldots, \frac{\partial h}{\partial u_d}(u)\right)$$

is C^∞ because δ is a C^∞ density on X (6.6.1). This completes the proof of lemma 6.7.9. $\qquad\square$

6.7.15. Lemma. *Denote by ε_x the canonical density on the subspace $N_x X$ of the Euclidean space E, and identify $N_x X$ with $\{x\} \times N_x X$. If $f \in C_\Delta^{\text{int}}(NX)$ we have*

$$f|_{N_x X} \in C_{\varepsilon_x}^{\text{int}}(N_x X)$$

for δ-almost every $x \in X$, and the function

$$x \mapsto \int_{N_x X} f|_{N_x X} \varepsilon_x$$

is δ-integrable. In addition we have

6.7.16 $$\int_{NX} f\Delta = \int_X \left(\int_{N_x X} f\varepsilon_x\right)\delta.$$

Proof. Using partitions of unity we can reduce to the case of $f \in C_\Delta^{\text{int}}(NX)$ with support in $H(U)$, where (U, H) is a parametrization of the type introduced in 6.7.11. In this case 3.3.16 gives, in view of the fact that H is a diffeomorphism between $H(U)$ and $U \times \mathbf{R}^{n-d}$:

$$\int_{NX} f\Delta = \int_{H(U)} f\Delta = \int_{U \times \mathbf{R}^{n-d}} (f \circ H)H^*\Delta = \int_{U \times \mathbf{R}^{n-d}} (f \circ H)Z(u, t)\Delta_1,$$

where Δ_1 is the canonical density on $\mathbf{R}^n$ (cf. 6.7.13).

By 6.7.13 and 6.7.14 we have, keeping the same notation:

$$Z(u, t)\Delta_1 = z(u)\,|du_1 \wedge \cdots \wedge du_d \wedge dt_{d+1} \wedge \cdots \wedge dt_n|$$
$$= z(u)\delta_1\,|dt_{d+1} \wedge \cdots \wedge dt_n|,$$

whence

$$\int_{NX} f\Delta = \int_{U \times \mathbf{R}^{n-d}} (f \circ H)h^*\delta(e_1, \ldots, e_d)\,|dt_{d+1} \wedge \cdots \wedge dt_n|$$
$$= \int_U \left(\int_{\{u\} \times \mathbf{R}^{n-d}} (f \circ H)\,|dt_{d+1} \cdots \wedge \cdots \wedge dt_n|\right)h^*\delta$$

by Fubini's theorem (0.4.5.1); here

$$u \mapsto \int_{\{u\} \times \mathbf{R}^{n-d}} (f \circ H) |dt_{d+1} \wedge \cdots \wedge dt_n|$$

is defined almost everywhere, and is $(h^*\delta)$-integrable.

Given $u \in U$, the restriction $H|_{\{u\} \times \mathbf{R}^{n-d}}$ sends $\{u\} \times \mathbf{R}^{n-d}$ onto $\{x\} \times N_x X$, where $x = h(u)$, and transforms ε_x into $dt_{d+1} \wedge \cdots \wedge dt_n$. Thus

$$\int_{\{u\} \times \mathbf{R}^{n-d}} (f \circ H)|dt_{d+1} \cdots \wedge \cdots dt_n| = \int_{\{u\} \times \mathbf{R}^{n-d}} (f \circ H)(H|_{\{u\} \times \mathbf{R}^{n-d}})^* \varepsilon_x$$

$$= \int_{\{x\} \times N_x X} f \varepsilon_x,$$

which proves the existence of this integral for almost every x. It also proves that the function

$$x \mapsto \int_{\{x\} \times N_x X} f \varepsilon_x$$

is $(h^*\delta)$-integrable. We thus get

$$\int_{NX} f \Delta = \int_U \left(\int_{\{h(u)\} \times N_{h(u)} X} f \varepsilon_{h(u)} \right) h^* \delta.$$

Since (U, h) is a parametrization of X and f is supported in $H(U)$, the function

$$x \mapsto \int_{\{x\} \times N_x X} f \varepsilon_x$$

is supported in U, and the previous equality can also be written

$$\int_{NX} f \Delta = \int_X \left(\int_{N_x X} f \varepsilon_x \right) \delta. \qquad \square$$

6.7.17. The unitary normal bundle. NUX also has a canonical density Ψ. In fact, $NU_x X$ (6.7.6) is the unit sphere in Euclidean space for every $x \in X$, and as such it has a canonical density τ_x. Then Ψ is characterized by the condition

$$\Psi(\lambda_1, \ldots, \lambda_d, \nu_{d+1}, \ldots, \nu_n) = \delta\big(T_{(x,v)} p(\lambda_1), \ldots, T_{(x,v)} p(\lambda_d)\big) \tau_x(\nu_{d+1}, \ldots, \nu_n)$$

for any $v \in NU_x X$, any $\lambda_1, \ldots, \lambda_d \in T_{(x,v)} NX$ and any $\nu_{d+1}, \ldots, \nu_n \in T_v(NU_x X)$. We can write this condition more compactly as

6.7.18 $$\Psi = p^* \delta \wedge \tau.$$

The reader can demonstrate the following lemma, along the same lines as 6.7.15:

6.7.19. Lemma. *For every function $f \in C_\Psi^{\mathrm{int}}(NUX)$ we have*

$$\int_{NUX} f \Psi = \int_X \left(\int_{\{x\} \times NU_x X} f \tau_x \right) \delta. \qquad \square$$

6.7.20. Remark. Formulas 6.7.16 and 6.7.19 are particular cases of a general formula for integrals on a submersion. For the orientable case, see [Die69, vol. III, 16.24.8.1].

6.7.21. Orientation

6.7.22. Lemma. *If E is oriented, NX has a canonical orientation.*

Proof. Orient the vector spaces T_xX and N_xX as in 6.7.10.1, but subject to the condition that the induced orientation on $E = \theta_x(T_xX) \oplus N_xX$ is the given orientation. Define a local volume form

6.7.23 $$\Xi = p^*\beta \wedge q^*\alpha$$

on NX. If we replace β by $-\beta$, we also have to replace α by $-\alpha$, so that the orientation on E remains unchanged; thus the definition of Ξ does not change. □

The form Ξ is called the canonical volume form on NX, for the given orientation of E.

6.7.24. Corollary. *NX is orientable (whether or not E is).* □

6.7.25. Remarks

6.7.25.1. If we switch the orientation of E, the canonical form Ξ changes sign.

6.7.25.2. A parametrization H of NX is positively oriented if

$$\left\{ \frac{\partial h}{\partial u_1}(u), \dots \frac{\partial h}{\partial u_d}(u), \nu_{d+1}, \dots, \nu_n \right\}$$

is a positively oriented basis for E. (Here $\partial h/\partial u_j(u)$ is considered as an element of E, via $\theta_{h(u)}$).

6.7.26. Orientability of NUX. Since NUX is the boundary of the sub-manifold-with-boundary N^1X and NX is oriented, NUX is oriented by 5.3.36. Let $\textcircled{H}$ be the canonical volume form on NUX associated with this orientation. If we denote by ς the vector field pointing out of NUX (6.4.4), defined by

6.7.27 $$\varsigma(x, v) = \theta_v^{-1}(v) \in T_v(N_xX) \subset T_{(x,v)}(NX),$$

we have, by 6.4.5:

6.7.28 $$\textcircled{H} = \text{cont}(\varsigma)\Xi.$$

6.8. Volume of Tubes II

We know by 2.7.12 that if X is compact the map can : $N^\epsilon X \to \text{Tub}^\epsilon X$ defined by $\text{can}(x, v) = x + v$ is a diffeomorphism for small enough ϵ. In this case $\text{Tub}^\epsilon X$ is relatively compact in the Euclidean space E; which means that its volume with respect to the canonical density Δ_0 on E is finite. We now want to compute

6.8.1
$$\text{vol}(\text{Tub}^\epsilon X) = \int_{\text{Tub}^\epsilon X} \Delta_0 = \int_{N^\epsilon X} \text{can}^* \Delta_0.$$

Denote by Δ the canonical density on the normal bundle (6.7.9). We must write down the function $G \in C^\infty(NX)$ such that

6.8.2
$$\text{can}^* \Delta_0 = G\Delta.$$

Consider a parametrization $(H, U \times \mathbf{R}^{n-d})$ of the form introduced in 6.7.11. Denoting again by Δ_1 the canonical measure on $\mathbf{R}^n$, we see that there exists a function $S \in C^\infty(U \times \mathbf{R}^{n-d})$ such that

6.8.3
$$(\text{can} \circ H)^* \Delta_0 = S\Delta_1.$$

Thus we have

$$S\Delta_1 = H^*(\text{can}^* \Delta_0) = H^*(G\Delta) = (G \circ H)(H^*\Delta).$$

If Z denotes the function such that $H^*\Delta = Z\Delta_1$ (cf. 6.7.13), we conclude that

6.8.4
$$G \circ H = \frac{S}{Z}.$$

Here we know Z (6.7.14) and must compute S. If $\{e_1, \ldots, e_d, f_{d+1}, \ldots, f_n\}$ is the canonical basis on $\mathbf{R}^n$ we have $\Delta_1(e_1, \ldots, e_d, f_{d+1}, \ldots, f_n) = 1$, so

$$S(u, t) = \big((\text{can} \circ H(u, t))^* \Delta_0\big)(e_1, \ldots, e_d, f_{d+1}, \ldots, f_n)$$

for every $u \in U$ and $t \in \mathbf{R}^{n-d}$. But then we get from 6.7.11 and the definition of can:

$$(\text{can} \circ H)(u, t) = h(u) + \sum_{j=d+1}^{n} t_j \nu_j(u).$$

A calculation analogous to that of 6.7.14, involving the jacobian H', gives:

$$S(u, t) = \big(\text{can} \circ H(u, t)\big)^* \Delta_0(e_1, \ldots, e_d, f_{d+1}, \ldots, f_n)$$
$$= \Delta_0 \bigg(\frac{\partial h}{\partial u_1} + \sum_{j=d+1}^{n} t_j \frac{\partial \nu_j}{\partial u_1}, \ldots, \frac{\partial h}{\partial u_d} + \sum_{j=d+1}^{n} t_j \frac{\partial \nu_j}{\partial u_d}, \nu_{d+1}, \ldots, \nu_d \bigg).$$

(Since $\{\nu_{d+1}, \ldots, \nu_n\}$ is an orthonormal basis for $N_x X = \big(\theta(T_{h(u)} X)^\perp\big)$, we are actually identifying each $\nu_k \in E$ with its image in the tangent space under the isomorphism θ.) If we denote by

6.8.5
$$\cdot^T : e \mapsto e^T$$

the orthogonal projection from $E = \theta(T_x X) \oplus N_x X$ onto $\theta(T_x X)$, it follows from the definition of Δ (6.7.9) that

6.8.6

$$S(u,t) = \delta\left(\frac{\partial h}{\partial u_1}(u) + \sum_{j=d+1}^{n} t_j\left(\frac{\partial \nu_j}{\partial u_1}\right)^T, \ldots, \frac{\partial h}{\partial u_d}(u) + \sum_{j=d+1}^{n} t_j\left(\frac{\partial \nu_j}{\partial u_d}\right)^T\right).$$

6.8.7. Remark. As we pointed out in 6.7.10, some of these equalities involve identifying vectors in E, in its tangent space, and in the tangent spaces to NX and to X. For instance, in 6.8.6 the vectors

$$\frac{\partial h}{\partial u_k} + \sum_{j=d+1}^{n} t_j\left(\frac{\partial \nu_j}{\partial u_k}\right)$$

are actually in $T_{h(u)}X$, so they should be written

$$\frac{\partial h}{\partial u_k} + \sum_{j=d+1}^{n} t_j\theta^{-1}\left(\theta_{i(x)}\left(\frac{\partial \nu_j}{\partial u_k}\right)\right)^T$$

in the notation of 6.7.1. You can see why we allow ourselves some abuse in notation!

Putting together the calculations above, we get the following result:

6.8.8. Proposition. *If* h, H *and* ν_j $(d+1 \leq j \leq n)$ *are as in the proof of lemma 6.7.9 and* (u,t) *is a point in* $U \times \mathbf{R}^{n-d}$, *with* $x = h(u)$ *and* $v = \sum_{j=d+1}^{n} t_j\nu_j(u)$, *we have*

6.8.9

$$G(x,v) = \frac{\delta\left(\dfrac{\partial h}{\partial u_1}(u) + \sum_{j=d+1}^{n} t_j\left(\dfrac{\partial \nu_j}{\partial u_1}\right)^T, \ldots, \dfrac{\partial h}{\partial u_d}(u) + \sum_{j=d+1}^{n} t_j\left(\dfrac{\partial \nu_j}{\partial u_d}\right)^T\right)}{\delta\left(\dfrac{\partial h}{\partial u_1}(u), \ldots, \dfrac{\partial h}{\partial u_d}(u)\right).}$$

$\square$

6.8.10. Remark. In particular, for $t = 0$ and $v = 0$ we get $G(x,0) = 1$. We have proved theorem 2.7.10 again: for every $x \in X$, $T_{(x,0)}(\text{can})$ has maximal rank.

6.8.11. Corollary. *There exist functions* $W_i \in C^\infty(NX)$ $(i = 0, \ldots, d+1)$, *with* $W_0 = 1$, *such that*

$$G = \left|\sum_{i=0}^{d} W_i\right|$$

and, for every $x \in X$, *the map* $v \mapsto W_i(x,v)$ *is a homogeneous polynomial of degree* i *on the vector space* $N_x X$.

For an intrinsic definition of homogeneous polynomials on vector spaces, see [Car71, p. 80]. A naïve definition is that $W_i(x, v)$ has degree i in the coordinates of v expressed in any basis.

Proof. Let β be a local canonical volume from on X; thus $\delta = |\beta|$. Expanding the numerator of 6.8.9 and ordering the terms according to their degree in the t_j, and using the fact that β is a d-linear form, we get

6.8.12

$$\frac{\beta\left(\frac{\partial h}{\partial u_1}(u) + \sum\limits_{j=d+1}^{n} t_j \left(\frac{\partial \nu_j}{\partial u_1}\right)^T, \ldots, \frac{\partial h}{\partial u_d}(u) + \sum\limits_{j=d+1}^{n} t_j \left(\frac{\partial \nu_j}{\partial u_d}\right)^T\right)}{\beta\left(\frac{\partial h}{\partial u_1}(u), \ldots, \frac{\partial h}{\partial u_d}(u)\right)} = \sum_{i=0}^{d} W_i,$$

where W_i denotes the sum of terms of degree i with respect to the t_j. We evidently have $W_0 = 1$ and

$$G = \left|\sum_{i=0}^{d} W_i\right|,$$

and the W_i are homogeneous polynomials with respect to the coordinates t_j of $v \in N_x X$. $\square$

So far the W_i may depend on the parametrization, or even on β. Clearly they don't change if we replace β by $-\beta$. There remains to see that they don't depend on the parametrization, only on $(x, v) \in NX$.

We know (6.8.2) that $G(x, v)$ depends only on (x, v). We also have $G(x, v) > 0$ for $\|v\|$ small, because $G(x, 0) = 1$ and G is continuous; in this case the homogeneous components of the polynomial

$$G(x, v) = \sum_{i=0}^{d} W_i(x, v)$$

are determined by the polynomial [Car71, p. 85, corollary 6.3.2]. But then the components are determined for every v, and they only depend on (x, v). Finally, 6.8.12 shows that they are C^∞ functions. (To learn how the W_i are deduced from G, see the proof of theorem 6.3.1 in [Car71].)

To wrap up the preliminaries to the calculation of the volume of $\text{Tub}^\varepsilon(X)$, we will present a fancy interpretation for the last term W_d of G. The idea is to introduce the unit sphere $S(E) = \{z \in E : \|z\| = 1\}$ and the Gauss map $\gamma \in C^\infty(NU X; S(E))$ defined by

6.8.13 $\gamma(x, v) = v.$

If E is oriented, the sphere $S(E)$ has a canonical orientation as the boundary of $B_E(0, 1)$. Thus there is a canonical volume form

6.8.14 $\Sigma = \text{cont}(\nu)\Omega_0$

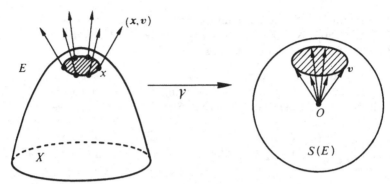

Figure 6.8.13

on $S(E)$, where Ω_0 is the canonical volume form on E (for the given orientation) and ν is the outward pointing unit vector field $\nu(z) = \theta_z^{-1}(z)$ (6.4.5). We have the following result:

6.8.15. Lemma. *If $\circledB$ is the canonical volume form on NUX (defined in 6.7.28), we have $\gamma^*(\Sigma) = W_d\circledB$, for either choice of orientation for E.*

Proof. Consider one point $(x, v) \in NUX$, with $x = h(u)$. Since $\|\theta(v)\| = 1$, we can assume that the basis $\{\nu_{d+1}(u), \ldots, \nu_n(u)\}$ of N_xX was chosen so that $v = \nu_{d+1}(u)$. If H is as in 6.7.11 we have $(x, v) = H(u, t)$ with $t = (1, 0, \ldots, 0)$.

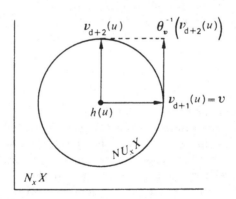

Figure 6.8.15

We assume also that H is positively oriented with respect to the given orientation of E, that is, that

$$\left\{ \theta\left(\frac{\partial h}{\partial u_i}(u) \right) \right\}_{i=1,\ldots,d} \cup \{\nu_j(u)\}_{j=d+1,\ldots,n}$$

is a positively oriented basis for E. Let $\frac{\partial h}{\partial u_i}(u)$ ($i = 1, \ldots, d$) and $\nu_j(u)$
($j = d+2, \ldots, n$) denote also the vectors in $T_{(x,v)}(NUX)$ corresponding
to the elements of the basis above. We know that

$$\left\{ \frac{\partial h}{\partial u_i}(u) \right\}_{i=1,\ldots,d} \cup \{\nu_j(u)\}_{j=d+2,\ldots,n}$$

is a basis for $T_{(x,v)}(NUX)$. Its orientation with respect to ℬ is $(-1)^d$,
because by 6.7.28 we're contracting with $\theta_v^{-1}(v) = \nu_{d+1}(u)$ (notice the
abuse in notation), and to drag this vector to the first slot takes d sign
flips. Thus we have

$$(\gamma^*\Sigma)(x,v) \left(\frac{\partial h}{\partial u_1}(u), \ldots, \frac{\partial h}{\partial u_d}(u), \nu_{d+2}(u), \ldots, \nu_n(u) \right)$$

$$= \Sigma(\gamma(x,v)) \left((T_{(x,v)}\gamma) \left(\frac{\partial h}{\partial u_1}(u) \right), \ldots, (T_{(x,v)}\gamma) \left(\frac{\partial h}{\partial u_d}(u) \right), \right.$$

$$\left. (T_{(x,v)}\gamma)(\nu_{d+1}(u)), \ldots, (T_{(x,v)}\gamma)(\nu_n(u)) \right).$$

But NUX, being a submanifold of NX, is also a submanifold of $E \times E$,
and so is $S(E)$. Thus we can compute $T_{(x,v)}\gamma$ using usual derivatives.

In the notation of 6.7.11, and taking into account the various identifications between tangent spaces, we get

6.8.16 $\qquad (T_{(x,v)}\gamma) \left(\frac{\partial h}{\partial u_i}(u) \right) = (\gamma \circ H)'(u,t)(e_i)$

and

$$(T_{(x,v)}\gamma)(\nu_j(u)) = (T_{(x,v)}\gamma)(u,t)(H'(f_j)) = (\gamma \circ H)'(f_j).$$

Since $\gamma(x,v) = v$, we can write

$$(\gamma \circ H)(w,s) = \sum_{j=d+1}^{n} s_j \nu_j(w),$$

where $(w,s) \in U \times \mathbf{R}^{n-d}$, the domain of H. Using the formula for H'
(6.7.11) to calculate $(\gamma \circ H)'(u,t)$, and bearing in mind that $t = (1,0,\ldots,0)$,
we get

$$(\gamma \circ H)'(u,t)(e_i) = \frac{\partial \nu_{d+1}}{\partial u_i}(u)$$

for $i = 1, \ldots, d$ and $(\gamma \circ H)'(u,t)(f_j) = \nu_j(u)$ for $j = d+2, \ldots, n$.

Now, since $\gamma(x,v) = v$, we get

$$(\gamma^*\Sigma)(x,v) = \Sigma(v) \left(\frac{\partial \nu_{d+1}}{\partial u_1}(u), \ldots, \frac{\partial \nu_{d+1}}{\partial u_d}(u), \nu_{d+2}(u), \ldots, \nu_n(u) \right),$$

or again, since $\Sigma = \text{cont}(\nu_{d+1}(u))\Omega_0$ by 6.8.14 and $\nu_{d+1}(u) = v$:

$$(\gamma^*\Sigma)(x,v) = \Omega_0 \left(\nu_{d+1}(u), \frac{\partial \nu_{d+1}}{\partial u_1}(u), \ldots, \frac{\partial \nu_{d+1}}{\partial u_d}(u), \ldots, \nu_{d+2}(u), \nu_n(u) \right)$$

$$= (-1)^d \Omega_0 \left(\frac{\partial \nu_{d+1}}{\partial u_1}(u), \ldots, \frac{\partial \nu_{d+1}}{\partial u_d}(u), \nu_{d+1}(u), \nu_n(u) \right).$$

Let β be an arbitrary local volume form on $\theta(T_x X)$. Since $E = \theta(T_x X) \oplus N_x X$ and $\{\nu_j(u)\}_{d+1 \leq j \leq n}$ is an orthonormal basis for $N_x X$, it follows from the definition of the canonical volume form on a Euclidean space (0.1.15.5) that

$$(6.8.15) \quad (\gamma^*\Sigma)(x,v) \left(\frac{\partial h}{\partial u_1}(u), \ldots, \frac{\partial h}{\partial u_d}(u), \nu_{d+2}(u), \ldots, \nu_n(u) \right)$$

$$= (-1)^d \beta \left(\left(\frac{\partial \nu_{d+1}}{\partial u_1}(u) \right)^T, \ldots, \left(\frac{\partial \nu_{d+1}}{\partial u_d}(u) \right)^T \right),$$

where $\cdot^T : E \to \theta(T_x X)$ was defined in 6.8.5.

On the other hand, if we set $\varsigma(x,v) = \theta_v^{-1}(v) = \nu_{d+1}(u)$ we have (cf. 6.7.27 and 6.7.28):

$$\oplus(x,v) \left(\frac{\partial h}{\partial u_1}(u), \ldots, \frac{\partial h}{\partial u_d}(u), \nu_{d+2}(u), \ldots, \nu_n(u) \right)$$

$$= \left(\text{cont}(\varsigma)\Xi \right)(x,v) \left(\frac{\partial h}{\partial u_1}(u), \ldots, \frac{\partial h}{\partial u_d}(u), \nu_{d+2}(u), \ldots, \nu_n(u) \right)$$

$$= (-1)^d \Xi \left(\frac{\partial h}{\partial u_1}(u), \ldots, \frac{\partial h}{\partial u_d}(u), \nu_{d+1}(u), \ldots, \nu_n(u) \right)$$

$$= (-1)^d \beta \left(\frac{\partial h}{\partial u_1}(u), \ldots, \frac{\partial h}{\partial u_d}(u) \right).$$

Comparing this with 6.8.17 we get

$$\textbf{6.8.18} \quad (\gamma^*\Sigma)(x,v) = \frac{\beta \left(\left(\dfrac{\partial \nu_{d+1}}{\partial u_1}(u) \right)^T, \ldots, \left(\dfrac{\partial \nu_{d+1}}{\partial u_d}(u) \right)^T \right)}{\beta \left(\dfrac{\partial h}{\partial u_1}(u), \ldots, \dfrac{\partial h}{\partial u_d}(u) \right)} \oplus(x,v).$$

Since we have taken $t = (1, 0, \ldots, 0)$ in 6.8.12, we get

$$\textbf{6.8.19} \quad W_d(x,v) = \frac{\beta \left(\left(\dfrac{\partial \nu_{d+1}}{\partial u_1}(u) \right)^T, \ldots, \left(\dfrac{\partial \nu_{d+1}}{\partial u_d}(u) \right)^T \right)}{\beta \left(\dfrac{\partial h}{\partial u_1}(u), \ldots, \dfrac{\partial h}{\partial u_d}(u) \right)},$$

which proves the lemma. $\square$

6.9. Volume of Tubes III

We propose to calculate the volume of $\mathrm{Tub}^\varepsilon X$ when ε is small enough that the canonical map is an embedding (2.7.12). With the notations introduced in sections 6.7 and 6.8,

$$\mathrm{vol}(\mathrm{Tub}^\varepsilon X) = \int_{N^\varepsilon X} \mathrm{can}^* \Delta_0 = \int_{N^\varepsilon X} G\Delta$$

or, by 6.7.15,

$$= \int_X \left(\int_{N_x^\varepsilon X} G(x,v)\varepsilon_x \right)\delta.$$

Since we're assuming that $\mathrm{can} : N^\varepsilon X \to \mathrm{Tub}^\varepsilon X$ is a diffeomorphism, G is non-zero on NX. In addition, on each $N_x^\varepsilon X$ we have

$$G(x,0) = 1 = \sum_{i=0}^{d} W_i(x,0),$$

since only the term $W_0(x,0)$ is non-zero. Thus $\sum_{i=0}^{d} W_i$ is strictly positive on the connected set $N_x^\varepsilon X$; by 6.8.11 we have

6.9.1
$$G(x,v) = \sum_{i=0}^{d} W_i(x,v)$$

on $N_\varepsilon X$, the sum being positive.

This leads to

$$\mathrm{vol}(\mathrm{Tub}^\varepsilon X) = \int_X \left(\sum_{i=0}^{d} \int_{N_x^\varepsilon X} W_i(x,v)\varepsilon_x \right)\delta,$$

so we now just have to find $\int_{N_x^\varepsilon X} W_i(x,v)\varepsilon_x$. To do this we use the same procedure as in 6.5.9, treating $N_x X$ as $\mathbf{R}^{n-d}$. Let $f : \mathbf{R}_+^* \times NU_x X \to N_x X \setminus 0$ be the diffeomorphism defined by $(r,v) \mapsto rv$. If τ_x denotes the canonical density on $NU_x X$ (which is the unit sphere in a Euclidean space E—cf. 6.7.17), we have

6.9.2
$$(f^* \varepsilon_x)(r,v) = r^{n-d-1}\, dr \wedge \tau_x,$$

whence

6.9.3
$$\int_{N_x^\varepsilon X} W_i(x,v)\varepsilon_x = \int_0^\varepsilon \int_{NU_x X} W_i(x, f(r,v))\, dr \wedge \tau_x.$$

But $W_i(x, f(r,v)) = W_i(x, rv) = r^i W_i(x,v)$ because W_i is homogeneous of degree i. Thus

$$\int_{N_x^\varepsilon X} W_i(x,v)\varepsilon_x = \int_0^\varepsilon r^{n-d-1+i}\, dr \int_{NU_x X} W_i(x,v)\tau_x,$$

or again

6.9.4
$$\int_{N_x^\varepsilon X} W_i(x,v)\varepsilon_x = \frac{\varepsilon^{n-d+i}}{n-d+1} \int_{NU_x X} W_i(x,v)\tau_x.$$

6.9.5. Remark. If i is odd we have

$$\int_{NU_x X} W_i(x,v)\tau_x = 0.$$

In fact,

$$\int_{NU_x X} W_i(x,v)\tau_x \int_{NU_x X} W_i(x,-v)\tau_x = (-1)^i \int_{NU_x X} W_i(x,v)\tau_x$$

by the definition of W_i and because τ_x is invariant under isometries.

For even i, we introduce the following definition:

6.9.6. Definition. Let X be a d-dimensional submanifold of an n-dimensional Euclidean vector space E. For every $i = 0, 1, \ldots, \left[\dfrac{d}{2}\right]$ we define the function $K_{2i} \in C^\infty(X)$ by

$$K_{2i}(x) = \int_{NU_x X} W_i(x,v)\tau_x.$$

This function is called the $2i$-th *Weyl curvature* of (E, X).

6.9.7. Relationship with 4.2.21. Let S be a surface (a two-dimensional submanifold of $\mathbf{R}^3$). In the notation of 4.2.21, we have (exercise 6.10.22):

$$\begin{cases} rt - s^2 \neq 0 \text{ at } m \in S \leftrightarrow K_2(m) \neq 0; \\ rt - s^2 > 0 \text{ at } m \in S \leftrightarrow K_2(m) > 0; \\ rt - s^2 < 0 \text{ at } m \in S \leftrightarrow K_2(m) < 0. \end{cases}$$

In fact we can say much more: $K_2(m)$ is the Gaussian curvature of S at m (see section 10.5). This is an immediate consequence of 6.9.15 and 10.6.2.2. The important thing about this curvature is that it only depends on the intrinsic, or riemannian, metric on S (cf. 10.3.1), and not on the particular embedding of S in $\mathbf{R}^3$ (cf. 10.5.3.2 and 10.6.2.1). Thus, for a tiny open piece of surface S we have

$$\text{vol}(\text{Tub}^\varepsilon S) = 2\varepsilon \, \text{area}(S) + \frac{4\pi}{\varepsilon^3} \int_S K_2(m)\, dm;$$

this volume only depends on the area of S and the integral of K_2 on S, and both quantities depend only on the intrinsic metric and not on the embedding. We will see even better results in 6.9.16, 7.5.5 and 11.7.1.

6.9.8. Important cultural digression. Just as we have seen that, for $d = 2$ and $n = 3$, the Weyl curvature K_2 depends only on the riemannian metric on X and not on the embedding, so it was proved by Hermann Weyl in the remarkable paper [Wey39] that, for any positive integers d, n and $i \leq d/2$ and any d-dimensional submanifold of $\mathbf{R}^n$, the function K_{2i} depends only on the riemannian metric on X. More precisely, K_{2i} is a universal i-th degree polynomial in R, the curvature tensor of X. When d is even, K_d is the integrand of a formula that generalizes the Gauss–Bonnet formula (11.7.1) to all even-dimensional riemannian manifolds. This formula, known as the Allendoerfer–Weyl–Fenchel–Gauss–Bonnet–Chern formula (cf. 7.5.7) was first proved by using Weyl's result, and then intrinsically by Chern.

See [CMS84] for more information on the K_{2i}, and [Kow80] for the volume of tubes.

6.9.9. Theorem. *Let X be a d-dimensional submanifold of an n-dimensional Euclidean vector space E. The volume of $\operatorname{Tub}^\varepsilon X$ is a polynomial in ε,*

$$\operatorname{vol}(\operatorname{Tub}^\varepsilon X) = \sum_{i=0}^{\lfloor d/2 \rfloor} a_{2i} \varepsilon^{n-d+2i},$$

where

$$a_{2i} = \frac{1}{n-d+2i} \int_X K_{2i} \delta.$$

In particular, $a_0 = \operatorname{vol} B_{n-d}(0,1) \cdot \operatorname{vol}(x)$.

Proof. We have

$$\operatorname{vol}(\operatorname{Tub}^\varepsilon X) = \sum_{i=0}^{d} \int_X \left(\frac{\varepsilon^{n-d+i}}{n-d+i} \int_{NU_x X} W_i(x,v)\tau_x \right) \delta,$$

which, together with 6.9.5 and 6.9.6, proves the first formula. For the second, we have

$$a_0 = \frac{1}{n-d} \int_X K_0(x)\delta.$$

But $K_0(x) = \int_{NU_x X} W_0(x,v)\tau_x = \int_{NU_x X} \tau_x$ by 6.8.11 (notice that we have $W_0(x,v) = 1$). $NU_x X$ is the unit sphere in $N_x X$; thus

$$\frac{K_0(x)}{n-d} = \operatorname{vol}(B_{n-d}(0,1))$$

and $a_0 = \operatorname{vol}(B_{n-d}(0,1)) \int_X \delta$ (cf. 6.5.6.1). Since $\varepsilon^{n-d} \operatorname{vol}(B_{n-d}(0,1)) = \operatorname{vol}(B_{n-d}(0,\varepsilon))$, we get

$$a_0 = \operatorname{vol}(X) \operatorname{vol}(B_{n-d}(0,\varepsilon)). \qquad \square$$

6.9.10. Corollary. *If X is a one-dimensional submanifold of a Euclidean space E we have*

$$\text{vol}(\text{Tub}^\varepsilon X) = \text{vol}(B_{n-1}(0, \varepsilon)) \, \text{leng}(X).$$

Proof. The only integer $\leq \left[\frac{1}{2}\right]$ is zero.   □

6.9.11. Examples

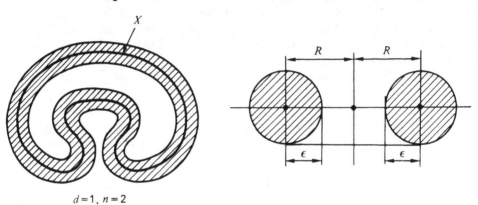

$d = 1, \, n = 2$

Figure 6.9.11

6.9.11.1. In $\mathbf{R}^2$ the only interesting case is $d = 1$, that is, X is a curve. Then corollary 6.9.10 says that

$$\text{area}(\text{Tub}^\varepsilon X) = 2\varepsilon \, \text{leng}(X).$$

6.9.11.2. Volume of the solid torus. If X is a circle in $\mathbf{R}^3$, $\text{Tub}^\varepsilon X$ is called a solid torus of revolution. If R is the radius of the circle, corollary 6.9.10 gives (see also 6.10.15):

$$\text{vol}(\text{Tub}^\varepsilon X) = 2\pi^2 R \varepsilon^2.$$

6.9.12. Corollary. *If X is an $(n-1)$-dimensional submanifold of an n-dimensional Euclidean space E we have*

$$\text{vol}(X) = \frac{1}{2} \frac{d}{d\varepsilon} \text{vol}(\text{Tub}^\varepsilon X)(0).$$

Proof. The first term in 6.9.9 is $a_0 \varepsilon$ and the others have degree greater than two in ε. Thus

$$\frac{d}{d\varepsilon} \text{vol}(\text{Tub}^\varepsilon X)(0) = a_0 = \text{vol}(X) \, \text{vol}(B_1(0, 1)).$$

But $\text{vol}(B_1(0, 1)) = 2$.   □

6.9.13. Example. Another proof for 6.5.6:

$$\text{vol}(S^d) = \frac{1}{2}\frac{d}{d\varepsilon}\big(\text{vol}(B_{d+1}(0, 1+\varepsilon)) - \text{vol}(B_{d+1}(0, 1+\varepsilon))\big)(0)$$
$$= \frac{1}{2}\text{vol}(B_{d+1}(0,1))\frac{d}{d\varepsilon}\big((1+\varepsilon)^{d+1} - (1-\varepsilon)^{d+1}\big)(0)$$
$$= (d+1)\,\text{vol}(B_{d+1}(0,1)).$$

6.9.14. Remark. For $X = S^d$ all the a_{2i} are non-zero:

$$\text{vol}(\text{Tub}^\varepsilon S^d) = \text{vol}(B_{d+1}(0, 1+\varepsilon)) - \text{vol}(B_{d+1}(0, 1+\varepsilon))$$
$$= B_{d+1}(0,1)\big((1+\varepsilon)^{d+1} - (1-\varepsilon)^{d+1}\big)$$
$$= 2B_{d+1}(0,1)\sum_{i=0}^{[d/2]}\binom{d+1}{2i}\varepsilon^{2i}.$$

Thus, in general, all the terms in 6.9.9 are non-zero.

We now give an interpretation for a_d when d is even.

6.9.15. Proposition. *If E is oriented, Σ denotes the canonical volume form on the unit sphere $S(E)$ and $\gamma : NUX \to S(E)$ the Gauss map, we have*

$$a_d = \frac{1}{n}\int_X K_d\delta = \frac{1}{n}\int_{NUX}\gamma^*\Sigma = \frac{1}{n}\int_{NUX}W_d\Psi,$$

where the orientation of NUX is chosen according to that of X.

Proof. We know by 6.7.19 that

$$\int_{NUX}W_d\Psi = \int_X\left(\int_{NU_xX}W_d\tau_x\right) = \int_X K_d\delta = na_d.$$

Since $\Psi = |\circledB|$, 6.8.15 gives

$$\int_{NUX}W_d\Psi = \int_{NUX}W_d\circledB = \int_{NUX}\gamma^*\Sigma. \qquad \square$$

6.9.16. Corollary. *If X is a two-dimensional submanifold of an n-dimensional Euclidean space E, we have*

$$\text{vol}(\text{Tub}^\varepsilon X) = \varepsilon^{n-2}\text{vol}(B_{n-2}(0,1))\,\text{area}(X) + \frac{\varepsilon^n}{n}\int_{NUX}\gamma^*\Sigma. \qquad \square$$

We shall see in the next chapter that $\int_{NUX}\gamma^*\Sigma$ is equal to $\text{vol}(S^{n-1})$ times an integer, and that this integer is exactly $\chi(X)$, the Euler characteristic of X. This provides a complete solution to the problem of the volume of tubes in the case of surfaces (dim $X = 2$). It is surprising that $\text{vol}(\text{Tub}_\varepsilon X)$ depends only on ε, area(X) and $\chi(X)$, and not on the embedding. See also 11.7.1.

6.10. Exercises

6.10.1. Let D be a compact submanifold-with-boundary of $\mathbf{R}^2$ and $f : \mathbf{R}^2 \to \mathbf{R}$ a C^2 function which vanishes on ∂D. Prove the formula

$$\int_D f\left(\frac{\partial^2 f}{\partial x^2} + \frac{\partial^2 f}{\partial y^2}\right) dx \wedge dy = -\int_D \left(\left(\frac{\partial f}{\partial x}\right)^2 + \left(\frac{\partial f}{\partial y}\right)^2\right) dx \wedge dy.$$

Deduce that if $\dfrac{\partial^2 f}{\partial x^2} + \dfrac{\partial^2 f}{\partial y^2} = 0$ on D then $f|_D = 0$.

6.10.2. The Hopf fibration and the Hopf invariant.

6.10.2.1. Let $b : \mathbf{R}^4 \to \mathbf{R}^3$ be the map

$$(x, y, z, t) \mapsto \left(p = 2(xz + yt), q = 2(-xt + yz), r = -x^2 - y^2 + z^2 + t^2\right)$$

and a the restriction of b to $S^3 \subset \mathbf{R}^4$. Let

$$\mu = (p\,dq \wedge dr + q\,dr \wedge dp + r\,dp \wedge dq)|_{S^2}$$

be the canonical volume form on $S^2 \subset \mathbf{R}^3$.

(a) Does $a \in C^\infty(S^3; S^2)$?
(b) Compute $\lambda = a^*\mu$. $\bigl($Answer: $4(dx \wedge dy + dz \wedge dt)|_{S^3}$. The otherwise cumbersome calculation can be shortened by a judicious use of the formulas

$$(x^2 + y^2 + z^2 + t^2)|_{S^3} = 1$$
$$(x\,dx + y\,dy + z\,dz + t\,dt)|_{S^3} = 0.\bigr)$$

(c) Show that $\lambda = 2d\theta$, where $\theta = (-y\,dx + x\,dy - t\,dz + z\,dt)|_{S^3}$. Compute $\lambda \wedge \theta$ and $\int_{S^3} \lambda \wedge \theta$.

6.10.2.2. Now let $f : S^3 \to S^2$ be an arbitrary C^∞ map, $\beta \in \Omega^2(S^2)$ a two-form on S^3, and $\alpha = f^*\beta$.

(a) Show that there exists $\xi \in \Omega^1(S^3)$ such that $d\xi = \alpha$.
(b) For a fixed f, we consider all forms $\beta \in \Omega^2(S^2)$ such that $\int_{S^2} \beta = 1$ and all forms $\xi \in \Omega^1(S^3)$ such that $d\xi = \alpha$. Show that $\int_{S^3} \alpha \wedge \xi$ does not depend on β or on ξ, only on f. We denote this integral by $\gamma(f)$, and call it the *Hopf invariant* of f.
(c) What is $\gamma(f)$ equal to when f is not surjective?
(d) Compute $\gamma(a)$ for a as in 6.10.2.1.

In fact γ is always an integer [Gre67, p. 151].

6.10.3. Prove 6.3.7 by showing that if $\xi(x) \neq 0$ for every $x \in \overline{B}(0, 1)$ we can find a map f contradicting 6.3.5.

6.10.4. Prove 6.3.5, 6.3.7 and 6.3.8 assuming just continuity, not differentiability. (Hint: Use Weierstrass's theorem to approximate continuous functions on compact sets by polynomials.)

6.10.5. Let X and Y be compact manifolds. Construct a natural map

$$F: \sum_{p+q=r} R^p(X) \otimes R^q(Y) \to R^r(X \times Y),$$

where $\otimes$ denotes the tensor product. Prove that f is injective.

6.10.6. Let D be the set of complex numbers z such that $\operatorname{Im} z > 0$, considered as an open subset of $\mathbf{R}^2$. Let G be the set of homographies $z \mapsto \dfrac{az+b}{cz+d}$, where a, b, c, d are real numbers satisfying $ad - bc \neq 0$.

(a) Show that $f(D) = D$ for every $f \in G$.

(b) Show that if ω is a two-form on D satisfying $f^*\omega = \omega$ for every $f \in G$, ω is proportional to $\dfrac{dx \wedge dy}{y^2}$ (where, as usual, we're setting $z = x+iy$).

(c) Let Δ be a compact submanifold-with-boundary of D such that $\partial\Delta$ is the union of three arcs of circle AB, BC and CA with centers on the line $y = 0$. Prove that

$$\int_\delta \frac{dx \wedge dy}{y^2} = \pi - (A + B + C),$$

where A, B and C are the angles of the "triangle" ABC.

6.10.7. Let D be a submanifold-with-boundary of an oriented Euclidean vector space, and $\nu(x)$ the unit normal vector point out at $x \in \partial D$. Prove that there exists $\varepsilon > 0$ such that $x + t\nu(x) \in \mathring{D}$ for every $t \in \,]{-\varepsilon}, 0[$.

6.10.8. Let n be the north pole of the sphere $S^2 \subset \mathbf{R}^3$, and $s : S^2 \setminus \{n\} \to \mathbf{R}^2$ the corresponding stereographic projection. Find $s(x, y, z)$ explicitly. If $\omega_0 = dx \wedge dy$ is the canonical volume form on $\mathbf{R}^2$ and σ the canonical volume form on S^2, find $s^*\omega_0$ as a function of σ.

6.10.9. Compute the volume of S^d using the map

$$\left[-\frac{\pi}{2}, \frac{\pi}{2}\right]^{n-1} \times [-\pi, \pi] : (\theta^1, \ldots, \theta^n) \mapsto (\varsigma^1, \ldots, \varsigma^{n+1}) \in \mathbf{R}^{n+1},$$

where

$$\varsigma_1 = \sin\theta^1,$$
$$\varsigma_2 = \cos\theta^1 \sin\theta^2,$$
$$\vdots$$
$$\varsigma_n = \cos\theta^1 \cos\theta^2 \cdots \cos\theta^{n-1} \sin\theta^n,$$
$$\varsigma_{n+1} = \cos\theta^1 \cos\theta^2 \cdots \cos\theta^{n-1} \cos\theta^n.$$

6.10.10. Prove 6.5.4 in detail.

6.10.11. Calculate the length and area of the hypocycloids and epicycloids studied in 8.7.17.

6.10.12. Consider the two surfaces
$$EP = \{(x, y, x^2 + y^2) : x, y \in \mathbf{R}\},$$
$$HP = \{(x, y, x^2 - y^2) : x, y \in \mathbf{R}\},$$
called *elliptic* and *hyperbolic paraboloids*, respectively. Let D be a compact submanifold-with-boundary of the plane. Show that, if $p : \mathbf{R}^3 \to \mathbf{R}^2$ is the projection onto the xy-plane, we have
$$\mathrm{vol}\big(EP \cap p^{-1}(\mathring{D})\big) = \mathrm{vol}\big(HP \cap p^{-1}(\mathring{D})\big).$$
Compute this volume explicitly when $D = \overline{B}(0, 1)$.

6.10.13. Find the center of mass of the circular sector
$$\{(r\cos\theta, r\sin\theta) : 0 \le r \le R, -T \le \theta \le T\}$$
as a function of $R \in \mathbf{R}_+^*$ and $T \in]0, \pi[$.

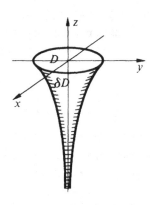

6.10.14. The paradox of the funnel. Given a real number α, consider the set $D = \{(x, y, z) \in \mathbf{R}^3 : z < 0, x^2 + y^2 \le (-z)^{2\alpha}\}$. Compute $\mathrm{vol}(D)$ and $\mathrm{area}(\partial D)$. Prove that if $\alpha \in [-1, -\frac{1}{2}[$ we have $\mathrm{vol}(D) < \infty$ and $\mathrm{area}(D) = \infty$. This means that to paint ∂D one should need an infinite amount of paint, but to fill D (and *a fortiori* paint ∂D) a finite amount is enough!

Figure 6.10.14

6.10.15. First Guldin theorem. Let $H^+ = \{(t, z) \in \mathbf{R}^2 : t > 0\}$, let $S^1 \subset \mathbf{R}^2$ be the unit circle and set
$$f : H^+ \times S^1 \ni \big((t, z), u\big) \mapsto (tu, z) \in \mathbf{R}^2 \times \mathbf{R} = \mathbf{R}^3$$
(*cylindrical coordinates*).
(a) Show that f is a diffeomorphism between $H^+ \times S^1$ and $\mathbf{R}^3 \setminus Z$, where Z is the z-axis $x^{-1}(0) \cap y^{-1}(0)$.
(b) Let ω_0, σ and τ be the canonical volume forms on $\mathbf{R}^3$, S^1 and $\mathbf{R}^2$, respectively (where $\tau = dt \wedge dz$ is restricted to H^+). Prove that $f^*\omega = -t(\tau \wedge \sigma)$.
(c) Let D be a compact submanifold-with-boundary of H^+, and consider the associated solid of revolution $f(D \times S^1) \subset \mathbf{R}^3$. Prove that
$$\mathrm{vol}\big(f(D \times S^1)\big) = 2\pi\delta \, \mathrm{area}(D),$$
where δ is the distance from the center of mass of D to the t-axis. Use this to find the volume of a torus of revolution and the center of mass of a half-disc.

6.10.16. Second Guldin theorem

(a) Let C be a compact, one-dimensional submanifold of a Euclidean affine space, and δ its canonical density. As in 6.5.14, show that

$$\frac{\int_C \mathrm{Id}_E\, \delta}{\int_C \delta}$$

defines a point in the affine space E, which we call the *center of mass* of C.

(b) With the same notation as 6.10.15, let $C \subset H^+$ be a compact, one-dimensional submanifold of H^+. Show that $f(C \times S^1)$ is a two-dimensional submanifold of $\mathbf{R}^3$ and that

$$\mathrm{area}\big(f(C \times S^1)\big) = 2\pi\delta\, \mathrm{leng}(C),$$

where δ is the distance from the center of mass of C to the t-axis. As applications, find the area of a torus of revolution and the position of the center of mass of a semicircle.

6.10.17. Let X be a compact, d-dimensional submanifold of $\mathbf{R}^n$ and $f : \mathbf{R}^n \to \mathbf{R}^n$ a homothety of ratio λ. Show that

$$\mathrm{vol}\big(f(X)\big) = \lambda^d \mathrm{vol}(X).$$

6.10.18. Let V and W be submanifolds of Euclidean spaces E and F, respectively. Show that

$$\mathrm{vol}(V \times W) = \mathrm{vol}(V) \times \mathrm{vol}(W).$$

6.10.19. Calculate the volume of $P^d(\mathbf{R})$ for the density derived from the canonical density on S^d by proposition 6.6.8.

6.10.20. Compute the area of the Möbius strip (5.9.11).

6.10.21. Compute the area of the image of S^2 under the map

$$(x, y, z) \mapsto \big(x^2, y^2, z^2, \sqrt{2}\, yz, \sqrt{2}\, zx, \sqrt{2}\, xy\big).$$

6.10.22. Prove 6.9.7.

6.10.23. Compute $\mathrm{vol}\big(\partial(\mathrm{Tub}^\varepsilon X)\big) = \mathrm{can}(NU^\varepsilon X)$ for ε small enough.

6.10.24. Compute

$$\int_{S^d} x_1^{\alpha_1} \ldots x_{d+1}^{\alpha_{d+1}}\, \sigma,$$

where σ is the canonical measure on $S^d \subset \mathbf{R}^{d+1}$, the x_i are the standard coordinate functions on $\mathbf{R}^{d+1}$ (restricted to S^d) and the α_i are positive integers. If you're familiar with the Γ-function, find this integral for α_i real.

6.10.25. Determine the center of mass of the half-ball
$$\left\{ (x_1, \ldots, x_n) \in \mathbf{R}^n : x_n \geq 0, \sum_{i=1}^n x_i^2 \leq 1 \right\}.$$
What can you say about the last coordinate of the center of mass when n tends to infinity?

6.10.26. Prove 6.5.9 without using coordinates.

6.10.27. Calculate the volume of a *spherical zone*, that is, the portion of a sphere in $\mathbf{R}^3$ comprised between two parallel planes. In particular, the lateral area of a spherical zone only depends on its thickness.

6.10.28. Calculate the volume of the solid bounded by two cylinders of revolution of same radius whose axes intersect at an angle α. Calculate the volume of the solid bounded by three cylinders of revolution of same radius and mutually orthogonal axes.

6.10.29. Calculate the volume of the solid bounded between a paraboloid of revolution and a plane not parallel to its axis.

6.10.30. Calculate the volume and lateral area of a *cylindrical wedge* (figure 6.10.30).

Figure 6.10.30

6.10.31. Calculate the volume of *Viviani's window*, the set of points of $\mathbf{R}^3$ defined by
$$\left\{ (x, y, z) \mid x^2 + y^2 + z^2 \leq 1 \text{ and } x^2 + y^2 \leq x \right\}.$$

6.10.32. Formula of the three levels. Let K be a compact set in a three-dimensional Euclidean space, and assume that, for $a \leq z \leq b$, the area $S(z)$ of the section $K \cap H(z)$ of K by a plane with a fixed z-coordinate is a polynomial in z, of degree three or less. Show that the volume of z between $H(a)$ and $H(b)$ is given by the formula of the three levels:
$$\frac{b-a}{6} \left(S(a) + 4S\left(\frac{a+b}{2}\right) + S(b) \right).$$

This gives another proof for the volume of a spherical zone (6.10.27). Apply also to truncated cones. Show that the condition of the statement is always satisfied when K is bounded on the sides by a ruled surface (in particular, if K is a polytope).

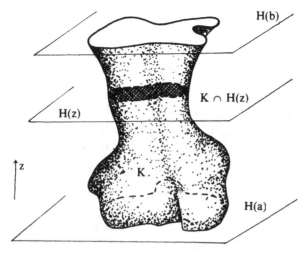

Figure 6.10.32

6.10.33. Find the area of the ellipsoid of revolution with equation

$$\frac{x^2}{a^2} + \frac{y^2}{a^2} + \frac{z^2}{c^2} = 1.$$

6.10.34. Compute the area of the open sets bounded by the curves defined in exercise 8.7.17.

CHAPTER 7

Degree Theory

Using a local lemma, we show that the d-th de Rham group of an oriented, compact, connected d-dimensional manifold is canonically isomorphic to $\mathbf{R}$ (7.2.1). From this fundamental fact we deduce Moser's theorem, which says that two volume forms whose integral is the same are conjugate under a diffeomorphism.

But the most important consequence of the isomorphism $R^d(X) \simeq \mathbf{R}$ is that it allows us to associate to any differentiable map $f : X \to Y$ between oriented, compact, connected manifolds of the same dimension a real number, called the degree of f, defined as the ratio between the integrals of certain differential forms (7.3.1(i)). The fecundity of this concept lies in that it can be reached in a completely different, geometric way: by counting, with appropriate signs, the number of inverse images of a regular value of f (7.3.1(ii)). In particular the degree is an integer; it is also invariant under continuous deformations (7.4.3).

From this we deduce a number of consequences bearing on vector fields on the sphere (7.4.6), the linking number of two curves (7.4.7), and the local behavior of vector fields near an isolated singularity (7.4.15). We also get a formula about vector fields on the unit ball pointing inward at the boundary (7.4.18).

The end of the chapter is devoted to the calculation of the last term in the formula giving the volume of a tube, in the even-dimensional case. This term turns out to be, on the one hand, a fixed scalar times the degree of the normal Gauss map of the submanifold, and, on the other hand, a fixed scalar times the Euler characteristic of the manifold (7.5.4). The equality between these two quantities is exactly the Gauss–Bonnet formula. All of this implies that the volume of a tube around a surface can be explicitly calculated (7.5.5).

Our first goal is to prove theorem 7.2.1: if X is an oriented, compact, connected d-manifold, the de Rham group $R^d(X)$ is canonically isomorphic to $\mathbf{R}$. We then give applications of this result.

> In this chapter everything is of class C^∞.

7.1. Preliminary Lemmas

7.1.1. Lemma. *Let $Q =]0,1[^d$ be the open unit cube in $\mathbf{R}^d$. If $\beta \in \Omega^d(\mathbf{R}^d)$ is a d-form such that $\operatorname{supp}\beta \subset Q_d$ and $\int_{Q_d}\beta = 0$, there exists $\gamma \in \Omega^{d-1}(\mathbf{R}^d)$ such that $\operatorname{supp}\gamma \subset Q_d$ and $\beta = d\gamma$.*

7.1.2. Remarks

7.1.2.1. Poincaré's lemma (5.6.1) applies to Q_d because $d\beta = 0$ (recall that β is a d-form on $\mathbf{R}^d$). Thus it is clear that there exists $\gamma \in \Omega^{d-1}(\mathbf{R}^d)$ such that $\beta = d\gamma$, but we don't know yet that $\operatorname{supp}\gamma \subset Q_d$.

7.1.2.2. In fact this inclusion isn't true if $\int_{Q_d}\beta$ is non-zero. Indeed, by Stokes' theorem (6.2.1 and remark in 6.2.2), we have

$$\int_{\partial\overline{Q}_d}\gamma = \int_{\overline{Q}_d}d\gamma = \int_{\overline{Q}_d}\beta = \int_{Q_d}\beta,$$

the last equality arising from $\operatorname{supp}\beta \subset Q_d$. By if $\operatorname{supp}\gamma \subset \partial\overline{Q}_d$, the first integral is zero.

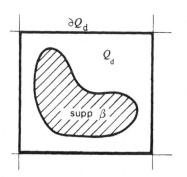
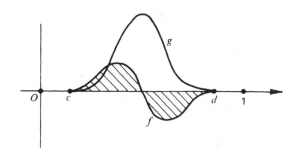

Figure 7.1.2

7.1.3. Proof of 7.1.1. We use induction on d. Let us first establish the case $d = 1$. If $\beta \in \Omega^1(\mathbf{R})$ has support in $]0,1[$, we can write β in the form $\beta = f(t)\,dt$, with $f \in C^\infty(]0,1[;\mathbf{R})$. We are also assuming that

$\int_{Q_1} \beta = \int_0^1 f(t) \, dt = 0$. Finding $\gamma \in \Omega^0(\mathbf{R}) = C^\infty(\mathbf{R})$ such that $d\gamma = \beta$ is the same as finding a function g such that $f(t) \, dt = g'(t) \, dt$.

The answer is

7.1.4
$$g(t) = \int_0^t f(s) \, dx;$$

g is defined for all t because f is continuous, and clearly $g'(t) = f(t)$. There remains to show that g has support in $]0, 1[$. The support of f is a closed subset of $]0, 1[$, that is, supp $f = [c, d]$ with $0 < c \leq d < 1$. Thus $g(t) = 0$ for $t \leq c$. But we also have $g(t) = 0$ for $t \geq d$:

$$g(t) = \int_0^t f(s) \, ds = \int_0^d f(s) \, ds = \int_0^1 f(s) \, ds = 0.$$

In order to take the induction step we will strengthen the induction assumption as follows:

7.1.5. (H_d) *If $\beta \in \Omega^d(\mathbf{R}^d)$ satisfies* supp $\beta \subset Q_d$ *and $\int_{Q_d} \beta = 0$, there exists $\gamma \in \Omega^{d-1}(\mathbf{R}^d)$ such that* supp $\gamma \subset Q_d$ *and $\beta = d\gamma$. In addition, if β depends on a parameter $\lambda \in \Lambda$, where λ is a finite-dimensional vector space and $(x, \lambda) \mapsto \beta(x, \lambda)$ is C^∞ on $\mathbf{R}^d \times \Lambda$, we can make γ depend on λ in such a way that $(x, \lambda) \mapsto \gamma(x, \lambda)$ is C^∞.*

Clearly (H_1) is true, because if $(t, \lambda) \mapsto \beta(t, \lambda) = f(t, \lambda) \, dt$ is C^∞ on $\mathbf{R} \times \Lambda$, the function

$$g(t, \lambda) = \int_0^t f(s, \lambda) \, ds$$

is C^∞ on $\mathbf{R} \times \Lambda$ and we can differentiate under the integral sign (0.4.8).

To show that (H_{d-1}) implies (H_d), we write $Q_d = Q_{d-1} \times]0, 1[\subset \mathbf{R}^d$ and define the maps

7.1.6 $p : Q_d \to Q_{d-1} : (x_1, \ldots, x_{d-1}, x_d) \mapsto (x_1, \ldots, x_{d-1})$;

7.1.7 $i_t : Q_{d-1} \to Q_{d-1} \times \{t\} : (x_1, \ldots, x_{d-1}) \mapsto (x_1, \ldots, x_{d-1}, t)$;

7.1.8 $j_t : Q_{d-1} \times \{t\} \to Q_d$ (inclusion).

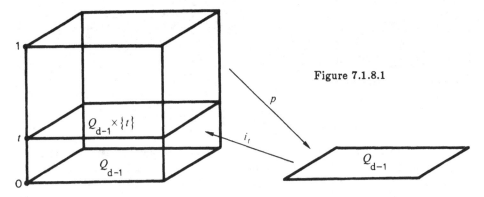

Figure 7.1.8.1

The idea of the proof is to decrease the dimension by dividing β by dx_d, restricting it to $Q_{d-1} \times \{t\}$ and transferring it to Q_{d-1}, this for each $t \in [0,1]$. In the process we introduce one additional parameter: Λ is replaced by $\mathbf{R} \times \Lambda$. The form we obtain on Q_{d-1} does not have zero integral, but we subtract its average and call the difference π_t. Then we apply (H_d) to π_t to get forms γ_t such that $\pi_t = d\gamma_t$, and we pull back the γ_t to $Q_{d-1} \times \{t\}$. The form obtained by combining all the pull-backs, after a slight correction, will be our γ.

The calculation below is done in coordinates for simplicity; we are in fact omitting a great number of occurrences of p^*, i_t^* and j_t^*.

First fix a reference form $\sigma \in \Omega^{d-1}(\mathbf{R}^{d-1})$ such that $\operatorname{supp}(\sigma) \subset Q_{d-1}$ and $\int_{Q_{d-1}} \sigma = 1$. Now assume that $\beta \in \Omega^d(\mathbf{R}^d)$ is a form satisfying the conditions in 7.1.5, and write

$$\beta(x, \lambda) = f(x, \lambda)\, dx_1 \wedge \cdots \wedge dx_d,$$

with $f \in C^\infty(Q^d \times \Lambda; \mathbf{R})$.

Write $x = (x_1, \ldots, x_{d-1}, t)$, and consider, for each $t \in]0,1[$, the form $\pi_t \in \Omega^{d-1}(\mathbf{R}^{d-1})$ given by

$$\pi_t(x_1, \ldots, x_{d-1}) = f(x_1, \ldots, x_{d-1}, t; \lambda)\, dx_1 \wedge \cdots \wedge dx_{d-1} - g(t; \lambda)\sigma,$$

where

$$g(t; \lambda) = \int_{Q_{d-1}} f(x_1, \ldots, x_{d-1}, t; \lambda)\, dx_1 \wedge \cdots \wedge dx_{d-1}.$$

Since $\int_{Q_{d-1}} \sigma = 1$ by assumption, we have $\int_{Q_{d-1}} \pi_t = 0$. By construction, $\pi_t \in \Omega^{d-1}(\mathbf{R}^{d-1})$ varies C^∞ with t and λ. In addition,

$$\operatorname{supp} \pi_t \subset \left(p(\operatorname{supp} f) \cap Q_{d-1} \times \{t\}\right) \cup \operatorname{supp} \sigma \subset Q_{d-1},$$

so, by an application of (H_{d-1}) with $\mathbf{R} \times \Lambda$ in lieu of Λ, there exists $\gamma_t \in \Omega^{d-2}(\mathbf{R}^{d-1})$ such that $d\gamma_t = \pi_t$ and γ_t is supported in Q_{d-1}.

Now we construct a form γ', dependent on λ, as follows:

$$\gamma'(x_1, \ldots, x_{d-1}, t; \lambda) = p^*(\gamma_t) \wedge dx_d.$$

If $\gamma_t(x_1, \ldots, x_{d-1}; \lambda) = \sum u_t(x_1, \ldots, x_{d-1}; \lambda)\, dx_1 \wedge \cdots \wedge \widehat{dx_i} \wedge \cdots \wedge dx_{d-1}$, we have

$$\gamma_t'(x_1, \ldots, x_{d-1}, t; \lambda) = \sum u_t(x_1, \ldots, x_{d-1}; \lambda)\, dx_1 \wedge \cdots \wedge \widehat{dx_i} \wedge \cdots \wedge dx_d$$

(for notational simplicity, we denote the derivative of the i-th coordinate function by dx_i, both on $\mathbf{R}^d$ and on $\mathbf{R}^{d+1}$). Formally we can write

$$d(\gamma') = d\left(p^*(\gamma_t) \wedge dx_d\right) = d\left(p^*(\gamma_t)\right) \wedge dx_d = p^*\left(d(\gamma_t)\right) \wedge dx_d = p^*(\pi_t) \wedge dx_d.$$

Bearing in mind the identifications we've made,

$$p^*(\pi_t) \wedge dx_d = f(x_1, \ldots, x_{d-1}, t; \lambda)\, dx_1 \wedge \cdots \wedge dx_d - g(t; \lambda)\, \sigma \wedge dx_d,$$

that is,

$$\beta = d\gamma' + g(t; \lambda)\sigma(x_1, \ldots, x_{d-1}) \wedge dx_d.$$

If we now show that the term $g(t; \lambda)\sigma(x_1, \ldots, x_{d-1}) \wedge dx_d$ is exact, we will have *formally* shown that β is exact, that is, that there is $\gamma \in \Omega^{d-1}(\mathbf{R}^d)$ such that $\beta = d\gamma$. To do this, set

$$\xi(x_1, \ldots, x_{d-1}, t; \lambda) = (-1)^{d-1} \left(\int_0^t g(s; \lambda) \, dx \right) \sigma.$$

Since σ is a $(d-1)$-form on $\mathbf{R}^{d-1}$, it is closed, and we have

$$\begin{aligned} d\xi(x_1, \ldots, x_{d-1}, t; \lambda) &= (-1)^{d-1} g(t, \lambda) \, dx_d \wedge \sigma \\ &= g(t; \lambda)\sigma(x_1, \ldots, x_{d-1}) \wedge dx_d, \end{aligned}$$

where $g(t, \lambda) \, dx_d$ is the derivative of $t \mapsto \int_0^t g(s) \, ds$.

But this result is as yet merely formal, in the sense that we still have to show that $\gamma = \gamma' + \xi$ is C^∞, that it has support in Q_{d-1} and that it varies C^∞ with λ. The first and last properties can be easily checked using (H_{d-1}) and going over the preceding calculation: the only non-trivial operation we've performed is integration of functions, and for that we have theorem 0.4.8, which allows differentiation under the integral sign.

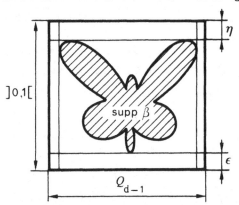

Figure 7.1.8.2

To see that γ has support in Q_d, notice first that each γ_t has support in Q_{d-1}, so γ' has support in $Q_{d-1} \times \mathbf{R}$. But β has support in Q_d, so there exist $\varepsilon, \eta > 0$, with $\varepsilon < 1 - \eta$, such that $f(x_1, \ldots, x_{d-1}, t) = 0$ for any $x_1, \ldots, x_{d-1} \in \mathbf{R}$ and $t \notin [\varepsilon, 1 - \eta]$. For such t, also, $\gamma_t = 0$, which shows that $\operatorname{supp} \gamma' \subset Q_d$.

As for ξ, we will show that the map $t \mapsto \int_0^t g(s; \lambda) \, ds$ is zero for $t \notin [\varepsilon, 1 - \eta]$; this is where the condition $\int_{Q_d} \beta = 0$ will come in. By Fubini's theorem and the construction of g we have, for $t \geq 1 - \eta$:

$$
\begin{aligned}
\int_0^t g(s; \lambda)\, ds &= \int_0^t \left(\int_{Q_{d-1}} f(x_1, \ldots, x_{d-1}, t; \lambda)\, dx_1 \wedge \cdots \wedge dx_{d-1} \right) dt \\
&= \int_0^1 \left(\int_{Q_{d-1}} f(x_1, \ldots, x_{d-1}, t; \lambda)\, dx_1 \wedge \cdots \wedge dx_{d-1} \right) dt \\
&= \int_{Q_d} f(x_1, \ldots, x_d; \lambda)\, dx_1 \wedge \cdots \wedge dx_d \\
&= \int_{Q_d} \beta = 0.
\end{aligned}
$$

Then

$$
\operatorname{supp} \xi \subset \operatorname{supp}\left(t \mapsto \int_0^t g(s; \lambda)\, ds \right) \times Q_{d-1} \subset [\varepsilon, 1 - \eta] \times Q_{d-1} \subset Q_d,
$$

which concludes the proof of the lemma. $\qquad \square$

7.1.9. Lemma. *If X is a compact, connected, d-dimensional manifold, the de Rham group $R^d(X)$ has dimension at most one.*

Proof. Let $\{(U_i, \phi_i, \psi_i)\}$ be a partition of unity on X, satisfying the condition that each $\phi_i(U_i) = Q_d$; this can always be achieved by modifying ϕ_i if necessary. Since X is compact, we can also assume that the partition of unity is finite, that is, $X = \bigcup_{i=1}^n U_i$.

Fix a d-form α_0 with support in U_1 and such that

7.1.10
$$
\int_{\mathbf{R}^d} (\phi_1^{-1})^* \alpha_0 = 1.
$$

We are going to show that, for every $\alpha \in \Omega^d(X)$, there exists a scalar k and a form $\mu \in \Omega^{d-1}(X)$ such that $\alpha = k\alpha_0 + d\mu$. Then α is homologous to $k\alpha_0$, and it follows that $\dim\big(R^d(X)\big) \leq 1$.

Since $\alpha = \sum_{i=1}^n \psi_i \alpha$ and the property we are trying to establish is additive, we can restrict ourselves to the case $\operatorname{supp} \alpha \subset U_i$, for i fixed. Take $m \in U_1$ and $n \in U_i$. Since X is connected, there exists $\gamma \in C^0\big([0,1]; X\big)$ such that $\gamma(0) = m$ and $\gamma(1) = n$ (2.2.13). The image $\gamma\big([0,1]\big)$ is covered by some of the U_j, say $U_{j_1} = U_1, U_{j_2}, \ldots, U_{j_k} = U_i$; we assume moreover that $U_{j_{r+1}} \cap U_n \neq \emptyset$ for $r = 1, \ldots, k - 1$.

Now choose forms $\alpha_{j_r} \in \Omega^d(X)$, with $\operatorname{supp} \alpha_{j_r} \subset U_{j_r} \cap U_{j_{r+1}}$ and

$$
\int_{\mathbf{R}^d} (\phi_{j_r}^{-1})^* \alpha_{j_{r-1}} = 1
$$

for $r = 1, \ldots, k - 1$. Also set $\alpha_{j_0} = \alpha_0$.

On U_{j_r} we have two forms $\alpha_{j_{r-1}}$ and α_{j_r} such that $(\phi_{j_r}^{-1})^* \alpha_{j_{r-1}}$ and $(\phi_{j_r}^{-1})^* \alpha_{j_r}$ have support in Q_d. The integral

$$
c_r = \int (\phi_{j_r}^{-1})^* \alpha_{j_r},
$$

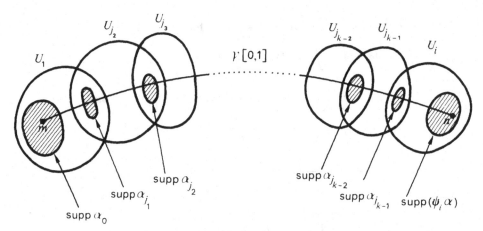

Figure 7.1.9

is a real number, and the form $\alpha_{j_r} - c_r \alpha_{j_{r-1}}$ satisfies $\operatorname{supp}(\phi_{j_r}^{-1})^*(\alpha_{j_r} - c_r \alpha_{j_{r-1}}) \subset Q_d$ and

$$\int (\phi_{j_r}^{-1})^*(\alpha_{j_r} - c_r \alpha_{j_{r-1}}) \subset Q_d.$$

By 7.1.1 there exists $\beta_r \in \Omega^{d-1}(\mathbf{R}^d)$ with support in $\subset Q_d$ and satisfying

$$(\phi_{j_r}^{-1})^*(\alpha_{j_r} - c_r \alpha_{j_{r-1}}) = d\beta_r.$$

We now set $\theta_r = \phi_{j_r}^*(\beta_r)$, which is possible because $\phi_i(U_i) = Q_d$ and $\operatorname{supp} \beta_r \subset Q_d$. On X we have

$$\alpha_{j_r} - c_r \alpha_{j_{r-1}} = d\theta_r.$$

Thus we have shown the existence of forms $\theta_1, \ldots, \theta_k$ and of scalars $c_1, \ldots, c_k$ such that

$$\alpha_{j_1} - c_1 \alpha_0 = d\theta_1,$$
$$\alpha_{j_2} - c_2 \alpha_{j_1} = d\theta_2,$$
$$\vdots$$
$$\alpha_{j_{k-1}} - c_{k-1} \alpha_{j_{k-2}} = d\theta_{k-1},$$
$$(\alpha) - c_k \alpha_{j_{k-1}} = d\theta_k.$$

In particular,

$$\alpha - (c_1 \ldots c_k)\alpha_0 = d(\theta_k + c_k \theta_{k-1} + \cdots + c_k c_{k-1} \ldots c_l \theta_{l-1} + \cdots + c_k \ldots c_2 \theta_1),$$

which shows that α and $(c_1 \ldots c_k)\alpha_0$ are homologous. $\square$

7.2. Calculation of $R^d(X)$

7.2.1. Theorem. *Let X be a compact, connected, d-dimensional manifold. The de Rham group $R^d(X)$ is canonically isomorphic to $\mathbf{R}$ if X is orientable, and trivial otherwise.*

Proof. We know by 7.1.9 that $R^d(X)$ has dimension at most one. If $F^d(X)$ denotes the set of closed d-forms (5.4.3), we have $F^d(X) = \Omega^d(X)$ because X has dimension d.

If X is oriented, we can consider by map

$$I : F^d(X) \ni \alpha \mapsto \int_X \alpha \in \mathbf{R},$$

defined because X is compact (6.1.3). But if $\alpha = d\beta$, Stokes' theorem implies that

$$\int_X \alpha = \int_X d\beta = \int_{\partial X} \beta = 0$$

because X has empty boundary. Thus I factors through $B^d(X)$ to give another map $\overline{I} : F^d(X)/B^d(X) = R^d(X) \to \mathbf{R}$, which is canonically associated to X (with a fixed orientation) and a vector space homomorphism. This map is surjective: if ω is a volume form on X (which exists because X is orientable), we have $\int_X \omega > 0$, so $\overline{I}(\overline{\omega}) > 0$, where $\overline{\omega}$ is the class of ω in $R^d(X)$. Since $R^d(X)$ has dimension at most one, $\overline{I}$ is also injective, and consequently an isomorphism.

If X is non-orientable, we consider the canonical double cover $p : \tilde{X} \to X$ (5.3.27). We know by 5.3.29 that $\tilde{X}$ is connected because X is connected and non-orientable. Now take $\omega \in \Omega^d(X)$. On $\tilde{X}$ we have the antipodal map $s : \tilde{X} \to \tilde{X}$, which satisfies $p \circ s = p$ and hence $p^*(\omega) = s^*(p^*\omega)$. By 5.3.31.1 and 6.1.4.5 we can write

$$\int_{\tilde{X}} p^*(\omega) = \int_{\tilde{X}} s^*(p^*(\omega)) = -\int_{\tilde{X}} p^*(\omega),$$

which implies $\int_{\tilde{X}} p^*\omega = 0$.

Since $\tilde{X}$ is orientable, there exists by 7.2.2 a $(d-1)$-form λ such that $p^*\omega = d\lambda$. Then

$$\mu = \frac{\lambda + s^*\lambda}{2}$$

is again a $(d-1)$-form on $\tilde{X}$, this time invariant under s^* (since $s \circ s = \text{Id}$). From $p^*\omega = d\lambda$ we get

$$s^*(p^*\omega) = s^*(d\lambda) = d(s^*\lambda) = (p \circ s)^*\omega = p^*\omega = d\lambda,$$

so

$$p^*\omega = \frac{d(\lambda + s^*\lambda)}{2} = d\mu.$$

By lemma 5.3.9 there exists $\underline{\mu} \in \Omega^{d-1}$ such that $p^*\underline{\mu} = \mu$. Then $p^*\omega = d\mu = d(p^*\underline{\mu}) = p^*(d\underline{\mu})$ and, since p is regular, $\omega = d\underline{\mu}$. Since $\omega = \Omega^d(X)$ was arbitrary, we have proved that $R^d(X) = 0$. $\qquad\square$

7.2.2. Corollary. *If X is an orientable, compact, connected d-dimensional manifold, we have*

$$B^d(X) = \{\alpha \in \Omega^d(X) : \exists \beta \text{ such that } d\beta = \alpha\} = \{\alpha \in \Omega^d(X) : \int_X \alpha = 0\}.$$

Proof. If I is as above, we have $\{\alpha \in \Omega^d(X) : \int_X \alpha = 0\} = I^{-1}(0)$, and this coincides with $B^d(X)$ because $\overline{I}$ is an isomorphism. $\qquad\square$

7.2.3. Theorem (Moser). *Let X be an oriented, compact, connected d-dimensional manifold. If α and β are volume forms on X such that $\int_X \alpha = \int_X \beta$, there exists $f \in \mathrm{Diff}(X)$ such that $\beta = f^*\alpha$.*

This amounts to saying that the only invariant of volume forms under diffeomorphisms is their volume.

Proof. Let α and β be volume forms. For every $t \in [0,1]$, set $\alpha_t = (1-t)\alpha + t\beta$, still a volume form. Since $\int_X(\beta - \alpha) = 0$ there exists $\gamma \in \Omega^{d-1}(X)$ such that $d\gamma = \beta - \alpha$ (7.2.2). By 5.3.16, there exists for every t a vector field $\xi(t)$, defined by $\mathrm{cont}(\xi(t))\alpha_t = -\gamma$; this field is C^∞ in t.

Let F be the global flow of $\xi(t)$; F is defined on $[0,1]$ because X is compact (cf. 3.5.6 to 3.5.14). Working as if calculating the derivative of a linear map, we can write

$$\frac{\partial F_t^*\alpha_t}{\partial t}(t) = \frac{\partial F_s^*\alpha_t}{\partial s}(t) + F_t^*\left(\left(\frac{\partial\alpha_s}{\partial s}\right)(t)\right) = \frac{\partial F_s^*\alpha_t}{\partial s}(t) + F_t^*(\beta - \alpha).$$

But 5.5.10 says that, for every closed form ω,

$$\frac{\partial F_s^*\alpha_t}{\partial s}(t) + F_t^*\big(d(\mathrm{cont}\,\xi(t)(\omega))\big),$$

so that

$$\frac{\partial F_t^*\alpha_t}{\partial t} = F_t^*(\beta - \alpha) + F_t^*\big(d(-\gamma)\big) = 0,$$

given that $\beta - \alpha = d\gamma$. Thus $F_t^*\alpha_t$ does not depend on t, and is equal to $F_0^*\alpha = \alpha$. In particular, $F_1^*\beta = \alpha$, and $\beta = f^*\alpha$ for $f = (F_1)^{-1}$. $\qquad\square$

7.3. The Degree of a Map

7.3.1. Fundamental theorem. *Let X and Y be oriented, compact, connected manifolds of same dimension d, and $f \in C^\infty(X; Y)$ a map. There exists an integer, called the degree of f and denoted by $\deg(f)$, such that:*
(i) *if $\omega \in \Omega^d(Y)$ we have*

$$\int_X f^*\omega = \deg(f) \int_Y \omega;$$

(ii) *if y is a regular value for f we have*

$$\deg(f) = \sum_{x \in f^{-1}(y)} \operatorname{sgn}(J_x(f)).$$

In particular, if $f^{-1}(y) = \emptyset$ the degree is zero.

For the definition of $\operatorname{sgn}(J_x(f))$, see 7.3.2.2.

7.3.2. Remark. If y is a regular value, f is a local covering and $f^{-1}(y)$ is finite (4.1.5 and 4.1.6). Thus the sum in (ii) is well-defined.

Proof. Let $f^* : R^d(Y) \to R^d(X)$ be the homomorphism associated to $f \in C^\infty(X; Y)$ (cf. 5.4.6). Because of the assumptions on X and Y, 7.2.1 gives canonical isomorphisms between their de Rham groups and $\mathbf{R}$; thus we can define $\overline{f}^* : \mathbf{R} \to \mathbf{R}$ in such a way that the following diagram commutes:

7.3.2.1
$$
\begin{array}{ccc}
R^d(X) & \xleftarrow{\ f^*\ } & R^d(Y) \\
\Big\downarrow{\scriptstyle \overline{I}_X} & & \Big\downarrow{\scriptstyle \overline{I}_Y} \\
\mathbf{R} & \xleftarrow[\ \overline{f}^*\ ]{} & \mathbf{R}.
\end{array}
$$

Since $\overline{f}^*$ is linear, it must be of the form $t \mapsto kt$ for some $k \in \mathbf{R}$. But according to 7.3.2.1, if $\overline{\omega} \in R^d(Y)$ and $\omega \in \Omega^d(Y)$ represents $\overline{\omega}$, we have

$$\int_X f^*\omega = \deg(f) \int_Y \omega.$$

That $\deg(f)$ is an integer will follow from the proof of (ii). By Sard's theorem (4.3.6) f has regular values y. Applying 4.1.5 we can find $V \in O_y(Y)$ such that, if $f^{-1}(y) = \{x_1, \ldots, x_n\}$, the inverse image $f^{-1}(V)$ is a disjoint union of open sets $U_i \in O_{x_i}(X)$, restricted to which f is a diffeomorphism. We can assume V to be connected. Choose $\omega \in \Omega^d(Y)$ such that $\operatorname{supp} \omega \subset V$ and $\int_Y \omega \neq 0$. We have $\operatorname{supp}(f^*\omega) \subset f^{-1}(V)$, so that

$$\int_X f^*\omega = \sum_{i=1}^n \int_{U_i} f^*\omega.$$

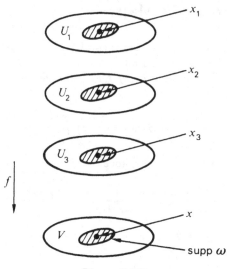

Figure 7.3.2

Now $f|_{U_i}$ is a diffeormorphism between U_i and a connected set V, so it either preserves or reverses orientation (the orientations of U_i and V being the ones induced from X and Y). This choice can be translated in terms of $T_{x_i}f$, so it only depends on x_i. Set

7.3.2.2 $$\text{sgn}(J_x(f)) = \begin{cases} +1 & \text{if } T_xf \text{ preserves orientation,} \\ -1 & \text{if } T_xf \text{ reverses orientation.} \end{cases}$$

Thus, by 6.1.4.4 and 6.1.4.5, we get

$$\int_{U_i} f^*\omega = \text{sgn}(J_{x_i}(f)) \int_Y \omega,$$

$$\int_X f^*\omega = \left(\sum_{x \in f^{-1}(y)} \text{sgn}(J_{x_i}(f)) \right) \int_Y \omega = k \int_Y \omega,$$

whence $\deg(f) = k = \sum_{x \in f^{-1}(y)} \text{sgn}(J_{x_i}(f))$ is an integer.

If $f^{-1}(y) = \emptyset$, there exists $V \in O_y(Y)$ such that $f^{-1}(V) = \emptyset$; if ω has support in V and $\int_Y \omega \neq 0$ we get

$$\int_X f^*\omega = \int_X 0 = 0,$$

and again here $\deg(f) = 0 = \sum_{x \in f^{-1}(y)} \text{sgn}(J_x(f))$.   □

7.3.3. Corollary. *Let X and Y be oriented, compact, connected manifolds of same dimension d, and $f \in C^\infty(X; Y)$ a map. If y and z are regular values the cardinals of $f^{-1}(y)$ and $f^{-1}(z)$ are congruent modulo 2.*

Proof. By 7.3.1(ii) we have $\deg f \equiv \#(f^{-1}(y)) \pmod 2$, and similarly for z. $\qquad\square$

7.3.4. Proposition. *Let X, Y and Z be oriented, compact, connected manifolds of same dimension d, and $f \in C^\infty(X;Y)$, $g \in C^\infty(Y;Z)$ maps. We have $\deg(g \circ f) = \deg g \times \deg f$.*

Proof. See exercise 7.8.2. $\qquad\square$

7.3.5. Examples

7.3.5.1. If f is not surjective, $\deg f = 0$.

7.3.5.2. The degree of the identity map Id_X is one.

7.3.5.3. Let X be the circle, Y the curve in figure 7.3.5 and $f : X \to Y$ the radial projection. The points $y_1, y_3, y_5 \in Y$ are regular values, and $f^{-1}(y_1)$, $f^{-1}(y_3)$ and $f^{-1}(y_5)$ have 1, 3 and 5 elements, respectively.

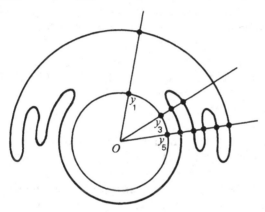

Figure 7.3.5

7.3.5.4. Let $X = Y = S^d$. The degree of the antipodal map $s = -\mathrm{Id}_{S^d}$ is $(-1)^{d+1}$. Indeed, if ω is the canonical volume form on S^d, we have $s_d^*\omega = (-1)^{d+1}\omega$ (cf. the proof of 5.3.18), so that

$$\int_{S^d} s^*\omega = (-1)^{d+1}\int_{S^d} \omega.$$

7.3.5.5. Let $X = Y = S^1 \subset \mathbf{R}^2$, and identify $\mathbf{R}^2$ with $\mathbf{C}$. The degree of $f : S^1 \ni z \mapsto z^n \in S^1$ is n.

7.3.6. Remarks

7.3.6.1. If we switch the orientation of both X and Y the degree does not change.

7.3.6.2. If X is differentiable and $f \in \mathrm{Diff}(X)$, the degree is $+1$ if f preserves orientation and -1 if it reverses it.

7.3.7. Definition and example. Let E be an oriented, two-dimensional Euclidean space, $\alpha \in C^\infty(S^1; E)$ a curve and $a \in E$ a point not in the image of α. The *winding number* of α with respect to a is the degree of the map μ defined by

$$\mu(t) = \frac{\alpha(t) - a}{\|\alpha(t) - a\|}.$$

See 7.6.7 for further discussion.

7.3.8. Remark. By 3.4.1 this applies to all compact one-dimensional submanifolds of the plane.

7.4. Invariance under Homotopy. Applications

7.4.1. Definition. Let X and Y be manifolds. Two maps $f, g \in C^\infty(X; Y)$ are said to be *homotopic* if there exists a *homotopy* between f and g, that is, a map $F : [0, 1] \times X \to Y$ such that:

(i)   For every $t \in [0, 1]$ the map $F_t : x \mapsto F(t, x)$ is in $C^\infty(X; Y)$;
(ii) The map $TF : [0, 1] \times TX \to TY$ defined by $(TF)(t, x) = T_x F_t$ is continuous;
(iii) $F_0 = f$ and $F_0 = g$.

7.4.2. Remark. Conditions (i) and (ii) are certainly satisfied if F can be extended to some $\overline{F} \in C^\infty(]-\varepsilon, 1 + \varepsilon[\times X; Y)$ for some $\varepsilon > 0$.

7.4.3. Theorem. *Let X and Y be oriented, compact, connected manifolds of same dimension. If $f, g \in C^\infty(X; Y)$ are homotopic we have $\deg f = \deg g$. (In fact, if F is a homotopy between f and g the degree of F_t does not depend on t.)*

Proof. Take $\omega \in \Omega^d(Y)$ such that $\int_Y \omega \neq 0$. By 7.3.1(i) we have

$$\deg F_t = \frac{\int_X F_t^* \omega}{\int_X \omega}$$

for every t, and this number is an integer. If we can show that the map $t \mapsto \int_X F_t^* \omega$ is continuous, the same will be true for $t \mapsto \deg(F_t)$, implying that $\deg(F_t)$ is constant. But this follows from 5.2.10.6 and 6.1.4.11. $\square$

7.4.4. Corollary. *If X is an oriented, compact, connected manifold, any map $f \in C^\infty(X)$ homotopic to the identity Id_X is surjective.*

Proof. Apply 7.3.5.1, 7.3.5.2 and 7.4.3. $\square$

7.4.5. First application: vector fields on spheres

7.4.6. Theorem. *There exists a nowhere vanishing vector field ξ on the sphere S^d if and only if d is odd.*

Proof. First we show the existence of such a vector field for odd d. This is done by identifying $\mathbf{R}^{2d}$ with $\mathbf{R}^2 \times \cdots \times \mathbf{R}^2 = \mathbf{C} \times \cdots \times \mathbf{C} = \mathbf{C}^d$ and by setting

$$\xi(z_1,\ldots,z_d) = \theta^{-1}_{(z_1,\ldots,z_d)}(iz_1,\ldots,iz_d).$$

Here $(z_1,\ldots,z_d) \in S^{2d-1} \subset \mathbf{R}^{2d} = \mathbf{C}^d$. It is clear that $\xi \neq 0$ everywhere.

Now we study the case that d is even. Assume that ξ is a nowhere vanishing vector field on S^d, and consider the field η given by

$$\eta(x) = \frac{\xi(x)}{\|\xi(x)\|}$$

for every $x \in S^d$. Next, define $F \in C^\infty(\mathbf{R} \times S^d; S^d)$ by

$$F(t,x) = (\cos \pi t)x + (\sin \pi t)\theta_x(\eta(x)).$$

Since $\|x\| = \|\eta(x)\| = 1$ and $(x \mid \theta_x(\eta(x))) = 0$, we do have $\|F(t,x)\| = 1$ for very $t \in [0,1]$ and $x \in S^d$. The conditions in 7.4.1 are satisfied (by 7.4.2), so the maps $F_0 = \mathrm{Id}_{S^d}$ and $F_1 = -\mathrm{Id}_{S^d}$ are homotopic. By 7.4.3, their degrees should be equal; but this is impossible for d even, since $\deg(\mathrm{Id}_{S^d}) = 1$ and $\deg(-\mathrm{Id}_{S^d}) = (-1)^{d+1}$ by 7.3.5.4. $\square$

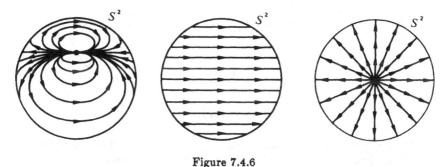

Figure 7.4.6

For instance, every vector field on S^2 must vanish somewhere. More concretely, it is impossible to comb a hairy sphere without leaving a cowlick. There exist vector fields on S^2 vanishing at one point only (figure 7.4.6, left).

7.4.7. Second application: linking number of two curves

7.4.8. Definition. From here through 7.4.14 we will understand by a *curve* in $\mathbf{R}^3$ a C^∞ embedding f of S^1 in $\mathbf{R}^3$. A *pair of curves* $\{f,g\}$ will consist of two curves f and g such that $f(S^1) \cap g(S^1) = \emptyset$.

A *homotopy between pairs or curves* $\{f, g\}$ and $\{f', g'\}$ will consist of two homotopies F and G (in the sense of 7.4.1), where F is a homotopy between f and f' and G a homotopy between g and g', subject to the condition that $\{F_t, G_t\}$ is a pair of curves for every $t \in [0, 1]$. Such a homotopy between pairs of curves will be denoted by $\{F, G\}$.

7.4.9. Definition. Let $\{f, g\}$ be a pair of curves. The *linking number of* $\{f, g\}$, denoted by $\mathrm{link}(f, g)$, is the degree of the map $\mu \in C^\infty(S^1 \times S^1; S^2)$ defined by

$$\mu(t, s) = \frac{f(t) - f(s)}{\|f(t) - f(s)\|}.$$

7.4.10. We will denote by $\{f_0, g_0\}$ the pair of curves in $\mathbf{R}^3$ consisting of the circles

$$f_0 = \big\{(x, y, z) : z = 0 \text{ and } (x - 2)^2 + y^2 - 1 = 0\big\},$$
$$g_0 = \big\{(x, y, z) : z = 0 \text{ and } (x + 2)^2 + y^2 - 1 = 0\big\}.$$

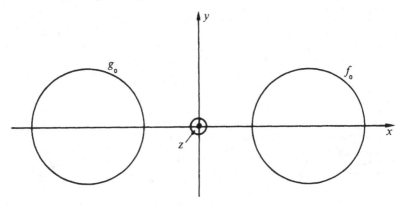

Figure 7.4.10

7.4.11. Definition. A pair of curves $\{f, g\}$ is *unlinked* if it is homotopic to the pair $\{f_0, g_0\}$ defined in 7.4.10. It is *linked* otherwise.

7.4.12. Theorem. *If $\{f, g\}$ and $\{f', g'\}$ are homotopic pairs of curves, we have*

$$\mathrm{link}(f, g) = \mathrm{link}(f', g').$$

If $\mathrm{link}(f, g) \neq 0$ *the pair* $\{f, g\}$ *is linked.*

Proof. Let $\{f, g\}$ and $\{f', g'\}$ be homotopic pairs of curves, and $\{F, G\}$ a homotopy between them. For fixed t, we associate to the pair $\{F_t, G_t\}$ the map $\mu_t : S^1 \times S^1 \to S^2$ defined by

$$\mu_t(x, y) = \frac{F_t(x) - F_t(y)}{\|F_t(x) - F_t(y)\|}.$$

The map $\mu : [0,1] \times S^1 \times S^1 \to S^2$ defined by $\mu(t,x,y) = \mu_t(x,y)$ satisfies the conditions in 7.4.1. Thus μ is a homotopy, and $\deg(\mu_t)$ is a constant by 7.4.3. In particular, $\mathrm{link}(f,g) = \deg(\mu_0) = \deg(\mu_1) = \mathrm{link}(f',g')$.

For the second part, we show that if f and g are unlinked their linking number is zero. Unlinkedness means that $\{f,g\}$ is homotopic to $\{f_0,g_0\}$, so the two linking numbers are the same. But the linking number of $\{f_0,g_0\}$ is zero: since f_0 and g_0 are contained in the xy-plane, the image of μ (cf. 7.4.9) also lies in the xy-plane, and in particular cannot be the whole of S^2. By 7.3.5.1 this implies $\deg(\mu) = 0$. $\square$

7.4.13. Example. Consider the circles in figure 7.4.13: f and g lie in mutually perpendicular planes and their centers are on the line common to those planes. We have $(f,g) = \pm 1$, and in particular f and g are linked.

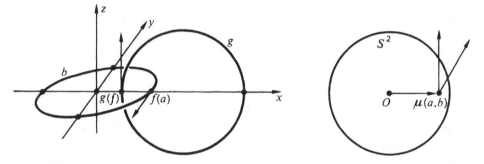

Figure 7.4.13

To calculate $\mathrm{link}(f,g)$, place the axes so that the centers lie on the x-axis and f and g are contained in the xy- and xz-planes, respectively. Now consider the map $\mu : S^1 \times S^1 \to S^2$ from 7.4.9. If $e_1 = (1,0,0) \in S^2$ and $(a,b) \in \mu^{-1}(e_1)$, the vector $f(a) - g(b)$ is parallel to the positive x-axis, so $f(a)$ and $g(b)$ must be placed as shown in the figure. Thus $\mu^{-1}(e_1)$ consists of one point. But e_1 is a regular value, because $T_{(a,b)}(S^1 \times S^1) \simeq T_a S^1 \times T_b S^1$ and, since $f'(a)$ and $g'(b)$ are perpendicular,

$$\big(T_{(a,b)}\mu\big)\big(T_{(a,b)}(S^1 \times S^1)\big) = T_{\mu(a,b)}S^2.$$

It follows from 7.3.1(ii) that $\deg(\mu) = \pm 1$. $\square$

7.4.14. Warning. The converse of the second statement in 7.4.12 is false. For example, in figure 7.4.14 both pairs of curves have linking number zero, but the pair on the right is linked (exercise 7.8.5).

For a sleight-of-hand involving linking numbers, see 7.8.11 and [Hil].

7.4.15. Third application: index of singularities. Let U be open in $\mathbf{R}^d$ and $\xi \in C^\infty(U; \mathbf{R}^d)$ a vector field on U (in the elementary sense, cf. 1.2.1). A point $\mu \in U$ is a *singularity* if $\xi(m) = 0$. It is an *isolated*

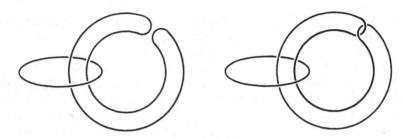

Figure 7.4.14

singularity if, in addition, $\xi(x) \neq 0$ for every $x \neq m$ in a neighborhood V of m in U. We will denote by $S(m, \varepsilon)$ the sphere of center m and radius ε in $\mathbf{R}^d$.

7.4.16. Proposition and definition. *Let m be an isolated singularity of ξ, and $V \in O_m(U)$ a neighborhood such that $\xi(x) \neq 0$ for $x \in V$ distinct from m. The degree of the map $f_\varepsilon : S^{d-1} \to S^{d-1}$, defined for all $\varepsilon > 0$ such that $S(m, \varepsilon) \subset V$ by the formula*

$$x \mapsto \frac{\xi(m + \varepsilon x)}{\|\xi(m + \varepsilon x)\|},$$

does not depend on ε. This number is called the index of ξ at m, and denoted by $\operatorname{ind}_m \xi$.

Proof. Let ε and η be such that $S(m, \varepsilon)$ and $S(m, \eta)$ are contained in V, define $F : [0, 1] \times S^{d-1} \to S^{d-1}$ by $F(t, x) = f_{(1-t)\varepsilon + t\eta}(x)$. Since $\xi \in C^\infty(U; \mathbf{R}^d)$ and ξ does not vanish on $V \setminus \{m\}$, there exists $\alpha > 0$ such that, for every $t \in]-\alpha, 1 + \alpha[$, the denominator of f_t does not vanish for any x. By 7.4.2 the map $F(t, x)$ is a homotopy between $F_0 = f_\varepsilon$ and $F_1 = f_\eta$, and 7.4.3 wraps up the proof. □

7.4.17. Examples

7.4.17.1. The origin is an isolated singularity of $\xi = \operatorname{Id}_{\mathbf{R}^d}$, of index one.

7.4.17.2. By 7.3.5.4, the index of the origin in $\xi = -\operatorname{Id}_{\mathbf{R}^d}$ is $(-1)^d$.

7.4.17.3. Let $\xi(x, y) = (-y, x)$ in $\mathbf{R}^2$. The index of the origin is one.

7.4.17.4. Let $\xi(x, y) = (x^2 - y^2, 2xy)$ in $\mathbf{R}^2$. The index of the origin is two.

  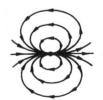

Figure 7.4.17

We are now in a position to generalize 6.3.7 in three ways: the dimension can be arbitrary, the field ξ can have singularities and it does not have to be normal at the boundary.

7.4.18. Theorem. *Let $\xi \in C^\infty(\mathbf{R}^d; \mathbf{R}^d)$ be a vector field having only isolated singularities in $B(0,1)$. Assume, moreover, that ξ points inward at the boundary, that is, $(x \mid \xi(x)) < 0$ for $x \in S^{d-1}$. Then there are only finitely many singularities $x_1, \ldots, x_n$ in $B(0,1)$, and*

$$\int_{i=1}^{n} \mathrm{ind}_{x_i} \xi = (-1)^d.$$

Proof. Since ξ points inward, it has no singularities on S^{d-1}. This means the singularities are isolated points in the compact set $\overline{B}(0,1)$, and consequently finite in number. For the same reason there is an open neighborhood U of $\overline{B}(0,1)$ in $\mathbf{R}^d$ such that ξ does not vanish on $U \setminus \{x_1, \ldots, x_n\}$. We can also find pairwise disjoint closed balls $D_i = \overline{B}(x_i, \varepsilon_i)$ $(i = 1, \ldots, n)$ such that ξ does not vanish on $B(x_i, \varepsilon_i) \setminus \{x_i\}$.

We will now apply Stokes' theorem to the submanifold-with-boundary

$$D = \overline{B}(0,1) \setminus \bigcup_{i=1}^{n} \mathring{D}_i.$$

But first notice that the map $f : x \mapsto \dfrac{\xi(x)}{\|\xi(x)\|} \in S^{d-1}$ is defined and C^∞ on $U \setminus \{x_1, \ldots, x_n\}$, which is a neighborhood of D. Thus, denoting by σ the canonical volume form on S^{d-1}, we have, by Stokes' formula (6.3.8):

$$0 = \int_D d(f^*\sigma) = \int_{S^{d-1}} f^*\sigma - \sum_{i=1}^{n} \int_{S(x_i, \varepsilon_i)} f^*\sigma.$$

For each i we have $\int_{S(x_i, \varepsilon_i)} f^*\sigma = \mathrm{ind}_{x_i} \xi$ (7.4.16), whence

$$\int_{S^{d-1}} f^*\sigma = \sum_{i=1}^{n} \mathrm{ind}_{x_i} \xi.$$

To calculate the integral on the left-hand side, set

$$F(t, x) = \frac{(1-t)\xi(x) - tx}{\|(1-t)\xi(x) - tx\|}.$$

Then there exists $\alpha > 0$ such that F is C^∞ on $]-\alpha, 1 + \alpha[\times S^{d-1}$: indeed, $(x \mid \xi(x)) < 0$ implies that, for all $x \in S^{d-1}$, the denominator of F is non-zero for t in an open interval containing $[0,1]$ (see figure 7.4.18, right).

By 7.4.2 this means that F is a homotopy between $F_0 = f|_{S^{d-1}}$ and $F_1 = -\mathrm{Id}_{S^{d-1}}$. But then 7.3.5.4 and 7.4.3 give

$$\int_{S^{d-1}} f^*\sigma = \int_{S^{d-1}} (-\mathrm{Id})^*\sigma = (-1)^d. \qquad \square$$

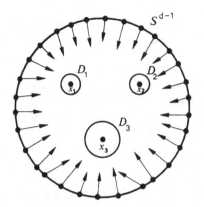 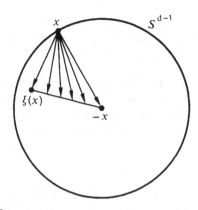

Figure 7.4.18

7.4.19. Corollary. *If a vector field $\xi \in C^{\infty}(\mathbf{R}^d; \mathbf{R}^d)$ points inward along S^{d-1}, it has at least one singularity in $B(0,1)$.*   □

See also section 7.7 for a discussion on arbitrary manifolds.

7.5. Volume of Tubes and the Gauss–Bonnet Formula

Let E be an n-dimensional Euclidean space and X a compact, connected, d-dimensional submanifold of E. Denote by $S(E)$ the unit sphere in E, by NUX the unitary normal bundle of X (6.7.6) and by $\gamma : NUX \to S(E)$ the Gauss map defined by $\gamma(x, v) = v$ (6.8.13). If we orient E, the sphere $S(E)$, being the boundary of $B_E(0,1)$, gets a canonical orientation, as does NUX (6.7.22 and 6.7.26). It then makes sense to consider the degree of $\gamma : NUX \to S(E)$. This does not depend on the orientation chosen for E.

The formulas in 6.9.15 take the following aspect:

$$\int_X K_d \delta = \int_{NUX} W_d \Psi = \int_{NUX} \gamma^* \Sigma = \deg(\gamma) \int_{S(E)} \Sigma = \text{vol}\, S(E) \cdot \deg \gamma.$$

To calculate $\deg(\gamma)$ we use the following trick: To each $(x, v) \in NUX$, associate the function $f_{(x,v)} \in C^{\infty}(X)$ defined by

7.5.1 $f_{(x,v)} : y \to (v \mid y).$

We have seen in 4.1.4.2 that $y \in X$ is a critical point of $f_{(x,v)}$ if and only if $v \in N_y X$. Thus x is a critical point of $f_{(x,v)}$, for every v.

The main tools in the sequel will be Morse theory (4.2.11 and 4.2.24) and the following lemma:

7.5.2. Lemma. *Let $(x, v) \in NUX$. The critical point x of $f_{(x,v)}$ is non-degenerate if and only if the Gauss map $\gamma : NUX \to S(E)$ is regular at (x, v). If, in addition, d is even, we have*

$$\text{sgn}(J_{(x,v)}\gamma) = (-1)^{\text{ind}_x f_{(x,v)}}.$$

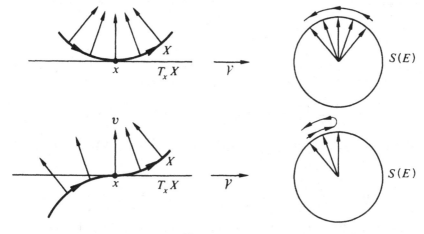

Figure 7.5.2

This result is easy to visualize when X is a plane curve. In figure 7.5.2, $f_{(x,v)}$ is the height function (coordinate parallel to v), and x is a critical point for it. The critical point is non-degenerate if the curvature of X there is non-zero (see 8.2.2.15, for example), in which case γ is regular at (x, v). Conversely, if x is an inflection point, for instance, the image of γ in $S(E)$ has a stationary point, that is, γ is no longer regular. The reader can verify the formula for $\text{sgn}(J_{(x,v)}\gamma)$. Section 8.5 makes this discussion of plane curves more rigorous.

Proof. Take $(x, v) \in NUX$, and a parametrization (U, h) of X at x. Let $(U \times \mathbf{R}^{n-d}, H)$ be a parametrization of NUX at (x, v) of the kind defined in 6.7.11 and satisfying the following conditions (where the notation is borrowed from 6.7.11):

(i) $h(u) = x$;

(ii) $\left\{ \dfrac{\partial h}{\partial u_1}(u), \ldots, \dfrac{\partial h}{\partial u_d}(u) \right\}$ is an orthonormal basis for $T_x X$, positively oriented for some local volume form β;

(iii) $\nu_{d+1}(u) = v$;

(iv) the matrix

$$\left(\frac{\partial^2 (f_{(x,v)} \circ h)}{\partial u_i \partial u_j}(u) \right) \quad (1 \le i, j \le d)$$

is diagonal. This can be achieved by applying the reduction of quadratic forms in the orthogonal group [Dix68, 36.9.4].

By definition 4.2.11, $f_{(x,v)}$ is non-degenerate at (x, v) if and only if

$$\prod_{i=1}^{d} \frac{\partial^2}{\partial u_i^2}(f_{(x,v)} \circ h)(u) \neq 0.$$

But $f_{(x,v)} \circ h(u) = (v \mid h(u))$. By the construction of H, we have

$$\left(\nu_{d+1}(w) \,\Big|\, \frac{\partial h}{\partial u_i}(w)\right) = 0$$

for all $w \in U$ and $i = 1, \ldots, d$. Differentiating with respect to u_j (the j-th coordinate in $\mathbf{R}^d \supset U$) and evaluating at u, we obtain

$$0 = \left(\frac{\partial \nu_{d+1}}{\partial u_j}(u) \,\Big|\, \frac{\partial h}{\partial u_i}(u)\right) + \left(\nu_{d+1}(u) \,\Big|\, \frac{\partial^2 h}{\partial u_i \partial u_j}(u)\right).$$

If $\cdot^T$ denotes the orthogonal projection from E onto $\theta(T_x X)$ (cf. 6.8.5), this becomes

$$0 = \left(\left(\frac{\partial \nu_{d+1}}{\partial u_j}\right)^T(u) \,\Big|\, \frac{\partial h}{\partial u_j}(u)\right) + \left(v \,\Big|\, \frac{\partial^2 h}{\partial u_i \partial u_j}(u)\right).$$

But differentiating $(v \mid w)$ with respect to w and evaluating at u we get

$$\left(v \,\Big|\, \frac{\partial^2 h}{\partial u_i \partial u_j}(u)\right) = \frac{\partial^2}{\partial u_i \partial u_j}(v \mid h(u)).$$

Substituting above,

$$\left(\left(\frac{\partial \nu_{d+1}}{\partial u_j}\right)^T(u) \,\Big|\, \frac{\partial h}{\partial u_j}(u)\right) = -\frac{\partial^2}{\partial u_i \partial u_j}(f_{(x,v)} \circ h)(u).$$

By assumptions (ii) and (iv), we can decompose $\left(\dfrac{\partial \nu_{d+1}}{\partial u_j}\right)^T(u)$, obtaining

7.5.3
$$\left(\frac{\partial \nu_{d+1}}{\partial u_i}\right)^T(u) = -\frac{\partial^2}{\partial u_i^2}(f_{(x,v)} \circ h)(u)\frac{\partial h}{\partial u_i}(u).$$

Thus the local form β of condition (ii) satisfies

$$\beta\left(\left(\frac{\partial \nu_{d+1}}{\partial u_1}\right)^T(u), \ldots, \left(\frac{\partial \nu_{d+1}}{\partial u_d}\right)^T(u)\right) = (-1)^d \prod_{i=1}^{d} \frac{\partial^2(f_{(x,v)} \circ h)}{\partial u_i^2}(u),$$

again because we're working with an orthonormal, oriented basis. Since $\gamma^*(\Sigma) = W_d \circledS$ (cf. 6.8.15) and, by 6.8.19,

$$W_d(s, v) = \beta\left(\left(\frac{\partial \nu_{d+1}}{\partial u_1}\right)^T(u), \ldots, \left(\frac{\partial \nu_{d+1}}{\partial u_d}\right)^T(u)\right),$$

we conclude that (x, v) is a non-degenerate critical point of $f_{(x,v)}$ if and only if γ is regular at (x, v).

If, in addition, d is even, the factor $(-1)^d$ disappears, and we obtain

$$\text{sgn}(J_{(x,v)}\gamma) = \text{sgn}(W_d(x,v)) = \text{sgn} \prod_{i=1}^{d} \frac{\partial^2(f_{(x,v)} \circ h)}{\partial u_i^2}(u) = (-1)^{\text{ind}_z f_{(x,v)}}.$$

$\square$

7.5.4. Theorem. *Let X be a compact, d-dimensional submanifold of an n-dimensional Euclidean space E. If d is even, the degree of the Gauss map γ is equal to $\chi(X)$, the Euler characteristic of X, and the Gauss-Bonnet formula holds:*

$$\int_X K_d \delta = \chi(X) \, \text{vol}(S^{n-1}).$$

Proof. By Sard's theorem (4.3.6), the map $\gamma : NUX \to S(E)$ has regular values; for such a regular value $v \in S(E)$ the inverse image $\gamma^{-1}(v) = \bigcup_{i=1}^{k}\{(x_i, v)\}$ is finite.

Denote by $f_v : X \to \mathbf{R}$ the map $y \mapsto (v|y)$. We have $f_v \in C^\infty(X)$, and y is a critical point of f_v if and only if $v \in N_y X$, if and only if $\gamma(y, v) = v$, if and only if y is one of the x_i. Thus $\{x_1, \ldots, x_k\}$ is the set of critical points of f_v. Since v is a regular value for γ, it follows form 7.5.2 that each x_i is a non-degenerate critical point of f_v. Applying Morse's theorem (4.2.24.4) we get

$$\chi(X) = \sum_{h=0}^{d} (-1)^h C_h(f_v),$$

where $C_h(f_v)$ is the number of critical points of f_v with index h. Thus

$$\chi(X) = \sum_{h'=0}^{h' \leq d/2} C_{2h'}(f_v) - \sum_{h''=0}^{h'' \leq d/2} C_{2h''+1}(f_v)$$
$$= \#\{x_i : \text{ind}_{x_i} f_v \text{ is even}\} - \#\{x_i : \text{ind}_{x_i} f_v \text{ is odd}\}.$$

Since d is even, we again get from lemma 7.5.2

$$\chi(X) = \#\{x_i : \text{sgn}(J_\gamma(x_i, v)) = +1\} - \#\{x_i : \text{sgn}(J_\gamma(x_i, v)) = -1\}.$$

But this is exactly the degree of γ, by 7.3.1(ii).

The equality $\int_X K_d \delta = \chi(X) \, \text{vol}(S^{n-1})$ follows from the formulas we recalled at the beginning of this section, since $\text{vol}(S(E)) = \text{vol}(S^{n-1})$. $\square$

7.5.5. Corollary. *Let X be a compact, d-dimensional submanifold of an n-dimensional Euclidean space E, and assume d even. The last term in the formula for $\text{vol}(\text{Tub}^\varepsilon X)$ (6.9.9) is*

$$\frac{\varepsilon^n}{n} \, \text{vol}(S^{n-1})\chi(X).$$

In particular, if X is a surface (i.e., $d = 2$), we have

$$\text{vol}(\text{Tub}^\varepsilon X) = \varepsilon^{n-2} \text{vol}(B_{n-2}(0,1)) \text{area}(X) + \frac{\varepsilon^n}{n} \text{vol}(S^{n-1})\chi(X),$$

so this volume only depends on n, ε, $\text{area}(X)$ and $\chi(X)$, and not on the embedding of X in E.   □

7.5.6. Examples

7.5.6.1. Take $X = S^2 \subset \mathbf{R}^3$. We have $\chi(S^2) = 2$ (recall the calculation of de Rham groups in 5.7.1), so

$$\text{vol}(\text{Tub}^\varepsilon X) = 8\pi\varepsilon + \frac{8\pi}{3}\varepsilon^3.$$

We get the same number by calculating the volume of the spherical shell $\text{Tub}^\varepsilon X = B_3(0, 1 + \varepsilon) - B_3(0, 1 - \varepsilon)$:

$$\text{vol}(\text{Tub}^\varepsilon X) = \frac{4\pi}{3}\left((1 + \varepsilon)^3 - (1 - \varepsilon)^3\right) = \frac{4\pi}{3}(2\varepsilon^3 + 6\varepsilon).$$

7.5.6.2. Let E be Euclidean of arbitrary dimension n, and $X = T^2$. By 5.8.1 we have $\chi(X) = 0$, so $\text{vol}(\text{Tub}^\varepsilon X) = \text{vol}(B_{n-2}(0, \varepsilon)) \text{area}(X)$.

7.5.6.3. We admit without proof that the Euler characteristic of a torus with g holes (4.2.24.3) is $2(1 - g)$. If we embed such a surface X is $\mathbf{R}^3$ we get

$$\text{vol}(\text{Tub}^\varepsilon X) = 2\varepsilon \text{area}(X) - \frac{8\pi\varepsilon^3}{3}(1 - g).$$

Notice that this number is less than the product $2\varepsilon \text{area}(X)$.

7.5.7. Remark. The Gauss–Bonnet formula holds also for (abstract) riemannian manifolds, for a suitably defined integrand. The right integrand turns out to be a kind of curvature, a function of x that depends only on the riemannian structure. (When X is embedded in n-dimensional Euclidean space, the curvature equals $K_d/\text{vol}(S^{n-1})$.) This formula is important because it affords an invariant $\int_X K_d\delta$ that does not depend on a particular embedding of X. For a proof, see [KN69, vol. II, p. 318].

7.6. Self-Maps of the Circle

In algebraic topology one studies the degree of *continuous* maps between oriented, compact, connected, *topological* manifolds. In the particular case of differentiable objects, this theory coincides with the one developed above [Gre67, p. 125]. We will only pursue this topic in the case $d = 1$, because we will need some of the results in the next chapter (9.1.11 and 9.4).

Compact, one-dimensional manifolds are homeomorphic to S^1 (see 3.4.1 for the C^1 case). Thus we can restrict ourselves to maps from S^1 into itself. In this case the study is elementary because we have the covering map $p : \mathbf{R} \to S^1$ (2.4.3) and the cohomology (de Rham groups) can be replaced by $\mathbf{Z}$ acting on $\mathbf{R}$.

In fact we take the covering map p to be

7.6.1 $p : \mathbf{R} \ni t \mapsto (\cos t, \sin t) \in S^1 \subset \mathbf{R}^2.$

If σ is the canonical length form on S^1, we have $p^*\sigma = dt$, the canonical length form on $\mathbf{R}$ (cf. 5.3.17.2 and 5.2.8.4).

7.6.2. Lemma. *Let $I = [a, b] \subset \mathbf{R}$, $f \in C^0([a, b]; S^1)$ and $u \in p^{-1}(f(a))$. There exists a unique $\overline{f} \in C^0(I; \mathbf{R})$ such that $p \circ \overline{f} = f$ and $\overline{f}(a) = u$. We say that $\overline{f}$ is a lifting of f.*

$$\begin{array}{ccc} & & \mathbf{R} \\ & \overline{f} \nearrow & \big\downarrow \\ [a, b] & \xrightarrow{\ f\ } & S^1 \end{array}$$

Proof. This is a particular case of the following result (itself a very particular case of the theorem on the lifting of homotopies—see [Gre67, p. 18]):

7.6.3. Lemma. *For $F \in C^0([0, 1] \times [a, b]; S^1)$ and $u \in p^{-1}(F(0, a))$, there exists a unique $\overline{F} \in C^0([0, 1] \times [a, b]; \mathbf{R})$ such that $p \circ \overline{F} = F$ and $\overline{F}(0, a) = u$.*

Proof. We take care of uniqueness first. If $\overline{F}$ and $\overline{F}'$ satisfy the desired conditions, we have

$$\overline{F}' - \overline{F} \in C^0([0, 1] \times [a, b]; \mathbf{R})$$

and $p \circ \overline{F}' = p \circ \overline{F}$, so for every $t \in [0, 1]$ and every $s \in [a, b]$ the difference $(\overline{F}' - \overline{F})(t, s)$ is an integral multiple of 2π. Since $\overline{F}' - \overline{F}$ is continuous and has values in a discrete topological space, it is locally constant, and constant because $[0, 1] \times [a, b]$ is connected. Thus

$$(\overline{F}' - \overline{F})(t, s) = (\overline{F}' - \overline{F})(0, a) = u - u = 0.$$

To show existence, notice that F, being continuous on a compact set, is uniformly continuous. Thus there exists $\varepsilon > 0$ such that, for every

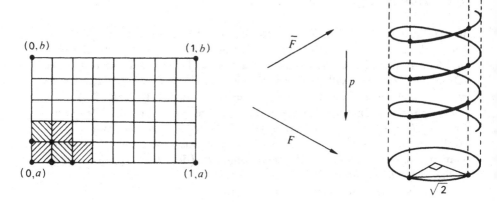

Figure 7.6.3

$x, y \in [0,1] \times [a,b]$ satisfying $\|x - y\| \le \varepsilon$, the distance between $F(x)$ and $F(y)$ in the metric induced from $\mathbf{R}^2$ is no more than $\sqrt{2}$, or a quarter-circle.

We now divide up $[0,1] \times [a,b]$ into rectangles of diameter less than ε. The restriction of F to each little rectangle can be lifted to a map onto $\mathbf{R}$, because its image is smaller than a quarter-circle in S^1 (2.4.3). We order the rectangles so that each intersects the previous one, then lift F one rectangle at a time, choosing the lifting on each rectangle to agree with the lifting on the previous one at one common point. By the uniqueness part above, the two partial liftings agree all along the common boundary. The continuity of $\overline{F}$ follows locally, by expressing $\overline{F}$ as a composition of F with the local inverse of a covering map. $\square$

Now let $f \in C^0(S^1; S^1)$ be an arbitrary continuous map. Consider the first S^1 as $\mathbf{R}/L\mathbf{Z}$, with $L > 0$, and identify $f \in C^0(S^1; S^1)$ with a map $f \in C^0(\mathbf{R}; S^1)$, periodic of period L. Take $t \in \mathbf{R}$, $u \in p^{-1}(f(t))$ and use 7.6.2 to define $\overline{f}$ associated with the restriction of f to $[t, t+L]$ (thus $\overline{f}(t) = u$). Since

$$p\big(\overline{f}(t+L)\big) = f(t+L) = f(t) = p\big(\overline{f}(t)\big),$$

we have $\overline{f}(t+L) - \overline{f}(t) \in 2\pi\mathbf{Z}$. Then:

7.6.4. Theorem and definition. *In the notation above, the integer k such that $\overline{f}(t+L) - \overline{f}(t) = 2k\pi$ is independent of $t \in \mathbf{R}$ and $u \in p^{-1}(f(t))$. This number is called the degree of $f \in C^0(S^1; S^1)$ and denoted by $\deg(f)$. If $f \in C^\infty(S^1; S^1)$ this definition coincides with the one in section 7.3.*

Proof. First we show that k does not depend on u. Let $u' \in p^{-1}(f(t))$. The map $\overline{f}'$ defined by

$$\overline{f}'(s) = \overline{f}(s) + u' - u$$

is a lifting of f such that $\overline{f}'(t) = \overline{f}(t) + u' - u = u + u' - u = u'$, and by uniqueness is the only such lifting. Since $\overline{f}' - \overline{f}$ is a constant, we have

$$\overline{f}'(t + L) - \overline{f}(t + L) = \overline{f}'(t) - \overline{f}(t),$$

whence

$$\overline{f}'(t + L) - \overline{f}'(t) = \overline{f}(t + L) - \overline{f}(t).$$

Next, k does not depend on t. For $\overline{f}$ is a lifting of the restriction of f to $[t, t+L]$. If we apply lemma 7.6.2 to the restriction of f to $[t+L, t+2L]$, with $u = \overline{f}(t + L)$, we can extend $\overline{f}$ into a continuous lifting of the restriction of f to $[t, t + 2L]$. Continuing in this way we get a lifting $\overline{f} \in C^0(\mathbf{R}; \mathbf{R})$ of $f \in C^0(\mathbf{R}; S^1)$. Then $s \mapsto \overline{f}(s + L) - \overline{f}(s)$ is constant, because it has values in $2\pi\mathbf{Z}$.

Finally we show that this definition is compatible with section 7.3. Take $f \in C^\infty(S^1; S^1)$, identify it with $f : \mathbf{R} \to S^1$ periodic of period L, and construct the lifting $\overline{f} : \mathbf{R} \to \mathbf{R}$ as in the previous paragraph:

7.6.4.1

$$
\begin{array}{ccc}
\mathbf{R} & \xrightarrow{\ \overline{f}\ } & \mathbf{R} \\
{\scriptstyle p'}\downarrow & \searrow{\scriptstyle f} & \downarrow{\scriptstyle f} \\
\mathbf{R}/L\mathbf{Z} = S^1 & \xrightarrow[f]{} & S^1 = \mathbf{R}/2\pi\mathbf{Z}.
\end{array}
$$

Since $f = p \circ \overline{f}$ we get

$$f^*\sigma = (p \circ \overline{f})^*\sigma = \overline{f}^*(p^*\sigma) = \overline{f}^*(dt)$$

by 5.2.8.4. But 2.5.23.2 and 2.5.17.3 give

$$\overline{f}^*(dt)(1_t) = dt\big((T\overline{f})(1_t)\big) = \frac{d\overline{f}}{dt},$$

the right-hand side expressing the usual derivative of $\overline{f} : \mathbf{R} \to \mathbf{R}$. Still identifying $f : S^1 \to S^1$ with $f : \mathbf{R} \to S^1$ we have

$$\int_{S^1} f^*\sigma = \int_t^{t+L} f^*\sigma = \int_t^{t+L} \frac{d\overline{f}}{dt}\,dt = \overline{f}(t + L) - \overline{f}(t) = 2\pi\deg(f).$$

This number is also $\deg(f)\int_{S^1}\sigma$ in the sense of section 7.3. $\qquad\square$

7.6.5. Theorem (invariance under homotopy). *If $F : [0, 1] \times S^1 \to S^1$ is continuous, the degree of the sections $F_s : x \mapsto F(s, x)$ does not depend on s. In particular, $\deg F_0 = \deg F_1$.*

Proof. Consider F as an element of $C^0\big([0, 1] \times \mathbf{R}; S^1\big)$, periodic of period L with respect to the second variable. By lemma 7.6.3 we can lift F to a map $\overline{F}$ on $[0, 1] \times [0, L]$. Then

$$2\pi\deg F_s = \overline{F}_s(L) - \overline{F}_s(0) = \overline{F}(s, L) - \overline{F}(s, 0) \in 2\pi\mathbf{Z}.$$

The map $s \mapsto \overline{F}(s, L) - \overline{F}(s, 0)$ is continuous and has values in a discrete space. By connectedness it is constant. $\qquad\square$

7.6.6. Proposition. *For any* $f, g \in C^0(S^1; S^1)$ *we have* $\deg(g \circ f) = \deg(g) \deg(f)$.

Proof. See exercise 7.8.19.   □

7.6.7. Example. Let γ be a loop in $\mathbf{R}^2$, that is, $\gamma \in C^0([a, b]; \mathbf{R}^2)$ satisfies $\gamma(a) = \gamma(b)$. We can think of γ as an element of $C^0(\mathbf{R}; \mathbf{R}^2)$ periodic of period $b - a$, hence also an element of $C^0(S^1; \mathbf{R}^2)$.

Now take $x \in \mathbf{R}^2$ not in the image of γ, and consider the map

$$\tilde{\gamma} : S^1 \ni t \mapsto \frac{\gamma(t) - x}{\|\gamma(t) - x\|} \in S^1.$$

7.6.8. Definition. The degree of $\tilde{\gamma}$ is called the *winding number* of x around γ.

The winding number plays a fundamental role in the theory of functions of one complex variable. It also figures prominently in the proof of Jordan's theorem (9.1.11 and 9.2).

7.6.9. The converse of theorem 7.6.5 is true (exercise 7.8.20). The two together are equivalent to saying that the fundamental group of $\pi_1(S^1)$ is isomorphic to $\mathbf{Z}$ [Gre67, p. 13].

7.7. Index of Vector Fields on Abstract Manifolds

7.7.0 | In this section all objects are C^∞. |

7.7.1. Definition. Let X and Y be manifolds. A homotopy $F : [0, 1] \times X \to Y$ is called an *isotopy* if F_t is a diffeormorphism for every $t \in [0, 1]$. In this case we say that the diffeomorphisms F_0 and F_1 are *isotopic*.

Isotopy is an equivalence relation.

7.7.2. Push-forwards. Let X, Y be manifolds and $f : X \to Y$ a diffeomorphism. If ξ is a vector field on X, the *push-forward* of ξ is the vector field $f_* \xi$ on Y defined by

$$f_* \xi = Tf \circ \xi \circ f^{-1}.$$

7.7.2.1

$$
\begin{array}{ccc}
TX & \xrightarrow{Tf} & TY \\
{\scriptstyle \xi}\big\uparrow{\scriptstyle p} & {\scriptstyle f_*\xi} & {\scriptstyle p}\big\uparrow \\
X & \xleftarrow[f^{-1}]{} & Y
\end{array}
$$

If $f \in \mathrm{Diff}(X; Y)$ and $g \in \mathrm{Diff}(Y; Z)$ we have $(g \circ f)_* = g_* \circ f_*$.

A point $x \in X$ is called a *singularity* of a vector field ξ if $\xi(x) = 0$. A singularity x is said to be *isolated* if $\xi(y) \neq 0$ for every $y \in U \setminus \{x\}$, where U is a neighborhood of x in X.

7.7.3. Lemma (invariance of the index under diffeomorphisms). *Let U and U' be open subsets of $\mathbf{R}^n$ and $f : U \to U'$ a diffeomorphism. If ξ is a vector field on U having an isolated singularity at $x_0 \in U$ and $\eta = f_* \xi$ is its push-forward, $f(x_0)$ is an isolated zero of η and*

$$\text{ind}_{x_0} \xi = \text{ind}_{f(x_0)} \eta.$$

Thanks to this lemma the following definition makes sense:

7.7.4. Definition. Let X be a manifold and ξ a vector field on X having an isolated singularity at x_0. The *index* of ξ at x_0, denoted by $\text{ind}_{x_0} \xi$, is the integer $\text{ind}_{\phi(x_0)}(\phi_* \xi)$, where (U, ϕ) is a chart at x_0.

Indeed, for another chart (V, ψ) at x_0 we have $\psi \circ \phi^{-1} \in \text{Diff}\big(\phi(U); \psi(V)\big)$, and the lemma implies that $\text{ind}_{\psi(x_0)}(\psi_* \xi) = \text{ind}_{\phi(x_0)}(\phi_* \xi)$.

In order to prove 7.7.3 we will need an auxiliary result:

7.7.5. Lemma. *If $U \subset \mathbf{R}$ is open and star-shaped at the origin, and $f \in \text{Diff}\big(U; f(U)\big)$ takes the origin into itself, f is isotopic to the identity.*

Proof. Define $F : [0, 1] \times U \to \mathbf{R}^n$ by

$$F(t, x) = \begin{cases} f(tx)/t & \text{for } t \in \;]0, 1], \\ f'(0)(x) & \text{for } t = 0. \end{cases}$$

Each F_t, for $t \in [0, 1]$, is clearly a diffeomorphism. We have to check that the conditions in 7.4.1 are satisfied; but this follows from writing $f = \sum_i x_i g_i$ as in 4.2.13, the g_i being vector-valued functions this time.

Thus f is isotopic to $f'(0) \in L(\mathbf{R}^n; \mathbf{R}^n)$ and $f'(0)$ preserves orientation. Now the linear group $\text{Isom}^+(\mathbf{R}^n; \mathbf{R}^n)$ of orientation-preserving isomorphisms of $\mathbf{R}^n$ is connected (see exercise 7.8.22, for example), hence path-connected because it is a submanifold of $\mathbf{R}^{n^2}$ (cf. 2.2.13). A path in $\text{Isom}^+(\mathbf{R}^n; \mathbf{R}^n)$ gives an isotopy between $f'(0)$ and Id_U. $\square$

Proof of lemma 7.7.3. Take first the case that f preserves orientation. We can assume that $x_0 = 0$ and $f(0) = 0$, and also that U is star-shaped (by restriction). By 7.7.5 there is an isotopy F between f and Id_U, that is, a family $F_t \in \text{Diff}\big(U; F_t(U)\big)$ $\big(t \in [0, 1]\big)$. Referring to 7.4.16, we can find, by uniform continuity and by the compactness of $[0, 1]$ and of spheres, a positive number ε such that $\eta_t = (F_t)_* \xi$ does not vanish on $B(0, \varepsilon) \setminus 0$ for any $t \in [0, 1]$. Now it suffices to apply definition 7.4.16 and the invariance of the degree under homotopy to see that $\text{ind}_0 \eta_t$ is constant, and in particular

$$\text{ind}_0 f_* \xi = \text{ind}_0 \eta_0 = \text{ind}_0 \eta_1 = \text{ind}_0 \xi.$$

If f reverses orientation, we reduce to the previous case by replacing ξ by $\rho_*\xi$ and f by $f \circ \rho^{-1}$, where ρ is an arbitrary hyperplane reflection in $\mathbf{R}^n$. Since $\operatorname{ind}_0(\rho_*\xi) = \operatorname{ind}_0 \xi$ and $f_*\xi = (f \circ \rho^{-1})_*(\rho_*\xi)$, we're done. $\square$

7.7.6. Cultural digression. Now that we know how to define the index of vector fields on arbitrary manifolds, we can ask whether there exists an analog of theorem 7.4.18 for manifolds. There is, in fact, the following result, which we will not prove:

7.7.6.1. Theorem. *If X is a compact manifold and ξ a vector field on X having only isolated singularities $x_1, \ldots, x_n$, then*

$$\sum_{i=1}^{n} \operatorname{ind}_{\xi_i} \xi = \chi(X),$$

where $\chi(X)$ denotes the Euler characteristic of X.

This theorem still holds for manifolds-with-boundary, as long as the vector field points outward all along the boundary. We have already seen two particular cases of this theorem:

(i) 7.4.18. This is the case $X = \overline{B}(0,1)$, so $\chi(X) = 1$ by 4.2.22.2 and 5.6.3. The factor $(-1)^d$ comes from our having assumed that ξ points inward: to apply 7.7.6.1 we have to flip signs, which multiplies the indices by $(-1)^d$.

(ii) 7.4.5. This is the case $X = S^d$, so $\chi(X) = 1 + (-1)^d$ by 4.2.22.2 and 5.7.1.

The proof of 7.7.6.1, as presented in [Mil69, p. 32–41], should pose no difficulty to the reader. Essential use is made of tubular neighborhoods (section 2.7).

7.7.7. An application of 7.7.5. Here is an application that will be useful in section 9.8, and which allows the quick determination of the index in certain cases:

7.7.8. Proposition. *Let ξ be a vector field on $U \in O(\mathbf{R}^n)$ and x_0 an isolated singularity of ξ. If $\xi'(x_0) \in \operatorname{Isom}(\mathbf{R}^n; \mathbf{R}^n)$, we have*

$$\operatorname{ind}_{x_0} \xi = \begin{cases} 1 & \text{if } J_{x_0}(\xi) > 0, \\ -1 & \text{if } J_{x_0}(\xi) < 0. \end{cases}$$

Here ξ is seen as a map $U \to \mathbf{R}^n$ and J denotes the usual jacobian (0.2.8.9).

Proof. Just notice that, by 0.2.22, ξ is a diffeomorphism on a neighborhood of 0. Thus there exists an isotopy F between ξ and Id_U if ξ preserves orientation, that is, if $J_{x_0}\xi > 0$. Each F_t can be considered as a vector

Figure 7.8.6

field having an isolated singularity at x_0, and by the invariance of the degree under homotopy we get

$$\operatorname{ind}_{x_0} \xi = \operatorname{ind}_{x_0} F_0 = \operatorname{ind}_{x_0} F_1 = \operatorname{ind}_{x_0} \operatorname{Id}_U = 1.$$

If ξ reverses orientation, we reduce the the previous case by considering $\xi \circ \phi$, where ϕ is a hyperplane reflection.   □

7.8. Exercises

All objects are C^∞ and all manifolds oriented, unless we say otherwise.

7.8.1. Let X be an oriented, compact, d-dimensional manifold (not necessarily connected). Calculate $R^d(X)$.

7.8.2. Prove 7.3.4 in two ways.

7.8.3. Let X be a compact manifold, Y a compact, connected manifold of the same dimension, and $f : X \to Y$ a map that preserves orientation at all regular points. Prove that if f is not surjective it has no regular points.

7.8.4. Let f and g be maps from X into S^d such that $\| f(x) - g(x) \| < 2$ for every $x \in X$, where $\| \cdot \|$ denotes the Euclidean norm in $\mathbf{R}^{d+1}$. Prove that f and g are homotopic.

Prove that if $\dim X < d$ every map $X \to S^d$ is homotopic to a constant map.

7.8.5. Compute the linking number of the two pairs of curves in figure 7.4.14.

7.8.6. Compute the linking number of the pair shown in figure 7.8.6, where the coil winds around the circle n times.

7.8.7. Generalize 7.4.18 to the case where we have spherical holes in $B(0,1)$ and the field points into the manifold along the holes.

7.8.8. Let $f : S^d \to S^d$ be a map such that $\deg(f) \neq (-1)^{d+1}$. Show that f leaves at least one point fixed.

7.8.9. Prove that every map $S^d \to T^d$ has degree zero. Prove that if X is a compact, connected, d-dimensional manifold such that $R^k(X) = 0$ for some $0 < k < d$, every map $X \to T^d$ has degree zero.

7.8.10. Let S^2 be the sphere with equation $x_1^2 + x_2^2 + x_3^2 = 1$ in $\mathbf{R}^3$. Denote by s^+ and s^- the stereographic projections onto the $x_1 x_2$-plane centered at $(0,0,1)$ and $(0,0,-1)$, respectively, and set $z = x_1 + ix_2$. To every polynomial $P : \mathbf{C} \to \mathbf{C}$, associate a map $\overline{P} : S^2 \to S^2$ defined by

$$\begin{cases} \overline{P}(x) = (s^+)^{-1}\big(P(s^+(x))\big) & \text{if } x \neq (0,0,1), \\ \overline{P}(0,0,1) = (0,0,1). \end{cases}$$

(a) Show that $\overline{P}$ is C^∞.
(b) Show that $\overline{P}$ has degree n if $P(z) = z^n$.
(c) If Q is a polynomial of degree n, $\overline{Q}$ is homotopic to the map $\overline{P}$ considered in (b). Deduce a proof of the fundamental theorem of algebra (every polynomial over $\mathbf{C}$ has a root).

7.8.11. Let M be the Möbius band, the image of the following map from $[0, 4\pi] \times [0, 3/4]$ into $\mathbf{R}^3$:

$$(\theta, r) \mapsto \left(\cos\theta + r\cos\theta \cos\frac{\theta}{2}, \ \sin\theta + r\sin\theta \cos\frac{\theta}{2}, \ r\sin\frac{\theta}{2} \right).$$

Find the linking number between the two curves $r = \frac{1}{4}$ and $r = \frac{3}{4}$. Make a drawing (this is more instructive than the traditional cut-and-paste construction—cf. [34], p. 294–297).

7.8.12. Let $f : S^d \to S^d$ be differentiable. Show that in the each of the following cases f takes at least one pair of diametrically opposite points into diametrically opposite points:
(a) d is even and $\deg(f) \neq 0$;
(b) $\deg(f)$ is odd.

7.8.13. Let f and g be a pair of curves. Assume there exists an open neighborhood $U \subset \mathbf{R}^2$ of $\overline{B}(0,1)$ and an embedding $F : U \to \mathbf{R}^3$ such that $F|_{S^1} = f$ and $F\big(B(0,1)\big) \cap g(S^1)$ is a finite set $\{x_1, \ldots, x_n\}$. Assume moreover that

$$T_{x_i}\big(F(B(0,1))\big) \cap T_{x_i}\big(g(S^1)\big) = \{0\}$$

for each x_i. Define $\operatorname{sgn}(x_i)$ to be $+1$ if the union of a positively oriented basis for $T_{x_i} F\big(B(0,1)\big)$ with a positively oriented basis for $T_{x_i}\big(g(S^1)\big)$ gives a positively oriented basis for $\mathbf{R}^3$, and -1 otherwise.

Prove that $\operatorname{link}(f,g) = \sum_i \operatorname{sgn}(x_i)$. (Hint: consider pairwise disjoint little disks C_i around each $F^{-1}(x_i) \subset B(0,1)$, and apply Stokes' theorem to the submanifold-with-boundary

$$\left(\overline{B}(0,1) \setminus \bigcup_i C_i\right) \times S^1 \subset U \times S^1.$$

Then show that for g arbitrary and f a tiny circle around a point of $g(S^1)$ the linking number of f and g is ± 1, where the sign is to be determined.)

7.8.14. The traveler in equilibrium. Consider in $\mathbf{R}^3$ a uniform gravitational field. A truck moves in the xy-plane, and its position is a C^∞ function of time. Attached to the floor of the truck is one end of a bar AB of uniform density. Prove that for some initial position of the bar, its free end B remains above the floor at all times.

7.8.15. Ampère's theorem. Let f and g be a pair of curves in the sense of 7.4.7, except that here f and g are periodic maps from $\mathbf{R}$ into $\mathbf{R}^3$, of period 2π.

(a) Express $\operatorname{link}(f,g)$ in terms of the coordinates of f and g and their derivatives.

(b) Assume that f is an electric circuit traversed by an electric current of uniform intensity i. The magnetic field created by this circuit at a point p is

$$i \int_0^{2\pi} \frac{f(t) - p}{\|f(t) - p\|^3} \times f'(t)\, dt.$$

Calculate the "circulation" of the magnetic field along g, thus proving Ampère's theorem.

(c) Assume that $g\big([0, 2\pi]\big)$ is the oriented boundary of a surface that does not intersect $f\big([0, 2\pi]\big)$. Prove that $\operatorname{link}(f,g) = 0$.

7.8.16. Let E and F be Euclidean spaces, an identify E with $E \times \{0\} \subset E \times F$. Show that if X is a submanifold of E, the value of K_d^X for X considered as a submanifold of $E \times F$ is proportional to the value of K_d^X for X considered as a submanifold of E, and that the proportionality constant does not depend on X.

7.8.17. Calculate $\operatorname{vol}(\operatorname{Tub}^\varepsilon X)$ for ε small enough for Veronese's surface (6.10.21).

7.8.18. Find $\operatorname{vol}\big(\partial(\operatorname{Tub}^\varepsilon X)\big)$ when $\dim X = 2$ (see 6.10.23).

7.8.19. Prove that $\deg(g \circ f) = \deg(g)\deg(f)$ for $f, g \in C^0(S^1; S^1)$.

7.8.20. Prove that maps $f, g \in C^0(S^1; S^1)$ having the same degree are homotopic.

7.8.21. Let U be a neighborhood of the origin in $\mathbf{R}^n$ and $f \in C^\infty(U)$ such that $f'(0) = 0$. Prove that if the origin is a non-degenerate critical point of f, it is an isolated singularity of the vector field $\hat{f} : x \mapsto (f'(x))^\sharp$, where $\sharp$ is the canonical isomorphism between $(\mathbf{R}^n)^*$ and $\mathbf{R}^n$. Find a formula relating $\mathrm{ind}_0 \, f$ and $\mathrm{ind}_0 \, \hat{f}$.

7.8.22. Prove that $\mathrm{Isom}^+(\mathbf{R}^n; \mathbf{R}^n)$ is connected.

7.8.23. Let X be a compact, connected manifold and δ, δ' densities on X such that $\int_X \delta = \int_X \delta'$. Is there a diffeomorphism f of X such that $f^*\delta' = \delta$?

CHAPTER 8

Curves: The Local Theory

In this chapter we study arcs, that is, immersions of open intervals of **R** into finite-dimensional affine or vector spaces (8.1.1). We define points of an arc and several important objects associated with them: the tangent, the osculating plane and the concavity (section 8.2).

If the ambient space is Euclidean we can also define arclength (section 8.3) and curvature (section 8.4). In the case of an oriented plane the curvature can be given a sign. No other invariants are then necessary: plane arcs are characterized, up to a rigid motion, by their curvature as a function of the arclength (8.5.7).

For arcs in three-dimensional Euclidean space we define another invariant, the torsion. Arcs in three-space are characterized, up to a rigid motion, by their curvature and torsion as functions of the arclength (8.5.7).

In fact we have to distinguish (8.1.4) between parametrized arcs (immersions of an open interval into space) and geometric arcs (equivalence classes of such immersions). All the preceding notions have to be introduced twice, but for brevity's sake we have sometimes left one or the other variant to the reader.

8.0. Introduction

For various reasons—kinematics, for example—it is desirable to extend the idea of a curve to include mathematical objects more general than one-dimensional submanifolds of $\mathbf{R}^n$.

In this chapter E will be a finite-dimensional vector space, and A an affine space, with underlying vector space $\vec{A}$. We are going to define a curve, or parametrized arc, as a C^p immersion of an open interval $I \subset \mathbf{R}$ in E. If $f \in C^p(I; E)$, with $p \geq 1$, is an immersion, we know by 0.2.24 that for every $t \in I$ there exists $J \in O_t(I)$ such that $f(J)$ is a one-dimensional submanifold of E. Thus we allow double or multiple points (figure 8.0), but we do not consider as curves maps that are not immersions, even if C^∞, because their study cannot be either simple or systematic, and in particular their singularities cannot be classified.

Figure 8.0

8.0.1. Examples of non-curves

8.0.1.1
$$f(t) = \begin{cases} (0, e^{1/t}) & \text{if } t < 0, \\ (0, 0) & \text{if } t = 0, \\ (e^{-1/t}, 0) & \text{if } t > 0. \end{cases}$$

This is a C^∞ function from $\mathbf{R}$ into $\mathbf{R}^2$ which is not an immersion at $t = 0$. Its image is the show on the left in figure 8.0.1.

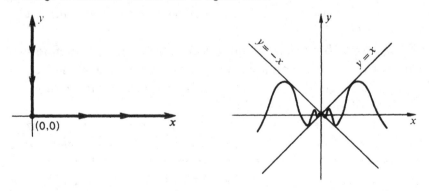

Figure 8.0.1

On the right we have the image of $f \in C^{\infty}(\mathbf{R}; \mathbf{R}^2)$ defined by

8.0.1.2
$$f(t) = \begin{cases} (e^{-1/t}, e^{-1/t} \sin e^{1/t}) & \text{if } t > 0, \\ (0,0) & \text{if } t = 0, \\ (e^{1/t}, e^{1/t} \sin e^{-1/t}) & \text{if } t > 0. \end{cases}$$

This image is bounded between the lines $y = \pm x$, and it has infinitely many "tangents" at $(0,0)$. Thus it is not an immersion at $t = 0$.

8.0.2. Remarks

8.0.2.1. The theory of C^0 curves is even more complicated: think of the Peano curve, for example. (This kind of thing doesn't occur for $p \geq 1$: by 0.4.4.5 the image of $f \in C^p(I; E)$ has measure zero if $\dim E \geq 2$.) With the restriction that $f \in C^0([a, b]; E)$ be a homeomorphism onto its image it is possible to develop a fairly rich, if lenghty, theory [BM70].

8.0.2.2. Singularities of maps that are not necessarily immersions can still be classified in the following cases: class C^ω (that is, real analytic), generic maps in class C^p, and algebraic curves.

8.0.2.3. Some authors call a parametrized arc any pair (I, f), with $I \subset \mathbf{R}$ an interval and $f \in C^p(I; E)$ ($p \geq 1$). In this terminology a point t is called regular if f is an immersion at t. But for us a parametrized arc will be regular everywhere (8.1.1).

8.1. Definitions

8.1.1. Definition. A *parametrized arc* of class C^p in a finite-dimensional vector space E is a pair (I, f), where $I \subset \mathbf{R}$ is an open interval and $f : I \to E$ is a C^p immersion (that is, $f'(t) \neq 0$ for all t).

8.1.2. Remarks

8.1.2.1. That (I, f) is an immersion implies that every $t \in I$ has a neighborhood J such that $f(J)$ is a one-dimensional submanifold of E and $f|_J$ is a diffeomorphism onto its image. But in general $f(I)$ is not an submanifold, even if f is injective (cf. 2.1.5).

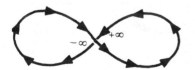

Figure 8.1.2

8.1.2.2. Even if $f(I)$ is a submanifold of E, it may not be parametrized by f in the sense of 2.1.8, because f is generally not injective: take the circle defined by $(\mathbf{R}, f)$, where $f(t) = (\cos t, \sin t)$, for instance. But if f is injective and $f(I)$ is a submanifold, f is a global parametrization for $f(I)$; this follows from 0.2.24.

8.1.3. Example. If E and F are vector spaces, $\Phi : E \to F$ is an isomorphism and (I, f) is a parametrized arc in E, the pair $(I, \Phi \circ f)$ is a parametrized arc in F, called the *image* of (I, f) under Φ.

8.1.4. Definition. Two parametrized C^p arcs (I, f) and (J, g) in E are said to be *equivalent* if there exists $\theta \in \mathrm{Diff}(I; J)$ such that $f = g \circ \theta$. A C^p *geometric arc* in E is an equivalence class of parametrized arcs.

It's clear that this is an equivalence relation. If C is a geometric arc and $(I, f) \in C$, we will generally say that (I, f) is a *parametrization* of C.

8.1.4.1

8.1.5. Remarks

8.1.5.1. Different geometric arcs can have the same image. (Of course different parametrizations of the same geometric arc always have the same image.) For example, take $I = \mathbf{R}$, $J =]-\pi, 3\pi[$ and $f(t) = g(t) = (\cos t, \sin t)$. Clearly $f(I) = g(J) = S^1$, but (I, f) and (J, g) cannot be equivalent: the inverse images of $(1, 0)$, for examples, have different cardinality. In the language of 8.1.8, all points of (I, f) have infinite multiplicity, whereas the points of (J, g) have multiplicity one or two.

Here are some more examples of different geometric arcs having the same image:

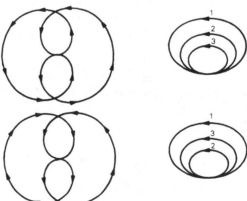

Figure 8.1.5

8.1.6. Example. If E and F are vector spaces, $\Phi : E \to F$ is an iso-morphism and C is a geometric arc in E, the *image* of C under Φ is the geometric arc represented by $(I, \Phi \circ f)$, where (I, f) is a parametrization of C.

8.1.7. Definition. A *point* in a geometric arc C is an equivalence class of triples (I, f, t), where $(I, f) \in C$ and $t \in I$, for the relation "$(I, f, t) \sim (J, g, s)$ if and only if there exists $\theta \in \text{Diff}(I; J)$ such that $f = g \circ \theta$ and $s = \theta(t)$." If m is a point in C, represented by (I, f, t), its image, denoted by $\textcircled{m}$, is the point $f(t) \in E$. The set of such images is called the *image* of C.

8.1.7.1

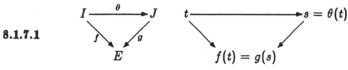

Sometimes the same point $p \in E$ is the image of several points $m \in C$; but if (I, f) and (J, g) are equivalent, we have $f^{-1}(p) = \theta(g^{-1}(p))$, so the sets $f^{-1}(p)$ and $g^{-1}(p)$ have the same cardinality. Thus:

8.1.8. Definition. The *multiplicity* of a point $M \in C$ is the number of points in $f^{-1}(\textcircled{m})$. A point is said to be *simple* (*multiple, double, triple*) if its multiplicity is 1 ($> 1, 2, 3$).

8.1.8.1. A geometric arc can have multiple points even if its image is a submanifold (8.1.5.1).

8.1.8.2. The multiplicity can be infinite (8.1.5.1), but not uncountable, because, by 8.1.1.1, $f^{-1}(\textcircled{m})$ is a discrete set.

8.1.8.3. A simple point is determined by its image. But a multiple point of C is best thought of as a *branch* of C going through its image.

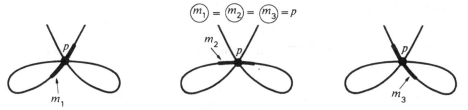

Figure 8.1.8

We have introduced arcs (or curves) to generalize the notion of a one-dimensional submanifold; now let's pin down the relationship between the two.

8.1.9. Proposition. *Every connected one-dimensional submanifold of a vector space E has a canonically associated geometric arc.*

Proof. Let $d \geq 1$ be the dimension of E and V a connected, one-dimensional C^p submanifold of E (with $p \geq 1$). We know by 3.4.1 that V is diffeomorphic to $\mathbf{R}$ or to S^1. According to the proof of that result, more can be said in each case:

If V is diffeomorphic to $\mathbf{R}$, there exists by 3.4.6 a global parametrization of V by arclength, that is, a pair (I, f) with $f \in \mathrm{Diff}(I; V)$. Two such parametrizations by arclength are equivalent in the sense of 8.1.4: if $g \in \mathrm{Diff}(J; V)$, we have $g^{-1} \circ f \in \mathrm{Diff}(I; J)$. Thus the two parametrizations define the same geometric arc, which we canonically associate to V.

If, on the other hand, V is diffeomorphic to S^1, it has a periodic parametrization by arclength $f : \mathbf{R} \to V$ (cf. 3.4.6). Let $(\mathbf{R}, f)$ and $(\mathbf{R}, g)$ be two such parametrizations. Since $f : \mathbf{R} \to V$ is periodic, we have a commutative diagram

8.1.9.1

$$
\begin{array}{ccc}
 & & \mathbf{R} \\
 & {\scriptstyle p}\swarrow & \big\downarrow{\scriptstyle f} \\
\mathbf{R}/L\mathbf{Z} & \underset{\theta^{-1}}{\overset{\theta}{\rightleftarrows}} & V,
\end{array}
$$

where L is the period. But $p : \mathbf{R} \to \mathbf{R}/L\mathbf{Z}$ is a covering map, so section 7.6 applies to show that every map from an interval into $\mathbf{R}/L\mathbf{Z}$ can be lifted, and uniquely so if we prescribe the image of one point. Transferring by θ we conclude the existence and uniqueness of liftings for the covering $f : \mathbf{R} \to V$.

In particular we can lift $g : \mathbf{R} \to V$ to $\bar{g} : \mathbf{R} \to \mathbf{R}$:

8.1.9.2

$$
\begin{array}{ccc}
 & & \mathbf{R} \\
 & {\scriptstyle \bar{g}}\nearrow & \big\downarrow{\scriptstyle f} \\
\mathbf{R} & \underset{g}{\longrightarrow} & V;
\end{array}
$$

we choose $\bar{g}$ satisfying $\bar{g}(t) = 0$, where t is a point such that $g(t) = f(0)$. Similarly, we can lift f by g to get $\bar{f} : \mathbf{R} \to V$

8.1.9.3

$$
\begin{array}{ccc}
 & & \mathbf{R} \\
 & {\scriptstyle \bar{f}}\nearrow & \big\downarrow{\scriptstyle g} \\
\mathbf{R} & \underset{f}{\longrightarrow} & V
\end{array}
$$

with $\bar{f}(0) = t$. Now $\bar{g} \circ \bar{f} : \mathbf{R} \to \mathbf{R}$ satisfies $f \circ (\bar{g} \circ \bar{f}) = g \circ \bar{f} = f$, that is, the diagram

8.1.9.4

$$
\begin{array}{ccc}
 & & \mathbf{R} \\
 & {\scriptstyle \bar{g} \circ \bar{f}}\nearrow & \big\downarrow{\scriptstyle f} \\
\mathbf{R} & \underset{f}{\longrightarrow} & V
\end{array}
$$

commutes. In addition, $(\bar{g} \circ \bar{f})(0) = \bar{g}(t) = 0$. Since $\mathrm{Id}_{\mathbf{R}}$ also satsifies $f \circ \mathrm{Id}_{\mathbf{R}} = f$ and $\mathrm{Id}_{\mathbf{R}}(0) = 0$, it follows from the uniqueness of liftings, applied to 8.1.9.4, that $\bar{g} \circ \bar{f} = \mathrm{Id}_{\mathbf{R}}$. One similarly shows that $\bar{f} \circ \bar{g} = \mathrm{Id}_{\mathbf{R}}$; in particular, $\bar{g} : \mathbf{R} \to \mathbf{R}$ is a diffeomorphism. This means that f and g are equivalent parametrizations (recall that $f \circ \bar{g} = g$), and hence define the same geometric arc, which we canonically associate with V. $\square$

8.1.10. Orientation. If $\theta : I \to J$ is a diffeomorphism θ' has constant sign because $\theta'(t) \neq 0$ for all t and I is connected. Let $\mathrm{Diff}^+(I; J)$ be the set of increasing diffeomorphisms (those for which $\theta'(t) > 0$), and $\mathrm{Diff}^-(I; J)$ be the set of decreasing diffeomorphisms.

8.1.11. Definition. Two parametrized arcs (I, f) and (J, g) are said to be *strictly equivalent* if there exists $\theta \in \mathrm{Diff}^+(I; J)$ such that $f = g \circ \theta$. An *oriented geometric arc* is a strict equivalence class of parametrized arcs.

Thus a geometric arc gives rise to exactly two oriented geometric arcs. A parametrized arc gives rise to one oriented geometric arc, to which it belongs.

If V is a connected, oriented, one-dimensional submanifold of E, the correspondence given by proposition 8.1.9 agrees with the orientation, if we take orientation-preserving parametrizations.

8.2. Affine Invariants: Tangent, Osculating Plan, Concavity

> In this section everything is of class C^1 at least.

8.2.1. The tangent

8.2.1.1. Definition. Let (I, f) be a parametrized arc in a vector space E (or an affine space A). The *tangent* to (I, f) at a point $t \in I$ is the vector line $\mathbf{R}f'(t)$ in E (or the affine line $f(t) + \mathbf{R}f'(t)$ in A).

8.2.1.2. Remarks. We have identified $f'(t) \in L(\mathbf{R}; E)$ with the vector $f'(t) \cdot 1$. Since f is an immersion, $f'(t)$ is non-zero.

8.2.1.3. Example. If $\Phi : E \to F$ is an isomorphism and (I, f) is an arc in E, the tangent to $(I, \Phi \circ f)$ at t is the image under Φ of the tangent to (I, f) at t.

This notion of tangency can be extended to geometric arcs because it is preserved under equivalence. Indeed, let C be a geometric arc and m a point in C. If (I, f, t) and (J, g, s) represent m, there exists $\theta \in \mathrm{Diff}(I; J)$ such that $f = g \circ \theta$ and $s = \theta(t)$ (definition 8.1.7). Then $f'(t) = (g \circ \theta)'(t) = g'(\theta(t))(\theta'(t)) = \theta'(t)g'(s)$, where $\theta'(t) \neq 0$ because θ is a diffeomorphism. This shows that $\mathbf{R}f'(t) = \mathbf{R}g'(s)$.

8.2.1.4. Definition. Let C be a geometric arc in a vector space E (or an affine space A) and m a point in C, represented by the triple (I, f, t). The vector line $\mathbf{R}f'(t)$ (or the affine line $f(t) + \mathbf{R}f'(t)$) depends only on C and m. It is called the *tangent* to C at m, and denoted by $\mathrm{tang}_m C$.

8.2.1.5. If C is a geometric arc in E and $\Phi : E \to F$ is an isomorphism, we have

$$\text{tang}_{\Phi(m)}(\Phi \circ C) = \Phi_{\text{tang}_m C}.$$

Thus the tangent is twice deserving of being called an invariant: it does not depend on the parametrization, and it is preserved by isomorphisms (and by affine maps, in the case of the affine definition).

8.2.1.6. Orientation. If C is an oriented geometric arc (8.1.11) and $m \in C$ is represented by two triples (I, f, t) and (J, g, s), with $\theta \in \text{Diff}(I; J)$ such that $f = g \circ \theta$ and $\theta(t) = s$, we have $\theta'(t) > 0$, so the line $\mathbf{R}f'(t) = \mathbf{R}g'(s)$ inherits the same orientation from $f'(t)$ and from $g'(s)$. Thus the orientation of $\text{tang}_m C$, where C is an oriented geometric arc, does not depend on the parametrization.

It is common and convenient to use arrows to indicate the orientation of an arc and its tangent:

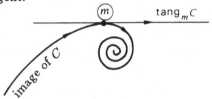

Figure 8.2.1.6

8.2.1.7. Relationship with the geometric definition. Let E be a vector space. The set of vector lines in E can be identified with the projective space $P(E)$, the quotient of $E \setminus 0$ by the equivalence relation "$x \sim y$ if and only if $y = kx$ for some $x \in \mathbf{R}^*$." We consider $P(E)$ with its quotient topology and denote the canonical projection by $p : E \setminus 0 \to P(E)$. (See 2.4.12.2 for the relation of $P(E)$ with the unit sphere, in the case that E is Euclidean.) Let (I, f) be a parametrized arc and $t \in I$; by 8.1.1.1, $f(s) \neq f(t)$ for s close enough but distinct from t.

8.2.1.8. Theorem. *The limit* $\lim_{s \to t} p(f(s) - f(t))$ *in* $P(E)$ *exists and equals* $p(\mathbf{R}f'(t))$.

Proof. As already observed, the map

$$\theta : s \mapsto \frac{f(s) - f(t)}{s - t}$$

is defined for s close enough but distinct from t, and it has a limit as s approaches t because f is of class C^1. This limit is $f'(t)$, which proves the theorem by the properties of the quotient topology. $\square$

8.2.1.9. For a similar property involving geometric arcs, see exercise 8.7.1.

In the rest of this section everything is of class C^2 at least.

8.2.2. The osculating plane

8.2.2.1. Definition. A parametrized arc (I, f) is said to be *biregular* at t if $f'(t)$ and $f''(t)$ are linearly independent (the derivatives being taken in the ordinary sense).

8.2.2.2. Definition. Let (I, f) be a parametrized arc in a vector space E (or an affine space A). The *osculating plane* to (I, f) at a biregular point t is the vector plane $\mathbf{R}f'(t) + \mathbf{R}f''(t)$ (or the affine plane $f(t) + \mathbf{R}f'(t) + \mathbf{R}f''(t)$).

8.2.2.3. Examples. If (I, f) is a *plane arc*, that is, if $f(I)$ is contained in some vector plane $V \subset E$, the osculating plane at every biregular point is V. The converse is trivial.

8.2.2.4. Proposition. *If $\Phi : E \to F$ is an isometry, Φ preserves biregularity and osculating planes.* $\qquad\qquad\qquad\qquad\qquad\qquad\qquad\qquad\qquad\qquad\qquad\qquad$ $\square$

Consider a point m in a geometric arc C, and let (I, f) and (J, g) be two parametrizations of C, with $\theta \in \text{Diff}(I; J)$ of class C^2 such that $f = g \circ \theta$. If m is represented by the triples (I, f, t) and (J, g, s), we have

8.2.2.5 $$f'(t) = \theta'(t)g'(s),$$

8.2.2.6 $$f''(t) = \theta''(t)g'(s) + \theta'^2(t)g''(s).$$

Since θ is a diffeomorphism, $\theta'(t) \neq 0$, and (I, f) is biregular at t if (J, g) is biregular at s. In addition, $\mathbf{R}g'(s) + \mathbf{R}g''(s) = \mathbf{R}f'(t) + \mathbf{R}f''(t)$. This justifies the following definition:

8.2.2.7. Definition. Let m be a point of a geometric arc C, represented by the triple (I, f, t). We say that C is *biregular* at m if the parametric arc (I, f) is biregular at t. The *osculating plane* of C at m, denoted by $\text{osc}_m C$, is the osculating plane of (I, f) at t.

See 8.0.2.3 for the origin of the word "biregular."

8.2.2.8. Example. If $\Phi \in \text{Isom}(E; F)$ we have $\text{osc}_{\Phi(m)}(\Phi \circ C) = \Phi(\text{osc}_m C)$. An analogous result holds for geometric arcs in affine spaces.

8.2.2.9. Geometric definition. If we can give the set of vector planes in E a canonical topology, we can interpret the osculating plane to a parametric arc (I, f) at t, for t biregular, as the limit of the plane $\mathbf{R}f'(t) + \mathbf{R}(f(s) - f(t))$ as $s \neq t$ approaches t. This is very easy if $\dim E = 3$ (exercise 8.7.3).

8.2.2.10. The concavity. Formula 8.2.2.6 and the fact that $\theta'^2(t) \geq 0$ show that the half-planes $\mathbf{R}f'(t) + \mathbf{R}_+ f''(t)$ and $\mathbf{R}g'(s) + \mathbf{R}_+ g''(s)$ coincide; whence the following definition:

8.2.2.11. Definition. Let m be a biregular point in a geometric arc C, represented by a triple (I, f, t). The *concavity* of C at m is the half-plane $\mathbf{R}f'(t) + \mathbf{R}_+ f''(t)$, which we denote by $\text{conc}_m C$. (In affine spaces we set $\text{conc}_m C = f(t) + \mathbf{R}f'(t) + \mathbf{R}_+ f''(t)$.)

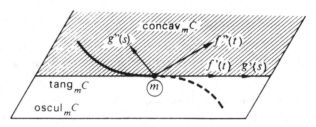

Figure 8.2.2.11

8.2.2.12. If $\Phi \in \text{Isom}(E; F)$ we have $\text{conc}_{\Phi(m)}(\Phi \circ C) = \Phi(\text{conc}_m C)$.

In the case of plane curves, the concavity is especially relevant:

8.2.2.13. Definition. Consider a plane curve, expressed in polar coordinates, and a point on the curve distinct from the origin. We say that at this point the curve *is turning towards the origin* if the origin belongs to the open affine half-plane of concavity.

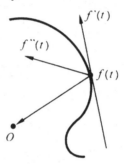

Figure 8.2.2.13

Here is the calculation that determines whether or not a curve $t \mapsto \rho(t)e^{it}$ is turning towards the origin. The origin will be in the concavity if the bases $\{-f(t), f'(t)\}$ and $\{f''(t), f'(t)\}$ have the same orientation, that is, if the determinants $\omega_0(-f, f')$ and $\omega_0(f'', f')$ have the same sign, where ω_0 is the canonical area form on $\mathbf{R}^2$ (example 6.4.2). The derivatives are

$$f' = \rho'e^{it} + \rho i e^{it},$$
$$f'' = (\rho'' - \rho)e^{it} + 2\rho' i e^{it}.$$

Since $\omega_0(e^{it}, ie^{it}) = 0$, we get

$$\omega_0(-f, f') = -\rho^2,$$
$$\omega_0(f'', f') = -(\rho^2 + 2\rho'^2 - \rho\rho''),$$

whence the following criterion:

8.2.2.14. Proposition. *At a biregular point distinct from the origin, a curve* $t \mapsto \rho(t)e^{it}$ *is turning toward the origin if* $\rho^2 + 2\rho'^2 - \rho\rho'' > 0$. $\qquad\square$

8.2.2.15. Proposition (local convexity). *A plane geometric arc is locally contained in its concavity at any biregular point. More precisely, if C is a geometric arc in an affine plane, $(I, f) \in C$ and $m \ni (I, f, t)$ is biregular with open half-plane of concavity H, there exists an open neighborhood $J \subset I$ of t such that $f(s) \in H$ for every $s \in J \setminus \{t\}$.*

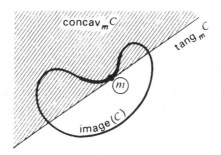

Figure 8.2.2.15

Proof. By assumption, $u = f'(t)$ and $v = f''(t)$ form a basis for $\vec{A}$, the vectorialization of A. Applying Taylor's formula to f at t, we get

$$f(t + h) = f(t) + hu + \frac{h^2}{2}v + o(h^2),$$

that is,

$$f(t + h) - f(t) = h(1 + o(h))u + \frac{h^2}{2}(1 + o(1))v.$$

This means the v-coordinate of $f(t + h) - f(t)$ is positive for $h \neq 0$ sufficiently small. $\qquad\square$

8.2.2.16. Remarks. The biregularity condition is essential, as shown by figure 8.2.2.16.1.

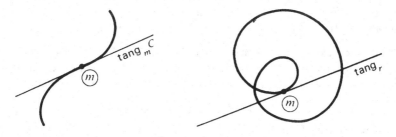

Figure 8.2.2.16.1   Figure 8.2.2.16.2

Figure 8.2.2.16.2 shows that, in general, 8.2.2.15 cannot be globally true. Finding conditions for the global validity of 8.2.2.15 is a typical *globalization problem*. It will be solved in section 9.6, where we will show

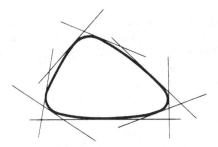

Figure 8.2.2.16.3

among other things that if C is closed, simple and everywhere biregular, it lies in $\mathrm{conc}_m C$ for every C (figure 8.2.2.16.3).

8.3. Arclength

From now on we assume that E (or A) is a *Euclidean* vector (or affine) space. Dealing with a richer structure, we can develop a correspondingly richer theory, and find more invariants.

Let C be a C^1 geometric arc in E. If $(I, f) \in C$ is an embedding, the image $V = f(I)$ has a canonical density δ (6.6.1), and (I, f) is a parametrization of V by arclength (3.4.3) if $\delta(f(t))(T_t f(1_t)) = 1$, or, equivalently, if $\|f'(t)\| = 1$, since we have identified $f'(t)$ with $T_t f(1_t)$. This leads to the following generalization:

8.3.1. Definition. A *parametrization by arclength* for a geometric arc C in a Euclidean space E is any parametrization $(I, f) \in C$ such that $\|f'(t)\| = 1$ for every $t \in I$.

8.3.2. Proposition. *Every geometric arc admits parametrizations by arclength. If (I, f) is a parametrization by arclength, any other such is of the form $t \mapsto f(t + a)$ or $t \mapsto f(-t + a)$, where a is an arbitrary real number.*

Proof. For existence, we proceed as in the proof of lemma 3.4.4. Take $(J, g) \in C$ and $s \in J$. If (I, f) is to be a parametrization by arclength, where $f = g \circ \lambda$ and $\lambda : I \to J$ is a diffeomorphism, we must have

$$\|f'(t)\| = \|f'(\lambda(t))\| |\lambda'(t)| = 1.$$

Thus we fix $a \in J$ and set

$$\theta(t) = \int_a^t \|g'(s)\| \, ds;$$

θ is invertible because $\theta'(t) = \|g'(t)\| > 0$. The desired parametrization by arclength (I, f) is then given by $I = \theta(J)$ and $f = g \circ \theta^{-1}$. To see this, notice that $\lambda = \theta^{-1} : I \to J$ is clearly a diffeomorphism and that

$$\|f'(s)\| = \|g'(\lambda(s))\| = \|g'(t)\| \frac{1}{\theta'(t)} = \frac{\|g'(t)\|}{\|g'(t)\|} = 1,$$

where $t = \lambda(s)$.

Uniqueness: let (I, f) and (J, g) be parametrizations by arclength and $\lambda : I \to J$ a diffeomorphism such that $f = g \circ \lambda$. Then

$$\|f'(t)\| = 1 = \|g'(\lambda(t))\| \|\lambda'(t)\| = |\lambda'(t)|,$$

whence $\lambda'(t) = \pm 1$. Since λ' is continuous on the connected open set I, its sign is constant, and we get $\lambda(t) = \pm t + a$. $\square$

8.3.3. Remark. Requiring that $f(t_0) = m$, for a fixed point m of C, restricts the choice of a parametrization by arclength to two possibilities.

8.3.4. Invariance under isometries. If $\Phi \in \mathrm{Isom}(E; F)$ and (I, f) is a parametrization of C by arclength, $(I, \Phi \circ f)$ is a parametrization of $\Phi \circ C$ by arclength.

8.3.5. Justification for the word "arclength" (cf. 6.5.3). If (J, g) is an arbitrary parametrized arc and $[a, b] \subset J$, the length of g between a and b is defined as

8.3.6

$$\sup \left\{ \sum_{i=0}^{n-1} \|g(t_{i+1}) - g(t_i)\| : n \in \mathbf{N}, a = t_0 < t_1 < \cdots < t_{n-1} < t_n = b \right\}.$$

It can be shown [Dix68, chapter 53] that this number is equal to

$$\int_a^b \|g'(t)\| \, dt.$$

This shows that if (I, f) is a parametrization by arclength and if $a, b \in I$, the length of (I, f) from a to b is $f(b) - f(a)$.

8.3.7. Definition. The *length* of a parametrized arc (I, f) from a to b, where $a, b \in I$, is the integral $\int_a^b \|f'(t)\| \, dt$.

8.3.8. It can be shown directly (without appeal to 8.3.6) that if C is a geometric arc parametrized by $(I, f) \in C$, the integral

$$\int_I \|f'(t)\| \, dt$$

does not depend on the parametrization. In any case, we can use 8.3.6 to introduce the following definition:

8.3.9. Definition. Let m and n be points in a geometric arc C, represented by the triples (I, f, a) and (J, g, b), respectively. The *length of the arc mn* of C is the integral

$$\int_a^b \| f'(t) \| \, dt.$$

8.3.10. If $\Phi \in \text{Isom}(E; F)$ is an isometry and m, n are points in an arc C, the length of the arc $\Phi(m)\Phi(n)$ of $\Phi \circ C$ is equal to the length of the arc mn of C.

8.3.11. Definition. Let C be an oriented geometric arc in a Euclidean space E. The *unit tangent vector to C* at a point m, represented by the triple (I, f, t), is the vector

$$\tau(m) = \frac{f'(t)}{\| f'(t) \|}.$$

8.3.12. Remarks

8.3.12.1. If (I, f) is a parametrization by arclength, $\tau(m) = f'(t)$.

8.3.12.2. Switching the orientation of C multiplies $\tau(m)$ by -1.

8.3.12.3. The unit tangent vector is invariant under isometries.

8.4. Curvature

Let m be a point in a geometric arc C. Let f and g be parametrizations by arclength for C, with $f = g \circ \theta$, and m a point represented by the triples (I, f, t) and (J, g, s). By 8.3.2 we have $\theta'(t) = \pm 1$, so $f''(t) = g''(s)$.

8.4.1. Definition. Let m be a point in a geometric arc C, represented by (I, f, t). If (I, f) is a parametrization by arclength, the quantity $\| f''(t) \|$, which depends only on C and m, is called the *curvature of C at m*, and denoted by $K_m C$.

8.4.2. Remarks.

8.4.2.1. For an explanation of the word "curvature," see note following formula 8.5.6.

8.4.2.2. We need not give a name to the invariant $f''(t)$, because it is completely determined by the ones introduced above. Indeed, $f'(t)$ is

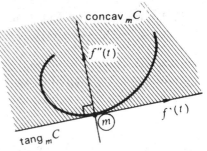

Figure 8.4.2

a unit vector, so $f'(t)$ and $f''(t)$ are orthogonal for all t, and $f''(t)$ lies in $\mathrm{conc}_m\, C$ (or else it is zero). Since $\left\| f''(t) \right\|$ is the curvature, $f''(t)$ is known. But see section 8.5 for the case of oriented curves.

The next two results are immediate:

8.4.3. Proposition. *A geometric arc C is biregular at m if and only if $K_m C \neq 0$.* $\qquad\square$

8.4.4. Proposition. *If $\Phi \in \mathrm{Isom}(E; F)$ is an isometry and C is a geometric arc in E, we have $K_{\Phi(m)}(\Phi \circ C) = K_m C$.* $\qquad\square$

8.4.5. The kinematic point of view. Let (J, g) be a parametrized arc in a Euclidean space E, and C the geometric arc determined by it. We associate to each point $m = g(s)$ the positive real number

8.4.6
$$v(s) = \left\| g'(s) \right\|$$

and call it the *scalar velocity*, or *speed of motion*, at m. If C is oriented (not necessarily by (J, g)) we let the *(algebraic) velocity* at m be the vector

8.4.7
$$g'(s) = v(s)\tau\big(g(s)\big),$$

where $\tau\big(g(s)\big)$ is the unit tangent vector to C at m (8.3.11).

Thus, if (I, f) is a parametrization by arclength and $g = f \circ \theta$ with $\theta \in \mathrm{Diff}(J; I)$, we have $g'(s) = v(s)f'(t) = f'(t)\theta'(s)$, where $s = \theta(t)$. This shows that $v = \theta'$.

8.4.8. Definition. The *curvature* of a parametrized arc (J, g) at $s \in J$ is the curvature of the associated geometric arc at the point m represented by (J, g, s). This number is denoted by

8.4.9
$$K(s) = K_{g(s)} C.$$

8.4.10. Calculation of the curvature. It is often too complicated, or even impossible, to find an explicit parametrization by arclength in order to compute the curvature. Here we develop a formula (8.4.13.1) for the curvature as a function of an arbitrary parametrization.

Let (I, f) be a parametrization by arclength of a geometric arc C, associated with the given parametrized arc (J, g), and let $\theta \in \mathrm{Diff}(J; I)$ be such that $g = f \circ \theta$. At a point $m = g(s) = f(t)$, the vector $g''(s)$ is the *acceleration* of the motion given by (J, g). Setting $\theta(s) = t$, we get from 8.2.2.6:

$$g''(s) = \theta'^2(s)f''(t) + \theta''(s)f'(t).$$

If m is biregular, we set

8.4.11
$$f''(t) = K_m C\, \nu(m),$$

thus defining the *principal normal
vector* $\nu(m)$ (since $K_m C \neq 0$; if the
point is not biregular $\nu(m)$ is not defined). This vector has length one,
by definition of $K_m C$, and is normal
to the curve (cf. 8.4.2.1).

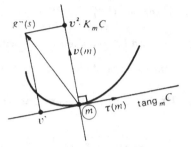

Setting $v = \theta'$, we can write the
acceleration $g''(s)$ as a sum of *intrinsic components*, in the sense that τ
and ν only depend on the geometric
arc and its orientation:

Figure 8.4.11

8.4.12
$$g''(s) = (v^2 K_m C)\nu(m) + v'\tau(m).$$

To compute $K_m C$ we just have to combine 8.4.12 with

8.4.13
$$g'(s) = v\tau(m)$$

and take the cross product (0.1.15.7), obtaining

8.4.13.1
$$K_m C = \frac{\|g'(s) \times g''(s)\|}{v^3} = \frac{\|g'(s) \times g''(s)\|}{\|g'(s)\|^3}.$$

8.4.14. Examples

8.4.14.0. Practical interpretation. The normal component $v^2 K_m C$ of the
acceleration represents the centripetal force, that is, the force that must be
applied to a unit mass, moving with velocity v, to ensure that its motion
will follow curve C. For a train in motion, for example, this force should
not change abruptly, lest the train be derailed or the tracks deformed. So
what we want is that K be continuous, or equivalently, that C be of class
C^2.

Whence the following problem in the design of railroads or highways:
how to connect two straight line segments D_1 and D_2 by means of a curve
C, in such a way that the curvature of C is continuous (and thus zero at
m_1 and m_2). It is also desirable that the maximum of $K_m C$ along C be as
small as possible. Finding such a curve is not trivial: circles and parabolas,
for example, are excluded because their curvature is non-zero everywhere.
Many families of suitable curves have been proposed, and the literature on
the subject is immense. See [Ali84, p. 294], for example.

8.4.14.1. In an affine Euclidean plane E, a circle of radius r and center a can
be parametrized by $g(t) = a + re^{it}$. Then $g'(t) = ire^{it}$ and $g''(t) = -re^{it}$,
which gives
$$\|g'(t) \times g''(t)\| = \|-r^2(ie^{it} \times e^{it})\| = r^2,$$
and $K_m C = r^2/r^3 = r^{-1}$ for all m. Thus a circle of radius r has constant
curvature $1/r$.

8.4.14.2. The curvature in polar coordinates. Consider a parametric arc in $\mathbf{R}^2$ defined by $g(t) = \rho(t)e^{it}$ (cf. 8.2.2.14). We have

$$g'(t) = \rho'(t)e^{it} + \rho(t)ie^{it},$$
$$g''(t) = \left(\rho''(t) - \rho(t)\right)e^{it} + 2\rho'(t)ie^{it},$$

whence

$$\|g' \times g''\| = |\rho^2 + 2\rho'^2 - \rho\rho''|.$$

But $v = \|g'\| = \sqrt{\rho^2 + \rho'^2}$, so

8.4.14.3
$$KC = \frac{|\rho^2 + 2\rho'^2 - \rho\rho''|}{(\rho^2 + \rho'^2)^{3/2}}.$$

For instance, the parametrized arc $\rho(t) = 1 + 2\cos t$ gives

$$K_{g(t)}C = \frac{3(3 + 2\cos t)}{(5 + 4\cos t)^{3/2}}.$$

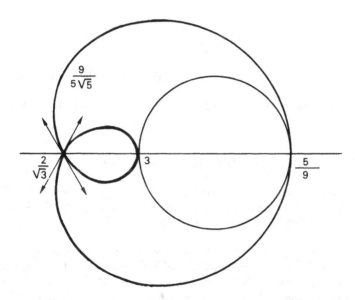

Figure 8.4.14.3

Setting $x = -\cos t$, we can write

$$K(x) = \frac{3(3 - 2x)}{(5 - 4x)^{3/2}}, \qquad \frac{\partial K}{\partial x} = \frac{12(2 - x)}{(5 - 4x)^{5/2}} > 0.$$

Thus, for $x \in [-1, 1]$, that is, $\theta \in [0, \pi]$, the curvature grows strictly monotonically from $\frac{5}{9}$ to 3, taking the values $\frac{9}{5\sqrt{5}}$ for $\theta = \pi/2$ and $\frac{2}{\sqrt{3}}$ for $\theta = 2\pi/3$. This curve is an example of a *Pascal limaçon*.

8.4.14.4. The circular helix. The parametrized arc in $\mathbf{R}^3$ given by $g : t \mapsto$ $(\cos t, \sin t, kt)$ satisfies $K_{g(t)} = (1 + k^2)^{-1}$ for all t.

8.4.15. Definition. The *radius of curvature* of C at a biregular point m is $R_m C = \dfrac{1}{K_m C}$. The *center of curvature* of C at m is the point $m + R_m C\, \nu(m)$ (8.4.11). The *osculating circle* to C at m is the circle of radius $R_m C$ centered at this point.

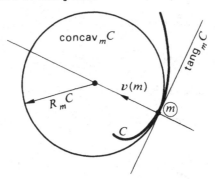

Figure 8.4.15

See exercise 8.7.4 for the geometric significance of the osculating circle.

8.4.16. For a metric definition of the curvature, see exercise 8.7.13.

8.5. Signed Curvature of a Plane Curve

In this section C denotes an oriented geometric arc in an oriented Euclidean plane E. (In any event we can orient E and C in order to apply the theory below; this turns out to be useful sometimes.)

8.5.1. In an oriented Euclidean plane E we have the notion of a rotation by $\pi/2$: it is the linear map $i : E \to E$ such that $\{x, ix\}$ is a positively oriented orthonormal basis for any unit vector $x \in E$. If we choose an isomorphism between E and $\mathbf{C}$ this rotation coincides with multiplication by i; but such an isomorphism is not canonical, whereas $i : E \to E$ is.

A point m of an oriented geometric arc C has an associated unit tangent vector $\tau(m)$ (8.3.11), given by $\tau(m) = f'(t)$ if (I, f) is a parametrization by arclength of C (with a compatible orientation) and $m = f(t)$. We can then form the vector $i\tau(m)$, normal to C at m. In general, $i\tau(m)$ doesn't have to equal the vector $\nu(m)$ introduced in 8.4.11; it can also equal $-\nu(m)$.

Now consider the real number $k_m C$ such that $f''(t) = k_m C\, i\tau(m)$. Since $f''(t)$ does not depend on the parametrization by arclength (I, f), we can introduce the following concept:

8.5.2. Definition. Let (I, f) be a parametrization by arclength of a geometric arc C, and $m = f(t)$ a point in C. The *signed curvature* of C at m is the real number $k_m C$ such that $f''(t) = k_m C \, i\tau(m)$.

Notice that here, contrary to 8.4.10, the normal vector $i\tau$ exists whether or not m is biregular; that's because we're dealing with a plane curve.

8.5.3. Local form. Figure 8.5.3 shows the sign of the curvature in the various possible cases. (The plane is oriented in the usual way, and the curve according to the arrows.)

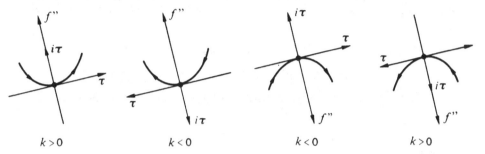

Figure 8.5.3

8.5.4. Elementary properties

8.5.4.1. Obviously $K_m C = |k_m C|$.

8.5.4.2. If Φ is an affine *rigid motion* of E (that is, an *orientation-preserving* isometry), we have

$$k_{\Phi(m)}(\Phi \circ C) = k_m C,$$

where $\Phi \circ C$ is oriented by Φ and C.

8.5.4.3. If we switch the orientation of E *or* the orientation of C, the signed curvature is multiplied by -1.

8.5.4.4. At a biregular point, the concavity (8.2.2.11) is the half-plane lying to the left or to the right of $\tan g_m C$, depending on whether $k > 0$ or $k < 0$. (Here $\tan g_m C$ is oriented by the orientation of C.)

We now propose to show (8.5.7) that, up to a rigid motion, there exists a unique geometric arc whose signed curvature is a specified function of the arclength (counting from some origin). We start by establishing a formula (8.5.6) that will be useful now and later. Let (I, f) be a parametrization by arclength of an arc C. The map $f' : I \to E$ has values in the unit circle

$$S(E) = \{x \in E : \|x\| = 1\}$$

of E, and $f'(t) = \tau(t)$, the unit tangent vector. Let $S(E)$ have the canonical orientation inherited from E, and call σ the length form on $S(E)$. Then

8.5.5 $\tau^* \sigma = k_m C \, dt.$

In fact, we have $(\tau^*\sigma)(1_t) = \sigma\big(\tau(t)\big)\big((T_t\tau)(1_t)\big)$, by the definition of τ^*. If ω_0 is the canonical area form on E, it follows from the calculation in 6.4.5 that $\sigma(t) = \mathrm{cont}(\tau(t))\omega_0$, hence:

$$
\begin{aligned}
(\tau^*\sigma)(1_t) &= \big(\mathrm{cont}(\tau(t))\omega_0\big)\big(\theta^{-1}(\tau'(t))\big) \\
&= \omega\big(\tau(t), \tau'(t)\big) = \omega\big(f'(t), f''(t)\big) \\
&= \omega\big(\tau(t), k_m C \, i\tau(t)\big) = k_m C,
\end{aligned}
$$

since $\omega_0(\tau, i\tau) = 1$ by the definition of i.

Now fix an arbitrary origin x on the circle $S(E)$, and consider the covering map $p : \mathbf{R} \to S(E)$ (in essence identical to the map $\mathbf{R} \to S^1$), defined by

8.5.5.1 $p : \mathbf{R} \ni t \mapsto \cos t\, x + \sin t\, ix.$

As in section 7.6, we have $p^*\sigma = dt$ and, if $\bar{\tau} : I \to \mathbf{R}$ is a lifting of $\tau : I \to S(E)$, we get $\tau^*\sigma = \dfrac{d\bar{\tau}}{dt}dt$, just as in the last part of the proof of 7.6.4.

8.5.5.2
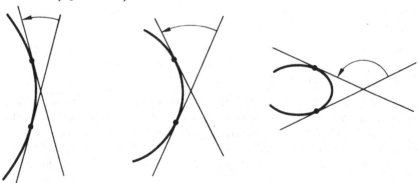

Comparing $\tau^*\sigma = \dfrac{d\bar{\tau}}{dt}dt$ with 8.5.5, we obtain

8.5.6 $k_m C = \dfrac{d\bar{\tau}}{dt}.$

Intuitively speaking, the curvature indicates how fast the tangent turns as we move along the curve. The faster the tangent turns, the greater the curvature (figure 8.5.6).

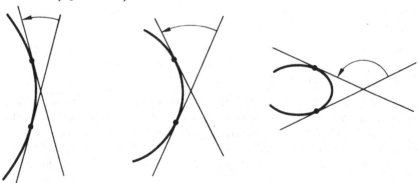

Figure 8.5.6

8.5.7. Theorem. *Let $I \subset \mathbf{R}$ containing the origin, $c : I \to \mathbf{R}$ a continuous function, and a,b elements of E with $\|b\| = 1$.   There exists a unique*

geometric arc C in E having a parametrization by arclength (I, f) such that $f(0) = a$, $f'(0) = b$ and $k_{f(t)}C = c(t)$ for every $t \in I$. In other words, a plane curve is determined, up to a rigid motion, by its signed curvature, and, conversely, the signed curvature can be any predetermined function of arclength.

Proof. If (I, f) is such a parametrization, we'll have $f' = \tau \in C^1(I; S(E))$, and a lifting $\bar{\tau} \in C^1(I; \mathbf{R})$ of τ will satisfy $\dfrac{d\bar{\tau}}{dt} = c$, by 8.5.6. Working backwards, we can define $\bar{\tau}$ by

$$\bar{\tau}(t) = \bar{\tau}(0) + \int_0^t c(u)\, du = b + \int_0^t c(u)\, du,$$

according to the requisite conditions. Then we set $\tau = p \circ \bar{\tau} = f'$, and integrate to obtain f:

$$f(t) = a + \int_0^t p\big(\bar{\tau}(u)\big)\, du. \qquad \qquad \square$$

See 8.7.7 for explicit examples.

8.5.8. Remark. The analog of 8.5.7 with k replaced by K is false, because of non-biregular points. The two curves in figure 8.5.8, for example, have the same unsigned curvature as a function of arclength, but they cannot be taken into one another by a rigid motion.

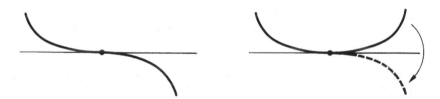

Figure 8.5.8

8.6. Torsion of Three-Dimensional Curves

In this section C will be an oriented geometric arc in an oriented three-dimensional Euclidean space E. We will show that, if C is of class C^3 and biregular, it is characterized by two functions of the arclength: the curvature and the torsion.

Let (I, f) be a parametrization by arclength of C, and $m = f(t)$ a point of C. We associate to m the unit tangent vector $\tau(m) = f'(t)$, and τ is of class C^{p-1} if f is of class C^p.

If m is biregular, we can also consider the principal normal vector $\nu(m)$ (8.4.11). If C is everywhere biregular, we have

$$\nu(t) = \frac{f''(t)}{\|f''(t)\|},$$

so that ν is of class C^{p-2}.

In the remainder of this chapter all arcs will be assumed biregular.

Since E is oriented, we can associate to each point m the vector $\beta(m)$ (and also $\beta(t)$, where $f(t) = m$) such that $\{\tau(m), \nu(m), \beta(m)\}$ is a positively oriented orthonormal basis for E. This condition can also be written (cf. 0.1.16):

8.6.1 $\beta(m) = \tau(m) \times \nu(m).$

Then β is of class C^{p-2}.

8.6.2. Definition. The vectors $\nu(m)$ and $\beta(m)$ are called the *principal normal* and *binormal* vectors, respectively.

We saw in 8.4.11 that, if $f(t) = m$,

8.6.3 $\tau'(t) = K_m C \, \nu(m),$

with $K \in C^{p-2}(I; \mathbf{R})$. Now assume that C is of class C^3 at least. Then we can differentiate the relations $\|\beta(t)\|^2 = 1$ and $(\beta'(t) \mid \tau(t)) = 0$ to obtain

$$(\beta(t) \mid \beta'(t)) = 0,$$
$$(\beta'(t) \mid \tau(t)) = (-\beta(t) \mid \tau'(t)) = (-\beta(t) \mid K_m C \, \nu(t))$$
$$= -K_m C \, (\beta(t) \mid \nu(t)) = 0.$$

Thus $\beta'(t)$ is a multiple of $\nu(t)$, and we let $T_m C$ be the real number such that

8.6.4 $\beta'(t) = T_m C \, \nu(m).$

The function $t \mapsto T(t) = T_{f(t)}(C)$ is of class C^{p-3}.

8.6.5. Definition. The real number $T_m C$ defined by 8.6.4 is called the *torsion* of C at m.

Differentiating the relations $\|\nu^2\| = 1$, $(\nu \mid \tau) = (\nu \mid \beta) = 0$, we get

$$(\nu' \mid \nu) = 0,$$
$$(\nu' \mid \tau) = (\nu \mid \tau') = -K,$$
$$(\nu' \mid \beta) = -(\nu \mid \beta') = -T.$$

Thus

8.6.6 $\nu'(t) = -K_m C \, \tau(m) - T_m C \, \beta(m).$

Formulas 8.6.3, 8.6.4 and 8.6.6 are called the *Frenet formulas*, and the basis $(\tau(m), \nu(m), \beta(m))$ the *Frenet frame*. Matrix lovers will delight in this one:

$$
\begin{array}{c}
\begin{array}{ccc} \tau' & \nu' & \beta' \end{array} \\
\begin{array}{c} \tau \\ \nu \\ \beta \end{array}
\left(\begin{array}{ccc}
0 & -K & 0 \\
K & 0 & T \\
0 & -T & 0
\end{array} \right).
\end{array}
$$

8.6.7. Remark. If K is zero, the torsion cannot be defined. Consider, for example, the following parametrized arc:

$$
f(t) = \begin{cases}
(t, e^{-1/t}, 0) & \text{if } t > 0, \\
(0, 0, 0) & \text{if } t = 0. \\
(t, 0, e^{1/t}) & \text{if } t < 0,
\end{cases}
$$

All points are biregular and have zero torsion, except for $t = 0$. Trying to extend the torsion to $t = 0$ by continuity is no good, because zero torsion everywhere is a characteristic of plane curves (8.6.12.1 and 8.6.15), and our the curve lies half in the xy-plane and half in the xz-plane.

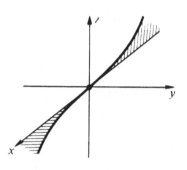

Figure 8.6.7

8.6.8. Elementary properties

8.6.8.1. If we change the orientation of E or the orientation of C, the torsion is multiplied by -1.

8.6.8.2. If Φ is a rigid motion, $T_{\Phi(m)}(\Phi \circ C) = T_m C$.

We now derive a formula for the torsion starting from an arbitrary parametrization, as we did for the curvature in 8.4.13.1.

8.6.9. Definition. Let (U, g) be a parametrized arc in an oriented, three-dimensional Euclidean space E, and C the associated geometric arc. The *torsion of* (U, g) *at* $g(s)$ (or at s) is the real number $T_{g(s)} C$.

8.6.10. To calculate the torsion, consider first a parametrization of C by arclength, say (I, f). With the notations above, and the abbreviations K and T for the curvature and torsion of C, we can write the Frenet formulas as follows:

8.6.10.1
$$
\begin{aligned}
f'(t) &= \tau(t), \\
f''(t) &= \tau'(t) = K_{f(t)} C \, \nu(t) = K(t)\nu(t), \\
f'''(t) &= K'\nu + K\nu' = K'\nu - KT\beta - K^2\tau.
\end{aligned}
$$

It follows that the mixed product (f', f'', f''') (cf. 0.1.16) is equal to $-K^2 T$.

In the case of an arbitrary parametrization (J, g), we take $\theta \in \mathrm{Diff}(J; I)$ such that $g = f \circ \theta$, and write

$$g'(t) = \theta' f',$$
$$g''(t) = \theta'' f' + \theta'^2 f'',$$
$$g'''(t) = \theta''' f' + 2\theta'' \theta' f'' + \theta'^3 f'''.$$

The mixed product (g', g'', g''') now equals

$$(g', g'', g''') = \theta'^6 (f', f'', f''') = -K^2 T \theta'^6 = -K^2 T v^6.$$

Since $\|g' \times g''\| = \|\theta'^3 f' \times f''\| = K v^3$, we get

8.6.10.2
$$T = -\frac{(g', g'', g''')}{\|g' \times g''\|^2}.$$

8.6.11. Examples

8.6.11.1. The circular helix (cf. 8.4.14.4). Here $g(t) = (a \cos t, a \sin t, bt)$, so that

$$K = \frac{a}{a^2 + b^2}, \qquad T = -\frac{b}{a^2 + b^2}.$$

8.6.11.2. If C is a plane curve, β is fixed and $T = 0$.

8.6.11.3. For each n, consider the parametrized arc P_n in $\mathbf{R}^3$ with equation

$$P_n(t) = \left(\left(\frac{1}{\sin \frac{2\pi}{n}} + \cos t \right) \cos \frac{t}{n} - \frac{1}{\sin \frac{2\pi}{n}}, \sin t, \left(\frac{1}{\sin \frac{2\pi}{n}} + \cos t \right) \sin \frac{t}{n} \right).$$

We claim that for n big enough the torsion of P_n is everywhere non-zero.

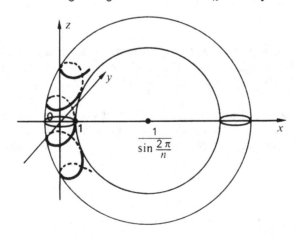

Figure 8.6.11.3

Since P_n is periodic of period $2\pi n$, and adding 2π to t is equivalent to rotating the image by $\dfrac{2\pi}{n}$ around the line $z = x = \dfrac{1}{\sin\frac{2\pi}{n}} = 0$, we can restrict ourselves to the interval $[0, 2\pi]$. It is easy to see that $\lim_{n\to\infty} P_n$ is the curve $P(t) = (\cos t, \sin t, t/2\pi)$, that this limit is uniform on $[0, 2\pi]$, and that all the derivatives of P_n converge uniformly to those of P on $[0, 2\pi]$. Thus the torsion of P_n converges to the torsion of P, which is given by $-\dfrac{2\pi}{1 + 4\pi^2}$ (8.6.11.1); this shows that $T < 0$ for large enough n.

P_n is a *toroidal coil*, the torus having radii $\dfrac{1}{\sin\frac{2\pi}{n}}$ and 1. As n tends to infinity, P_n becomes a cylindrical coil, or circular helix.

8.6.12. Geometric meaning of the torsion

8.6.12.1. Since $\beta(t)$ is perpendicular to the osculating plane, β' indicates how fast the osculating plane $\mathrm{osc}_m\, C$ turns around the tangent as we move along the curve. Thus the torsion says how far C is from being a plane curve. In particular, if C is a plane curve, the osculating plane is fixed, so $\beta' = 0$ and $T = 0$. Conversely, if T is identically zero, we have $\beta'(t) = 0$ by 8.6.4, and $\beta(t) = \tau(t) \times \nu(t)$ is constant, so $\tau(t)$ is perpendicular to the fixed vector $\beta(t)$. But $f'(t) = \tau(t)$ is perpendicular to $\beta(t)$, so $\beta(t) = $ constant implies $\big(f(t) \mid \beta(t)\big) = $ constant, and C is a plane curve.

8.6.12.2. Local position of a curve with respect to the Frenet frame. Formulas 8.6.10.1 imply that the Taylor series of C, up to order three, is

$$f(t) - f(t_0) = (t - t_0)\tau + \frac{(t - t_0)^2}{2}K\nu$$
$$+ \frac{(t - t_0)^3}{6}(-K^2\tau + K'\nu - KT\beta) + o(t - t_0)^3,$$

where (τ, ν, β) is the Frenet frame of C at $m_0 = f(t_0)$. Separating the components, we get

$$f(t) - f(t_0) = (t - t_0)\left(1 - K^2\frac{(t - t_0)^2}{6} + o(t - t_0)^2\right)\tau$$
$$+ \frac{(t - t_0)^2}{2}\left(K + K'\frac{t - t_0}{3} + o(t - t_0)\right)\nu$$
$$+ \frac{(t - t_0)^3}{6}(-KT\beta + o(t - t_0)^3)\beta.$$

Figure 8.6.12.3 show the projections of the curve looks on the three coordinate planes of the Frenet frame:

In particular, we see that if $T \neq 0$, the curve crosses its osculating plane. The sign of the torsion is apparent from the projection on the $\tau\beta$-plane; for the helix in 8.6.11.1, it is negative (figure 8.6.12.4).

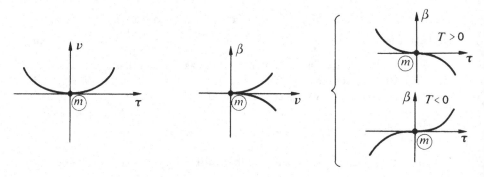

Figure 8.6.12.3

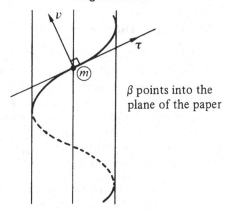

β points into the
plane of the paper

Figure 8.6.12.4

8.6.13. Fundamental theorem. *Let $I \subset \mathbf{R}$ be an interval, $c \in C^1(I; \mathbf{R}_+^*)$
and $d \in C^0(I; \mathbf{R})$ functions and u, v, w elements of E with $\|v\| = \|w\| =
1$. There exists a unique oriented geometric arc C in E, of class C^3 and
necessarily biregular, having a parametrization by arclength (I, f) such that
$f(0) = u$, $f'(0) = v$,*

$$\frac{f''(0)}{\|f''(0)\|} = w, \qquad and \qquad \left.\begin{array}{r} K_{f(t)}C = c(t) \\ T_{f(t)}C = d(t) \end{array}\right\} \quad for\ all\ t \in I.$$

Proof. Uniqueness will follow from the proof of existence below, but it can
also be shown by a neat elementary argument that deserves to be included.
Assume (I, f) and $(I, \underline{f})$ are two parametrizations by arclength satisfying
the conditions of the theorem, and let $\{\tau, \nu, \beta\}$ and $\{\underline{\tau}, \underline{\nu}, \underline{\beta}\}$ be their Frenet
frames. Form the C^1 function

$$\alpha = (\tau \mid \underline{\tau}) + (\nu \mid \underline{\nu}) + (\beta \mid \underline{\beta});$$

by formulas 8.6.3, 8.6.4 and 8.6.6, its derivative is identically zero:

$$\alpha'(t) = (K_m C \nu \mid \underline{\tau}) + (\tau \mid K_m C \underline{\nu}) + (-K_m C \tau - T_m C \beta \mid \underline{\nu})$$

$$+ (\nu \mid -K_m C \tau - T_m C \beta) + (T_m C \nu \mid \beta) + (\beta \mid T_m C \nu) = 0.$$

Thus $\alpha(t) = \alpha(0) = 3$. But if a and $\underline{a}$ are unit vectors we have $(a \mid \underline{a}) \leq \|a\|\|\underline{a}\| = 1$, with equality if and only if $a = \underline{a}$; thus $\alpha(t) = 3$ implies $\tau(t) = \underline{\tau}(t)$, $\nu(t) = \underline{\nu}(t)$ and $\beta(t) = \underline{\beta}(t)$, and, in particular, $f'(t) = \tau = \underline{\tau} = \underline{f}'(t)$. Together with $f(0) = \underline{f}(0)$, this shows that $f = \underline{f}$.

Now for the proof of existence. Consider on E the differential equation

$$(X', Y', Z') = (cY, -cX - dZ, dY),$$

subject to the initial condition $(X(0), Y(0), Z(0)) = (v, w, v \times w)$. Since c and d are continuous and this system is linear, theorem 1.6.6 says that a C^1 solution exists, and is unique, on the whole of I. Let it be (X, Y, Z).

We claim that (X, Y, Z) is a positively oriented orthonormal basis, for every $t \in I$. For the map $\theta = (\theta_1, \theta_2, \theta_3, \theta_4, \theta_5, \theta_6) : I \to \mathbf{R}^6$ defined by

$$\theta(t) = \left(\|X\|^2, \|Y\|^2, \|Z\|^2, (X \mid Y), (X \mid Z), (Y \mid Z) \right)$$

satisfies a certain continuous, first-order differential equation: for instance, θ_1' is equal to $2c(Y \mid X) = 2c\theta_4$, and so on. Since the constant map $\eta(t) = (1, 1, 1, 0, 0, 0)$ clearly satisfies that same equation, we must have $\theta(t) = (1, 1, 1, 0, 0, 0)$ by 1.3.1.

Now set $f(t) = u + \int_0^t X(s) \, ds$. We have $f'(t) = X(t)$ and $\|X(t)\| = 1$. Thus $f(t)$ is a parametrization by arclength and, since $X' = cY$ with c and Y of class C^1, f is of class C^3. Clearly $X(t) = \tau(t)$ is the tangent to the arc C defined by that parametrization; in addition, $c(t)$ is the curvature of C because $\tau'(t) = X'(t) = c(t)Y(t)$ with $\|Y(t)\| = 1$. Now we've assumed $c(t) > 0$, so C is everywhere biregular; thus Y is the principal normal vector, $Z = X \times Y$ is the binormal and $Z'(t) = d(t)X(t)$, so $d(t)$ is the torsion. $\qquad\square$

8.6.14. Theorem 8.6.13 says that, up to rigid motions, there exists a unique curve having predetermined curvature and torsion, as functions of the arclength.

8.6.15. Caution. Biregularity is an essential hypothesis. Otherwise, not only does the torsion not exist (8.6.7), but even if we could extend it by continuity the uniqueness conclusion would fail. Consider the curves in 8.5.8, for example, which have same K and same $T = 0$.

8.6.16. Example. The only arcs with constant, non-zero curvature and torsion are the circular helices (8.6.11.1). This is a consequence of the uniqueness part of theorem 8.6.13, and of the fact that, for any K and T, we can find a helix with curvature K and torsion T, namely, $t \mapsto (a \cos t, a \sin t, bt)$ with

$$a = \frac{K}{K^2 + T^2}, \qquad b = -\frac{T}{K^2 + T^2}.$$

8.6.17. The keen reader may have sniffed out the following generalization of theorems 8.5.7 and 8.6.13: arcs in n-dimensional Euclidean space are characterized, up to rigid motions, by a system of $n - 1$ invariants. See [Spi79, vol. II, chapter 1].

8.7. Exercises

8.7.1. Given a geometric arc C and a point $m \in C$, define the sentence "n approaches m along C". Show that $\lim_{n \to m} p(\textcircled{n} - \textcircled{m}) = \mathrm{tang}_m C$ (see 8.2.1.8).

8.7.2. What are the parametrized arcs having no biregular points?

8.7.3. Let (I, f) be a parametrized arc in E, and $t \in I$ a point such that (I, f) is biregular at t.

(a) Let E be three-dimensional and Euclidean. Give the set P of planes (two-dimensional vector subspaces) in E the topology obtained by identifying P with the real projective plane, via the map that takes a plane to its normal direction. Show that, as s approaches t, the plane spanned by the vectors $f'(t)$ and $f(s) - f(t)$ approaches the osculating plane to (I, f) at t. (First you have to show that, for s close enough to t, the vectors $f'(t)$ and $f(s) - f(t)$ are linearly independent.)

(b) Prove the same result in arbitrary dimension, this time giving P the topology defined in 2.8.8 (which coincides with the one in part (a) in dimension three).

8.7.4. The osculating circle

(a) Let m be a biregular point in a geometric arc C. Show that, as n approaches m along C (see 8.7.1), the circle containing n and tangent to C at m approaches the osculating circle to C at m. (Define a suitable topology on the set of circles in the plane; or, if you can't, just study the limit of the center of the circle as n approaches m.)

(b) Show that the osculating circle to C at m can also be obtained as the limit of the circles going through three distinct points on C that tend toward m.

(c) Assume that the curvature of C (for some parametrization by arc-length) is a strictly increasing of the parameter. Show that the points whose parameter is greater than (resp. less than) the parameter at m lie strictly inside (resp. outside) the osculating circle at m.

(d) Let S be a circle of radius r and center O in E, and consider the map $i : E \setminus \{O\} \to E \setminus \{O\}$ that takes a point $x \in E \setminus \{O\}$ to the point $i(x)$ on the half-line Ox such that $\|d(O, i(x))\| = 1/d(O, x)$ (such a map is called an *inversion*, cf. [Ber87, section 10.8]). Show that osculating circles are preserved by i.

8.7.5. Let E be an oriented Euclidean plane, C an oriented geometric arc of class C^2, and (I, f) a parametrization of C by arclength.

(a) When is it possible to find a scalar function α such that $g = f + \alpha\nu$ defines a geometric arc D whose tangent at $g(t)$ is normal to C at $f(t)$, for all $t \in I$? Show that D does not depend on the orientation of C or of E. We call D the *evolute* of C.

(b) Conversely, given D, find the curves C of which D is the evolute. Study the case that D is a circle or a logarithmic spiral (8.7.16).

(c) Let C' a subarc of C not containing vertices (9.7.1). What is the relative position of two osculating circles to C'?

8.7.6. State and prove an analog to theorem 8.5.7 concerning the unsigned curvature $K_m C$, assuming that $K_m C$ does not vanish.

8.7.7. With the notation of theorem 8.5.7, find the curves which satisfy:

(a) $c^2(s) + s^2 = 1$;
(b) $c(s) = 1/s$;
(c) $c(s) = ks$.

8.7.8. Let (I, f) and (I, g) be parametrizations by arclength for geometric arcs C and D, respectively, and assume that $K_{f(t)} C \leq K_{g(t)} D$ for all $t \in I$. Show that, if $t, t' \in I$ are close enough, we have

$$d\big(f(t), f'(t)\big) \geq d\big(g(t), g(t')\big).$$

8.7.9. Define an *osculating sphere* to an arc in $\mathbf{R}^3$, and determine that sphere.

8.7.10. Helices. Let C be an oriented, biregular arc in $\mathbf{R}^3$. Prove that the four conditions below are equivalent:

(i)   the tangent makes a constant angle with a fixed direction;
(ii) the principal normal is parallel to a fixed plane;
(iii) the binormal makes a constant angle with a fixed direction;
(iv) T and K are proportional.

Show that under these conditions C has a parametrization of the form

$$t \mapsto f(T) + (t - t_0)\alpha,$$

where f is a plane curve parametrized by arclength and α is a vector perpendicular to the plane of that curve.

An arc satisfying the conditions above is called a *helix*.

8.7.11. We say that two C^∞, p-dimensional submanifolds M and N of $\mathbf{R}^d$ have a *contact of order k or more* at x if there exist parametrizations (U, f) and (U, g) of M and N, respectively, such that $0 \in U$, $f(0) = g(0) = x$ and $f^{(n)}(0) = g^{(n)}(0)$ for $n = 1, \ldots, k$. We say that M and N have a *contact of* order k if they have a contact of order k or more but do not have a contact of order $k + 1$ or more. We say that a p-dimensional submanifold M and a q-dimensional submanifold N, with $q \leq p$, have a contact of order k at x if there exists a q-dimensional submanifold N' of N having a contact of order k with M at x. (The reader should check that these definitions do not depend on the parametrizations.)

(a) Show that these notions are invariant under diffeomorphisms. Define the order of contact between two geometric arcs, and between a geometric arc and a submanifold.

(b) Show that a plane curve and its osculating circle at a biregular point have a contact of order three or more, and of order four or more if and only if $k' = 0$ at the contact point. Show that a curve in $\mathbf{R}^3$ and its osculating plane at a biregular point have a contact of order three or more, and of order four or more if and only if $T = 0$ at the contact point.

(c) Let C be a plane curve in $\mathbf{R}^3$ and i an inversion (8.7.4(d)) whose pole does not lie on C. Using part (b), show that if m is a point of C where $k' = 0$, the torsion of $i(C)$ at $i(m)$ is zero. Prove the same result by direct calculation (using Taylor series, for example).

8.7.12. Show that if a curve in $\mathbf{R}^3$, with nowhere vanishing curvature K and torsion T, is to be contained in a sphere of radius r, we must have

$$ r^2 = \left(\frac{1}{K} \right)^2 + \left(\left(\frac{1}{K} \right)' \frac{1}{T} \right)^2, $$

where the derivative is taken with respect to arclength. The converse is also true, but difficult; see [Won72].

8.7.13. Menger curvature. Let $f \to C$ be a curve of class C^2 in a Euclidean space. Let x, y, z be distinct points of f, and set

$$ K(x, y, z) = \frac{\sqrt{(xy + yz + zx)(xy + yz - zx)(xy - yz + zx)(xy - yz + zx)}}{xy \cdot yz \cdot zx}, $$

where xy, for example, denotes the Euclidean distance from x to y. Show that, as y and z approach x along the curve, $K(x, y, z)$ tends towards the curvature of f at x. Find examples of curves of class C^1 for which $K(x, y, z)$ does not have a limit, or becomes infinite, as y and z approach x.

8.7.14. Draw the curve whose polar equation (8.4.14.2) is $\rho(t) = -\log(1 - \sin t)$. Do the bottle and the pebble have the same curvature radius at their contact point?

8.7.15. Find the curvature radius at the origin of the curve with equation

$$ x^{15} + y^{13} + x^3 y^4 + x^2 - y^2 + x + 199y = 0. $$

8.7.16. Logarithmic spirals. What are the plane curves whose tangent makes a constant angle with the line joining the foot of the tangent to a fixed point? What are the plane curves whose radius of curvature is proportional to arclength? Write the polar equation of such curves, called *logarithmic spirals* [Ber87, 9.6.9].

Can a logarithmic spiral coincide with its evolute (8.7.5)?

8.7.17. Hypocycloids and epicycloids

8.7.17.1. Definition. A *hypocycloid* (resp. *epicycloid*) is the set C of points of the Euclidean plane occupied by a given point on a circle Γ' that rolls without sliding inside (resp. outside) a fixed circle Γ of commensurable radius (figure 8.7.17.1). Study the shape of C according to the ratio between the radii. How many cusps does C have?

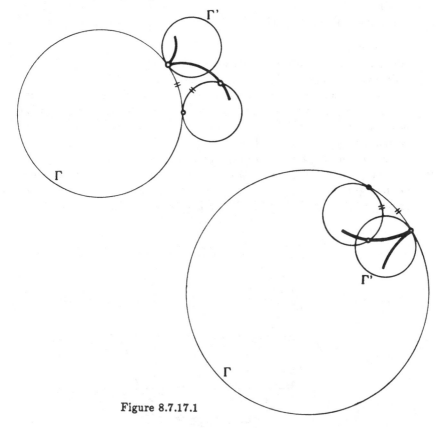

Figure 8.7.17.1

8.7.17.2. Equivalent definitions. Let Σ be the unit circle of $\mathbf{R}^2 = \mathbf{C}$, and let $r \neq 0$ be rational. Show that the envelope of the lines $D(\theta)$ joining the points $e^{i\theta}$ and $e^{ir\theta}$, for $\theta \in \mathbf{R}$, is a hypo- or epicycloid. Discuss what kind

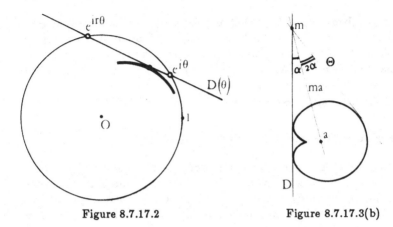

Figure 8.7.17.2   Figure 8.7.17.3(b)

it is and how many cusps it has, according to the value of r. Are all hypo- and epicycloids obtained in this way?

8.7.17.3. Examples

(a) What does the figure look like when Γ' has half the radius of Γ and rolls inside Γ? (It is called *Lahire's cogwheel.*)

(b) If Γ' and Γ have the same radius and Γ' rolls outside Γ, one obtains a Pascal limaçon, called a *cardioid*. Let a and D be a point and a line in the plane, with $a \notin D$; show that the envelope of the family of lines Θ defined by $m \in \Theta$ and $\widehat{D\Theta} = 3(\widehat{ma, D})$, for $m \in D$, is a cardioid, whose *center* is, by definition, the point a.

(c) Show that the caustic of a plane spheric mirror (that is, the envelope of the light rays reflected from a beam parallel to the axis) is a piece of a two-cusped epicycloid (called a *nephroid*).

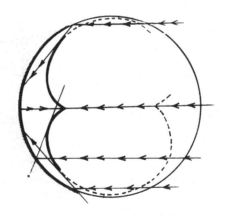

Figure 8.7.17.3(c)

(d) Show that horizontal projections of *spherical helices* with a vertical axis (that is, curves on the sphere whose tangent makes a constant angle α with the axis) are epicycloids for appropriate values of α.

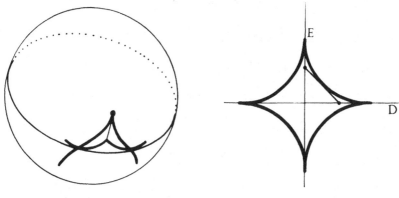

Figure 8.7.17.3(d)   Figure 8.7.17.3(e)

(e) Show that the envelope of a segment of fixed length whose endpoints describe two orthogonal lines is a four-cusped hypocycloid (called an *astroid*).

8.7.17.4. Properties. Show that the evolute (8.7.5) of a hypo- or epicycloid is similar to the original curve. Show that the arclength s (computed from an appropriated starting point) and the curvature K of a hypo- or epicycloid satisfy a relation of the form

$$as^2 + bK^{-2} = c,$$

for a, b, c constant. Conversely, what are the curves satisfying such a relation? Find the total length of a hypo- or epicycloid.

8.7.17.5. For more information on hypo- and epicycloids, see [Lem67]. See also [Zwi63], a very agreeable text on plane curves an their links with mechanics, optics and electricity; in particular, see chapter XXI, where the connection between epicycloids and cogwheel design is discussed. Finally, [LA74, pp. 413–435] gives an analytic presentation of the various cycloids, including a derivation of the shape of the carter of the Wankel engine.

8.7.18. The lemniscate

(a) Prove that the locus of the points whose distance to two fixed points m and n of the plane has a constant product is a curve whose polar equation is $\rho^2 = 2a^2 \cos 2t$, where the origin is the midpoint of the segment mn and a is the half the distance between m and n. This curve is called *Bernoulli's lemniscate*.

(b) Prove the properties indicated in figure 8.7.18, about the direction of the tangent and the center of curvature. Study possible converses.

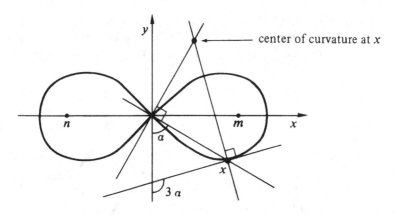

Figure 8.7.18

8.7.19. The curvature radius of the ellipse. Let the normal at m to the ellipse $\dfrac{x^2}{a^2} + \dfrac{y^2}{b^2} - 1 = 0$ intersect the x-axis at n. If d is the distance between m and n, show that the curvature radius of the ellipse at m is given by

$$R = \frac{a^2 d^3}{b^4}.$$

For applications, see 10.6.6.6.2 and 11.18.

8.7.20. Center of curvature of a catenary. Catenaries are the shapes taken by chains hanging between two fixed points. Their equation, up to translation, is $y = a \cosh(x/a)$. Show that the center of curvature of a catenary has the property shown in figure 8.7.20. Study a converse. For an application, see 10.6.6.6.3.

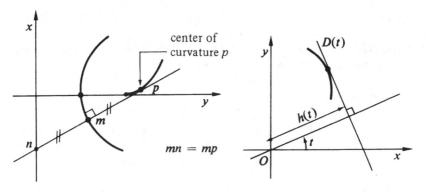

Figure 8.7.20   Figure 8.7.21

8.7.21. Plane curves in Euler form. Define a curve C as the envelope of the lines

$$D(t) = \{(x, y) \in \mathbf{R}^2 : \cos t\, x + \sin t\, y = h(t)\},$$

where $h : \mathbf{R}/2\pi\mathbf{Z} \to \mathbf{R}_+^*$. Show that the curvature of C at the point where C touches $D(t)$ is given by $h(t) + h''(t)$.

8.7.22. Schur's comparison theorem. Let (I, f) and (I, g) be parametrizations by arclength for geometric arcs C and D in $\mathbf{R}^2$ and $\mathbf{R}^3$, respectively. Assume that $k_{f(t)} > 0$ and $K_{f(t)}C \le K_{g(t)}D$ for all $t \in I$. Show that, if $t, t' \in I$ are close enough, we have

$$d\big(f(t), f(t')\big) \ge d\big(g(t), g(t')\big),$$

the inequality being strict unless D is the image of C under a plane isometry. Give a physical interpretation for this result.

Deduce that, among all closed curves with curvature not exceeding $1/R$, the shortest is the circle of radius R.

8.7.23. Bertrand curves. A curve C in $\mathbf{R}^3$ is called a *Bertrand curve* if it is possible to find another curve D with the following property: for each parameter value the principal normals to the two curves coincide (i.e. are the same line in $\mathbf{R}^3$). Show that C is a Bertrand curve if and only if its curvature K and torsion T satisfy a linear relation $aK + bT = 1$, where a and b are constants. Show that if C has two distinct such Bertrand partners D_1 and D_2, it has infinitely many, and is necessarily a circular helix. Show that if m and n are points on C and D, respectively, parametrized by the same number, and p and q are the corresponding centers of curvature, the *cross-ratio* $(pm/pn)/(qm/qn)$ is a constant. (Here pm, for instance, is the signed distance from p to m along the common normal; cf. [Ber87, chapter 6].)

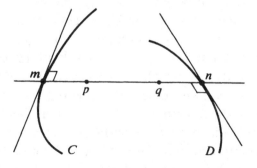

Figure 8.7.23

CHAPTER 9

Plane Curves: The Global Theory

In this chapter we will study immersions of the circle in the plane, called closed curves, and embeddings of the circle in the plane, called simple closed curves. Naturally, the interesting objects are equivalence classes of such immersions or embeddings.

The study of simple closed curves starts with Jordan's theorem (9.2.1), which states that the complement of such a curve has two connected components, one of which, called its interior, is compact. Then comes the isoperimetric inequality, which relates the area of such a curve with the area of its interior (9.3.1). Later we have the theorem of the four vertices (9.7.4) and a global convexity property (9.6.2).

For arbitrary closed curves we introduce the turning number (section 9.4), which says by how much the tangent vector to a curve turns when we follow it once. The theorem of Whitney–Grauenstein (9.4.8) asserts that two curves having the same turning number are homotopic. The turning tangent theorem (9.5.1) says that the turning number of a simple closed curve is ±1. We conclude with a formula (9.8.1) that relates the number of inflection points, the number of double points and the number of double tangents of a closed curve.

The differentiability class p is assumed to be at least one, but is otherwise unspecified. Some of the results can be established in class C^0, but this complicates the proofs a lot.

9.1. Definitions

Recall that every compact, connected, one-dimensional manifold is diffeomorphic to the circle (3.4.1).

9.1.1. Definition. A *simple closed curve* C in a finite-dimensional vector space E is a compact, connected, one-dimensional submanifold of E. We say that C is *oriented* if it is oriented as a manifold.

9.1.2. Since C is diffeomorphic to S^1, there is an equivalent formulation: a simple closed curve in E is an equivalence class of embeddings (S^1, f) of S^1 into E under the equivalence relation "$(S^1, f) \sim (S^1, g)$ if and only if $g^{-1} \circ f \in \text{Diff}(S^1)$."

9.1.3. By proposition 8.1.9 (second case) we have yet a third definition: a simple closed curve in E is a geometric arc C in E having a representative $(\mathbf{R}, f)$ that is periodic of period L, for some $L > 0$, and injective on $[0, L[$.
   This leads to the following, more general, definition:

9.1.4. Definition. A C^p *closed curve* C in E is an equivalence class of C^p immersions of S^1 into E under the equivalence relation "$(S^1, f) \sim (S^1, g)$ if and only if there exists $\theta \in \text{Diff}^p(S^1)$ such that $f = g \circ \theta$."

9.1.5. Definition. An *oriented closed curve* C in E is an equivalence class of immersions of S^1 into E under the equivalence relation "$(S^1, f) \sim (S^1, g)$ if and only if there exists $\theta \in \text{Diff}(S^1)$ such that $f = g \circ \theta$."

9.1.6. Remark. By just repeating the proof of 8.1.9 (second case) one comes up with an equivalent definition:

9.1.7. A closed curve in E is a geometric arc C in E having a representative $(\mathbf{R}, f)$ that is periodic of period L, for some $L > 0$.

   As expected, a simple closed curve canonically determines a closed curve.

9.1.8. Examples

9.1.8.1. Notice that the image $f(\mathbf{R}) = V$ does not determine a unique closed curve, because the curve can be drawn several times—the number L

in 9.1.7 is not necessarily the shortest period. For example, take the closed curves $f_n : S^1 \to \mathbf{R}^2$ given by $\overline{f}_n : t \mapsto (\cos n\pi t, \sin n\pi t)$, for $N \in \mathbf{Z}^*$. They all have the same image $f_n(S^1) = S^1 \subset \mathbf{R}^2$, but they are distinct for distinct values of $|n|$ (and, as oriented closed curves, for distinct values of n as well). See also 9.4.5.1.

9.1.8.2. Figure 8.1.5 also shows pairs of distinct curves with the same image.

9.1.8.3. Figure 9.1.8.3 shows the curve $f : S^1 \to \mathbf{R}^2$ given by $(\cos t, \sin t) \mapsto (\cos t, \sin 2t)$, a particular case of a *Lissajous figure*:

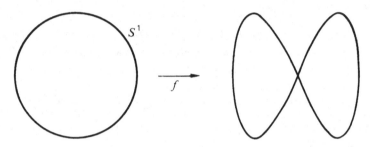

Figure 9.1.8.3

9.1.8.4. Let D be a compact submanifold-with-boundary of the Euclidean plane E. The boundary ∂D of D is made up of a finite number of simple closed curves of E, by 3.4.1, 5.3.35. If E is oriented, the components of ∂D are canonically oriented by 5.3.36.

Notice that, even if D is connected, ∂D may not be (figure 9.1.8.4). In fact, if D is connected, ∂D is connected if and only if D is simply connected. This follows from the uniformization theorem, for example [Car63, p. 188].

Figure 9.1.8.4

9.1.9. Definition. A *point* in a closed curve C in E is an equivalence class of triples (S^1, f, m), under the relation "$(S^1, f, m) \sim (S^1, g, n)$ if and only if there exists $\theta \in \mathrm{Diff}(S^1, S^1)$ such that $f = g \circ \theta$ and $n = \theta(m)$." The point $f(m) = g(n) \in E$ is called the *image* of the corresponding point of C, and denoted by $\textcircled{m}$. The *multiplicity* of a point m in C is the number of elements of any of the sets $f^{-1}\big(f(m)\big)$, for $(S^1, f) \in C$.

In example 9.1.8.1, the multiplicity of an arbitrary $x \in S^1$ is $|n|$; this shows again that curves with different $|n|$ are distinct.

9.1.10. We will now generalize the notion of index introduced in 7.3.7 and 7.6.8. Let C be an oriented closed curve in an oriented Euclidean plane E, and call $S(E)$ the unit circle in E, which is canonically oriented (beginning of section 7.5). For $m \in E$, define a map $f_m : S^1 \to S(E)$ by

$$f_m(s) = \frac{f(s) - m}{\|f(s) - m\|}.$$

The degree of this map depends only on C, because if $(S^1, f) \sim (S^1, g)$, the orientation-preserving diffeomorphism θ such that $f = g \circ \theta$ gives $f_m = g_m \circ \theta$, whence $\deg f_m = \deg g_m \circ \deg \theta = \deg f_m$ by 7.3.4 and 7.3.6.2.

9.1.11. Definition. The degree of the map above is called the *index of m with respect to C*, and denoted by $\mathrm{ind}_m C$.

The index does not depend on the choice of a Euclidean structure (see exercise 9.9.1).

9.1.12. Proposition. *The map $m \mapsto \mathrm{ind}_m C$ is constant on each connected component of the complement $E \setminus C$ of C.*

Proof. Take $m, n \in E \setminus C$ and $(S^1, f) \in C$, and consider the maps

$$f_m : s \mapsto \frac{f(s) - m}{\|f(s) - m\|} \quad \text{and} \quad f_n : s \mapsto \frac{f(s) - n}{\|f(s) - n\|}.$$

If m and n lie on the same component of $E \setminus C$, we use 2.2.13 to connect the two with a path $\gamma : [0, 1] \to E \setminus C$ (figure 9.1.12). This gives a homotopy $F : [0, 1] \times S^1 \to S(E)$ between f_m and f_n, as follows:

$$F(t, s) = \frac{f(s) - \gamma(t)}{\|f(s) - \gamma(t)\|}.$$

But by 7.6.5, homotopic maps have the same degree. $\square$

9.1.13. Proposition. *Any closed curve C in Euclidean space admits periodic parametrizations by arclength.*

Proof. Apply 8.3.2 and 8.1.9. $\square$

Figure 9.1.12

All such parametrizations have the same period, which we denote by leng(C).

9.1.14. Example. The curve f_n in 9.1.8.1 has period $2\pi|n|$.

9.1.15. Remark. More generally, it is possible to integrate along C: Let $(\mathbf{R}, f)$ be a parametrization by arclength of C, periodic of period L, and associate to every function $g : C \to \mathbf{R}$ the function $\tilde{g} : [0, L] \to \mathbf{R}$ such that $\tilde{g}(t) = g(m)$ for $m = f(t) \in C$. The integral of g is $\int_0^L \tilde{g}(t)\,dt$. This gives a measure on C.

9.2. Jordan's Theorem

9.2.1. Theorem. *Let C be a simple closed curve of class C^2 in an affine plane E. The complement $E \setminus C$ has exactly two components, exactly one of which is bounded. We denote the bounded component by C_{int} and the other by C_{ext}. The closures $\overline{C}_{\text{int}}$ and $\overline{C}_{\text{ext}}$ are submanifolds-with-boundary of E and C is the boundary of each of them. We have the following two criteria:*

(i) *$\overline{C}_{\text{int}}$ is compact and $\overline{C}_{\text{ext}}$ isn't;*

(ii) *$\text{ind}_x C = \pm 1$ if $x \in C_{\text{int}}$ and $\text{ind}_x C = 0$ if $x \in C_{\text{ext}}$.*

9.2.2. Definition. We call C_{int} the *interior* of C and C_{ext} its *exterior*. Points in C_{int} and C_{ext} are said to be *inside* and *outside* C, respectively.

9.2.3. Remarks

9.2.3.1. Jordan's theorem is true in class C^0 [Die69, vol. I, chapter 9, app. 4.2].

9.2.3.2. As shown by figure 9.2.3, this result is not obvious, even intuitively, if the curve is complicated: for example, is x inside or outside C?

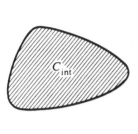

Figure 9.2.3

9.2.3.3. What submanifolds-with-boundary of E are obtained in this way? Exactly those that are simply connected (cf. 9.1.8.4).

9.2.4. Proof of Jordan's theorem. We orient C and E, and give E a Euclidean structure.

9.2.5. Lemma. *Let m be a point on a simple closed curve C and $u, v \neq m$ points on the normal to C at m, one on each side of m, and close enough to m. Then $|\mathrm{ind}_u C - \mathrm{ind}_v C| = 1$. If u is the point to the left of m, C being oriented, then $\mathrm{ind}_u C - \mathrm{ind}_v C = 1$.*

Proof. Using theorem 2.1.2(iv), draw a rectangle R around m so that C is a graph within R. If u an v are in R (this is the meaning of "close enough" in the statement), they are not in the image of C. Now deform C by the homotopy shown in figure 9.2.5.1; the result is a parametrized arc D of class C^0 only, but $\mathrm{ind}\,D$ is still defined (9.1.11). And, by 7.6.5, we have $\mathrm{ind}_u C = \mathrm{ind}_u D$ and $\mathrm{ind}_v C = \mathrm{ind}_v D$. Similarly, the index doesn't change if u and v move along the normal without crossing C (9.1.12).

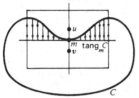
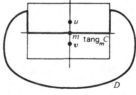

Figure 9.2.5.1

Now fix a and b on the straight part of D, and set u and v at a distance ε from m. For $p \in D \setminus [a, b]$, the angle between $\vec{pu}$ and $\vec{pv}$ approaches zero uniformly as ε decreases; this is immediate, because the distance pu, say, is bounded below by $\inf(bm, am)$.

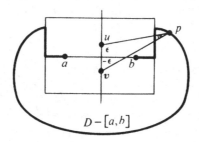

Figure 9.2.5.2

Choose a parametrization h of D, with period $L > 1$ and such that $h(0) = a$ and $h(1) = b$, and define the maps $f, g : \mathbf{Z}/L\mathbf{Z} \to S(E)$ by

$$f : t \mapsto \frac{h(t) - u}{\|h(t) - u\|} \quad \text{and} \quad g : t \mapsto \frac{h(t) - v}{\|h(t) - v\|}.$$

By definition, $\mathrm{ind}_u D = \deg f$ and $\mathrm{ind}_v D = \deg g$. Also, if $\overline{f}$ is a lifting of f to $\mathbf{R}$ (cf. 8.5.5.1 and 8.5.5.2), the degree of f is the integer k such that $\overline{f}(t + L) - \overline{f}(t) = 2k\pi$, for any t (7.6.4), and similarly for $\overline{g}$.

Now take ε small enough that the angle between $\vec{pu}$ and $\vec{pv}$ is less than some fixed $\eta < \pi$, for every $p \in D \setminus [a, b]$. If $\overline{f}$ and $\overline{g}$ are chosen so that $\left| \overline{f}(1) - \overline{g}(1) \right| < \eta$, we will have $\left| \overline{f}(t) - \overline{g}(t) \right| < \eta$ for all t, since $\overline{f}$ and $\overline{g}$ are continuous and the projection $\mathbf{R} \to S(E)$ is injective on any interval of length π. In particular, $\left| \overline{f}(L) - \overline{g}(L) \right| < \eta$.

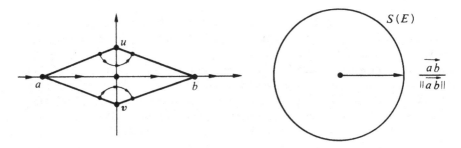

Figure 9.2.5.3

Fix as the origin on $S(E)$ the direction $\vec{ab}$. For ε small, we can take $\overline{f}(0)$ close to $-\pi$ and $\overline{g}(0)$ close to π. The behavior of $\overline{f}$ and $\overline{g}$ on $[0, 1]$ is then given by figure 9.2.5.4; in particular, we see that $\overline{f}(1)$ and $\overline{g}(1)$ are

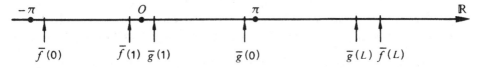

Figure 9.2.5.4

close to zero, so $\left|\overline{f}(1) - \overline{g}(1)\right| < \eta$. By the previous paragraph, this implies $\left|\overline{f}(L) - \overline{g}(L)\right| < \eta$.

Now the difference $\operatorname{ind}_u C - \operatorname{ind}_v C$ is given by

$$\left(\overline{f}(L) - \overline{f}(0)\right) - \left(\overline{g}(L) - \overline{g}(0)\right) = \left(\overline{f}(L) - \overline{g}(L)\right) - \left(\overline{f}(0) - \overline{g}(0)\right),$$

so it must be close to $\overline{g}(0) - \overline{f}(0)$, which is close to 2π. Since the index is an integer multiple of 2π, we must have $\operatorname{ind}_u C - \operatorname{ind}_v C = 2\pi$, proving the formula.

The formula in the oriented case also follows from this construction. □

Now for the proof of Jordan's theorem proper. By theorem 2.7.12, we can choose $\varepsilon > 0$ such that the map can : $N^\varepsilon C \to \operatorname{Tub}^\varepsilon C$ (section 2.7) is an embedding. But $N^\varepsilon C \setminus C \times \{0\}$ has exactly two connected components S_1 and S_2: indeed, $N_x C$ has a *canonical* orientation (6.7.22), so the positive components of $N_x C$ for x ranging over C agree, and their union is a component of $N^\varepsilon C \setminus C \times \{0\}$. Since $\operatorname{can}|_{N^\varepsilon C}$ is a diffeomorphism and takes $C \times \{0\}$ into C, we see that $\operatorname{Tub}^\varepsilon C \setminus C$ has two components: $\operatorname{can}(S_1) = T_1$ and $\operatorname{can}(S_2) = T_2$.

Let $\Omega_1, \dots, \Omega_k, \dots$ be the components of $E \setminus C$. Clearly T_1 and T_2 are contained in components of $E \setminus C$; but they cannot be in the same component because the indices of C with respect to points in T_1 and T_2 are different, by lemma 9.2.5. Thus we can assume $T_1 \subset \Omega_1$ and $T_2 \subset \Omega_2$.

Since C is closed, each Ω_i is closed, and its frontier is contained in C; the frontier is also non-empty (otherwise $\Omega_i = E$), so we can take $x \in C$ such that every neighborhood of x intersects Ω_i. But $\operatorname{Tub}^\varepsilon(C) = T_1 \cup C \cup T_2$ is a neighborhood of x, intersecting only Ω_1 and Ω_2; thus $E \setminus C$ has no other connected components. This also shows that C is the frontier of both Ω_1 and Ω_2.

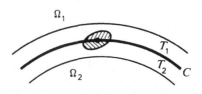

Figure 9.2.5.5

We now show that Ω_1 and Ω_2 have the properties stated in 9.2.1. First, $\overline{\Omega}_1$ and $\overline{\Omega}_2$ are submanifolds-with-boundary, because if f is a local equation for C, we locally have $\overline{\Omega}_1 = f^{-1}\left(]-\infty, 0]\right)$ and $\overline{\Omega}_2 = f^{-1}\left([0, +\infty[\right)$. Next, one of the components, which we call C_{ext}, is unbounded, because $\overline{\Omega}_1 \cup \overline{\Omega}_2 = E$. But C is contained in some ball $B(0, a)$, by compactness; since $E \setminus$

$B(0, a)$ is connected, the other component C_{int} has to be entirely contained in $B(0, a)$, which means it's bounded and relatively compact. Finally, to show property (ii), notice that we can contract C to a constant curve via a homotopy that fixes $E \backslash B(0, a)$. For any $m \notin B(0, a)$, this gives a homotopy between f_m and a constant map (cf. 9.1.10 and 9.1.11); by 7.4.1, this says that $\text{ind}_m C = 0$. By 9.1.12, the same is true for $m \in C_{\text{ext}}$. Finally, 9.2.5 gives $\text{ind}_m C = \pm 1$ for $m \in C_{\text{int}}$.   □

9.2.6. Corollary. *A simple closed curve in an oriented affine plane has a canonical orientation, for which* $\text{ind}_x C = 1$ *for all* $x \in C_{\text{int}}$.

In other words: when walking forward along C, the interior of C lies to the left.

Proof. The orientation is chosen so that $C = \partial C_{\text{int}}$ is canonically oriented as the boundary of a submanifold-with-boundary (theorem 5.3.36). With the notation of lemma 9.2.5, this gives $u \in C_{\text{int}}$ and $v \in C_{\text{ext}}$. Since $v \in C_{\text{ext}}$ implies $\text{ind}_v C = 0$, we get $\text{ind}_u C = 1$ by lemma 9.2.5.   □

We now give a practical criterion to determine whether $x \in C_{\text{int}}$ or $x \in C_{\text{ext}}$. For another criterion, see the end of the proof of 9.5.1.

9.2.7. Lemma. *Let C be a one-dimensional submanifold of an affine plane A, with underlying vector space E; both C and A are oriented. Take $x \in A \backslash C$, and consider the map $f : C \to S(E)$ defined by*

$$f(n) = \frac{\overrightarrow{xn}}{\|\overrightarrow{xn}\|}.$$

A point $m \in C$ is a critical point of f if and only if the line xm is tangent to C at m.

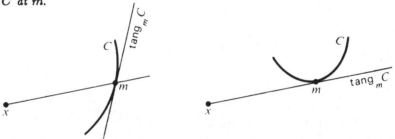

Figure 9.2.7

Proof. Writing $f = F \circ i$, where $i : C \to A$ is the inclusion and $F : A \to E$ is given by the same formula as f, we see that $x \in C$ is a critical point of f if $T_x C$ is taken into zero by the derivative of F. This derivative is given by

$$F'(m)(z) = \frac{z}{\|m - x\|} - \frac{(m - x \mid z)(m - x)}{\|m - x\|^{3/2}},$$

for $z \in E$ and $m \in A$, and it is easy to see that it vanishes if and only if $m - x$ and z are proportional. Thus $F'(m)(T_x C) = 0$ is equivalent to $m - x \in T_m C = \text{tang}_m C$. $\qquad\qquad\qquad\qquad\qquad\qquad\qquad\qquad\qquad\square$

By a corollary of Sard's theorem (4.3.6), the map f has regular values; let $u \in S(E)$ be one. The half-line D_x with origin x and direction u is nowhere tangent to C, by lemma 9.2.7, and intersects C in a finite number of points, by the proof of 4.1.5. Thus, by criterion (ii) in theorem 9.2.1, saying that $x \in C_{\text{int}}$ is the same as saying that $\text{ind}_x C = \pm 1$, which implies that D_x intersects C an odd number of times.

9.2.8. Practical criterion. *Let C be a simple closed curve in an affine plane A, and x a point not in C. There exist half-lines originating at x that are nowhere tangent to C. Such a half-line intersects C in k points, where k is finite, and*

$$x \in C_{\text{int}} \Leftrightarrow k \text{ is odd},$$
$$x \in C_{\text{ext}} \Leftrightarrow k \text{ is even}.$$

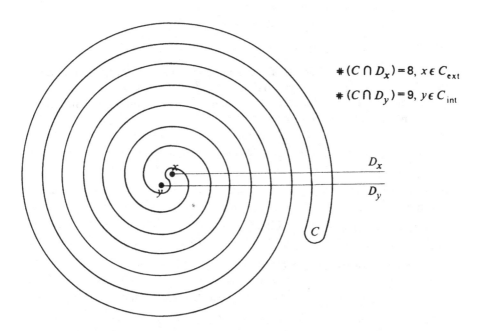

$$\#(C \cap D_x) = 8, \, x \in C_{\text{ext}}$$
$$\#(C \cap D_y) = 9, \, y \in C_{\text{int}}$$

Figure 9.2.8

9.2.9. Remark. Corollary 4.3.5 to Sard's theorem says in fact that "almost all" half-lines are non-tangent. Such half-lines are said to be *transversal* to C. See [AR67] for the important notion of transversality.

9.2.10. Remark. See the proof of 9.5.1 for another characterization of C_{int}.

9.3. The Isoperimetric Inequality

In this section E is a Euclidean plane and C a simple closed curve in E, of class C^p. Associated with C are its length and the area of C_{int}, which is finite because $\overline{C}_{\text{int}}$ is compact (6.5.2).

9.3.1. Theorem. *Every simple closed curve C of class C^p in a Euclidean plane E satisfies the inequality*

$$\text{leng}^2(C) \geq 4\pi \, \text{area}(C_{\text{int}}).$$

Furthermore, equality holds if and only if C is a circle.

9.3.2. Lemma (Wirtinger's inequality). *Let $f \in C^1(\mathbf{R})$ be periodic of period 2π, and such that $\int_0^{2\pi} f(t)\, dt = 0$. Then*

$$\int_0^{2\pi} f'^2(t)\, dt \geq \int_0^{2\pi} f^2(t)\, dt,$$

and equality holds if and only if $f(t) = a\cos t + b\sin t$, with $a, b \in \mathbf{R}$.

Proof. (See [Rud74, p. 94–97] for the necessary background.) The restriction $f|_{[0,2\pi]}$, being continuous, belongs to $L^2([0, 2\pi])$, the Hilbert space of square-integrable functions on $[0, 2\pi]$, and hence can be approximated in the L^2 norm by its Fourier series

$$\frac{a_0}{2} + \sum_{n=1}^{\infty} (a_n \cos nt + b_n \sin nt).$$

Since we've assumed $\int_0^{2\pi} f(t)\, dt = 0$, we have $a_0 = 0$, and Parseval's theorem gives

$$\|f\|^2 = \sum_{n \geq 1} a_n^2 + b_n^2 = \int_0^{2\pi} f^2\, dt.$$

The Fourier series of f' is given by

$$\sum_{n=1}^{\infty} \alpha_n \cos nt + \beta_n \sin nt = \sum_{n=1}^{\infty} n(b_n \cos nt - a_n \sin nt);$$

this can be seen by performing the integration of $\alpha_n = \int_0^{2\pi} f'(t) \cos nt\, dt$ and $\beta_n = \int_0^{2\pi} f'(t) \sin nt\, dt$ by parts. Applying Parseval again, we get

$$\|f'\|^2 = \int_0^{2\pi} f'^2\, dt = \sum_{n \geq 1} n^2(a_n^2 + b_n^2) \geq \|f\|^2,$$

with equality if and only if $a_n^2 + b_n^2 = 0$ for all $n > 1$. $\square$

9.3.3. Proof of the isoperimetric inequality. We can assume, by applying a homothety, that C has length 2π. Let $h = (f, g)$ be a parametrization by arclength (9.1.13), and set the coordinate axes so that the x-axis goes through the center of mass of $\overline{C}_{\text{int}}$, which is compact (6.5.14 and 9.2.1(i)); this is the same as saying that $\int_0^{2\pi} f(t)\, dt = 0$. Since h is a parametrization by arclength, we have

$$\int_0^{2\pi} (f'^2 + g'^2)\, dt = \int_0^{2\pi} dt = 2\pi = \text{leng}(C).$$

On the other hand, Stokes' theorem gives

$$\text{area}(C_{\text{int}}) = \int_{C_{\text{int}}} dx \wedge dy = \int_{\overline{C}_{\text{int}}} dx \wedge dy = \int_C x\, dy = \int_0^{2\pi} f g'\, dt,$$

because $d(x\, dy) = dx \wedge dy$. Subtracting, we get

$$2\big(\pi - \text{area}(C_{\text{int}})\big) = \int_0^{2\pi} (f'^2 + g'^2 - 2f g')\, dt$$

$$= \int_0^{2\pi} (f'^2 - f^2)\, dt + \int_0^{2\pi} (f - g')^2\, dt.$$

The first integral is non-negative by lemma 9.3.2, and so is the second. Thus $2\,\text{area}(C_{\text{int}}) \geq 2\pi$ and, since $\text{leng}(C) = 2\pi$, we finally get

$$\text{leng}^2(C) \geq 2\pi \cdot 2\,\text{area}(C_{\text{int}}).$$

For equality to hold it is necessary that $\int_0^{2\pi}(f'^2 - f^2)\, dt = 0$ and $\int_0^{2\pi}(f - g')^2\, dt = 0$. The first condition gives $f(t) = a\cos t + b\sin t$, by lemma 9.3.2; the second gives $g'(t) = f(t)$ almost everywhere, and, by continuity, $g(t) = a\sin t - b\cos t + c$. Thus (f, g) is the parametrization of a circle. $\square$

9.3.4. Remarks.

9.3.4.1. The isoperimetric inequality is valid in class C^1 and even C^0 [Fed69, p. 278].

9.3.4.2. Jordan's theorem has a counterpart in higher dimension: a compact, connected, codimension-one submanifold V of $\mathbf{R}^n$ is the boundary of a compact submanifold-with-boundary $\overline{V}_{\text{int}}$, and in particular orientable [Gre67, p. 81].

9.3.5. Thus we can define $\text{vol}(V)$ and $\text{vol}(\overline{V}_{\text{int}})$ (cf. 6.5.1), and there is a version of the isoperimetric inequality relating the two quantities (6.6.9). This is harder to prove than the plane version.

9.4. The Turning Number

In this section everything is assumed of class C^1. Let C be an oriented closed curve in an oriented Euclidean plane E, represented by an immersion $(S^1, f) \in C$ (9.1.4). The *unit tangent map* $\tau_f : S^1 \to S(E)$, defined by

9.4.1
$$\tau_f(t) = \frac{f'(t)}{\|f'(t)\|}$$

is continuous, and its degree does not depend on the choice of f: if $(S^1, f) \sim (S^1, g)$ with $f = g \circ \theta$, where $\theta \in \text{Diff}(S^1)$ preserves orientation, $\tau_f = \tau_g \circ \theta$. Hence the following definition makes sense:

9.4.2. Definition. The *turning number* of C, denoted by $\text{turn}(C)$, is the degree of the map τ_f in 9.4.1.

9.4.3. Intuitively, the turning number tells how many times the (oriented) tangent turns around, as we go along the curve once.

9.4.4. Remarks

9.4.4.1. This definition does not depend on the choice of a Euclidean structure.

9.4.4.2. If the orientation of C or of E is switched, $\text{turn}(C)$ changes sign.

9.4.5. Examples

9.4.5.1. If C_n is the curve defined in 9.1.8.1, we have $\text{turn}(C_n) = n$.

9.4.5.2. The closed curves in figure 9.4.5.2 have turning number 2 and $n+1$, respectively (if the plane is oriented canonically).

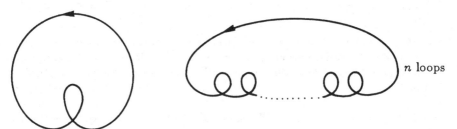

Figure 9.4.5.2

9.4.5.3. The closed curve in 9.1.8.3 (figure 9.4.5.3) has turning number zero, because τ is not surjective (7.3.5.1); it misses the shaded part of the circle.

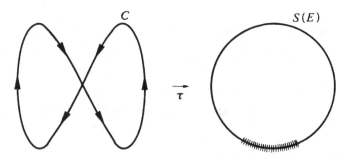

Figure 9.4.5.3

9.4.6. Definition. Two closed curves C and D in E, of class C^1, are said to be *homotopic* if there exists a homotopy $F \in C^0([0,1] \times S^1 \to E)$ such that, $F_0 = C$, $F_1 = D$ and $F_t : x \mapsto F(t,x)$ is a C^1 immersion for all $t \in [0,1]$.

In other words, we want each F_t to be a closed curve.

9.4.7. Proposition. *Two homotopic closed curves in an oriented Euclidean plane E have the same turning number.*

Proof. Homotopic maps have the same degree (7.6.5).   □

9.4.7.1. If we hadn't required each F_t to be an immersion, the proposition would be false. For instance, in figure 9.4.7, the turning number is 1 for F_0 and 0 for F_1, by 9.4.5.1 and 9.4.5.3.

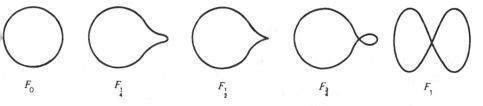

Figure 9.4.7

9.4.7.2. It follows that a circle cannot be turned inside out by a continuous sequence of immersions, since the turning number would change from 1 to -1. The situation is different for a sphere in $\mathbf{R}^3$: see 11.11.1.

9.4.8. Theorem (Whitney–Grauenstein). *Two closed curves C and D having the same turning number are homotopic.*

Before proving the theorem we indicate some of its consequences.

9.4.9. Corollary. *Every closed curve in E is homotopic to one of the following:*

(i) $S(E)$ *if* $\mathrm{turn}(C) = 1$;

(ii) $S(E)$, *with orientation reversed, if* $\mathrm{turn}(C) = -1$;

(iii) C_n *(cf. 9.1.8.1) or the curve in figure 9.4.5.2, if* $|\mathrm{turn}(C)| > 1$;

(iv) *the curve in figure 9.4.5.3, if* $\mathrm{turn}(C) = 0$.   □

Another, very nice consequence, involves the total curvature. Let C be a closed curve and $f \in C^1(\mathbf{R}; E)$ a parametrization of C by arclength, with period $L = \mathrm{leng}(C)$ (cf. 9.1.13). Assume C and E are oriented.

9.4.10. Definition. The *total (signed) curvature* of C is the integral $\int_0^L k(f(t))\, dt$, where $k(f(t))$ is the signed curvature of C at $f(t)$ (8.5.2).

By 9.1.13, this integral is independent of the parametrization.

9.4.11. Corollary. *The total curvature of C is equal to $2\pi\,\mathrm{turn}(C)$. In particular, it is a homotopy invariant.*

Proof. By 8.5.6 and 7.6.4 we have

$$\int_0^L k(f(t))\, dt = \int_0^L \frac{d\bar{\tau}}{dt}\, dt = \bar{\tau}(L) - \bar{\tau}(0) = 2\pi \deg \tau = 2\pi\,\mathrm{turn}(C). \quad \square$$

9.4.12. Examples. Here is the total curvature of some curves:

2π -2π 0 0 $2(n+1)\pi$

Figure 9.4.12

9.4.13. Proof of theorem 9.4.8. Let $f, g \in C^1(\mathbf{R}; E)$ be parametrizations by arclength of C and D, respectively. We may assume that f and g are periodic of period 2π, that is, C and D have length 2π; otherwise we can replace C, for example, by the curve $\tilde{C}$ defined by

$$\tilde{f} : t \mapsto \frac{2\pi}{L} f\left(\frac{Lt}{2\pi}\right),$$

which is easily seen to be parametrized by arclength, homotopic to C and of length 2π. We may also assume $f(0) = g(0) = 0$.

By assumption, f' and g', considered as maps from S^1 into $S(E)$, have the same degree (9.4.2). Define a homotopy $H \in C^0([0,1] \times \mathbf{R}; S^1)$ between f' and g' such that each H_λ is periodic of period 2π: fix a covering map $p : \mathbf{R} \to S(E)$ (cf. 8.5.5.1), take liftings $\overline{f}'$ and $\overline{g}'$ of f' and g', and average them as follows:

9.4.14 $\overline{H}_\lambda(s) = (1 - \lambda)\overline{f}'(s) + \lambda\overline{g}'(s).$

Then set $H = p \circ \overline{H}$.

Now define a family of parametrized arcs, for $\lambda \in [0, 1]$, by

9.4.15 $$F_\lambda(t) = \int_0^t H(\lambda, s)\, ds - \frac{t}{2\pi} \int_0^{2\pi} H(\lambda, s)\, ds = F(\lambda, t);$$

we will show that F is a homotopy between C and D. Clearly $F_\lambda \in C^1(\mathbf{R}; E)$, and F_λ is periodic of period 2π because H_λ is:

$$F_\lambda(t + 2\pi) = \int_0^{2\pi} H(\lambda, s)\, ds + \int_{2\pi}^{t+2\pi} H(\lambda, s)\, ds - \int_0^{2\pi+t} H(\lambda, s)\, ds$$
$$= \int_0^t H(\lambda, s)\, ds - \frac{t}{2\pi} \int_0^{2\pi} H(\lambda, s)\, ds = F_\lambda(t).$$

We also have

$$F_0(t) = \int_0^t H(0, s)\, ds - \frac{t}{2\pi} \int_0^{2\pi} H(0, s)\, ds$$
$$= \int_0^t f'(s)\, ds - \frac{t}{2\pi} \int_0^{2\pi} f'(s)\, ds$$
$$= f(t) - f(0) - \frac{t}{2\pi}\big(f(2\pi) - f(0)\big) = f(t),$$

and $F_1(t) = g(t)$ follows analogously.

There remains to show that F_λ is an immersion for every λ (9.4.6), that is, that

9.4.16 $$F_\lambda'(t) = H(t, \lambda) - \frac{1}{2\pi} \int_0^{2\pi} H(\lambda, s)\, ds \neq 0$$

for all t and λ. By construction, $\|H(t, \lambda)\| = 1$, so

$$\left\| \frac{1}{2\pi} \int_0^{2\pi} H(\lambda, t)\, dt \right\| \leq \frac{1}{2\pi} \int_0^{2\pi} \|H(\lambda, t)\|\, dt = 1.$$

But, as the reader should check, equality can only take place if $H(\lambda, t)$ is a positive multiple of a fixed vector x_0, for every t; and this, in turn, implies that $H(\lambda, t)$ is constant. We now prove that this cannot be the case.

If H_λ is constant, so is $\overline{H}_\lambda$. But it follows from 9.4.14 and the definition of the turning number that $\overline{H}_\lambda(2\pi) - \overline{H}_\lambda(0) = (1-\lambda)\,\mathrm{turn}(C) + \lambda\,\mathrm{turn}(D) = \mathrm{turn}(C)$, so this cannot happen if $\mathrm{turn}(C) \neq 0$. Now assume $\mathrm{turn}(C) = 0$; then $\overline{f}'$ is periodic, so it has a minimum, say at t_0; similarly $\overline{g}'$ has a minimum at t_1. Reparametrize C and D so that $t_0 = t_1 = 0$. Then $\overline{H}_\lambda = (1 - \lambda)\overline{f}' + \lambda\overline{g}'$ cannot be constant unless $\overline{f}'$ and $\overline{g}'$ are. But $\overline{f}'$ cannot be a constant: if $f'(a) = a$ for all t, then $f(t) = f(0) + at$ is not periodic. $\qquad\square$

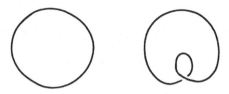

Figure 9.4.18.2

9.4.17. Remarks.

9.4.17.1. Theorem 9.5.1 below gives the value of turn(C) for a simple closed curve.

9.4.17.2. Theorem 9.4.8 can be considered as a classification theorem: the equivalence classes for the homotopy equivalence relation are in one-to-one correspondence with **Z**. There exists a similar, but much more difficult, result for biregular curves in three-space (see section 8.6), due to Feldman [Fel68]; this time we get only two classes, represented by the curves in figure 9.4.18.2.

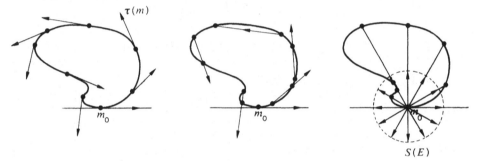

Figure 9.5.2.1

9.5. The Turning Tangent Theorem

9.5.1. Theorem. *The turning number of any simple closed curve is ± 1. If the curve is oriented by Jordan's theorem* (9.2.6) *its turning number is 1.*

9.5.2. Corollary. *The total curvature* (9.4.11) *of a simple closed curve oriented by Jordan's theorem is 2π.*   □

Proof. The idea of the proof is neat (figure 9.5.2.1). It consists in constructing a homotopy between the unit tangent map on the left, whose degree is turn(C), and the "sweep map" on the right, which is clearly of degree one: if you stand on the curve and turn so as to keep facing a moving point as it goes around the curve one, you will have turned 180°; repeating the process facing away from the point makes a full turn. The formalization, however, is quite long and technical.

As in 9.4.13, we can assume that C has length 2π. Let f be a parametrization of C by arclength (9.1.13), periodic of period 2π and compatible with the orientation given by corollary 9.2.6.

Choose $m_0 \in C$ such that C lies entirely on one side of $\text{tang}_{m_0} C$. Any point of C minimizing the function $(e\,|\,\cdot)$, where $e \in E$ is a fixed vector, will do; such points exist since C is compact. We can assume that $m_0 = f(0)$; in addition, we choose $f'(0) = \tau(m_0)$ as the origin for $S(E)$.

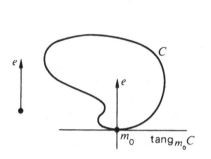

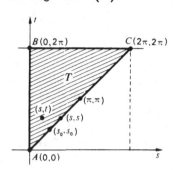

Figure 9.5.2.2   Figure 9.5.2.3

Let T be the triangle $\{(s,t) \in \mathbf{R}^2 : 0 \le s \le t \le 2\pi\}$ (figure 9.5.2.3), and define $G : T \to S(E)$ by

9.5.3
$$G(s,t) = \begin{cases} \dfrac{f(t) - f(s)}{\|f(t) - f(s)\|} & \text{if } s \ne t \text{ and } (s,t) \ne (0,2\pi), \\ -f'(0) & \text{if } (s,t) = (0,2\pi), \\ f'(s) & \text{otherwise.} \end{cases}$$

Here is where we use the fact that C is a simple closed curve, because then $f(s) \ne f(t)$ for $s - t \notin 2\pi\mathbf{Z}$. Notice that the composition of G with the diagonal map $t \mapsto (t,t)$ is the unit tangent map f'. Now we deform the diagonal into two sides of T (figure 9.5.2.4), by means of a continuous one-parameter family of curves $H : [0,1] \times [0,2\pi] \to T$. More specifically, we require that $H_0(u) = H(0,u) = (u,u)$ for all $u \in [0,2\pi]$ and

$$H_1(u) = H(1,u) = \begin{cases} (0,2u) & \text{for } u \in [0,\pi], \\ (2(u - \pi), 2\pi) & \text{for } u \in [\pi,2\pi]. \end{cases}$$

We show below that G is continuous, so $F = G \circ H : [0,1] \times [0,2\pi] \to S(E)$ is continuous. In fact F can be seen as a family of *closed* curves, since $F_\lambda(2\pi) = F_\lambda(0)$ for all λ. At one end of the family is F_0, which, as already remarked, is just f'; at the other end is the map from figure 9.5.2.1, whose degree we will show to be one. By 7.6.5 this means that $\text{turn}(C) = \deg F_0 = 1$.

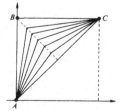

Figure 9.5.2.4

By definition,

$$F_1(0) = f'(0),$$

$$F_1(u) = \frac{f(2u)}{\|f(2u)\|} \quad \text{for all } u \in \,]0, \pi[,$$

$$F_1(\pi) = -f'(0),$$

$$F_1(u) = \frac{-f(2(u-\pi))}{\|-f(2(u-\pi))\|} \quad \text{for all } u \in \,]\pi, 2\pi[,$$

$$F_1(2\pi) = f'(0).$$

By our choice of m_0, C lies entirely to the left of $\text{tang}_{m_0} C$. Consider the covering map $p : \mathbf{R} \to S(E)$ such that $p(0) = f'(0)$ (cf. 8.5.5.1), and lift F_1 to a map $F_1 : \mathbf{R} \to \mathbf{R}$ such that $\overline{F}_1(0) = 0$.

For all $u \in [0, \pi]$, we have $F_1(u)$ to the left of $\text{tang}_{m_0} C$, which implies $0 \leq \overline{F}_1(u) \leq \pi$ and $\overline{F}_1(\pi) = \pi$ because $F_1(\pi) = -f'(0)$. Similarly, for all $u \in [\pi, 2\pi]$ we have $F_1(u)$ to the right of $\text{tang}_{m_0} C$; this implies $\pi \leq \overline{F}_1(u) \leq 2\pi$ and $\overline{F}_1(2\pi) = 2\pi$, since $\overline{F}_1(2\pi)$ is again $f'(0)$. By definition, $\deg(F_1) = (\overline{F}_1(2\pi) - \overline{F}_1(0))/2\pi = 1$.

We still must check that the orientation of C is the one given by Jordan's theorem; this amounts to showing that, at m_0, the interior of C lies to the left of $\text{tang}_{m_0} C$. But this is obvious, because the open half-plane to the right of $\text{tang}_{m_0} C$ is an unbounded connected set, not intersecting C, so it must be contained in the exterior of C (9.2.1(i)).

This completes the proof, modulo lemma 9.5.4. $\square$

9.5.4. Lemma. *The map $G : T \to S(E)$ defined in 9.5.3 is continuous.*

Proof. Since f is continuous and $f(s) \neq f(t)$ for $s - t \notin 2\pi\mathbf{Z}$, it is clear that G is continuous at (s, t) for $s \neq t$ and $(s, t) \neq (0, 2\pi)$. In studying the behavior of G along the diagonal and at $(0, 2\pi)$, we apply the technique used in Morse reduction (4.2.13). Since f is C^1 we can write

$$f(s) = f(t) + \int_0^1 \frac{d(f(t + \lambda(s - t)))}{d\lambda} \, d\lambda;$$

setting $\phi(s, t) = \int_0^1 (s - t) f'(t + \lambda(s - t)) \, d\lambda$, this becomes

$$f(s) - f(t) = (s - t)\phi(s, t),$$

with $\phi \in C^0(T; E)$, because f is C^1 and we're integrating over a compact set (0.4.8.2). Away from $(0, 2\pi)$ and the diagonal we have

$$G(s, t) = \frac{(t - s)\phi(t, s)}{\|(t - s)\phi(t, s)\|} = \frac{\phi(t, s)}{\|\phi(t, s)\|},$$

since $s \leq t$. For a point on the diagonal, say (s_0, s_0), we can write

$$\phi(s_0, s_0) = \int_0^1 f'(s_0 + \lambda(s_0 - s_0)) \, d\lambda = f'(s_0);$$

in particular, $\phi(s_0, s_0)$ is non-zero because $\|f'\| = 1$. Thus we have

$$\lim_{(s,t) \to (s_0,s_0)} G(s,t) = \lim_{(s,t) \to (s_0,s_0)} \frac{\phi(t,s)}{\|\phi(t,s)\|} = \frac{\phi(s_0, s_0)}{\|\phi(s_0, s_0)\|} = f'(s_0),$$

which proves continuity at (s_0, s_0). Finally, around $(0, 2\pi)$ we set $2\pi - t = t'$, whence

$$G(s,t) = G(s, 2\pi - t') = \frac{f(2\pi - t') - f(s)}{\|f(2\pi - t') - f(s)\|} = \frac{f(-t') - f(s)}{\|f(-t') - f(s)\|}$$

$$= \frac{(-t' - s)\phi(-t', s)}{\|(-t' - s)\phi(-t', s)\|} = -\frac{\phi(-t', s)}{\|\phi(-t', s)\|}$$

since $t' + s$ is positive. Thus

$$\lim_{(s,t) \to (0,2\pi)} G(s,t) = \lim_{(s,t)} \left(-\frac{\phi(-t', s)}{\|\phi(-t', s)\|} \right) = -\frac{\phi(0,0)}{\|\phi(0,0)\|} = -f'(0),$$

since $\phi(0,0) \neq 0$. This shows that G is continuous everywhere. $\square$

9.6. Global Convexity

In this paragraph the differentiability class is at least C^2.

9.6.1. Definition. The *total unsigned curvature* of a curve C is the integral

$$\int_0^L K_{f(t)} C \, dt,$$

where $([0, L], f)$ is a parametrization of C by arclength and $K_{f(t)} C$ is the curvature of C at $f(t)$ (8.4.1). This integral will be denoted by $\int_C K$.

9.6.2. Theorem. *Let C be a simple closed curve in a Euclidean plane E. The following properties are equivalent:*
(i)   *the signed curvature of C (for some choice of orientations) has constant sign;*
(ii) $\int_C K = 2\pi$;
(iii) *C is globally convex, that is, for every $m \in C$ the curve lies entirely to one side of its tangent $\tan g_m C$;*
(iv)   *C is the boundary of a convex submanifold-with-boundary of E.*

9.6.2.1. Note. In the case of surfaces in $\mathbf{R}^3$ there is a much stronger theorem: the conclusion is true even if the surface is *a priori* only assumed to be immersed. See 9.6.5.3 and 11.13.

9.6.3. Definition. A simple closed curve C in E satisfying the properties in theorem 9.6.2 is called *convex*.

Proof. (i)$\Leftrightarrow$(ii). If $k_m C$ is the signed curvature of C at m, we have $|k_m C| = K_m C$ (8.5.4.1). Thus

$$\int_C K = \int_C |k| \geq \left| \int_C k \right| = 2\pi$$

by 9.5.2. If k does not have constant sign, the inequality is strict because k is continuous.

(i)$\Rightarrow$(iii). By contradiction: assume that k has constant sign but there exists $m \in C$ such that C has points on both sides of $\mathrm{tang}_m\, C$. By compactness, choose points p and q on opposite sides of $\mathrm{tang}_m\, C$ whose distance to $\mathrm{tang}_m\, C$ is maximal (cf. the beginning of the proof of 9.5.1); then $\mathrm{tang}_p\, C$, $\mathrm{tang}_m\, C$ and $\mathrm{tang}_q\, C$ are parallel (figure 9.6.3.2).

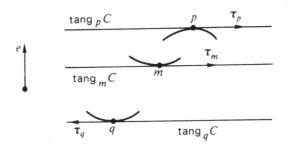

Figure 9.6.3.2

Since $\|\tau_m(C)\| = \|\tau_p(C)\| = \|\tau_q(C)\| = 1$, two of these vectors are equal, say $\tau_m(C) = \tau_p(C)$. Let f be a parametrization of C by arclength, periodic of period L, and let s, t be such that $f(s) = m$ and $f(t) = p$, with $s, t \in [0, L[$. If $\overline{\tau} : \mathbf{R} \to \mathbf{R}$ is a lifting of the unit tangent map $\tau : \mathbf{R} \to S(E)$, formula 8.5.6 says that, with appropriately chosen orientations, $k = \dfrac{d\overline{\tau}}{dt}$. Since k has constant, say positive, sign, $\overline{\tau}$ is non-decreasing. In addition $\overline{\tau}(L) = 2\pi$, if $\overline{\tau}$ is chosen so that $\overline{\tau}(0) = 0$, because $\mathrm{turn}(C) = 1$ by 9.5.1.

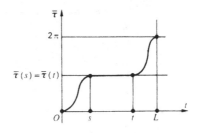

Figure 9.6.3.3

Since $p : \mathbf{R} \to S(E)$ is injective on $[0, 2\pi[$ and $\overline{r}([0, L[) = [0, 2\pi]$ by the above, the equality $r(s) = r(t)$ implies $\overline{r}(s) = \overline{r}(t)$. But $\overline{r}(s) = \overline{r}(t)$ and $\overline{r}$ non-decreasing imply $\overline{r}$ constant on $[s, t]$; thus r is constant on $[s, t]$, and $f(t) = f(s) + (t - s)f'(s)$, contradicting $p \neq \tan g_m C$.

(iii)$\Rightarrow$(iv). This is a general theorem on convexity: a compact subset of the plane having a support hyperplane at each point of its frontier is convex. But in our case we can give a more elementary proof.

Take $m \in C$ and let H_m be the closed hyperplane determined by $\tan g_m C$ and containing C. We claim that $D = \overline{C}_{\text{int}} \subset H_m$. By contradiction: let $y \in \overset{\circ}{D} \setminus H_m$. The half-line Δ originating at y and containing m is not entirely in D because D is compact; since $y \in \overset{\circ}{D}$ there exists a point $z \in \text{Fr}(\Delta \cap D) \subset \Delta \cap \text{Fr}\, D = \Delta \cap C$ such that y lies between m and z. Thus C is not entirely contained in H_m.

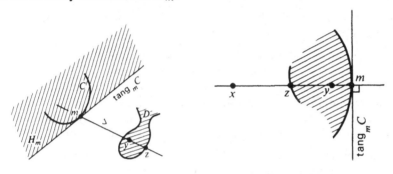

Figure 9.6.3.4   Figure 9.6.3.5

This is true for arbitrary $m \in C$, so $D \subset \bigcap_{m \in C} H_m$. We claim that, in fact,

9.6.4
$$D = \bigcap_{m \in C} H_m;$$

this implies that D is convex, being an intersection of convex sets.

By contradiction: take $x \in \bigcap_{m \in C} H_m \setminus D$. By compactness, choose a point $m \in C$ minimizing the function

$$C \ni n \mapsto d^2(s, n) = \|n - x\|^2 \in \mathbf{R}.$$

The tangent $\tan g_m C$ is orthogonal to the line Δ containing x and m, because the derivative of $\| \cdot \|^2(z)$ is $2(z|\cdot)$.

By assumption, C is in H_m, and in particular $\overset{\circ}{D}$ contains points y strictly between x and m. But, as before, we see that there would exist

$$z \in \text{Fr}(\Delta \cap D) \subset \Delta \cap \text{Fr}\, D = \Delta \cap C,$$

where z is between x and y, because $x \notin D$. But such a z would be closer to x than to m, that is, $d(x, z) < d(x, m)$, contradicting the definition of m.

(iv)$\Rightarrow$(i). At a point m of C, oriented by Jordan's theorem, the submanifold-with-boundary D lies to the left of $\text{tang}_m\, C$; since D is convex, C is also to the left of $\text{tang}_m\, C$. But then $k_m C \geq 0$, for if $k_m C < 0$, the local study made in 8.2.2.15 shows that C has some points to the right of $\text{tang}_m\, C$.   □

9.6.5. Remarks

9.6.5.1. If C is a closed curve in a Euclidean space E of arbitrary finite dimension, we can still define $\int_C K$, the total scalar curvature. It can be shown that $\int_C K \geq 2\pi$ and if equality holds C is plane and convex. This is in fact a very particular case of a general result valid for any compact manifold X immersed in E, with $\int_C K$ replaced by $\int_X |K_d|\, \delta$ (in the notation of 6.9.6 and 6.9.15). See [Kui70].

9.6.5.2. One also shows that if dim $= 3$ and C is knotted then $\int_C K > 4\pi$; see [Kui70].

9.6.5.3. As already remarked in 8.2.2.16, a (non-simple) plane closed curve can have strictly positive curvature everywhere and still not be globally convex. But this phenomenon only happens in two dimensions. If X is a compact codimension-one manifold, immersed in a Euclidean space of dimension ≥ 3, and if $K_d \geq 0$ (6.9.6), then X is necessarily embedded, and is the boundary of a convex submanifold-with-boundary: see 11.13.

9.6.5.4. For some results about $\int_C T$, the integral of the torsion of a closed curve in three-space, see exercise 9.9.6 and [Poh68].

9.7. The Four-Vertex Theorem

Here the differentiability class is at least C^3.

Let C be a closed curve in a Euclidean space E of arbitrary dimension, and f a parametrization of C by arclength. The curvature function $K : t \to K(t) = K_{f(t)} C$ being continuous and periodical, it has a maximum and a minimum; at these points $K' = 0$. The condition $K' = 0$ at a given point does not depend on the choice of a parametrization by arclength (cf. 8.3.2); so the following definition makes sense:

9.7.1. Definition. A *vertex* of a geometric arc is a point where $K' = 0$ (expressed in terms of a parametrization by arclength).

9.7.2. Remark. In fact, $m = g(t)$ is a vertex if the function $K : t \mapsto K_{g(t)} C$ satisfies $K'(t) = 0$, even if g is not a parametrization by arclength; indeed, C is locally a submanifold V of E, and the curvature is a function on V; saying that m is a vertex is the same as saying that m is a critical point of K.

9.7.3. Examples

9.7.3.1. The closed curve in figure 8.4.14.3 has exactly two vertices; for we have seen that, between 0 and π, its curvature is strictly increasing, and on the other hand $K'(0) = K'(\pi) = 0$.

9.7.3.2. The ellipse $f : t \mapsto (a\cos t, b\cos t)$ has curvature

$$K_{f(t)} = \frac{ab}{(a^2\sin^2 t + b^2\cos^2 t)^{3/2}},$$

by formula 8.4.13.1. It follows that $K' = 0$ if and only if $t = k\pi/2$. Thus the ellipse has exactly four vertices (in the curvature sense), which are the same as the vertices (in the quadric sense). This is one justification for the name "vertex".

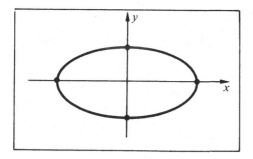

Figure 9.7.3.2

The introduction to section 9.7 shows that a closed curve has at least two vertices; example 9.7.3.1 shows it can have more.

9.7.4. Theorem (of the four vertices). *A convex simple closed curve in Euclidean plane has at least four vertices.*

9.7.5. Remark. The convexity hypothesis is not necessary, but the proof is much more difficult in the general case [BF58]. For another proof, due to Osserman, see exercise 9.9.7.

9.7.6. Proof. We know that C has at least two vertices. Suppose by contradiction that C has only two or three, say m_1, m_2 and m_3.

By assumption the derivative k' of the curvature can only change sign at one of the m_i, so we can assume that it is positive from m_1 to m_2 and negative from m_2 to m_3 and from m_3 to m_1. Take the line m_1m_2 for the x-axis, and let $\{e_1, e_2\}$ be the associated orthonormal basis.

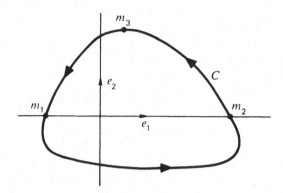

Figure 9.7.6

Let f be a parametrization of C by arclength, periodic of period L, where L is the length of C, and let k be the signed curvature function of C. (We orient E and C if necessary.) Introduce the vector-valued function

9.7.7 $$t \mapsto k'(t)f(t)$$

and the vector-valued integral

9.7.8 $$\int_0^L k'(t)f(t)\,dt$$

which can be integrated by parts:

$$\int_0^L k'(t)f(t)\,dt = \left[k(t)f(t)\right]_0^L - \int_0^L k(t)f'(t)\,dt = -\int_0^L k(t)f'(t)\,dt,$$

since f and k are periodic. Since f is a parametrization by arclength, $f'(t) = \tau(t)$ is the unit tangent vector, so

$$-k(t)\tau(t) = (i\tau)',$$

where i denotes a rotation by $\pi/2$ (definition 8.5.2).

Thus

$$\int_0^L k'(t)f(t)\,dt = \int_0^L (i\tau)'\,dt = i\big(\tau(L) - \tau(0)\big) = 0.$$

In particular, the second coordinate is zero, that is,

$$\left(\int_0^L k'(t)f(t)\,dt \mid e_2\right) = \int_0^L k'(f \mid e_2)\,dt = 0.$$

On the other hand, by convexity, the part of the curve between m_1 and m_2 is below the x-axis, and there $k' > 0$, so $k'(f \mid e_2) < 0$ from m_1 to m_2.

Similarly $(f \mid e_2) > 0$ and $k' < 0$ from m_2 to m_3 and from m_3 to m_1, which again implies $k'(f \mid e_2) < 0$. Thus we must have

$$\int_0^L k'(f \mid e_2)\, dt < 0,$$

contradiction.   □

9.7.9. Corollary. *Let C be a curve in $S^2 \subset \mathbf{R}^3$ that is the image under stereographic projection of a simple closed curve D of class C^3. Then C has at least four points where the torsion is zero.*

Proof. One can show that stereographic projection takes a vertex of D into a point of C with zero torsion (see 9.7.10.1).   □

9.7.10. Remarks

9.7.10.1. An informal argument to prove 9.7.9 is the following: saying that m is a vertex of D is the same as saying that the osculating circle to D at m has a contact of order ≥ 4 with D. By stereographic projection this still gives a circle having contact of order ≥ 4, which implies that the image point m' is a vertex of D and has zero torsion, since the torsion has to be zero at a point where the curve has contact of order four or more with a plane. This reasoning can be made rigorous if a theory of contact is developed; see 8.7.11 or [Lel63].

9.7.10.2. The theorem of the four vertices for a non-convex plane curve is proved exactly by demonstrating that a simple closed curve on the sphere S^2 has at least four points where the torsion vanishes.

9.7.10.3. Example 8.6.11.3 shows that a simple closed curve in $\mathbf{R}^3$ can has constantly non-zero torsion; see also exercise 9.9.5. But it can be shown that simple close curves in $\mathbf{R}^3$ satisfying certain simple conditions have at least two points where the torsion vanishes. See also exercise 9.9.6.

9.7.10.4. The theorem of the four vertices gives a necessary condition on a function $c \in C^0(S^1; \mathbf{R}_+^*)$ for there to exist a simple closed curve C and a parametrization $(S^1, f) \in C$ such that $k_{f(m)}C = c(m)$. This condition is also sufficient, but the converse is more difficult [Glu71]. Notice that the situation here is not the same as in the statement of theorem 8.5.7, where the curvature was a function of the arclength, in which case the curve is predetermined and generally does not close up. In fact, the proof in [Glu71] consists in modifying c by a diffeomorphism of S^1 in such a way that 8.5.7 gives a curve that closes up.

9.7.10.5. An analogous notion to that of a vertex for a plane curve is that of an *umbilic* for a surface in $\mathbf{R}^3$. See 11.7.4.

9.8. The Fabricius–Bjerre–Halpern Formula

9.8.1. Theorem. *Let C be a closed curve in an affine plane E, satisfying certain regularity conditions to be made precise later. Call*

N^+ *the set of double tangents to C at two points, having the same principal normal vector;*

N^- *the set of double tangents to C at two points, having opposite principal normal vectors;*

D *the set of double points of C;*

I *the set of inflection points of C.*

These four sets are finite and their cardinalities satisfy the following relation:

$$\#N^+ = \#N^- + \#D + \tfrac{1}{2}\#I.$$

Here are some examples:

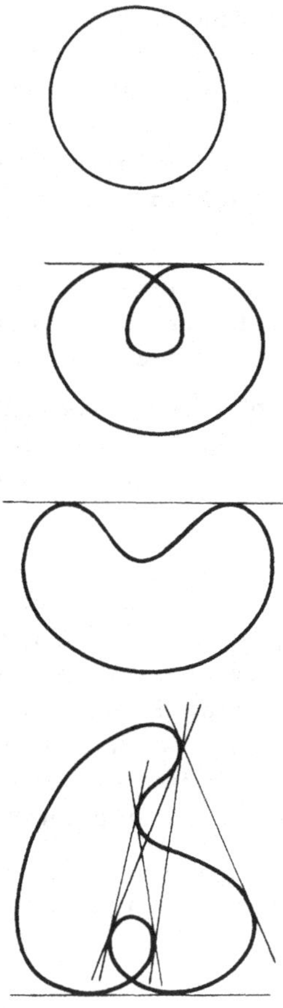

	N^+	N^-	D	I
	0	0	0	0
	1	0	1	0
	1	0	0	2
	4	2	1	2

Figure 9.8.1

9.8.1.1. One can ask whether there are other necessary conditions that $\#N^+$, $\#N^-$, $\#D$ and $\#I$ must satisfy. Such is not the case, as has been recently shown by Ozawa [Oza84]. He also found results on the number of planes thrice tangent to a compact curve in $\mathbf{R}^3$ [Oza85].

9.8.2. Elaboration. For simplicity, assume that C is C^∞ and that it's drawn in $\mathbf{R}^2$. Let L be the length of C and $f \in C^\infty(\mathbf{R}; \mathbf{R}^2)$ a parametrization of C by arclength, periodic of period L. To simplify the notation, let the signed curvature (8.5.2), the affine tangent (8.2.1.1) and the principal normal vector (8.4.11) at $f(s)$ be denoted by $k(s)$, T_s and $n(s)$, respectively.

9.8.3. A pair $\{t, s\}$ is called a *double tangent* if $f(t) \neq f(s)$ and $T_s = T_t$. A double tangent is called *regular* if $k(s)k(t) \neq 0$. We denote by N^+ and N^- the sets of regular double tangents such that $n(s) = n(t)$ and $n(s) = -n(t)$, respectively.

9.8.4. A pair $\{t, s\}$ is called a *double point* if $s \notin t + L\mathbf{Z}$ and $f(t) = f(s)$. A double point is called *regular* if $T_s \neq T_t$. The set of regular double points is denoted by D.

9.8.5. A point $\{s\}$ is called an *inflection point* if $k(s) = 0$. An inflection point is called *regular* if $k'(s) \neq 0$. The set of regular inflection points is denoted by I.

9.8.6. The regularity conditions missing from the statement of theorem 9.8.1 are the following: all double tangents, all double points and all inflection points are to be regular. From now on we assume these conditions to be fulfilled.

Proof. The key to the proof is the introduction of the vector field W : $\mathbf{R} \times \mathbf{R} \to \mathbf{R}^2$ defined by

9.8.7
$$W(t, s) = \big(u(t, s), v(t, s)\big) = \big(\sigma(f'(t), f(t) - f(s)), \sigma(f'(s), f(s) - f(t))\big),$$

where σ denotes the canonical volume form (given in $\mathbf{R}^2$ by the determinant of two vectors).

9.8.8. Study of the zeros of W. If $f(s) = f(t)$, we certainly have $W(t, s) = 0$. Otherwise $W(t, s) = 0$ implies that $f'(t)$ and $f'(s)$ are both proportional to $f(t) - f(s)$, that is, $\{t, s\}$ is a double tangent. The regularity conditions imply that $W'(t, s) \neq 0$ at any zero, so all the zeros are isolated and we can apply 7.7.8 to compute their indices.

At a double point, the jacobian of W is

$$J_{t,s}W = \begin{vmatrix} \sigma\big(f''(t), f(t) - f(s)\big) & \sigma\big(f'(t), -f'(s)\big) \\ \sigma\big(f'(s), -f'(t)\big) & \sigma\big(f''(s), f(s) - f(t)\big) \end{vmatrix} = \big(\sigma(f'(s), f'(t))\big)^2,$$

which is positive. Thus double points have index one.

At a double tangent, the jacobian is

$$J_{t,s}W = |k(t)||k(s)|\sigma\big(n(t), f(t) - f(s)\big)\sigma\big(n(s), f(s) - f(t)\big).$$

It is positive if $n(t) = -n(s)$, that is, if $\{t, s\} \in N^-$, and negative if $\{t, s\} \in N^+$.

In short, all the zeros of W are isolated, their set is the union $N^+ \cup N^- \cup D$ and their indices are

$$+1 \text{ for points in } N^- \text{ or } D,$$

$$-1 \text{ for points in } N^+.$$

Since W is doubly periodic and $[0, L] \times [0, L]$ is compact, there are only finitely many zeros, and we have shown that N^+, N^- and D are finite sets.

9.8.9. The idea now is, roughly speaking, to apply to W a theorem similar to 7.4.18, with the ball replaced by a parallelogram. One difficulty is that a parallelogram has corners, which we must smooth out in order to apply Stokes' theorem (cf. 6.2.2). Also, W will no longer be pointing inward at the boundary; in fact, the integral over the boundary is related to the number $\#I$ of inflection points.

9.8.10. From now on we fix $\varepsilon > 0$ so small that, on every interval $[t, t + \varepsilon]$, the map f is injective, has at most one inflection point, and the angle of f' changes by less than $\pi/2$. Such a number exists by the compactness of $[0, L]$ and our assumptions: f is an immersion, $\|f'\| = 1$ and the inflection points are isolated (9.8.5).

Under these conditions, $W(t, s) \neq 0$ if $t - \varepsilon < s < t + \varepsilon$. This is obvious for points in D; for those in N^+ and N^- it follows from an argument similar to the proof of 9.8.13.1. We then choose $t_0 \in [0, L]$ such that the boundary of the parallelogram

$$B = \big\{(t, s) : t_0 < t < t_0 + L, t + \varepsilon \leq s \leq t + L - \varepsilon\big\}$$

does not intersect $W^{-1}(0)$; this is possible because there are only finitely many values of t_0 to avoid (figure 9.8.10).

Since every double point or double tangent has exactly two representatives in B (namely, $\{t, s\}$ and $\{s, t + L\}$ if $t < s$), we can write

$$\sum_{x \in W^{-1}(0) \cap B} \operatorname{ind}_x W = 2(\#N^- + \#D - \#N^+).$$

9.8.11. We now approximate B by sets B_η ($\eta > 0$) with differentiable boundary. For instance, we can replace the corners of B by arcs of circle of radius η; the resulting sets are submanifolds-with-boundary of $\mathbf{R}^2$, with C^1 boundary (figure 9.8.11).

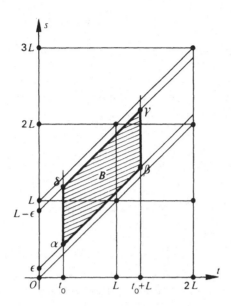

Figure 9.8.10

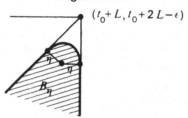

Figure 9.8.11

We now mimic the proof of 7.4.18. First we normalize W into a vector field $\dfrac{W}{\|W\|}$, which exists away from $W^{-1}(0)$, and take η small enough that $B_\eta \cap W^{-1}(0) = B \cap W^{-1}(0)$. Then we draw little disjoint circles B_i around the points of $W^{-1}(0)$, and set $B'_\eta = B_\eta \setminus \bigcup_i B_i$. As in 7.4.18, we get

$$0 = \int_{B'_\eta} d(Z^*\sigma) = \int_{\partial B'_\eta} Z^*\sigma = \int_{\partial B_\eta} Z^*\sigma - \sum_i \int_{\partial B_i} Z^*\sigma.$$

The last term equals 2π times the sum of the indices of points in $B_\eta \cap W^{-1}(0) = B \cap W^{-1}(0)$, which, according to 9.8.10, is equal to

$$2(\#N^- + \#D - \#N^+).$$

The term $\int_{\partial B_\eta} Z^*\sigma$ is continuous in η, so we can take the limit to get

$$4\pi(\#N^- + \#D - \#N^+) = \int_{\partial B} Z^*\sigma.$$

9.8.12. With $\alpha, \beta, \gamma, \delta$ as in figure 9.8.10, we have

$$\int_{\partial B} Z^*\sigma = \int_\alpha^\beta Z^*\sigma + \int_\gamma^\delta Z^*\sigma + \int_\beta^\gamma Z^*\sigma - \int_\alpha^\delta Z^*\sigma.$$

By periodicity, the last two terms cancel out. The first can be written in terms of $g : S^1 \to S^1$, the quotient modulo L of the map

$$t \mapsto \frac{W(t, t+\varepsilon)}{\|W(t, t+\varepsilon)\|};$$

we get $\int_\alpha^\beta Z^*\sigma = 2\pi \deg g$. Similarly, $\int_\gamma^\delta Z^*\sigma = -2\pi \deg h$ where h comes from the map

$$t \mapsto \frac{W(t, t+L-\varepsilon)}{\|W(t, t+L-\varepsilon)\|}.$$

But it's easy to see that

$$W(t, t+L-\varepsilon) = W(t, t-\varepsilon) = \tau W(t-\varepsilon, t),$$

where $\tau : \mathbf{R}^2 \to \mathbf{R}^2$ switches the coordinates; so $h(t) = \tau g(t-\varepsilon)$, and $\deg h = \deg \tau \deg g = -\deg g$. Putting it all together we get

$$\int_{\partial B} Z^*\sigma = 4\pi \deg g.$$

9.8.13. To find the degree of g, we will look at the quadrant of $\mathbf{R}^2$ that contains $g(t)$, and consequently $W(t, t+\varepsilon)$. Denote the four quadrants by $\Delta_1, \Delta_2, \Delta_3$ and Δ_4, in their conventional order.

9.8.13.1. Lemma. *We have:*
(i) $g(t) \in \Delta_3$ *if* $k(u) > 0$ *for all* $u \in \,]t, t+\varepsilon[$;
(ii) $g(t) \in \Delta_1$ *if* $k(u) < 0$ *for all* $u \in \,]t, t+\varepsilon[$;
(iii) $g(t) \notin \Delta_4$ *if there exists* $t \in \,]t, t+\varepsilon[$ *such that* $k(u) > 0$ *for all* $u \in \,]t, t'[$ *and* $k(u) < 0$ *for all* $u \in \,]t', t+\varepsilon[$;
(iv) $g(t) \notin \Delta_2$ *if there exists* $t \in \,]t, t+\varepsilon[$ *such that* $k(u) < 0$ *for all* $u \in \,]t, t'[$ *and* $k(u) > 0$ *for all* $u \in \,]t', t+\varepsilon[$.

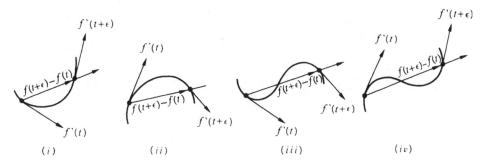

Figure 9.8.13.1

Proof. The proof is analogous in the four cases; we prove the third. By contradiction, assume that $g(t) \in \Delta_4$; this implies $\sigma\big(f'(t), f(t) - f(t+\varepsilon)\big) > 0$. Consider the maximum of the function $z \mapsto (z \mid e)$, where z is a non-zero vector perpendicular to $f(t + \varepsilon) - f(t)$; let s be the first point in $]t, t + \varepsilon[$ where this maximum is reached. Then $\big(f(s) \mid e\big) > 0$ and $f'(s)$ is parallel to $f(t + \varepsilon) - f(t)$. Since the direction of f' does not change by more than $\pi/2$ within $[t, t + \varepsilon]$ (9.8.10), we see that $f'(s)$ is a positive multiple of $f(t + \varepsilon) - f(t)$; this implies $s > t'$ by the definition of t' in condition (iii). A similar argument involving $\sigma\big(f'(t + \varepsilon), f(t + \varepsilon) - f(t)\big) < 0$ gives an s' such that $\big(f(s') \mid e\big) < 0$ and $s' < t'$; but $s < s'$ by construction, contradiction. $\qquad\square$

9.8.14. We have already seen (cf. 9.8.5) that inflection points are finite in number, and by construction (9.8.10) no two of them are less than ε apart. We now cut up the interval $[t_0, t_0 + L]$ into subintervals $[t_i, t_{i+1}]$ whose endpoints are inflection points. By the regularity assumption, if $k > 0$ on $[t_i, t_{i+1}]$ we have $k < 0$ on $[t_{i+1}, t_{i+2}]$. By choosing a different t_0, if necessary, we can assume that $k < 0$ on $[t_0, t_1]$.

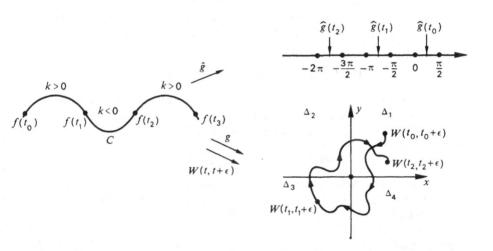

Figure 9.8.13.2

According to the lemma, $W(t, t + \varepsilon)|_{[t_0, t_1]}$ is a path that starts in Δ_1 and ends in Δ_3 without ever getting into Δ_2. This means that, if $\hat{g} : [t_0, t_0 + L] \to \mathbf{R}$ is a lifting of $g : [t_0, t_0 + L] \to S^1$ with $\hat{g}(t_0) \in [0, \pi/2]$, we will have $\hat{g}(t_1) \in [-\pi, -\pi/2]$. By the same token, $\hat{g}(t_2) \in [-2\pi, -3\pi/2]$, and so forth. More exactly, if $t_0 + L = t_i$, that is, if there are i inflection points, we will have

$$\hat{g}(t_i) = \left[-i\pi, -i\pi + \frac{\pi}{2}\right].$$

But $\hat{g}(t_i) = \hat{g}(t_0 + L) = \hat{g}(t_1) + 2\pi \deg g$; thus we conclude that $\deg g = -i/2 = -\#I/2$, which completes the proof of theorem 9.8.1. $\qquad\square$

9.9. Exercises

9.9.1. Prove that definition 9.1.11 does not depend on the choice of a Euclidean structure.

9.9.2. Compute the turning number of the four curves in figure 8.1.5.

9.9.3. Let C be an oriented, convex, simple closed curve of class C^2 in $\mathbf{R}^2 = \mathbf{C}$, and assume the curvature of C never vanishes.

(a) Show that C can be written in Euler form (see 8.7.21).

(b) Compute the curvature (8.7.21), the length and the area of C as a function of p and its derivatives.

(c) Let C' be another curve satisfying the same properties, and assume that, for any pair of points $m \in C$, $m' \in C'$, there exists a rigid motion f such that $f(m') = m$, $f(\tan g_{m'} C') = \tan g_m C$ and $f(C') \subset \bar{C}$. Prove that

$$\mathrm{leng}(C)\,\mathrm{leng}(C') \le 2\pi\big(\mathrm{area}(C_{\mathrm{int}})\,\mathrm{area}(C'_{\mathrm{int}})\big).$$

(d) Deduce from (c) that the radius R of the smallest circle Γ containing C in its interior and the radius r of the largest circle γ contained in the interior of C satisfy *Bonnesen's* inequality

$$\big(\mathrm{leng}(C)\big)^2 - 4\pi\,\mathrm{area}(C) \ge \pi^2(R^2 - r^2).$$

(The enterprising reader can show that Γ is unique, but γ is not; cf. [Ber87, 11.5.8].) Notice that the left-hand side is the expression occurring in the isoperimetric inequality.

9.9.4. Determine the Frenet frame, the curvature and the torsion of the curve defined by

$$t \mapsto \left(\frac{\cos t}{\sin t/2},\, 2\cos\frac{t}{2},\, \frac{\sqrt{3}}{\sin t/2}\right).$$

9.9.5. Consider the torus of revolution

$$T = \big\{(x, y, z) : (x^2 + y^2 + z^2 + a^2 - r^2)^2 - 4a^2(x^2 + y^2) = 0\big\},$$

obtained by rotating the circle of equations $(x - a)^2 + z^2 - r^2 = 0 = y$ around the z-axis. Set $r = a \sin \theta$.

(a) Find the curves on T that make a constant angle V will all *parallels of latitude* of T (that is, the curves having z and $x^2 + y^2$ fixed). (Hint: using polar coordinates, find their projections on the xy-plane.) Prove that for $V = \theta$ the curves thus obtained are circles, whose projection on the xy-plane are ellipses with a focus at the origin. More generally, if $\frac{\tan V}{\tan \theta}$ is an integer, one obtains simple closed curves.

(b) Deduce that there are curves on the torus with nowhere vanishing torsion.

9.9.6.

(a) Let f be a parametrization by arclength of an oriented geometric arc C in $\mathbf{R}^3$. Under what conditions are there real-valued functions θ and a such that the map

$$t \mapsto f(t) + \big(\cos\theta(t)\nu + \sin\theta(t)\beta\big)a(t)$$

defines a geometric arc whose tangent at $g(t)$ is normal to C at $f(t)$?

(b) Assume now that C is a closed spherical curve of length L. Use part (a) to show that

$$\int_0^L T\,dt = 0.$$

9.9.7. More curves in Euler form.

(a) Consider two plane curves in Euler form (8.7.21), defined by functions h and k. Assume that they turn toward the origin (8.2.2.13), and that they are tangent at $t = 0$, for example; thus $h(0) = k(0)$ and $h'(0) = k'(0)$. Show that if, for every t, the curvature of C at the point with parameter t is greater than or equal to the curvature of D at the point with parameter t, then C is contained in the interior of D.

(b) Let C be a compact convex plane curve, biregular and of class C^2. Let A (resp. a) be a point on C where the curvature is maximal (resp. minimal). Show that the osculating circle γ at a can roll all around C, always staying inside C_{int}, and the C can roll all around the osculating circle Γ at A.

(c) Osserman [Oss85] gives a very simple and natural proof for the theorem of the four vertices. Let C be a simple closed curve, assumed convex for simplicity. Let R be the radius of the smallest circle Γ containing C in its interior, and m, n consecutive contact points of C with Γ. Show that there exists a point p on C, between m and n, where the curvature is greater than $1/R$, as long as C and Γ do not coincide. Deduce that if Γ is tangent to C at n points, C has at least $2n$ vertices. Notice that generically $n \geq 3$; thus the generic simple closed curve has six vertices at least.

CHAPTER 10

A Brief Guide to the Local Theory of Surfaces in $\mathbf{R}^3$

The local study of curves in $\mathbf{R}^2$, in sections 4 and 5 of chapter 9, was simple: theorem 8.5.7 shows that a single invariant, the curvature (expressed as a function of the arclength), is enough to characterize such a curve. The fundamental reason for this simplicity, and one that remains true no matter what the dimension of the ambient space, is that the intrinsic geometry of curves is trivial; the metric given by the length of paths on a curve is always the same as the metric on some interval of $\mathbf{R}$. And the "shape", or "position," of a curve in $\mathbf{R}^2$ is specified by a mere scalar function, the curvature.

When it comes to surfaces in $\mathbf{R}^3$, however, things are quite different. For one thing, the intrinsic geometry of surfaces is already a subject in itself, since, in general, the Euclidean geometry of subsets of $\mathbf{R}^2$ can no longer describe the surface. For example, a sphere can never be isometric, even locally, to a plane open set. The study of the intrinsic geometry of surfaces constitutes riemannian geometry in two dimensions.

Another complication is that a scalar function is not sufficient to characterize the way a surface sits in space; we need a quadratic form, called the second fundamental form. (The first fundamental form is the riemannian structure, or intrinsic geometry.)

Our brief guide will consist of four parts. In the first we define the two fundamental forms and give numerous examples, all of them important from one or several points of view. Next we study all the phenomena associated exclusively with the first fundamental form, and then with the second. Finally the relationship between the two forms, which are not independent, is discussed.

Our subject is vast: already in 1896 Darboux published a two-thousand-page treatise, which was not exhaustive even then. So we had to make choices, according to personal taste. We hope the result will please the average reader, without apalling the specialists.

Bibliography: After [Dar72], which contains the old stuff— not all of which has been rephrased or generalized in contemporary language—the only work that contains most of the results mentioned here is [Spi79]. Do Carmo [Car76] covers the foundations thoroughly and well. Klingenberg [Kli78] packs a lot in a remarkably thin volume. The reader may find [Str61] interesting in that its scope is complementary to ours. Most of the global results we discuss can be found in [Hop83], which often covers them even more quickly. We mention [Car76], [Ste64], [ST67], [KN69], [Lel63], [Hic65], [Wal78], [Kli78], [Tho79] and [LS82] as general references on differential geometry that have good chapters on surfaces; [Val84] is old-fashioned but relevant.

In order not to encumber the reading, we omit references to classical definitions and results which are found in all textbooks and can be looked up in their indexes.

10.1. Definitions

10.1.1. In this chapter we study surfaces in $\mathbf{R}^3$, that is, two-dimensional submanifolds of $\mathbf{R}^3$ (considered as a Euclidean space). By the fundamental theorem 2.1.2, we can restrict ourselves to the following definitions:

10.1.2. Definition. *A (global) surface in $\mathbf{R}^3$ is a two-dimensional submanifold of $\mathbf{R}^3$. A (local) surface in $\mathbf{R}^3$ is a surface V that can be defined by a single chart.*

We recall from 2.2.3 what this condition means: there exists an open set $U \subset \mathbf{R}^2$ and an immersion $g \in C^\infty(U; \mathbf{R}^3)$ such that g is a homeomorphism between U and its image $V = g(U)$.

In the sequel we will not distinguish between local and global surfaces unless it is essential to do so. Some results won't hold unless we restrict the surface to an appropriately chosen open set; we leave this task to the reader. (Even in the presence of a guide one is not prevented from looking around.)

10.1.3. Corollary (cf. 2.2). *If $\phi, \psi : V \rightarrow \mathbf{R}^2$ are complete charts on a local surface, the images $\phi(V)$ and $\psi(V)$ are open subsets of $\mathbf{R}^2$ diffeomorphic under $\phi \circ \psi^{-1}$.* □

For pedagogical and even conceptual reasons (see examples 10.2.1.4, 10.2.2.5, 10.2.3.8 and 10.2.4, among others), it is well to introduce here the global objects that will occupy us in chapter 11:

10.1.4. Definition. *An immersed (global) surface in $\mathbf{R}^3$ is a two-dimensional abstract manifold X, together with an immersion $f \in C^\infty(X; \mathbf{R}^3)$.*

Contrary to the case of curves in chapter 9, which was simple because there exist only two one-dimensional abstract manifolds ($\mathbf{R}$ and the circle, but even the latter can be camouflaged by the use of periodic functions on $\mathbf{R}$), the situation here is more complex, because there exist many abstract surfaces (4.2.25).

10.2. Examples

Theorem 2.1.2 yields a local equivalence which shows that three natural classes of examples of local surfaces can be formulated: graphs, implicit equations and parametric representations. We follow this outline in this section.

10.2.1. Graphs. Every open set $U \subset \mathbf{R}^2$ and every $f \in C^\infty(U; \mathbf{R})$ gives a surface in $\mathbf{R}^3$ via its graph

$$\{(x, y, f(x,y)) : (x,y) \in U\}.$$

We will simply write $z = f$ or $z = f(x, y)$.

10.2.1.1. The sphere (cf. 2.1.6.2). We can take U to be the open disc $x^2 + y^2 < 1$ and $f = \sqrt{1 - x^2 - y^2}$. But we need six such patches to cover the whole of $S^2 \subset \mathbf{R}^3$.

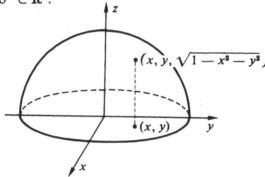

Figure 10.2.1.1

10.2.1.2. Quadrics. The two quadrics that can be written as graphs are the elliptic paraboloid $z = \dfrac{x^2}{a^2} + \dfrac{y^2}{b^2}$ and the hyperbolic paraboloid $z = \dfrac{x^2}{a^2} - \dfrac{y^2}{b^2}$. We could also consider the improper quadric $z = ax^2$. In each case U is the whole of $\mathbf{R}^2$.

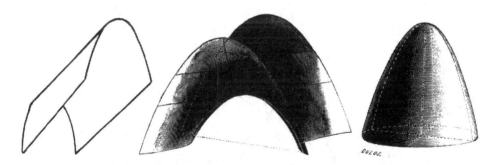

Figure 10.2.1.2

From Rouché and Comberousse, *Traité de Géométrie*, Paris, Gauthier-Villars, p. 486 and 492

10.2.1.3. Surfaces of translation (particular case; see also 10.2.3.1). Consider the graph of $z = f(x, y) = A(x) + B(y)$. This is a surface of translation

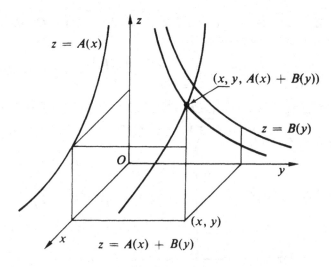

Figure 10.2.1.3.1

in two ways: with respect to x and with respect to y (figure 10.2.1.3.1).
See also example 10.2.3.1 below.

Such a surface can also be written in the form

$$z = \frac{A(x) + B(y)}{2},$$

that is, as the locus of the midpoint of a segment whose endpoints describe
two fixed curves:

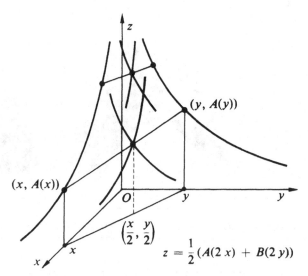

Figure 10.2.1.3.2

If A and B are defined on the whole of $\mathbf{R}$ the surface is defined on $\mathbf{R}^2$. On the other hand, *Scherk's surface* $z = \log \cos x - \log \cos y$ is only defined on the shaded region in figure 10.2.1.3.3(b). To see this just write z as the logarithm of a quotient.

Any surface of the form

$$z = b \log(\sqrt{\rho^2 + a^2} + \sqrt{\rho^2 - b^2}) + a \arctan\left(\frac{b\sqrt{\rho^2 + a^2}}{a\sqrt{\rho^2 - b^2}}\right) + a\theta + c,$$

where (ρ, θ) are polar coordinates on $\mathbf{R}^2$ and $a, b, c \in \mathbf{R}$, is also called a Scherk surface. These are the only helicoids that are minimal surfaces (10.6.9.2); see [Dar72, vol. I, p. 328], for a proof. For a study of minimal surfaces of translation, see [Dar72, vol. I, p. 406].

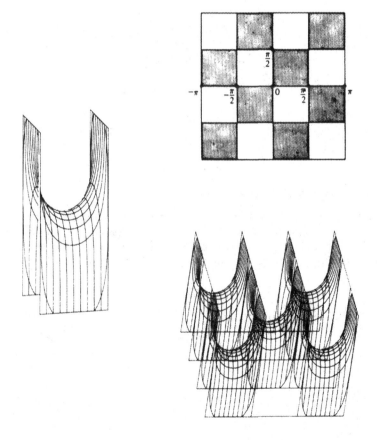

Figure 10.2.1.3.3

From M. P. do Carmo, *Differential Geometry of Curves and Surfaces*, Prentice-Hall, p. 209

10.2.1.4. Plücker's conoid. This surface is defined by $\dfrac{xy}{x^2 + y^2}$; here U is $\mathbf{R}^2$ minus the origin. See also 10.2.2.4. Plücker's conoid is a ruled surface. It has the property that the projections of any point $p \in \mathbf{R}^3$ onto the rules form a plane curve; this property characterizes Plücker's conoid among all ruled surfaces. In addition, this plane curve is a conic, and its projection onto the xy-plane is a circle. An alternative formulation of this property involves one-parameter families of rigid motions under which all points on a fixed rigid curve describe plane curves. For more information, see [Dar72, vol. I, p. 99].

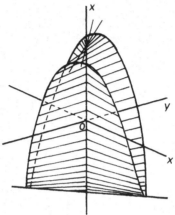

Figure 10.2.1.4

10.2.2. Implicitly defined surfaces

10.2.2.1. Cylinders. If a surface $V = F^{-1}(0)$ satisfies an equation involving only two coordinates we call it a *cylinder*. If $F(x, y) = 0$, for instance, we say that the cylinder is parallel to the z-axis. Examples are the improper quadrics (elliptic and hyperbolic cylinders) defined by

$$\frac{x^2}{a^2} \pm \frac{y^2}{x^2} - 1 = 0$$

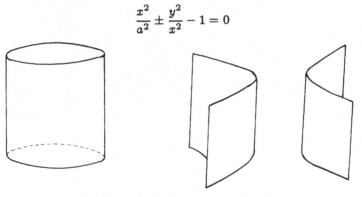

Figure 10.2.2.1

10.2.2.2. Planes. Their general equation is $ax + by + cz = d$.

10.2.2.3. Proper quadrics. Apart from the paraboloids in 10.2.1.2, these are as follows:

$$\frac{x^2}{a^2} + \frac{y^2}{b^2} + \frac{z^2}{c^2} - 1 = 0 \quad \text{(ellipsoid)},$$

$$\frac{x^2}{a^2} + \frac{y^2}{b^2} - \frac{z^2}{c^2} - 1 = 0 \quad \text{(one-sheet hyperboloid)},$$

$$\frac{x^2}{a^2} - \frac{y^2}{b^2} - \frac{z^2}{c^2} - 1 = 0 \quad \text{(two-sheet hyperboloid)}.$$

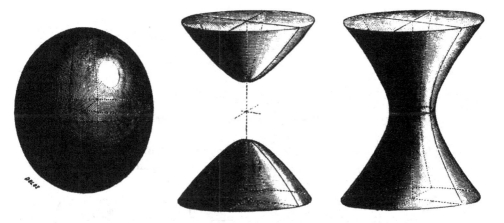

Figure 10.2.2.3.1

From Rouché and Comberousse, *ibid.*, p. 483, 490 and 493.

It so happens that, even if we're interested in only one such quadric, it is important to consider the one-parameter family of *confocal* quadrics

$$\frac{x^2}{a - \lambda} + \frac{y^2}{b - \lambda} + \frac{z^2}{c - \lambda} - 1 = 0,$$

where $a > b > c$ are fixed and λ can be in any of the intervals $]-\infty, c[$, $]c, b[$ and $]b, a[$. Within each interval the type of quadric remains the same. Generically, a point in $\mathbf{R}^3$ belongs to exactly three quadrics in this family, one of each type; the corresponding parameter values $\rho_0 \in]b, a[$, $\rho_1 \in]c, b[$ and $\rho_2 \in]-\infty, c[$ can be found by clearing denominators in the equation above and observing that the resulting numerator is a degree-three polynomial in λ. In other words, if $f(\lambda) = (a - \lambda)(b - \lambda)(c - \lambda)$ and $\phi(\lambda) = (\lambda - \rho_0)(\lambda - \rho_1)(\lambda - \rho_2)$, we have the identity

$$\frac{x^2}{a - \lambda} + \frac{y^2}{b - \lambda} + \frac{z^2}{c - \lambda} - 1 = \frac{\phi(\lambda)}{f(\lambda)}.$$

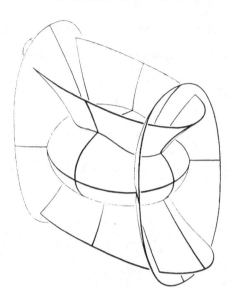

Figure 10.2.2.3.2

From Hilbert and Cohn-Vossen, *Geometry and the Imagination*, Chelsea, p. 23

This immediately leads to the important parametrization of each octant of $\mathbf{R}^3$ by (ρ_0, ρ_1, ρ_2); to find x, say, we multiply the equation above by $a - \lambda$ and take the limit as λ approaches a. The result is

$$(\rho_0, \rho_1, \rho_2) \mapsto \left(\sqrt{\frac{\phi(a)}{(a-b)(a-c)}}, \sqrt{\frac{\phi(b)}{(b-a)(b-c)}}, \sqrt{\frac{\phi(c)}{(c-a)(c-b)}} \right).$$

In particular, if $\rho_2 = 0$ we get the parametrization of an ellipsoid by *elliptic coordinates*:

$$(u, v) \mapsto \left(\sqrt{\frac{a(a-u)(a-v)}{(a-b)(a-c)}}, \sqrt{\frac{b(b-u)(b-v)}{(b-a)(b-c)}}, \sqrt{\frac{c(c-u)(c-v)}{(c-a)(c-b)}} \right).$$

An important consequence of these formulas is that this family of quadrics is *triply orthogonal*, that is, the tangent planes (section 2.5) to the three quadrics that meet at a point are pairwise orthogonal. This fact is most obviously stated by writing the element of length

$$dx^2 + dy^2 + dz^2 = 4 \sum_i \frac{\phi'(\rho_i)}{f(\rho_i)} d\rho_i^2.$$

10.2.2.4. Algebraic surfaces. If F is a polynomial, the surface defined by $F(x, y, z) = 0$ is said to be *algebraic*. If F is homogeneous, the surface is a *cone* with vertex at the origin. To be exact, we may have to remove the points where the derivative dF vanishes, since we're only guaranteed to get a surface where F is a submersion (theorem 2.1.2). But dF may be zero for an immersed surface: think of $x^2 - y^2 = 0$, for example.

The Plücker conoid is an algebraic surface, since it can be written in the form $z(x^2 + y^2) - xy = 0$; but then we have to add the z-axis to the previously described graph. We get an immersed surface, except at the points $(0, 0, \pm 1)$.

10.2.2.5. Enneper's surface. This surface, defined by

$$\left(\frac{y^2 - x^2}{2z} + \frac{2z^2}{9} + \frac{2}{3}\right)^3 = 6\left(\frac{y^2 - x^2}{4z} - \frac{1}{4}\left(x^2 + y^2 + \frac{8}{9}z^2\right) + \frac{2}{9}\right)^2,$$

has a way of cropping up unexpectedly, in seemingly unrelated contexts: see 10.2.3.6, 10.2.3.13 and 11.16.7. By direct calculation or by looking at the parametric representation in 10.2.3.6 one can easily check that Enneper's surface is immersed everywhere.

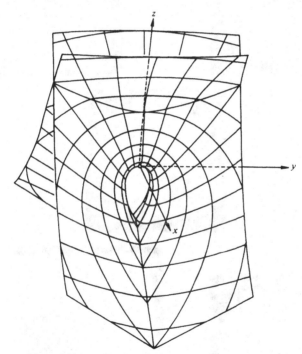

Figure 10.2.2.5

From do Carmo, *ibid.*, p. 205

10.2.2.6. Cyclids and Dupin cyclids. *Cyclids* are things of the form

$$\frac{x^2}{a_1 - \lambda} + \frac{y^2}{a_2 - \lambda} + \frac{z^2}{a_3 - \lambda} + \frac{(r^2 - R^2)^2}{4R^2(a_4 - \lambda)} - \frac{(r^2 - R^2)^2}{4R^2(a_5 - \lambda)} = 0,$$

where $a_1, \ldots, a_5, R$ are arbitrary, λ is a parameter, and $r^2 = x^2 + y^2 + z^2$. As λ changes, these surfaces form a triply orthogonal system (cf. 10.2.2.3). When at least two of a_1, a_2, a_3 are equal, we say that the surface is a *Dupin cyclid*. These surfaces too are loci of apparently unrelated properties: see 10.6.8.2(4), 11.21, [Ber87, 20.7], and [Dar17, chapter 6]. Figure 10.2.2.6 shows the appearance of some Dupin cyclids.

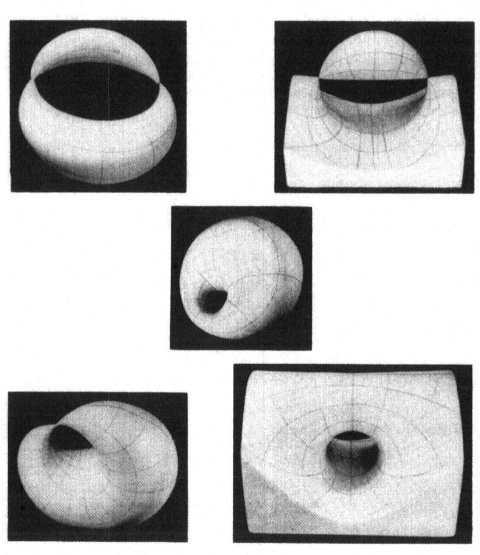

Figure 10.2.2.6
From Hilbert and Cohn-Vossen, *ibid.*, pp. 218–219

10.2.2.7. Wave surfaces. Consider two fixed lines in a Euclidean plane and a moving line, say a ruler, with two points marked on it. As the ruler moves subject to the condition that each of the marked points remains on one of the fixed lines, all the other points on the ruler describe ellipses.

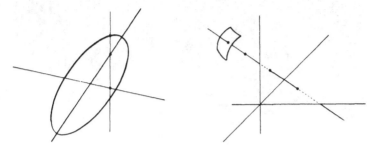

Figure 10.2.2.7.1

Now consider the analogous problem in space: given three fixed planes, pairwise orthogonal for simplicity, and a moving line with three points marked, move the line around so that each of the marked points always lies on the same plane. As the line takes all possible positions, every point in it describes an ellipsoid (or possibly an ellipse); this a mechanical way to generate an ellipsoid.

There exists a one-parameter family of *wave surfaces* having this two-parameter family of lines as its set of normals (3.5.15.5). Wave surfaces were first studied by Fresnel, who used them to describe the wavefronts of a punctual light source in a birefringent medium. They can be obtained from the corresponding ellipsoid by means of a geometric construction called "apsidal;" see [Sal74, p. 424], [Dar72, vol. IV, notes VII and VIII], and [RdC22, p. 496].

One particular wave surface has equation

$$\frac{x^2}{r^2 - a^2} + \frac{y^2}{r^2 - b^2} + \frac{z^2}{r^2 - c^2} - 1 = 0,$$

where $r^2 = x^2 + y^2 + z^2$. It has four cone points (singularities) and four tangent planes which touch it along whole circles. For details, see [Dar72, vol. IV, note VIII]. See figure 10.2.2.7.2.

10.2.2.8. Surfaces defined by $(x/a)^m + (y/b)^n + (z/c)^p = 1$ are called *tetrahedral.* One can also consider surfaces of the form $x^m y^n z^p = 1$. All of these are algebraic if the exponents are commensurable. Notice that, depending on whether or not m, n and p are positive, integers, or rationals, some parts of $F^{-1}(0)$ must be eliminated.

10.2.2.9. Surfaces of revolution. If $f(x, z)$ denotes a curve in the xz-plane, the surface of revolution obtained by rotating this curve around the z-axis will have equation $F(x, y, z) = f(\sqrt{x^2 + y^2}, z) = 0$. If $f^{-1}(0)$ doesn't have

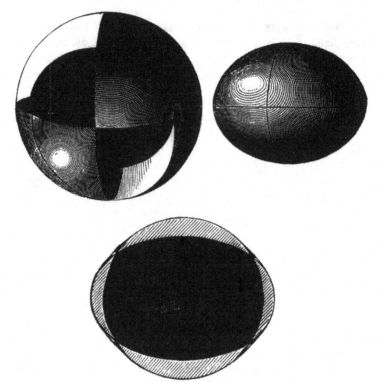

Figure 10.2.2.7.2

From Rouché and Comberousse, *ibid.*, p. 501 and 503

singularities, $F^{-1}(0)$ will be non-singular except possibly on the z-axis. Quadrics (10.2.2.3) for which at least two of the numbers a, b, c are equal are surfaces of revolution.

The *catenoid* is the surface of revolution defined by $\cosh z = \sqrt{x^2 + y^2}$; its name derives from the catenary (8.7.20). In example (3) of 10.6.6.6 we show that the catenoid is the only minimal surface of revolution.

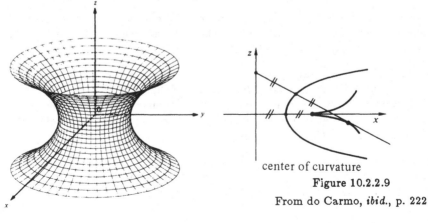

center of curvature

Figure 10.2.2.9

From do Carmo, *ibid.*, p. 222

10.2.2.10. The helicoid. This surface, defined by the equation $\tan\dfrac{z}{a}=\dfrac{y}{x}$, is obtained by moving a line perpendicular to the z-axis along the axis, rotating it as it moves (*screw motion*). The amount by which the line advances as it goes around the axis once is called the *pitch* of the helicoid; it equals $2\pi a$.

Figure 10.2.2.10

From Spivak, *A Comprehensive Introduction to Differential Geometry*, Publish or Perish, vol. 3, p. 248

10.2.2.11. The surface defined by $e^x+e^y+e^z=1$ is a surface of translation in two ways; see 10.2.3.1.

10.2.2.12. Parallel surfaces. If the gradient of F has constant norm, say

$$\left(\frac{\partial F}{\partial x}\right)^2+\left(\frac{\partial F}{\partial y}\right)^2+\left(\frac{\partial F}{\partial z}\right)^2=1,$$

any two surfaces of the form $F(x,y,z)=k$, for different but nearby values of k, are parallel. This means that each point of one lies at a constant distance form the other. For a converse, see 3.5.15.5.

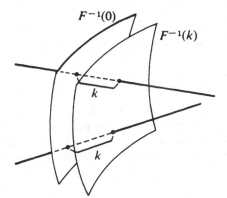

Figure 10.2.2.12

10.2.3. Parametrically defined surfaces

10.2.3.1. Surfaces of translation. We generalize 10.2.1.3 by considering two arbitrary curves F, G in $\mathbf{R}^3$, defined on the same interval I, and forming the map $(u, v) \mapsto F(u) + G(v)$ from $I \times I$ into $\mathbf{R}^3$. Its image is a surface of translation in two ways. Can it be a surface of translation in more than two ways? Sophus Lie gave a complete answer to this question; the general solution starts from a plane algebraic curve of degree four and four arbitrary points on it, and involves the corresponding abelian integrals [Dar72, vol. I, p. 159]. A very particular case is when this curve degenerates into two conics. Then the result always is, up to translation, the surface $e^x + e^y + e^z = 1$ mentioned in 10.2.2.11. Indeed, for arbitrary real numbers a, b, c, we can represent this surface by a very different-looking immersion:

$$(u, v) \mapsto \left(\log \frac{(u - a)(v - a)}{(a - b)(a - c)}, \log \frac{(u - b)(v - b)}{(b - c)(b - a)}, \log \frac{(u - c)(v - c)}{(c - a)(c - b)} \right).$$

10.2.3.2. General note. In each of our examples, we give a formula, but we do not check, at least not systematically, whether the formula really gives an immersion.

10.2.3.3. Spheres. Besides the parametrization $(x, y) \mapsto \pm\sqrt{1 - x^2 - y^2}$ mentioned in 10.2.1.1, we have the latitude-longitude chart (6.1.6):

$$(\phi, \theta) \mapsto (\cos \phi \sin \theta, \sin \phi \sin \theta, \cos \theta);$$

the Mercator projection, obtained by letting $u = \log \tan(\theta/2)$ and $v = \phi$:

$$(u, v) \mapsto (\operatorname{sech} u \cos v, \operatorname{sech} u \sin v, \tanh v);$$

and the stereographic projection from the north pole (see 2.8.7, 5.7 and [Ber87, 18.1.4]):

$$(u, v) \mapsto \left(\frac{4u}{u^2 + v^2 + 4}, \frac{4v}{u^2 + v^2 + 4}, \frac{2(u^2 + v^2)}{u^2 + v^2 + 4} \right).$$

10.2.3.4. Ellipsoids. See 10.2.2.3 for a very useful parametrization of ellipsoids.

10.2.3.5. Surfaces of revolution. If a curve in the xz-plane has parametric equation $v \mapsto (f(v), g(v))$, the surface of revolution obtained by rotating it around the z-axis (cf. 10.2.2.9) is

$$(u, v) \mapsto (\cos u \, f(v), \sin u \, f(v), g(v)).$$

For the catenoid we have $x = \cos u \cosh v$, $y = \sin u \cosh v$, $z = v$.

The *pseudosphere* or *Beltrami's surface*, which was probably first discovered by Minding, is the surface of revolution generated by the *tractrix* $v \mapsto (\operatorname{sech} v, v - \tanh v)$. The tractrix is an envelope of the catenary; it

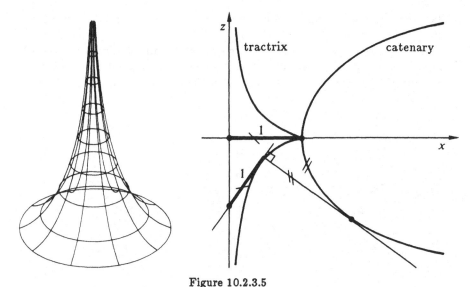

Figure 10.2.3.5

Left diagram taken from do Carmo, *ibid.*, p. 169

can also be characterized by the property of its tangent indicated in figure
10.2.3.5.

For other important surfaces of revolution, see 10.5.3.10, 10.6.9.6 and
11.10.2.

10.2.3.6. Weierstrass's formula for minimal surfaces. Let $U \ni 0$ be an open
set in $\mathbf{R}^2 = \mathbf{C}$ and $f(Z) = f(x + iy)$ an arbitrary holomorphic function on
U. Define an immersion into $\mathbf{R}^3$ by setting

$$x(u, v) = \operatorname{Re} \int_0^w (1 - Z^2) f(Z) \, dZ,$$

$$y(u, v) = \operatorname{Re} \int_0^w i(1 + Z^2) f(Z) \, dZ,$$

$$z(u, v) = \operatorname{Re} \int_0^w 2Z f(Z) \, dZ,$$

where the integral is taken along any path connecting the origin to $w =
u + iv \in \mathbf{C}$: see [Rud74]. (In general, this only makes sense locally; globally
we get a "multi-valued function." For a definition without integrals, see
[Dar72, vol. I, p. 340].) The surface thus obtained is minimal (10.6.9.2)
and, conversely, any minimal surface can be locally written in this form
(away from umbilics): see [Nit75, p. 144] or [Oss69, p. 64].

The simplest case of this construction, $f(Z) = 1$, gives Enneper's surface
(10.2.2.5) in the following representation:

$$(u, v) \mapsto \left(u - \frac{u^3}{3} + uv^2, v - \frac{v^3}{3} + vu^2, u^2 - v^2 \right).$$

Its symmetries are evident in this form: consider the maps $(u, v) \mapsto (-u, v)$, $(u, v) \mapsto (u, -v)$ and $(u, v) \mapsto (v, -u)$.

If we replace f by $e^{i\alpha} f$ we get a one-parameter family of minimal surfaces, all isometric. In the particular case $f = 1/Z^2$, this family includes the catenoid and the helicoid of pitch 2π (see 10.4.1.7 and 11.16.5).

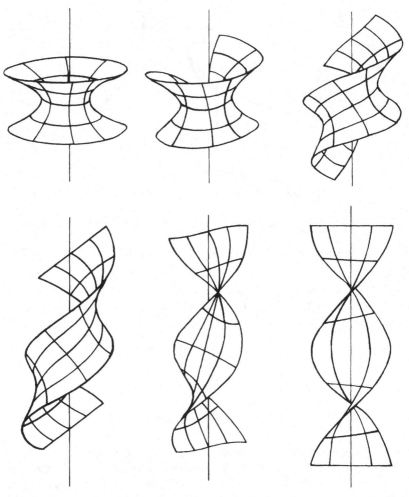

Figure 10.2.3.6
From Spivak, *ibid.*, p. 248 and 249

10.2.3.7. Ruled surfaces. A surface S is *ruled* if it is the union of a one-parameter family of straight lines, called its *rulings*. Roughly, we can write

10.2.3.8 $(u, v) \mapsto f(u, v) = m(u) + v\xi(u),$

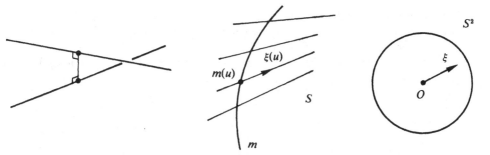

Figure 10.2.3.8.1

where m is a curve in $\mathbf{R}^3$ and the vector ξ describes a curve on the unit sphere S^2.

In addition to the elementary cylinders and cones, we have encountered the following examples of ruled surfaces: the helicoid (10.2.2.10), the hyperbolic paraboloid (10.2.1.2), the one-sheet hyperboloid (10.2.2.3) and Plücker's conoid (10.2.1.4).

Two parametrizations have special interest: the first consists in taking for m a curve orthogonal to the rulings, that is, $(m'\,|\,\xi) = 0$. The second, more subtle and unique, is given by the condition $(m'\,|\,\xi') = 0$. Its geometric interpretation is the following: two non-parallel lines in $\mathbf{R}^3$ determine a unique common perpendicular, a distance and an angle. Taking the limit as two rulings approach one another, we get, for each interval in which $\xi'(u) \neq 0$ (this is a reasonable restriction, since $\xi'(u) = 0$ for every u gives just a cylinder), a unique point on each ruling. These points form a curve, called the *line of striction*. The reader should determine the line of striction for the examples of ruled surfaces above.

Assume from now on that $(m'\,|\,\xi') = 0$ in 10.2.3.8, and also that u is the arclength of ξ, that is, $\|\xi'\| = 1$. A parametrization satisfying these conditions is called *standard*. The limit λ of the ratio between the angle and the distance between rulings that approach one another is the *parameter of distribution* of S; it is given by the mixed product $\lambda = (m', \xi', \xi)$.

The tangent plane to S is found by calculating the normal:

$$\frac{\partial f}{\partial u} \times \frac{\partial f}{\partial v} = m' \times \xi + v\xi' \times \xi = \lambda \xi' + v\xi' \times \xi.$$

This formula shows, among other things, that the cross-ratio (see [Ber87, chapter 6]) of the tangent planes at four points on the same ruling is equal to the cross-ratio of the four points.

A last example of a ruled surface: the *Möbius strip* (cf. 5.9.11):

$$(u, v) \mapsto \left(\left(2 - v\sin\frac{u}{2}\right)\sin u, \left(2 - v\sin\frac{u}{2}\right)\cos u, \cos\frac{u}{2}\right).$$

We remark that this equation cannot describe the surface that one obtains by carefully gluing the ends of a strip of paper after a half-twist: the paper construction yields, by rigidity, a surface locally isometric to the Euclidean plane, and the surface above does not have this property. See [Wun62] and [Hal77] for a discussion of the physical feasibility of a Möbius strip from a paper strip (the ratio between length and width must be large enough), and equations describing the result. See also 7.8.11 and [Hil], p. 294–297 for applications to magic and degree theory.

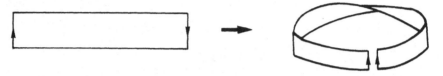

Figure 10.2.3.8.2

10.2.3.9. Developable surfaces. According to the formula above, the tangent plane is fixed along a ruling if and only if $\lambda = 0$, that is, ξ' is parallel to m'. A surface satisfying this condition is called *developable*; see 10.4.1.8 for an explanation of the name. By construction, a developable surface can be obtained as the union of the tangents to a given curve; also as the envelope of the set of osculating planes to a curve (10.2.3.12).

This curve is a singular locus for the developable surface, and coincides with its line of striction; we also call it its *cuspidal edge*. Where its torsion is not zero, one can describe things by saying that two sheets of the surface meet tangentially along the cuspidal edge.

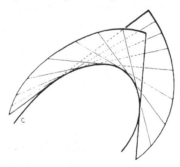

Figure 10.2.3.9

From W. Klingenberg, *A Course in Differential Geometry*, Springer, p. 57

10.2.3.10. Moulding surfaces. A natural generalization of ruled surfaces consists in taking a curve C in space and applying to it a one-parameter family of rigid motions. The simplest case of this procedure, leading to the so-called *molding surfaces*, is when C is a plane curve and the tangents to the trajectories of fixed points on C are everywhere perpendicular to the plane P containing C; kinematically, this means that P must roll on its envelope without sliding. Applying Frenet's formulas (section 8.6) we see that this condition is equivalent to saying that, in the plane P where C sits, C remains fixed with respect to the curve in P having the same curvature (as a function of arclength, cf. 8.5.7) as the curve in $\mathbf{R}^3$ which is the line of striction of the envelope of P.

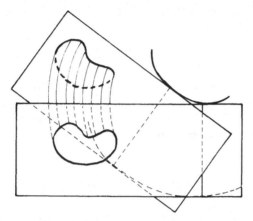

Figure 10.2.3.10

An easily visualizable particular case is when C reduces to a point, that is, the envelope of P is a cone (or a cylinder, if the point is at infinity). If P rotates around a fixed line, we obtain a surface of revolution; thus surfaces of revolution are also particular cases of molding surfaces. So are boundaries of tubes (10.2.3.12). For more information, see [Dar72, vol. I, p. 143 ff.] and [Val84, p. 244].

10.2.3.11. Parallel surfaces. The parametric representation of a surface parallel to $(u, v) \mapsto f(u, v)$ (10.2.2.12) is given by

$$(u, v) \mapsto f(u, v) + k \frac{\dfrac{\partial f}{\partial u} \times \dfrac{\partial f}{\partial v}}{\left\| \dfrac{\partial f}{\partial u} \times \dfrac{\partial f}{\partial v} \right\|}.$$

10.2.3.12. Envelopes of one-parameter families of surfaces. The practical procedure for finding the envelope of a one-parameter families of surfaces, one which gives a regular result in the generic case, consists in writing the family in the form $F(x, y, z, \lambda) = 0$ and in eliminating the parameter λ between the two equations $F(x, y, z, \lambda) = 0$ and $\frac{\partial F}{\partial \lambda}(x, y, z, \lambda) = 0$. If we manage to carry out the calculations explicitly, we get something of the form $G(x, y, z) = 0$. For a family of planes we get a developable surface (10.2.3.9).

The case of a family of spheres is particularly interesting. While each plane was tangent to its envelope along a line (intersection of infinitely close planes), here each sphere is tangent to the envelope along a circle.

Such surfaces will be characterized in 10.6.8.2.3 by their curvature lines. Notice that the envelope of spheres going through a common point is the image under inversion (0.5.3.1) of a developable surface.

Two cases are worth mention. First, if the radius of the sphere is constant (and small enough), the envelope is the boundary of a tube, as defined in section 2.7 (see figure 2.7.6.2).

Second, when is a surface the envelope of two distinct families of spheres? Notice that the spheres in one family must be tangent to those in the other, and since three spheres determine a one-parameter family of spheres tangent to the three, we see that each family consists of spheres tangent to the spheres in the other. It immediately follows that the locus of the centers (which consists of two parts, one for each family), is made up of two *focal conics*, that is, geometric limits of quadrics in a confocal family (10.2.2.3) as λ approaches b or c:

$$\frac{x^2}{a-c} + \frac{y^2}{b-c} - 1 = 0 = z, \qquad \frac{x^2}{a-b} - \frac{z^2}{b-c} - 1 = 0 = y.$$

(Two focal conics are contained in mutually perpendicular planes, and the foci of each are vertices of the other. One must also include the special cases of two parabolas, the focus of one being the vertex of the other, and of a circle and its axis.) The surfaces thus obtained are exactly the Dupin cyclids introduced in 10.2.2.6. The case of a circle and its axis gives a torus of revolution.

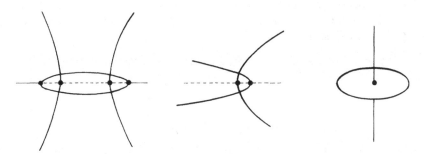

Figure 10.2.3.12

10.2.3.13. Envelopes of two-parameter families. We only discuss the case of planes. In $\mathbf{R}^2$ a one-parameter family of lines is best expressed—at least for some calculations—in Euler form, that is, by giving the distance of the line to the origin as a function of its direction (see exercise 9.9.3 and [Ber87, 11.8.12.3 and 12.12.1]). Calculations are simple because this description amounts to a real-valued function on the circle, and a circle can be parametrized as a line: for instance, if the distance from the origin is $p(\theta)$, the curvature of the envelope is $p + p''$ (see figure 11.19 or [Ber87, figure 12.12.14]).

For a two-parameter family of planes in $\mathbf{R}^3$ the corresponding description amounts to a function $p : S^2 \to \mathbf{R}$; as we have seen that there is no really simple parametrization for S^2, calculations are more involved. An analogue for the formula $K = p + p''$ will be studied in 11.19.1.

Wave surfaces (10.2.2.7) can be easily expressed as envelopes of a two-parameter family of planes; just take p to be

$$p = \frac{(a+k)u^2 + (b+k)v^2 + (c+k)w^2}{2},$$

where $u^2 + v^2 + w^2 = 1$ and k is a constant. In cartesian coordinates we have $x = (p-a)u$, $y = (p-b)v$ and $z = (p-c)w$.

The reader can verify that Enneper's surface (10.2.2.5) can be expressed as an envelope in a simple way: take two arbitrary point describing parabolas focal to one another (figure 10.2.3.12). The envelope of the planes equidistant to these two points is Enneper's surface.

10.2.3.14. String construction for ellipsoids. An ellipsoid can be mechanically generated as follows: a piece of string is kept taut so that it is always supported by two focal conics, as indicated in figure 10.2.3.14. The endpoint of the string, which depends on two parameters, describes an ellipsoid belonging to the family of homofocal quadrics (10.2.2.3) arising from the two focal conics. For a proof, see [Sal74, p. 450] or [Coo68, p. 198].

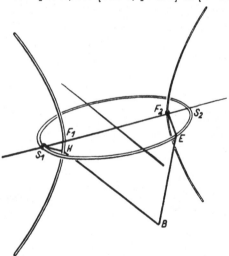

Figure 10.2.3.14

From Hilbert and Cohn-Vossen, *ibid.*, p. 20

10.2.4. Globally immersed surfaces. We present only some figures; most often, their equations are too complicated. See [Abr86]. Figures 10.2.4.1 and 10.2.4.2 both represent immersions of the torus T^2, the second being the standard one. Figure 10.2.4.3 is the well-known Boy's surface, an immersion of the projective plane $P^2(\mathbf{R})$; we know that $P^2(\mathbf{R})$ cannot be embedded in $\mathbf{R}^3$. Explicit formulas for Boy's surface, involving low-degree polynomials, can be found in [Ape86].

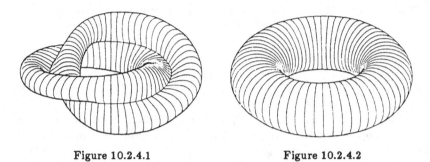

Figure 10.2.4.1   Figure 10.2.4.2

From U. Pinkall, *Regular Homotopy Classes of Immersed Surfaces*, Bonn, p. 5

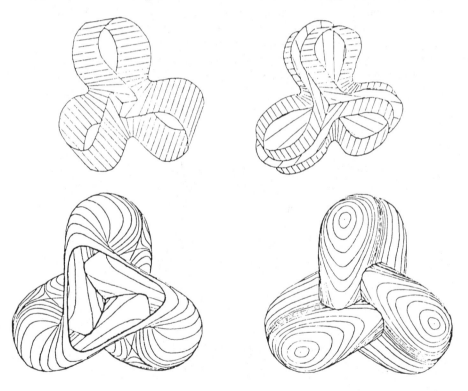

Figure 10.2.4.3

From U. Pinkall, *ibid.*, p. 10

10.3. The Two Fundamental Forms

10.3.1. The first fundamental form. Let V be a (local or global) surface in $\mathbf{R}^3$. What can the inhabitants of V learn about the geometry of their world, if they're allowed to move around on it and measure lengths, but not to step out into $\mathbf{R}^3$?

First let's see how to think of the geometry of V independently of $\mathbf{R}^3$. For each point $v \in V$ consider the tangent space $T_v V$ (2.5.3), with its Euclidean structure inherited from $\mathbf{R}^3$. The collection of these Euclidean structures g_v on the tangent bundle $TV = \bigcup_v T_v V$ (2.5.24) forms a *riemannian structure* on V (exercise 3.6.3), that is, the map $v \mapsto g_v$ is C^∞. Generalizing the language of differential forms introduced in section 5.2, we say that $v \mapsto g_v$ is a symmetric bilinear form on V; it is also positive definite. This is the so-called *first fundamental form*, which we denote by g or ds^2.

Let's state these definitions in a more precise form:

10.3.2. Definitions. Consider an arbitrary abstract differentiable manifold X. To each tangent space $T_x X$ we can associate the tensor product $(\bigoplus^p T_x) \oplus (\bigoplus^q T_x^* X)$ whose elements are the *tensors of type* (p,q) at x. The union of these vector spaces for all x forms a vector bundle over X, denoted by $T^{(p,q)}$. A *tensor of type* (p,q) on X is a C^∞ section of this bundle. A q-form, as defined in section 5.2, is in this nomenclature a skew-symmetric tensor of type $(0,q)$. A *riemannian structure* on X is a tensor of type $(0,2)$ that is symmetric and everywhere positive definite. A vector field on X is a tensor of type $(1,0)$. A field of endomorphisms on X is a tensor of type $(1,1)$, since $\text{End}(T_x X)$ and $T_x X \times T_x^* X$ are canonically identified for all x. All tensors of type $(0,q)$, including q-forms and riemannian structures, are contravariant: given a differentiable map $f : X \to Y$ and a tensor ω of type $(0,q)$ on Y, we have a *pullback* $f^*\omega$ on X given by formula 5.2.4. On the other hand, if $p > 0$, a tensor of type (p,q) cannot be pulled back any more than a vector field can (except under a diffeomorphism).

An immersed submanifold of $\mathbf{R}^n$ inherits a natural riemannian structure, called its *first fundamental form*.

Two abstract riemannian manifolds (X,g) and (Y,h) are called *isometric* if there exists a diffeomorphism $f : X \to Y$ such that $f^*h = g$.

10.3.3. The second fundamental form. This notion applies to local, global and immersed surfaces. For simplicity, let's consider the case $V \subset \mathbf{R}^3$. We also assume that V is oriented (locally this is always the case) and we denote by ν the canonical normal to V (see 6.4.3). Thus we have a differentiable map $\nu : V \to S^2$, called the *Gauss map*. Since, by construction, $T_v V = T_{\nu(v)} S^2$, the tangent map $T_v \nu$ is a linear map from $T_v V$ into itself, called the *Weingarten endomorphism*. This map is self-adjoint with respect to the first fundamental form g, and thus comes from a symmetric bilinear form on $T_v V$, called the *second fundamental form*, and denoted by

II_v:

$$\text{II}_v(x, y) = \big(T_v\nu(x) \,|\, y\big).$$

Do not confuse this with the *third fundamental form*, which is by definition the pullback under ν of the first fundamental form on S^2. We will not discuss it any further, since it is a linear combination of g and II.

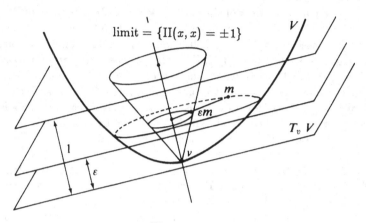

$\text{limit} = \{\text{II}(x, x) = \pm 1\}$

Figure 10.3.3

The second fundamental form gives the shape of V in $\mathbf{R}^3$. In order to determine it geometrically, we can cut V with planes parallel to T_vV, rescale the intersection curves, and take their limit as the plane approaches T_vV. If II has rank two, we obtain a limiting curve, the conic defined by $\text{II}(x, x) = \pm 1$; this is called the *Dupin indicatrix*. All of this results from 4.2.20; see also 10.6.4.

There is another, important geometric method for determining the second fundamental form, based on the notion of parallel surfaces (10.2.2.12). Let $V = F^{-1}(0)$ be the surface being considered and V_k the surface parallel to V at signed distance k (for a fixed orientation—the second fundamental form changes sign when we switch the orientation of V). Then figure 10.2.2.12 gives a diffeomorphism $\phi_k : V \to V_k$, at least for k small. If g_k denotes the first fundamental form of V_k, it can be shown that the second fundamental form of V is given by the derivative

10.3.3.1
$$\text{II} = \frac{d\big(\phi_k^*(g_k)\big)}{dk}(0),$$

where the notation $\phi_k^*(g_k)$ is borrowed from the end of 10.3.2. See 10.6.6.7 for a local version of this infinitesimal result.

10.4. What the First Fundamental Form Is Good For

10.4.1. The first fundamental form in coordinates

10.4.1.1. Assume that V is locally the image of an immersion $(u,v) \mapsto f(u,v)$, defined on an open set $U \subset \mathbf{R}^2$. The pullback f^*g is a riemannian structure on U, which can be written down in coordinates; we say that g has been expressed in coordinates (u,v). The standard notation is

$$f^*(ds^2) = E\, du^2 + 2F\, du\, dv + G\, dv^2,$$

where du^2, $du\,dv$ and dv^2 are tensor products of the one-forms du and dv. The functions E, F and G are given by

$$E = \left\| \frac{\partial f}{\partial u} \right\|^2, \qquad F = \left(\frac{\partial f}{\partial u} \,\middle|\, \frac{\partial f}{\partial v} \right), \qquad G = \left\| \frac{\partial f}{\partial v} \right\|^2.$$

These three functions completely determine the riemannian structure, but they are not determined by it. In fact (cf. 10.1.3 and 10.3.2) we can apply a diffeomorphic change of coordinates $(u,v) \mapsto (u',v')$ to obtain an isometric structure, which is in some sense the same. The new coefficients E', F' and G' of du'^2, $du'\,dv'$ and dv'^2 are found by plugging in

$$du = \frac{\partial u}{\partial u'}du' + \frac{\partial u}{\partial v'}dv', \qquad dv = \frac{\partial v}{\partial u'}du' + \frac{\partial v}{\partial v'}dv'$$

into the standard notation above. In the language of quadratic forms [Ber87, 13.1.3.8], the matrix $I = \left(\begin{smallmatrix} E & F \\ F & G \end{smallmatrix} \right)$ becomes $I' = \left(\begin{smallmatrix} E' & F' \\ F' & G' \end{smallmatrix} \right) = {}^tJIJ$, where J is the jacobian

$$\begin{pmatrix} \dfrac{\partial u}{\partial u'} & \dfrac{\partial v}{\partial u'} \\[2mm] \dfrac{\partial u}{\partial v'} & \dfrac{\partial v}{\partial v'} \end{pmatrix}.$$

Naturally, one tries to find coordinates in which the expressions for E, F and G are as simple as possible. This task is facilitated by a knowledge of the (intrinsic or extrinsic) geometry of V.

10.4.1.2. Graphs. Using the classical notation $p = \dfrac{\partial f}{\partial x}$, $q = \dfrac{\partial f}{\partial y}$ for the graph of a function $f : \mathbf{R}^2 \to \mathbf{R}$, we get

$$ds^2 = (1 + p^2)dx^2 + 2pq\, dx\, dy + (1 + q^2)dy^2.$$

10.4.1.3. The sphere (cf. 10.2.3.3). For the latitude-longitude chart we have

$$ds^2 = d\theta^2 + \cos^2\theta\, du^2;$$

for the Mercator projection,

$$ds^2 = \operatorname{sech}^2 u(du^2 + dv^2);$$

and for the stereographic projection,

$$ds^2 = \frac{4}{(x^2 + y^2 + 4)^2}(dx^2 + dy^2)$$

(see 11.2.2 for a similar formula in the case of negative curvature).

10.4.1.4. Surfaces of revolution. If $\sqrt{x^2 + y^2} = J(z)$ we have $ds^2 = (1 + J'^2)dz^2 + J^2 dv^2$. We can obtain an expression in geodesic coordinates (10.4.2) by introducing the *latitude* u, defined as the arclength along a meridian: we have $du = \sqrt{1 + J'^2}\,dz$, and

$$ds^2 = du^2 + J^2(z(u))dv^2.$$

The further substitution $du = J^2(z(u))dw$ gives an expression in conformal coordinates (10.4.2):

$$ds^2 = \psi(w)(dw^2 + dv^2).$$

For the catenoid we have $ds^2 = \cosh^2 v(dv^2 + du^2)$, or, with $\sigma = \sinh v$,

$$ds^2 = d\sigma^2 + (1 + \sigma^2)dv^2.$$

For Beltrami's surface:

$$ds^2 = du^2 + e^{2u}dv^2.$$

10.4.1.5. Enneper's surface. We find $ds^2 = (1 + u^2 + v^2)^2(du^2 + dv^2)$, which is also in conformal coordinates (10.4.2).

10.4.1.6. Ellipsoids. If (u, v) are elliptic coordinates (10.2.2.3) we get

$$ds^2 = \frac{u - v}{4}\left(\frac{u\,du^2}{(a - u)(b - u)(c - u)} - \frac{v\,dv^2}{(a - v)(b - v)(c - v)}\right).$$

Defining α and β by the integrals

$$d\alpha = \frac{\sqrt{u}\,du}{\sqrt{(a - u)(b - u)(c - u)}}, \qquad d\beta = \frac{\sqrt{-v}\,dv}{\sqrt{(a - v)(b - v)(c - v)}},$$

we can write

$$ds^2 = \frac{u - v}{4}(d\alpha^2 + d\beta^2).$$

10.4.1.7. Ruled surfaces. We start with an example, the helicoid of pitch 2π (10.2.2.10), whose equation is $\tan z = y/x$. Using the parametrization $(\theta, r) \mapsto (r\cos\theta, r\sin\theta, \theta)$, we get

$$ds^2 = dr^2 + (1 + r^2)d\theta^2.$$

To our great surprise, we verify that this helicoid and the catenoid (10.4.1.4) are isometric, at least locally. But they do not have the same shape in space: no isometry of $\mathbf{R}^3$ maps one into the other. From this phenomenon arises the need, first understood clearly by Gauss, to distinguish between the two fundamental forms, between the intrinsic and extrinsic geometries of a surface. There are numerous examples of intrinsic isometries: see 10.4.1.8 (developable surfaces) and especially 10.2.3.6, 10.5.3.10 and section 11.14.

For an arbitrary ruled surface, expressed in such a way that $(m' \mid \xi) = 0$, we get

$$ds^2 = dv^2 + \left(v^2\|\xi'\|^2 + 2v(m' \mid \xi') + \|m'\|^2\right)du^2.$$

If, on the other hand, we work with the standard parametrization (10.2.3.7), we get

$$ds^2 = dv^2 + 2(m' \mid \xi)\, du\, dv + \left(v^2 + \|m'\|^2\right)du^2.$$

10.4.1.8. Developable surfaces. Consider the developable S of a curve $C \in \mathbf{R}^3$, that is, the union of the tangents to C (10.2.3.9). We can use the previous formula with $\|m'\| = 1$, $\xi = m'$ and $\|\xi'\| = \rho(u)$, where $\rho(u)$ is the curvature of C as a function of arclength, to obtain

$$ds^2 = dv^2 + 2\, du\, dv + (1 + v^2\rho^2)du^2 = (dv + du)^2 + v^2\rho^2\, du^2,$$

which already shows that developable surfaces coming from curves with same curvature function are isometric.

However, we can get much more if we change coordinates. As we roll the osculating plane P to C around its envelope S without sliding, let the new coordinates (α, β) of a point $m(u) + vm'(u) \in S$ be its cartesian coordinates with respect to an orthonormal affine frame rigidly attached to P. By construction, these are the same as the coordinates of

$$r(u) + vr'(u)$$

with respect to the chosen frame, where $r : \mathbf{R} \to \mathbf{R}^2$ is the equation of the curve in P that has the same curvature $\rho(u)$ as C does (recall the discussion in 10.2.3.10 and the existence and uniqueness theorem 8.5.7). This says exactly that $d\alpha^2 + d\beta^2 = (dv + du)^2 + v^2\rho^2\, du^2$, so we finally get

$$ds^2 = d\alpha^2 + d\beta^2.$$

The striking thing about this equation is that it shows that the surface is isometric (at least locally, of course) to the Euclidean plane. In the degenerate case of cones and cylinders, when C is just a fixed point (at infinity in the case of cylinders), it is trivial to write down an explicit isometry. The burning question then is: does this local isometry property characterize developable surfaces? Turn to 10.6.6.2 and section 11.12 for the answer.

10.4.2. The various kinds of coordinates. Here are the classical names for some of the special cases we have met. A set of coordinates is called *orthogonal* if $F = 0$, that is,

$$ds^2 = E(u, v)du^2 + G(u, v)dv^2.$$

If, in addition, $E = G = C(u, v)$, the coordinates are called *conformal*, because the angle between two directions on the surface is then equal to their angle on the Euclidean uv-plane. This is desirable in plotting charts, especially for marine navigation, because it makes straight lines on the chart correspond to lines of constant heading. In cartography one also uses *area-preserving coordinates*, for which $EG = 1$ and $F = 0$. See [Ber87, 18.1.7 and 18.1.8], and the references therein.

A surface is in *geodesic polar coordinates* (cf. 10.5.1) if it is in the form $ds^2 = du^2 + J^2(u, v)dv^2$, that is, if $E = 1$ and $F = 0$. We talk about *Liouville coordinates* if

$$ds^2 = \big(A(u) + B(v)\big)\big(C(u)du^2 + D(v)dv^2\big);$$

the ellipsoid above (10.4.1.6) is an example of this type. Finally, we call *Chebyshev coordinates* those in which $E = G = 1$: they occur when we try to "dress" a surface with a fabric that can be deformed by a shearing motion, without changing the length of the threads.

Orthogonal coordinates can always be found; this is trivial. Conformal coordinates also always exist, but this is much harder to prove, especially when the functions involved are not real analytic; see, for example, the addendum to [Spi79, vol. IV]. Geodesic coordinates always exist (10.4.8). Liouville coordinates, on the other hand, only exist for a special class of surfaces, whose characterization can be found in [Dar72, vol. III, p. 16].

10.4.3. The metric of a local surface. The notions that we will introduce in the remainder of this section make sense for abstract riemannian manifolds of any dimension, and the definitions stay the same, but the two-dimensional case is by far the simplest. This not only in details, like the simple description of parallel transport (10.4.8.3), but from the very root of thing: in two dimensions, the fundamental invariant of surfaces is a scalar function, the Gaussian curvature K (section 10.5), whereas in higher dimension it is the curvature tensor, which is a quadrilinear form, a very complicated animal, not yet fully understood. Going from riemannian geometry in two dimensions to three or more is a huge leap, in terms of both technical complexity and essential depth.

So we're going to stick to surfaces V in $\mathbf{R}^3$, but consider them insofar as possible as abstract riemannian manifolds. Let $\big([a, b], f\big)$ be a curve on V. Its *length* is

$$\text{leng}(f) = \int_a^b \|f'(t)\| \, dt;$$

this number only depends on the first fundamental form. The *distance* between two points $v, w \in V$ is the infimum of the length of curves connecting them:

$$d(v, w) = \inf\big\{\text{leng } f : f \text{ is a curve in } V \text{ with } f(a) = v \text{ and } f(b) = w\big\}.$$

This defines the *intrinsic metric* of V. Unless V is a subset of a plane, this does not coincide with the metric induced from $\mathbf{R}^3$, though of course we have the trivial inequality $d_{\mathbf{R}^3}(v, w) \le d(v, w) = d_V(v, w)$. It is clear that d

is a metric, if $V \subset \mathbf{R}^3$; the case of an abstract two-dimensional riemannian manifold will be proved in 10.4.8.

10.4.4. Some natural questions. As with any metric space, we may want to:

(i)   calculate $d(v, w)$ explicitly;
(ii)   determine when there exists a curve of length $d(v, w)$ joining v and w (such curves are called *shortest paths*, or *segments* if there is no danger of confusion);
(iii) find out what the shortest paths look like;
(iv)   find out when there is a unique shortest path joining v and w.

Needless to say, these problems do not have an explicit solution in most cases, even those that appear simple at first sight. See 10.4.9.5, for example, for the case of ellipsoids. On the other hand, that are some very pretty and general qualitative results: for example, any two sufficiently closed points are connected by a unique segment (just as in Euclidean space). Segments are subsets of curves called geodesics, which will be defined in 10.4.5; in fact, it is by calculating the geodesics that one determines the segments. Geodesics, in turn, are found by considerations of parallel transport (10.4.6), a notion that has numerous applications and a strong heuristic interest.

In our study we will seem to wander away from V, but we will show that the notions of geodesics and parallel transport only depend on the first fundamental form of V, that is, on the abstract riemannian structure. For this reason they are automatically invariant under isometries. We remark here that the isometries of a riemannian manifold V (10.3.2) coincide with those of the metric space (V, d), where d is the intrinsic metric; this is not entirely obvious [Pal57].

10.4.5. Geodesics. Consider the problem of finding a curve $C = ([a, b], f)$ on V that has shortest length among all curves with same endpoints. The calculus of variations shows that a necessary condition (which is also sufficient if we just want the length to have zero derivative, instead of being minimal) is that the acceleration of C be everywhere perpendicular to V, that is, $f''(t) \in (T_{f(t)})^{\perp}$ for every $t \in [a, b]$. Since this implies that $\|f'(t)\|$ is constant, we may as well take this constant to be one.

This observation seems to go against our philosophy, in that it makes use of the external shape of V. In other words, we have wandered away from V into the ambient space, something we had agreed was forbidden. But, in fact, this criterion for geodesics only depends of the first fundamental form. This will be shown by introducing covariant derivatives—in some sense a generalization of the notion of geodesics—but should not come as a surprise in any case, since the original problem, minimizing the length of a path joining two points, only depends on the intrinsic metric, that is, on the first fundamental form.

There are several ways to obtain geodesics geometrically. For example, one can take two wheels of same size, connected by an axle, and roll them around V. In the limit, as the distance between the wheels tends to zero, the path they follow will be a geodesic. This operation is intrinsic, for it amounts to saying that the wheels follow paths of same length and constant distance to one another. Another idea is to stretch an elastic band between two points, making sure that it remains on V. (In $\mathbf{R}^3$ we can guarantee this by laying out the elastic band on the appropriate side of V, as long as the endpoints are close enough.)

10.4.6. Parallel transport. Let $C = \big([a,b], f\big)$ be a curve on V. We say that a vector field Z along C, such that $Z(t)$ is everywhere tangent to V, is *parallel* if its derivative is everywhere perpendicular to V, that is, $Z'(t) \in (T_{f(t)})^{\perp}$ for every $t \in [a,b]$. The fundamental result is the following:

10.4.6.1. Theorem. *Given a curve C on V and a vector $z \in T_{f(a)}V$, there exists a unique parallel vector field Z along C satisfying $Z(a) = z$. In addition, Z only depends on the first fundamental form.* $\square$

We say that the vectors $Z(t)$ are obtained from z by *parallel transport*. Parallel transport preserves inner products (that is, the first fundamental form).

Geodesics are then just self-parallel curves: their velocity vector field is parallel.

10.4.7. Covariant derivative and geodesic curvature. Parallel transport is not yet the most fundamental notion; behind it we find the idea of covariant derivative. Let Z be a vector field along C. We know how to differentiate Z along C, but the result is generally outside V, so it's not good enough. The right derivative on V turns out to be the orthogonal projection of $Z'(t)$ on $T_{f(t)}(V)$; it is called the *covariant derivative* of Z along C and denoted by $D_{f'}Z$. The *geodesic curvature* of a curve in V (or in an arbitrary riemannian manifold) is defined as $\|D_{f'}f'\|$; this number replaces the curvature defined in chapter 8.

The covariant derivative depends only on the first fundamental form. This is to be expected: by projecting Z' onto the tangent space to V we're doing all we can to remain in V. But the real, deep reason is that, given any riemannian manifold (X, g), we can associate to each pair of vector fields (ξ, η) on X a new vector field, called the *covariant derivative* of η with respect to ξ and denoted by $D_{\xi}\eta$, characterized by the following conditions: D_{ξ} preserves the scalar product g for every ξ, and $D_{\xi}\eta - D_{\eta}\xi = [\xi, \eta]$ for every ξ and η (see 2.8.17 and 3.5.15 for the definition of the brackets). This second condition is the counterpart for riemannian manifolds of the commutativity of partial derivatives in $\mathbf{R}^n$ (cf. section 4.2).

A direct proof of the invariance of covariant derivatives, but one that doesn't show what's actually going on, can be given by starting from the initial definition and showing that only E, F and G are involved. Existence and uniqueness derive from the existence and uniqueness of solutions of differential equations (1.4.5).

In order to write the covariant derivative in coordinates, one introduces certain coefficients called *Christoffel symbols*. These numbers can be calculated using just the first order derivatives of E, F and G.

Here is a formula for the geodesic curvature of a curve $t \mapsto (u(t), v(t))$ on a riemannian surface. As you can see, it is not simple, and we omit the proof, which can be found in [Lel63, p. 194] or [Str61, p. 131]. Assuming that the coordinates on V are orthogonal, that is, $F = 0$, we have

10.4.7.1
$$\frac{1}{\sqrt{Eu'^2 + Gv'^2}} \left(\frac{d\phi}{dt} + \frac{1}{2\sqrt{EG}} \left(\frac{\partial G}{\partial u} v' - \frac{\partial E}{\partial v} u' \right) \right),$$

where ϕ is defined by $\tan \phi = \sqrt{\dfrac{G}{E}} \dfrac{dv}{du}$.

If the surface is in geodesic coordinates, that is, $ds^2 = du^2 + J^2(u, v)dv^2$, and v is given as a function of u, we have

$$\frac{d\phi}{dt} = \frac{1}{1 + J^2 v'^2} \frac{d}{du} (J(u, v(u))v'),$$

and 10.4.7.1 simplifies to

10.4.7.2
$$\frac{1}{(1 + J^2 v'^2)^{3/2}} \left(\frac{\partial J}{\partial u} (2v' + J^2 v'^3) + \frac{\partial J}{\partial v} v'^2 + Jv'' \right).$$

10.4.8. Geodesics revisited; locally shortest paths. The equations above show that $D_{f'} f' = 0$, the equation characterizing a geodesic, is a second-order differential equation (not linear, but quadratic). This implies the following result:

10.4.8.1. Theorem. *Given $v \in V$ there exists $\varepsilon > 0$ such that, for every unit vector $\xi \in T_v V$, there exists a geodesic of V with initial speed ξ and defined at least on the interval $]-\varepsilon, \varepsilon[$. As ξ runs over the set of unit vectors in $T_v V$, the associated geodesics γ_ξ make up exactly the open ball $B(v, \varepsilon) = \{x \in V : d(x, v) < \varepsilon\}$. Furthermore, for $\varepsilon' < \varepsilon$ the restriction of γ_ξ to $[0, \varepsilon']$ is the shortest path on V joining its endpoints.* □

This shows that the topology of S as a surface coincides with its topology as a metric space.

10.4.8.2. Note. In general a geodesic cannot be extended to all of $\mathbf{R}$; think of the unit ball in $\mathbf{R}^2$, for example. Moreover, if a geodesic is extended too far it can cease to be the shortest path between two points; see section 11.1.

10.4.8.3. Note. In dimension two only, knowing the geodesics is equivalent to knowing the covariant derivative and parallel transport. Indeed, along a geodesic γ a vector is defined by its length and its angle with γ', the angle being oriented by continuity. Thus we have for each point of γ an orthonormal basis whose first vector if γ', and a vector field whose components in this basis are constant is parallel. Similarly, the covariant derivative of a vector field is given by the ordinary derivative of the components of the field in this basis.

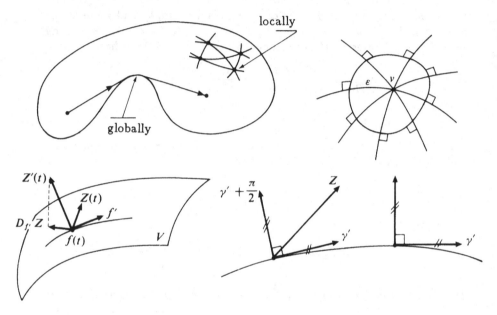

Figure 10.4.8

10.4.9. The geodesics of a few surfaces.

10.4.9.1. Spheres. Great circles are geodesics on a sphere, because their acceleration points toward the center. Since there are enough great circle to go through all points and in all directions, the uniqueness part of 10.4.8.1 says that there are no other geodesics. Hence all the questions in 10.4.4 can be completely answered: the distance between v and w is the real number $d(v, w) \in [0, \pi]$ defined by $\cos(d(v, w)) = (v \mid w)$, and two points $v, w \in S^2$ are connected by a unique shortest path, unless they happen to be *antipodal points*, i.e., $w = -v$.

10.4.9.2. Cylinders, cones and developable surfaces. If a surface is isometric to the Euclidean plane, its geodesics are just straight lines in the appropriate coordinate system (they are generally not straight in $\mathbf{R}^3$). The study of shortest paths requires a bit more care, however, because of global non-uniqueness: think of a cylinder or cone or revolution, for example.

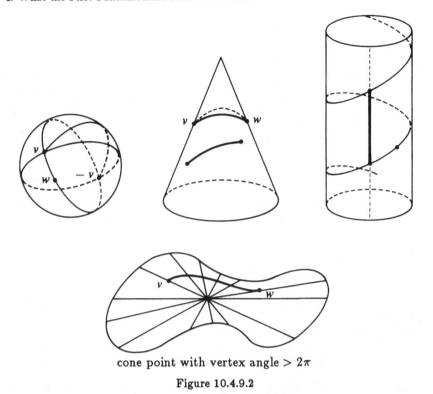

cone point with vertex angle $> 2\pi$

Figure 10.4.9.2

10.4.9.3. Surfaces of revolution. If we write $ds^2 = du^2 + J^2(u)dv^2$, where J is the distance to the z-axis (10.4.1.4), the quantity

$$J^2\big(u(t)\big)\frac{dv(t)}{dt}$$

is a *first integral*, that is, it remains constant along a geodesic (this is known as *Clairault's relation*). The reason is simple: the acceleration of a geodesic is normal to the surface, which here means that it intersects the z-axis, or that its projection on the xy-plane always goes through the origin. And it is well known in mechanics (Kepler's second law) that if a point moves under a central force its radius vector sweeps equal areas in equal time intervals.

It follows that the geodesics of a surface of revolution can be found by solving one integral. Even if the integration cannot be carried out explicitly, we're much better off than in the case of an arbitrary surface, since a degree-two second order differential equation is generally intractable.

It is easy to describe the geodesics qualitatively: they consist of all meridians, all equators (parallels of latitude where the meridians are parallel to the axis) and other curves whose distance to the axis can never fall below a certain threshold (determined by their initial point and direction) and

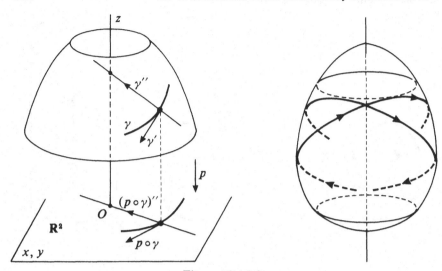

Figure 10.4.9.3

which, as a consequence, may weave back and forth between two parallels of latitude of same radius.

Surprisingly enough, there exist surfaces of revolution whose geodesics are all closed and which are not spheres (section 11.10). One of the tools in their study is Clairault's formula.

10.4.9.4. Surfaces in geodesic polar coordinates. If a surface is in the form $ds^2 = du^2 + J^2(u, v)dv^2$, it follows from 10.4.7.2 that its geodesics $\big(u, v(u)\big)$ satisfy

$$2\frac{\partial J}{\partial u}(v' + J^2 v'^3) + \frac{\partial J}{\partial v}v'^2 + Jv'' = 0.$$

10.4.9.5. Ellipsoids. An ellipsoid's geodesics can be completely determined, modulo the solution of one integral, thanks to the Liouville-type elliptic coordinates (10.4.1.6), which satisfy a counterpart of sorts for Clairault's relation (10.4.9.3):

$$\sqrt{\frac{(a-u)(b-u)(c-u)}{k-u}}u' \pm \sqrt{\frac{(a-v)(b-v)(c-v)}{k-v}}v' = 0,$$

where k is a constant. This gives the following qualitative behavior for the geodesics: each one-sheet hyperboloid H from the family of confocal quadrics defined in 10.2.2.3 has an associated one-parameter family of geodesics, all of which oscillate between the two curves where H intersects the ellipsoid E. These curves, as we shall see in 10.6.8.3, are curvature lines of E. Geometrically they are characterized by the fact that their tangents are tangent to H as well as to E. The two-parameter family of lines formed by these tangents as H varies satisfies the condition for the existence of surfaces having these lines as normals (3.5.15.5); see [Dar72, vol. II, p. 310] for an explicit calculation.

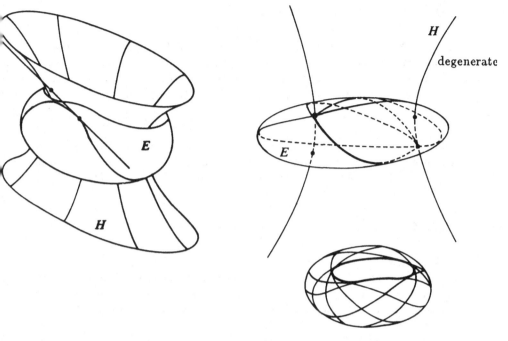

Figure 10.4.9.5

From H. Knörrer, Geodesics on the Ellipsoid, *Inventiones Math.*, Springer

When H degenerates into the focal hyperbola

$$\frac{x^2}{a-b} - \frac{z^2}{b-c} - 1 = 0 = y$$

(cf. 10.2.3.12) the intersection $H \cap E$ reduces to the four umbilical points
(10.6.4) of E. Every geodesic going through one of these points also goes
through the opposite one.

More about the geodesics of ellipsoids can be found in the reference above
and in [Dar72, vol. III, p. 14]; [Car76, p. 263–264]; [Kli82, p. 302 and ii];
[Kno80]; and [Sal74, p. 447 and neighboring pages].

10.4.9.6. Geodesic maps. A map between riemannian surfaces that takes
geodesics to geodesics does not have to be an isometry, not even up to a
scale factor. For example, take the central projection from a sphere into
any plane and use 10.4.9.1. However, only riemannian structures that can
be put in Liouville form (10.4.2) admits such geodesic maps: see [Dar72,
vol. III, p. 51]. Beltrami showed that every riemannian manifold that can
be geodesically mapped to the Euclidean plane is a sphere or a hyperbolic
plane: see [Car76, p. 296] or [Spi79, vol. IV, p. 26].

This problem belongs to a theory called *projective differential geometry*:
see [Lel82], [Ven79], [Car37b] and [Eis49, p. 131].

10.5. Gaussian Curvature

10.5.1. Normal and geodesic coordinates. A (local) surface is written in *normal coordinates* if all geodesics going through the origin of the coordinate plane are straight lines in that plane, parametrized by arclength. In normal coordinates the expansion of ds^2 at the origin is

10.5.1.1 $$ds^2 = dx^2 + dy^2 + K(x\,dy - y\,dx)^2 + o(x^2 + y^2),$$

where K is a real number, depending only on the metric and on the point taken as the origin. Its value is easily found if we know the expression of ds^2 in geodesic polar coordinates (10.4.2): comparing $ds^2 = dr^2 + J^2(r, \theta)$ with 10.5.1.1 and taking a limit we get

10.5.1.2 $$K = -\frac{1}{J(0,0)} \frac{\partial^2 J}{\partial r^2}(0,0).$$

We call K the *Gaussian curvature* (or *total curvature*, cf. 10.6.2) of the surface at the point in question. The Gaussian curvature is clearly invariant under riemannian isometries; in particular, a surface with $K \neq 0$ can never be isometric to the Euclidean plane, it being clear that the latter has zero curvature.

In fact, formula 10.5.1.1 says that the Gaussian curvature in some sense measures how far a surface is from being Euclidean. The inhabitants of V can calculate K at a point v by drawing circles of small radius ε around v (they know by experience what the shortest paths are) and measuring their circumference. The radius and length of a circle are related by the formula

10.5.1.3 $$\text{leng}(C(v, \varepsilon)) = 2\pi\varepsilon \left(1 - \frac{K(v)}{6}\varepsilon^3 + o(\varepsilon^3)\right).$$

This is called *Puiseux's formula*. The reader can prove a similar formula due to Diquet relating the area of the disc $B(v, \varepsilon)$ and its radius.

10.5.2. Surfaces of constant curvature. Figure 10.5.2 shows that the curvature of the sphere S^2 is everywhere equal to 1; more generally, the curvature of a sphere of radius R is R^{-2}:

Riemannian surfaces of constant curvature will play an important role in the sequel. We have already seen surfaces of zero and positive constant curvature. For surfaces of negative curvature, one uses formula 10.5.3.3 below to get

$$ds^2 = dr^2 + \frac{\sinh^2(\sqrt{-K}r)}{\sqrt{-K}}\,d\theta^2.$$

Compare this with $ds^2 = dr^2 + r^2\,d\theta^2$ for the Euclidean plane and $ds^2 = dr^2 + \sin^2 r\,d\theta^2$ for the unit sphere.

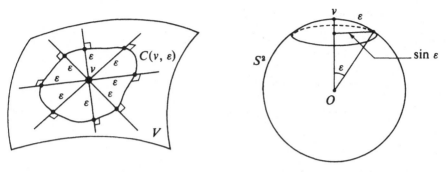

Figure 10.5.2

Formula 10.5.3.3 also shows that manifolds of same constant curvature are locally isometric. For a global statement, see 11.2.1. For embeddings in $\mathbf{R}^3$, see 10.5.3.10 and section 11.15.

10.5.3. Formulary for K. Let $(u, v) \mapsto f(u, v)$ be an immersed surface. A simple expression of K in coordinates involves, in addition to the coefficients E, F and G of the first fundamental form (10.4.1.1), the coefficients of the second fundamental form (10.6.6), traditionally denoted by L, $2M$ and N and given by

$$L = \frac{(f_{uu} \mid n)}{\sqrt{EG - f^2}}, \qquad M = \frac{(f_{uv} \mid n)}{\sqrt{EG - f^2}}, \qquad N = \frac{(f_{vv} \mid n)}{\sqrt{EG - f^2}},$$

where the subscripts denote partial differentiation and $n = f_u \times f_v$ is the (non-unit) normal to the surface. In this notation we have

10.5.3.1
$$K = \frac{LN - M^2}{EG - F^2}.$$

This will be proved in 10.6.6.

10.5.3.2. Gauss's formula. Formula 10.5.3.1 hides the fundamental fact, apparent from definition 10.5.1.1, that K only depends on the first fundament form of V. In proving his "Theorema egregium," Gauss accomplished a computational tour de force in expressing K explicitly as a function of E, F and G. His formula reads

$$4K(EG - F^2)^2 = \begin{vmatrix} 4F_{uv} - 2E_{vv} - 2G_{uu} & E_u & 2F_u - E_v \\ 2F_v - G_u & E & F \\ G_v & F & G \end{vmatrix} - \begin{vmatrix} 0 & E_v & G_u \\ E_v & E & F \\ G_u & F & G \end{vmatrix}.$$

For a relatively short proof, see [Str61, p. 112].

10.5.3.3. Geodesic coordinates. If the surface is in the form $ds^2 = du^2 + J^2(u, v)dv^2$, we get from 10.5.1.2 or 10.5.3.2 that

$$K = -\frac{1}{J}\frac{\partial^2 J}{\partial u^2}.$$

This is related with the notion of Jacobi fields (11.5.1).

10.5.3.4. Graphs. For a graph $(x, y, f(x, y))$, if p and q are as in 10.4.1.2 and r, s and t as in 4.2.20, we have

$$K = \frac{rt - s^2}{(1 + p^2 + q^2)^2}.$$

10.5.3.5. Enneper's surface. We have

$$K = -\frac{4}{(1 + u^2 + v^2)^2}.$$

10.5.3.6. Ruled surfaces. If λ is the parameter of distribution (10.2.3.7) we have

$$K = -\frac{\lambda^2}{(\lambda^2 + v^2)^2}.$$

Hence the Gaussian curvature of a ruled surface is always strictly negative, unless $\lambda = 0$, in which case it is zero along the corresponding generating line. In particular K is identically zero for a developable surface; this, together with 10.5.2, gives another proof for 10.4.1.8.

10.5.3.7. Ellipsoids. For the ellipsoid $\frac{x^2}{a^2} + \frac{y^2}{b^2} + \frac{z^2}{c^2} - 1 = 0$ we find the very nice formula $K = \frac{p^4}{a^2 b^2 c^2}$, where p denotes the distance from the origin to the tangent plane at the point (x, y, z).

10.5.3.8. Implicitly defined surfaces. Let V be the zero-set of $F : \mathbf{R}^3 \to \mathbf{R}$, and define a, b, c by the identity

$$\begin{vmatrix} F'' - \lambda I & F' \\ {}^t F' & 0 \end{vmatrix} = a + b\lambda + c\lambda^2,$$

where F' is the vector of partial derivatives of F, F'' its matrix of second derivatives and I the identity matrix. In this notation we have

$$K = \frac{a/c}{F_x^2 + F_y^2 + F_z^2}.$$

10.5.3.9. Surfaces of revolution. If V is given by $ds^2 = du^2 + J^2(u)dv^2$, formula 10.5.3.3 shows that K is independent of v: it is a function of the latitude only. This is to be expected, since K is invariant under isometries and revolutions are isometries of $\mathbf{R}^3$, hence also of V.

Furthermore, one can find all surfaces of revolution whose curvature is a given function of the latitude u just by integrating 10.5.3.3; once J is known one obtains the z-coordinate as a function of u by solving the equation

$$\frac{dz}{du} = \sqrt{1 - \left(\frac{\partial J}{\partial u}\right)^2},$$

an easy consequence of 10.4.1.4.

10.5.3.10. We do this for constant K. The case $K = 0$ is trivial: J is linear in u, and consequently in z. We simply get cones and cylinders.

If $K = 1$, the solution of 10.5.3.3 is $J = a \cos u$, and integrating 10.5.3.10 we get the following parametric equation for the meridians:

$$u \mapsto \left(r = a \cos u, \, z = \int_0^u \sqrt{1 - a^2 \sin^2 t} \, dt \right).$$

This gives three kinds of surfaces, depending on whether a is less than, greater than, or equal to zero; the latter case represents the sphere.

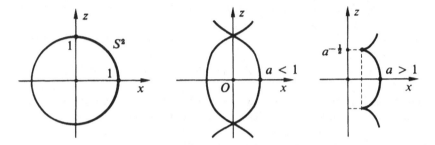

Figure 10.5.3.10.1

For $K = -1$, we also get three kinds of surface (figure 10.5.3.10.2), corresponding to three qualitatively distinct solutions for 10.5.3.3: $J = e^u$, $J = a \cosh u$ and $J = a \sinh u$. The equations for the meridians are:

$$u \mapsto \left(e^u, \int_0^u \sqrt{1 - e^{2t}} \, dt \right) \quad \text{(Beltrami's surface, see 10.2.3.5)},$$

$$u \mapsto \left(a \cosh u, \int_0^u \sqrt{1 - a \sinh^2 t} \, dt \right),$$

$$u \mapsto \left(a \sinh u, \int_0^u \sqrt{1 - a \cosh^2 t} \, dt \right).$$

10.5.4. To what extent does K determine the metric? In 10.4.1.7 we saw two different surfaces with the same intrinsic metric. In fact we know from 10.2.3.6 that there exist non-trivial one-parameter families of isometric surfaces. So we have to get accustomed to the idea that the first fundamental form does not at all determine the shape of a surface in $\mathbf{R}^3$. But we may suspect that K determines the metric of a riemannian surface, based on the fact that this is true if K is constant, and on the observation that K is a measure of non-Euclideanness, so to speak.

However, this is not the case either. To get counterexamples it is enough to take a metric with variable curvature and apply to a it a diffeomorphism that preserves the lines where the curvature is constant, the level curves of K. The result, as a rule, is not isometric to the original surface. The reader

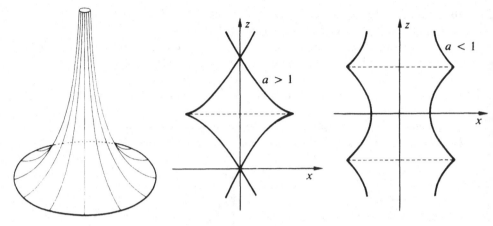

Figure 10.5.3.10.2

may consult [Car84] for a characterization of two-dimensional riemannian metrics by the curvature and the invariants derived therefrom.

10.5.5. Gaussian curvature and parallel transport

10.5.5.1. The key to the spectacular result we're about to discover is the fact that, in geodesic polar coordinates $ds^2 = dr^2 + J^2(r,\theta)d\theta^2$, the partial derivative $\dfrac{\partial J}{\partial r}(r,\theta)$ represents the geodesic curvature of the metric circle of radius r centered at the origin. See also 11.3.3.

10.5.5.2. Here we need to introduce the *canonical measure* dm of a riemannian manifold. In dimension two dm is equal to $\sqrt{EG - F^2}\, du\, dv$, where the coefficients of the first fundamental form are expressed in any set of coordinates. In our case, then, we can write $dm = J\, dr\, d\theta$.

10.5.5.3. Let D be a submanifold-with-boundary given, in our coordinates, by a positive function $r(\theta)$ defined on $[0, 2\pi]$. We calculate the integral over D of the Gaussian curvature K:

$$\mathcal{A} = \int_D K\, dm = \int_D KJ\, dr\, d\theta.$$

Using formula 10.5.3.3, the integral over r is trivial:

$$\mathcal{A} = -\int_D \frac{\partial^2 J}{\partial r^2}\, dr\, d\theta = \int_\theta \left(\frac{\partial J}{\partial r}\bigl(r(\theta)\bigr) - \frac{\partial J}{\partial r}(0) \right) d\theta.$$

Performing this calculation carefully (using 10.4.7.2, for instance), we get the following result:

10.5.5.4. Theorem (Gauss–Bonnet formula.) *Let D be a submanifold-with-boundary as above, except that we allow the boundary $C = \partial D$ to be only piecewise differentiable. Letting ρ be the signed geodesic curvature* of C (positive if C curves inward), ds the element of length along C and $\beta_1, \ldots, \beta_n$ the exterior angles of C at its corners, we have

$$A = \int_D K\, dm = 2\pi - \sum_{i=1}^{n} \beta_i - \int_C \rho\, ds. \qquad \square$$

A formal proof requires working in the unit tangent bundle UV, the three-dimensional manifold consisting of unit tangent vectors to V, and showing that the inverse image under the projection $UV \to V$ of the exterior two-form $K\, dm$, for a fixed orientation of V, is the exterior derivative of a one-form on UV that represents the geodesic curvature. The Gauss–Bonnet formula is then a consequence of Stokes' theorem. See [Ste64, p. 284] or the end of [Spi79, vol. III].

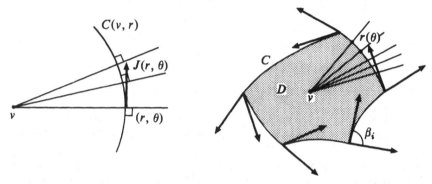

Figure 10.5.5.4

10.5.5.5. Applications. The most spectacular case is when D is a geodesic triangle $\mathcal{T}$, still contained within one geodesic coordinate patch. If the interior angles of $\mathcal{T}$ are A, B and C, the Gauss–Bonnet formula gives

$$\int_D K\, dm = A + B + C - \pi.$$

On S^2, for example, we recover *Girard's formula*: $\operatorname{area}(\mathcal{T}) = \alpha + \beta + \gamma - \pi$ [Ber87, 18.3.8.4]. On the Euclidean plane, we get that the sum of the angles of a triangle is π. On Beltrami's surface, where $K = -1$, we have $\operatorname{area}(\mathcal{T}) = \pi - \alpha - \beta - \gamma$; for a proof in this particular case, see 6.10.6. See also 11.2.2.

10.5.5.6. Geodesy. Formula 10.5.5.4 is fundamental in geodesy, the science of determining and mapping the shape of the earth. Let $\mathcal{T}$ be a geodesic triangle on a surface, with angles A, B, C, vertices m, n, p and sides of length a, b, c. Let A^*, B^*, C^* be the angles of the Euclidean triangle having sides a, b, c. The problem is to approximate the differences $A - A^*$, $B - B^*$ and $C - C^*$. Legendre knew that, on a sphere of curvature K,

10.5.5.7 $$A - A^* = \frac{\operatorname{area}(\mathcal{T})}{3} \cdot K + o(a^2 + b^2 + c^2),$$

and similarly for $B - B^*$ and $C - C^*$. In fact this formula holds as written for any surface, as long as K is measured at a point within the triangle. This means that, up to the second order, geodesic corrections cannot give or use any information on the variation of the curvature of the earth. Gauss, by means of very clever calculations, pushed the formula to the third order:

$$A - A^* = \text{area}(\mathcal{T})\frac{2K(m) + K(n) + K(p)}{12} + o(a^3 + b^3 + c^3).$$

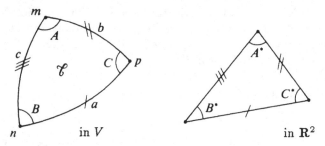

Figure 10.5.5.7

Very few books on the geometry of surfaces supply this formula. A proof for it can be found, together with other formulas including even higher order terms, in [Dar72, vol. III, p. 168–176]. A very thorough exposition of mathematical geodesy, involving elliptic functions, which give the geodesics of ellipsoids of revolution, is to be found in [Hal88]. Reference [Dom79] contains good historical remarks, and [CMS84] gives an application of formula 10.5.5.7 to modern geometry, as well as generalizations.

10.6. What the Second Fundamental Form Is Good For

The second fundamental form allows a quantitative study of the shape of V in the neighborhood of a point, extending the qualitative study in 4.2.20. It measures how far V is from being a plane: If II is identically zero, V is a plane, and vice versa. It is also used in calculating the curvature (in the sense of chapter 8) of curves in V.

Whenever surfaces are considered for practical applications, the second fundamental form is of paramount importance: in optics, it determines the caustics (see 10.6.8, [Ben86], [BW75], [Car37a], [Syn37] and [Mac49]) and the image one sees when one looks into the surface [Ben86]; in hydrostatics, it provides the notion of a *metacenter* [Cha53, vol. 2, chapter XIX], crucial to the design of ship hulls; in mechanics, it's used in the design of large smokestacks.

We recall that II switches sign when we change the orientation of V.

10.6.1. Curvature of curves on V. Let f be a curve drawn on V and parametrized by arclength. By the definition of II we see that if the angle between the normal f'' to f and the normal n to V is θ, the curvature $K(f) = \|f''\|$ is given by

10.6.1.1 $$K(f) = \text{II}(f', f') \cos \theta.$$

In particular, the curvature of a geodesic is $\text{II}(f', f')$. Notice that the geodesic curvature of f is $\text{II}(f', f') \sin \theta$.

Readers with a geometric bent may prefer to express things in terms of the center of curvature of f (8.4.15), which becomes the orthogonal projection on the osculating plane to f of the center of curvature of the geodesic of V tangent to f at the point in question. Yet another way to say this is that the component of f'' in the direction of n only depends on the direction of f'; this component is called the *normal curvature vector* of f.

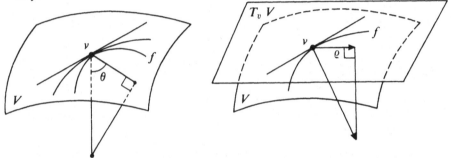

Figure 10.6.1

10.6.2. Total and mean curvatures. Principal directions. Since the first fundamental form is positive definite we can diagonalize II with respect to it (see [Ber87, 13.5], for example). The eigenvalues of the diagonalized form are those of the Weingarten endomorphism (10.3.3); they are called the *principal curvatures* of V at the point, and are generally denoted by k_1 and k_2. The eigenspaces are called the *principal directions*. One can also introduce the *principal curvature radii* $R_1 = k_1^{-1}$ and $R_2 = k_2^{-1}$. The determinant $k_1 k_2$ of II is called the *total curvature*, and the average $H = \frac{1}{2}(k_1 + k_2)$ (one-half the trace) is the *mean curvature*. Notice that the total curvature does not depend on the orientation, whereas the mean curvature does.

10.6.2.1. Theorema egregium. Gauss's stunning discovery was that the total curvature equals the curvature defined in 10.5.1! In our notation,

$$K = k_1 k_2.$$

This means, in particular, that the two fundamental forms are not independent; we will discuss this point further in section 10.7. Using the Gauss map (10.3.3) we can rewrite 10.6.2.1 as follows:

10.6.2.2 $\nu^*\omega = K\,dm,$

where ω denotes the oriented canonical measure on S^2 and dm the oriented measure on V (10.5.5.2), both being considered as volume forms. Thus the Gauss map preserves or reverses orientation depending on whether K is positive or negative.

If we recall 10.5.3.2 we see that $K = k_1 k_2$ remains invariant when V is deformed in $\mathbf{R}^3$, maintaining its intrinsic metric. Intuitively this can be justified by considering the case of polyhedra, as shown in figure 10.6.2 (for more details, see [Sto69, p. 141]) and the fact that the area of a polygon in S^2 can be expressed solely in terms of the sum of its vertex angles, without using the side lengths (this follows from Girard's formula 10.5.5.6; see also [Ber87, 18.3.8.5]).

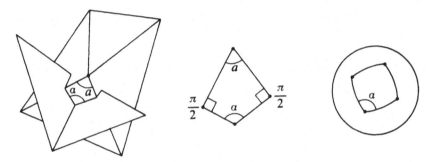

Figure 10.6.2.2

10.6.3. Application to the volume of tubes. Using 6.9.15 and 10.6.2.2, it is easy to see that what we called K_2 in 6.9.7 is exactly the total curvature K of the surface in $\mathbf{R}^3$. Thus formula 6.9.16 gives, if V is relatively compact, an exact value for the volume of the region $\mathrm{Tub}_\varepsilon\,V$ between two surfaces parallel to V at distance ε (10.2.2.12):

$$\mathrm{vol}(\mathrm{Tub}_\varepsilon\,V) = 2\varepsilon\,\mathrm{vol}(V) + \frac{2}{3}\pi\varepsilon^2 \int_V K d.$$

See also 11.20.2.

10.6.4. The osculating paraboloid. One might try to extend the notion of the osculating circle (8.4.15 and 8.7.4) to surfaces in $\mathbf{R}^3$. However, one cannot define an osculating sphere, because the radius of curvature of a geodesic that turns around a point $v \in V$ changes if $R_1 \neq R_2$. What can be defined is an *osculating paraboloid*; this may have been guessed from the qualitative study in 4.2.20. (In the plane we can define an osculating parabola, but circles are metrically more natural.)

Take orthonormal coordinates x, y, z in $\mathbf{R}^3$ coinciding, respectively, with the principal directions and the normal to V at v. Then the paraboloid

$z = k_1 x^2 + k_2 y^2$ will have with V a contact of order three, by the very definition of II. This will be an elliptic paraboloid if $K(v) > 0$ and a hyperbolic paraboloid if $K(v) < 0$ (see 10.2.1.2 and 4.2.20 for figures). If $K(v) = 0$ and either k_i is non-zero it will be a parabolic cylinder. If both k_i are zero, that is, if $II(v) = 0$, the tangent plane to V at v has a contact of third order with V. We know from 4.2.22 that in this case we cannot say anything about the local behavior of the surface.

Let's give names to these cases:

10.6.4.1. Definition. *Let v be a point on a surface V. We say that v is umbilical if $k_1 = k_2$ (possibly zero); elliptic if $K(v) > 0$; hyperbolic if $K(v) < 0$; parabolic if $K(v) = 0$ but II does not vanish; and planar if $II(v) = 0$. An asymptotic direction at V is one in which II vanishes.*

10.6.4.2. Remarks. The principal directions are well-determined and orthogonal, unless v is an umbilic. Asymptotic directions exist if and only if $K \leq 0$; at a parabolic point there is only one (counted twice), at a hyperbolic point two. At a planar point all directions are asymptotic.

It is to be suspected that integral curves of principal directions (called *curvature lines* of V) and those of asymptotic directions play an important role in the extrinsic geometry of V. See 10.6.8, for example. It is easy to see that a surface all of whose points are umbilics is a subset of a sphere.

Notice that the preceding discussion, contrary to the one in 4.2.20, is quantitative; but on the other hand it doesn't say anything about the intersection of V with $T_v V$.

10.6.5. Formulary for the second fundamental form. If $(u, v) \mapsto f(u, v)$ is an arbitrary immersion, the second fundamental form is traditionally written
$$II = L\,du^2 + 2M\,du\,dv + N\,dv^2.$$
By definition 10.6.2, the principal curvatures k_1 and k_2 are the maximum and minimum values of the quotient
$$\frac{L\,du^2 + 2M\,du\,dv + N\,dv^2}{E\,du^2 + 2F\,du\,dv + G\,dv^2} = \frac{L + 2M\lambda + N\lambda^2}{E + 2F\lambda + G\lambda^2},$$
where $\lambda = dv/du$. A bit of straightforward algebra shows that these extremal values must satisfy
$$(Ek_i - L) + (Fk_i - M)\lambda = (Fk_i - M) + (Gk_i - N)\lambda = 0 \qquad (i = 1, 2),$$
which in turn implies
$$\begin{vmatrix} Ek_i - L & Fk_i - M \\ Fk_i - M & Gk_i - N \end{vmatrix} = 0.$$
From this second-degree equation we get
$$K = k_1 k_2 = \frac{LN - M^2}{EG - F^2}, \qquad H = \frac{k_1 + k_2}{2} = \frac{1}{2}\frac{EN - 2FM + GL}{EG - F^2},$$

as well as k_1 and k_2.

10.6.6. Some explicit calculations

10.6.6.1. Graphs. In the notation of 10.5.3.4 we have

$$\text{II} = \frac{r\,dx^2 + 2s\,dx\,dy + t\,dy^2}{\sqrt{1 + p^2 + q^2}}, \qquad K = \frac{rt - s^2}{1 + p^2 + q^2},$$

$$H = \frac{1}{2}\frac{t(1 + p^2) - 2pqs + r(1 + q^2)}{(1 + p^2 + q^2)^{3/2}}.$$

10.6.6.2. Developable surfaces. Let's show that if K is identically zero V is locally a developable surface. In the graph notation, $K = 0$ implies $rt - s^2 = 0$, which says that the one-forms $dp = r\,dx + s\,dy$ and $dq = s\,dx + t\,dy$ are linearly dependent at each point. Where dp and dq are not both zero this implies the existence of an exact, nowhere vanishing one-form dh such that dp, dq and $d(z - px - qy) = -x\,dp - y\,dq$ are all multiples of dh. It follows that we can express p, q and $z - px - qy$ as functions of $h : \mathbf{R}^2 \to \mathbf{R}$:

$$p = \tilde{p} \circ h, \qquad q = \tilde{q} \circ h, \qquad z - px - qy = \tilde{w} \circ h.$$

Since $x\tilde{p}' + y\tilde{q}' + \tilde{w}' = 0$, this says exactly that V is the envelope of the one-parameter family of planes $z = \tilde{p}(h)x + \tilde{q}(h)y + \tilde{w}$.

For a more complete discussion, see section 11.12 and especially [Spi79, vol. V, p. 349 ff].

10.6.6.3. Enneper's surface. We have

$$\text{II} = 2(du^2 - dv^2), \qquad R_1 = -R_2 = \frac{1}{2}(u^2 + v^2 + 1)^2, \qquad H = 0.$$

Thus Enneper's surface has zero mean curvature: we say it is a *minimal surface* (10.6.9.2). The reader should check that the helicoid of pitch 2π and the surfaces in 10.2.3.6 are also minimal.

10.6.6.4. Implicit surfaces. If V is of the form $F^{-1}(0)$, we first calculate K and H to get k_1 and k_2. In the notation of 10.5.3.8 we have

$$H = -\frac{b/c}{2\sqrt{F_x^2 + F_y^2 + F_z^2}}.$$

If we want to determine the principal directions, we can take the bisectors of the asymptotic directions, which are the vectors v simultaneously satisfying $F'(v) = F''(v, v) = 0$, where $F' \in L(\mathbf{R}^3; \mathbf{R})$ and $F'' \in \text{Bilsym}(\mathbf{R}^3)$ are the first and second derivative forms of F [Spi79, vol. V, p. 204–205].

10.6.6.5. Ruled surfaces. In the standard parametrization (10.2.3.7) we find

$$\text{II} = \frac{1}{\sqrt{\lambda^2 + v^2}}\big((\lambda(m'' \mid \xi') + v(m'' + v\xi'', \xi', \xi))du^2 + 2\lambda\,du\,dv\big).$$

This, together with 10.6.5, implies 10.5.3.6. It can also be used to show that the only ruled minimal surface is the helicoid of pitch 2π.

10.6.6.6. Surfaces of revolution. We could use 10.4.1.4 and 10.6.6 to compute the curvatures of a surface of revolution, but let's instead work geometrically. First, by symmetry reasons, the principal directions are given by the tangents to meridians and parallels. Formula 10.6.1.1 shows then that one of the curvature radii is $R_1 = r/\sin\theta$, where r is the distance to the axis and θ is the angle between the normal and the axis. Since meridians are geodesics, the second curvature radius coincides with the radius of curvature of the meridian at the point, considered as a plane curve (8.4.15).

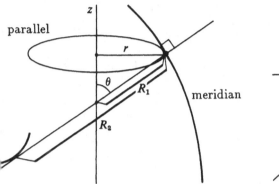

Figure 10.6.6.6.1

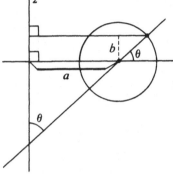

Figure 10.6.6.6.2

Here are some examples:

(1) The torus generated by the circle of radius b and centered at the point $(a, 0) \in \mathbf{R}^2$ (see figure 10.6.6.6.2):

$$K = \frac{\sin\theta}{b(a + b\sin\theta)}, \qquad H = \frac{a + 2b\sin\theta}{2b(a + b\sin\theta)}.$$

(2) The ellipsoid of revolution. Exercise 8.7.19 shows that k_1^3/k_2 is a constant, a pretty result that places ellipsoids of revolution among the so-called Weingarten surfaces (section 11.18).

(3) The catenoid is a minimal surface because of the property of the center of curvature of the catenary that can be seen on figure 10.6.6.6.3 (see 8.7.20). Since this property characterizes the catenary among plane curves, we see that the catenoid is the only minimal surface of revolution.

(4) Beltrami's surface has Gaussian curvature identically equal to -1; this comes from a geometric property of the tractrix (figure 10.6.6.6.4).

10.6.6.7. Parallel surfaces. The radii of curvature of a surface parallel to V at distance k are $R_1 - k$ and $R_2 - k$, where R_1 and R_2 are the radii of curvature of V. See 10.6.9.5 for an application. See also 10.3.3.

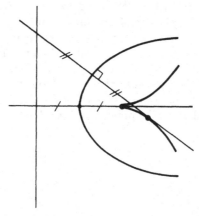

Figure 10.6.6.6.3

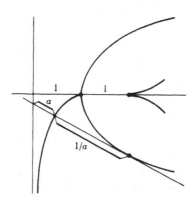

Figure 10.6.6.6.4

10.6.6.8. Arbitrary ellipsoids. If (u, v) are elliptic coordinates (see end of 10.2.2.3), one shows that

$$-R_1^2 = \frac{u^3 v}{abc}, \qquad -R_2^2 = \frac{uv^3}{abc}$$

(for this and other formulas see [Dar72, vol. II, p. 392]). In particular, along each line of curvature the ratio R_1^3/R_2 is constant. This generalizes (2) in 10.6.6.6. For the converse property, see [Dar72, vol. II, p. 415].

10.6.7. Geodesic torsion. As in section 8.6, we consider a biregular curve C, this time drawn on a given surface V. There are two natural orthonormal frames at a point in C: The Frenet frame, and the frame whose first two vectors are the tangent to C and the normal to V. If we call the corresponding normals ν_C and ν_V, respectively, the difference between the two bases is given just by the angle θ between ν_C and ν_V. This angle is sure to be something interesting, as is its derivative $d\theta/dt$, where t is the arclength along C. The right invariant turns out to be the function

$$\tau_g = \tau - \frac{d\theta}{dt},$$

called the *geodesic torsion* of C; here, of course, τ is the torsion of C considered as a curve in $\mathbf{R}^3$. Two important points: the geodesic torsion of C on V depends only on the tangent to C, and not, as the curvature in general does, on the normal to C. In fact, we have the formula

$$\tau_g = (k_1 - k_2) \sin \phi \cos \phi,$$

where ϕ is the angle that the tangent to C makes with the principal direction associated with k_2. As a consequence, curves of constant geodesic torsion coincide with lines of curvature.

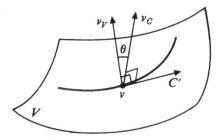

Figure 10.6.7

10.6.7.1. Note. If one wants to go a bit further in the study of surfaces, things can soon get pretty hairy. For instance, one may want to know the (ordinary) curvature and torsion of the asymptotic curves on a surface V, which arise naturally as the curves whose osculating plane is always tangent to V (this follows from 10.6.1.1). For the torsion the formula is simple:

$$\tau = \sqrt{-k_1 k_2} = \sqrt{-K}.$$

But the curvature is given by the following formula:

$$\frac{4(-R_1 R_2)^{7/8}}{(R_1 - R_2)^{3/2}} \left(\frac{\partial}{\partial s_1} \left(-\frac{R_2}{R_1^3} \right)^{1/8} \pm \frac{\partial}{\partial s_2} \left(-\frac{R_1}{R_2^3} \right)^{1/8} \right),$$

where the partial derivatives are taken with respect to arclength along the lines of curvature. See [Dar72, vol. II, p. 415], and compare with 10.6.6.8.

10.6.8. Lines of curvature, parallel surfaces and caustics. Figure 10.6.8 is very important, both in theory and in practice. It translates the following essential result:

10.6.8.1. Theorem. *A curve C in V is a line of curvature (10.6.4.2) if and only if the normals to V along C form a developable surface.* $\quad\square$

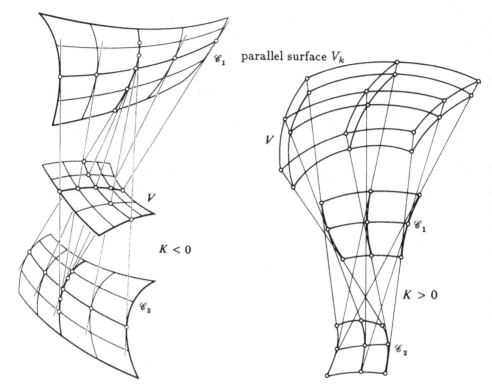

Figure 10.6.8

We recall that this condition is equivalent to saying that the normals form a surface that is locally a plane, a cylinder, a cone or the set of tangents to some curve in $\mathbf{R}^3$, which is the line of striction.

Figure 10.6.8 depicts the situation away from umbilics (10.6.4.1). In the neighborhood of an umbilic too many things can happen, and it is impossible to draw a general picture; but see [SG82] for a discussion of the generic case.

Here we just assume we're in a region free of umbilics, and that the normals to each curve form a tangent developable or a cone. Then the lines of curvature fall into two mutually orthogonal one-parameter familes $\mathcal{F}_1$ and $\mathcal{F}_2$, and the line of striction of the developables in each family describes a surfaces $\mathcal{C}_i$, for $i = 1, 2$, called a *caustic* or *focal surface* for V. The reason for the name is that if V is a wavefront, its normals are light rays, and the light rays would burn anyone rash enough to get near either caustic. *Astigmatism* occurs when the focal surface of the cornea does not degenerate into a single point. In the study of ship hulls, focal surfaces are the loci of the metacenters.

The following observations are trivial: parallel surfaces have the same caustics; the contact points of a normal with C_1 and C_2 are at distance R_1 and R_2, respectively, from the foot of the normal; whence 10.6.6.7, of which 10.3.3.1 is an infinitesimal version. All these results are part of the following important general philosophy for riemannian manifolds in general, and even for a fairly general class of metric spaces: the various curvatures of an object can be obtained by looking at its parallel objects (those at constant distance).

10.6.8.2. Examples.
(1) Moulding surfaces. The moving curve C (10.2.3.10) is a line of curvature and also a geodesic. Conversely, a surface having a family of geodesics which are also lines of curvature is necessarily a molding surface.
(2) Surfaces of translation. Not surprisingly, both caustics of a surface of translation are cylinders.
(3) The envelope of a one-parameter family of spheres (cf. 10.2.3.12). It is easy to see (figure 10.6.8.2) that, since each sphere is tangent to the surface along a circle, such a circle is a line of curvature, and the associated developable is a cone. Thus one caustic degenerates into a curve. Conversely, if one caustic is a curve, the surface is an envelope of spheres. Another equivalent condition is that the radius of curvature of each line of curvature is a constant. Recall the two important particular cases of envelopes of spheres: surfaces of revolution and boundaries of tubes.

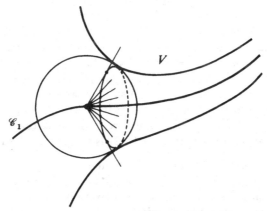

Figure 10.6.8.2.3

(4) Dupin cyclids come up again: by (3) and 10.2.3.12, a surface is a Dupin cyclid if and only both of its caustics are curves.

10.6.8.3. Additional properties. Each line of striction (or cuspidal edge, cf. 10.2.3.9) is a geodesics of the caustic it lies on. This is just because the normal to a caustic and the tangent to the associated line of curvature are parallel.

If V and W are surfaces that intersect along a line of curvature of V, say C, a necessary and sufficient condition for C to be a line of curvature of W is that the angle between V and W remain constant along C.

In particular, if three one-parameter families of curves form a triply orthogonal system, their pairwise intersections are lines of curvature for each of the surfaces. We have seen examples of triply orthogonal systems in 10.2.2.6 (cyclids) and in 10.2.2.3 (quadrics). The ellipsoid's lines of curvature are especially interesting: in addition to being connected with the geodesics, as mentioned in 10.4.9.5, they are "bifocal conics" with the umbilics as foci. This means that each line of curvature is the locus of points whose intrinsic distances to two non-antipodal umbilics add up to a constant. The difference of the distances is also constant, because, as we have mentioned, every geodesic starting from one umbilic reaches its antipodal, and this implies by 11.3.2 that the length of all these geodesics is the same. For a proof of all this, see [Sal74, p. 431 ff].

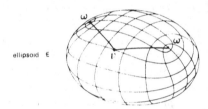

Figure 10.6.8.3

Curvature lines are preserved under inversion (8.7.4). This can be see as follows: inversions preserve angles [Ber87, 10.8.5], hence triply orthogonal systems; by the previous paragraph, this reduces the problem to showing that every surface can be included in a triply orthogonal system. This is guaranteed by the existence of parallel surfaces and figure 10.6.8.

Similarly, asymptotic lines are preserved under projective transformations [Ber87, chapter 4]. This comes from the fact that the definiting property of an asymptotic curve (10.6.4.2) is a contact of order three.

10.6.8.4. More advanced formulas. The reader will find in [Dar72, vol. III, p. 340–341], a table with just about every formula for the calculation of the various elements of the caustics. For example, here's a neat formula for the total curvature of the caustic C_1:

$$K_{C_1} = -(R_1 - R_2)^2 \frac{\partial R_1/\partial u}{\partial R_2/\partial u},$$

where u and v are the coordinates associated with the curvature lines.

As an application of this formula, we get the following necessary and sufficient condition for the existence of a universal relation between R_1 and

R_2 (cf. section 11.18 and example (2) in 10.6.6.6):

$$K_{C_1} K_{C_2} = (R_1 - R_2)^4.$$

The reader who is fond of delicate theorems should read [Dar72, vol. IV, p. 407]. This reference contains, among others, the following result: if all the lines of curvature in one family are plane curves and one of them is a circle, all of them are circles.

10.6.9. Significance of the mean curvature. For each point v on an oriented surface V, measure along the normal ν to V at v a distance $f(v)$. The points thus determined form a new surface, which we denote by V_f; for instance, if f is constant, V_f is parallel to V. Assume that V is relatively compact so the integrals will be finite. Then

10.6.9.1 $$\frac{d\bigl(\mathrm{area}(V_{tf})\bigr)}{dt}(0) = 2 \int_V f(v) H(v)\, dv,$$

where H is the mean curvature. This can be seen as follows: the element of area on V_{tf} is the square root ot the determinant of its first fundamental form. The derivative of a determinant involves the trace of the derivative of the form. Here this derivative equals II by 10.3.3.1, and $\mathrm{Tr}(\mathrm{II}) = 2H$ by definition.

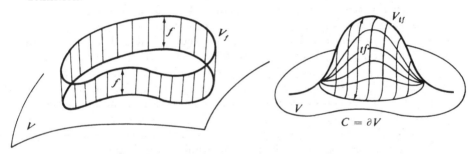

Figure 10.6.9

This again illustrates the philosophy introduced in 10.6.8.1, because for parallel surfaces the derivative of the area is measured by $\int_V H\, dv$.

Now fix a closed curve C in $\mathbf{R}^3$ and let's look for a surface V with boundary C and having the smallest possible area (that is, a *minimal surface*). A classical reasoning from the calculus of variations, applied to 10.6.9.1, shows the following result:

10.6.9.2. Theorem. *The mean curvature of a minimal surface is identically zero.* □

We have already encountered several examples of minimal surfaces. In graph notation the equation of a minimal surface is

$$(1 + q^2)r - 2pqs + (1 + p^2)t = 0.$$

10.6.9.3. A geometric interpretation of the condition $H = $ constant is that the coordinates associated with the lines of curvature of V are conformal. More simply, the Gauss map is conformal. This result will be fundamental in 11.16.6.

An interesting physical interpretation of the mean curvature is given by soap bubbles. Because of surface tension, the equilibrium position taken by a soap film is a surface whose mean curvature is the difference between the pressures on either side. In particular, a soap film with a given boundary will take the shape of a minimal surface with that boundary.

10.6.9.4. We will continue in 11.19.3 and 11.19.4 the study, initiated in 10.2.3.13, of surfaces whose total or mean curvature is prescribed as a function of the direction of the normal.

10.6.9.5. Remark. Here is an immediate consequence of 10.6.6.7: if V is a surface of constant total curvature, there exist at least two surfaces of constant mean curvature parallel to V. Conversely, a surface of constant mean curvature has at least one parallel surface of constant total curvature. For applications, see [Ser69].

10.6.9.6. Surfaces of revolution with constant mean curvature. By 10.6.6.6 we see that finding all surfaces of revolution whose mean curvature is a given function of the latitude amounts to finding all plane curves whose radius of curvature satisfies a certain condition.

The problem can be solved using calculus, but Delaunay has found a stunning geometric construction for all surfaces of revolution with constant mean curvature (see [Eel87] and [Ste]): Roll a conic along a straight line, without slippage. The curve described by a focus of the conic is the meridian of a surface of constant mean curvature (if the line is the axis of revolution).

Here are the special cases: an ellipse degenerated into a segment gives a sphere; a circle gives a cylinder; a parabola gives a catenoid.

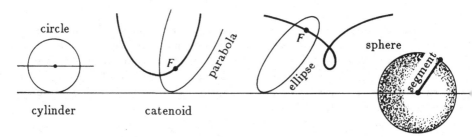

Figure 10.6.9.6

10.6.9.7. The mean curvature and solids of equal area. Here's a geometric interpretation for constant mean curvature: if $\Omega \subset \mathbf{R}^3$ is a compact submanifold-with-boundary bounded by some $\partial\Omega = S$, the volume of Ω will be largest, for a given area of S, exactly when S has constant mean curvature. This follows by applying the calculus of variations to formula 10.6.9.1. For more information, see section 11.20.

10.7. Links Between the two Fundamental Forms

We mentioned in 10.6.2.1 one relation between the two fundamental forms: the total curvature equals the determinant $k_1 k_2$ of the second fundamental form. Another relation is the *Codazzi–Mainardi equation* (or equations, if they are written out in coordinates). This equation involves the covariant derivative D, introduced in 10.4.7. If W is the Weingarten endomorphism (10.3.3) and ξ, η are arbitrary vector fields on V, it reads

10.7.1 $$D_\xi\big(W(\eta)\big) - D_\eta\big(W(\xi)\big) - W\big([\xi, \eta]\big) = 0.$$

To write the equation in coordinates one must resort to Christoffel symbols, the coefficients of D in coordinates (10.4.7). In any case we know that D is determined by the first fundamental form.

10.7.2. Special case: parametrization by lines of curvature. If (u, v) are the parameters associated with the lines of curvature, the expression of the Codazzi–Mainardi equation in coordinates is very simple:

$$\frac{\partial L}{\partial v} = H \frac{\partial E}{\partial v}, \qquad \frac{\partial N}{\partial u} = H \frac{\partial G}{\partial u}.$$

Here is a counterpart for theorem 8.6.13, on the existence and uniqueness of curves in $\mathbf{R}^3$ with curvature and torsion given as functions of arclength:

10.7.3. Theorem. *Let $U \subset \mathbf{R}^2$ be open. If $g = ds^2$ and II are differentiable quadratic forms on U, the first being positive definite, and if 10.6.2.1 and 10.7.1 are satisfied, there exists an immersed surface in $\mathbf{R}^3$, unique up to a rigid motion, having g and II for fundamental forms.*

Proof (outline). The idea is to work as in 8.6.13, looking for a moving frame that has nice properties like the Frenet frame. The only source of difficulties is that here, instead of an ordinary differential equation, we get a partial differential equation, with derivatives with respect to the coordinates u and v of U. Thus the problem is apparently more difficult. But the system is over-determined, and its integration is not difficult if the Frobenius compatibility conditions (3.5.15.4) are satisfied. And in our case these conditions amount exactly to 10.6.2.1 and 10.7.1! For details, see [Car76, p. 311]; [Spi79, vol. IV, p. 61]; [KN69, vol. II, p. 47] and [Sto69, p. 146]. □

We hadn't talked about the Codazzi–Mainardi equations before because they do not play a role in the statement of the results discussed so far, and, by and large, not even in their proofs. By contrast, they play a fundamental role in several global results, which are the object of the next chapter.

10.8. A Word about Hypersurfaces in $\mathbf{R}^{n+1}$

The reader may be surprised that, after having talked about curves in $\mathbf{R}^2$ and surfaces in $\mathbf{R}^3$, we stop. The reason for this is that the theory of surfaces in $\mathbf{R}^3$ is much richer than that of hypersurfaces in $\mathbf{R}^{n+1}$, for $n \geq 2$. Here's why:

We can still define the two fundamental forms, the first one being $g = ds^2$. By diagonalizing the second fundamental form with respect to the first we again define principal curvatures $k_1, \ldots, k_n$; naturally, there exist generalizations of the formulas above that relate the two fundamental forms. In particular, Gauss's theorem (10.6.2.1) can be generalized to say that, for every i and j, the product $k_i k_j$ is equal to the sectional curvature σ_{ij} of the plane spanned by the two principal directions associated with i and j, a quantity that only depends on the first fundamental form (see [Spi79, vol. IV, p. 61] or [KN69, vol. II, p. 23]).

This is such a strong relation that it implies that, in general, the second fundamental form is determined by the first. For, if $k_i k_j = k'_i k'_j$ for all pairs $i < j$, and if $k_i \neq 0$ for at least three values of i, it follows that $k'_i = k_i$ for every i or $k'_i = -k_i$ for every i. This change in sign corresponds to switching the surface's orientation.

Thus the common phenomenon of surfaces that are intrinsically isometric but are not equivalent under rigid motions of $\mathbf{R}^3$ has no counterpart here, except if the shape of the manifold in space is very degenerate: the rank of the second fundamental form must be everywhere less than three. The interested reader can consult [Spi79, vol. III, chapter 12].

For surfaces, on the other hand, problems involving the relationship between the two fundamental forms were the staple of many nineteenth-century geometers. We discuss a few of them in chapter 11; the reader can take a look at [Dar72], [Spi79, vol. III, chapter 12], and [Eis62].

The discussion above applies only to the local theory. The global geometry of hypersurfaces in $\mathbf{R}^{n+1}$ is rich in any dimension. For lack of space, we can only refer the reader to [Spi79, vol. III, chapter 12], [BZ86], [Lei80] and [Oli84].

CHAPTER 11

A Brief Guide to the Global Theory of Surfaces

Our aim is to understand complete, compact surfaces from the global point of view, posing problems and obtaining results parallel to the ones discussed in chapter 9. As we have seen in chapter 10, such problems fall naturally into two categories: the ones dealing with abstract riemannian manifolds (X, g) and the ones dealing with surfaces embedded or immersed in $\mathbf{R}^3$.

Thus the chapter is divided into two parts. In the first we study surfaces (X, g), perhaps defined by an embedding into $\mathbf{R}^3$, for their intrinsic properties. After discussing the problem of globally shortest paths, we tackle the following question: what can one say about the intrinsic geometry of (X, g) when given information about its natural invariant, the Gaussian curvature? The answer is given by a comparison theorem, which works in both directions: if $K \geq k$ (resp. $K \leq k$) the geometry of (X, g) is bounded above (resp. below) by the geometry of a simply connected manifold with constant curvature

k. Ultimately, it is the existence of this one-parameter family of universal objects that makes this result so satisfying.

The relationship between curvature and topology is discussed next, based on the Gauss–Bonnet formula. After that we extend to submanifolds-with-boundary of an arbitrary (X, g) the isoperimetric inequality proved in chapter 9 for the plane, with the consequent introduction of the isoperimetric profile of X. We also study isosystolic inequalities, which are in a sense isoperimetric inequalities for compact manifolds without boundary; the role of the boundary is played by the shortest closed geodesic of the surface.

By way of transition we study the problem, still almost entirely open, of realizing a surface (X, g) by an immersion or embedding into $\mathbf{R}^3$.

The second part starts with problems having to do with the Gaussian curvature K. Then come problems concerning the mean curvature H, which is also a natural invariant, since, together with K, it completely determines the second fundamental form. When H is identically zero we have the famous minimal surfaces; when H is a constant we also have interesting and strong results. We conclude with the isoperimetric inequality for submanifolds-with-boundary of $\mathbf{R}^3$ and their refinement, due to Minkowski.

In all these questions we have striven for completeness, in the sense of talking about all the important problems and results, and giving references when the subject is not discussed in the majority of classical books on surfaces.

We mention two important topics that are absent from our short guide: first, analysis on abstract riemannian surfaces, including the Laplace–Beltrami operator, the heat and wave equations, and vibrations. Two excelent references for this subject are [Ber85] and [Cha84]. Then, complex structures on surfaces: every oriented riemannian surface can be made into a Riemann surface, that is, a one-dimensional complex manifold. This result of paramount importance is the starting point for a huge field of studies, still very active; see [Wol85] and [Gun62], for instance.

PART I: INTRINSIC SURFACES

11.1. Shortest Paths

11.1.1. We saw in 10.4.8 that two sufficiently close points in (X, g) can be connected by a unique shortest path. Is this true globally? Evidently not in general; both uniqueness and existence fail. For instance, take the standard sphere and remove a point from it (or a bigger chunk, if you're hungry). Then two points on opposite sides of the hole and close enough to it will not be connected by a segment. On the other hand, antipodal points are connected by infinitely many segments.

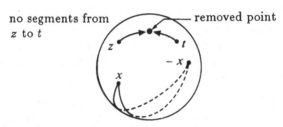

no segments from
z to t removed point

infinitely many shortest paths from x to $-x$

Figure 11.1.1.1

Now the punctured sphere is not a complete metric space, and certain geodesics cannot be extended to the whole real line. These two remarks are linked by the very important Hopf–Rinow theorem, which essentially says that the only thing that can prevent a geodesic from being extended indefinitely is a hole in the surface. To state this theorem, we must introduce some notation.

For any $x \in X$, define the *exponential map* at x, denoted by $\exp_x$, as the map that takes $\eta \in T_x X$ into the point $\gamma_{\eta/\|\eta\|}(\|\eta\|)$ reached by walking a distance $\|\eta\|$ along the geodesic $\gamma_{\eta/\|\eta\|}$ (10.4.8.1). In general, $\exp_x$ is not defined on the whole of $T_x X$; its maximal domain is a star-shaped subset of $T_x X$ whose frontier corresponds to points to which geodesics can no longer be extended (figure 11.1.1.2).

11.1.2. Theorem (Hopf–Rinow). *The following three conditions are equivalent for a riemannian manifold* (X, g):
(i) *for some x, the domain of $\exp_x X$ is the whole tangent space, that is, every geodesic emanating from x is defined on the whole real line;*
(ii) *for every x, the domain of $\exp_x X$ is the whole tangent space;*
(iii) (X, g) *is a complete metric space.*

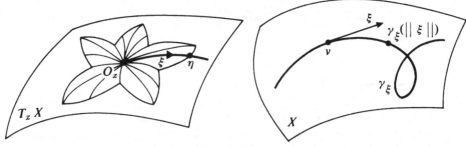

Figure 11.1.1.2

These conditions imply that any pair of points in X can be joined by at least one segment.   □

11.1.3. Notes. The last condition does not imply the other three, as you can see by taking an open disc in the Euclidean plane.

A surface that is a closed subset of $S \subset \mathbf{R}^3$ is obviously complete, but the converse is not true: take a cylinder over a spiral that wraps around a limit circle without ever touching it.

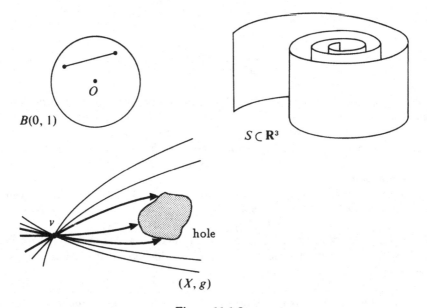

Figure 11.1.3

From now on all riemannian manifolds, including surfaces in $\mathbf{R}^3$, will be assumed complete unless we explicitly say otherwise.

11.1.4. The diameter. The *diameter* of a metric space (X, g), denoted by $\mathrm{diam}(g)$, is the supremum of the distance between any two points in X. The Hopf–Rinow theorem implies that $\mathrm{diam}(g)$ is finite if and only if X is compact (recall from 10.4.8.1 that the topology of X as a manifold X coincides the metric space topology arising from g).

11.2. Surfaces of Constant Curvature

We saw in 10.5.2 that spheres of radius R have constant Gaussian curvature equal to $1/R^2$; that the Euclidean plane has zero curvature; and that $\mathbf{R}^2$, with the metric given in polar coordinates by

$$ds^2 = dr^2 + \frac{\sinh^2 \sqrt{-k}r}{-k} d\theta^2,$$

where $k < 0$ is arbitrary, has constant curvature k. For each real number k, then, we have found one riemannian surface of constant curvature k, which we denote by $\mathbf{S}_k^2$. This notation is fully justified by the following result:

11.2.1. Theorem. *Every simply connected riemannian surface of constant curvature k is isometric to $\mathbf{S}_k^2$.*

Proof. Locally this comes from 10.5.3.3 (cf. 10.5.2); globally it is a consequence of simple connectedness (watch out for the case of spheres, which require two charts). □

11.2.2. There are two other useful models for $\mathbf{S}_k^2$ when $k < 0$. The first is the *Poincaré model*, given by the metric

$$ds^2 = \frac{4}{\left(1 + \frac{k}{4}(x^2 + y^2)\right)^2}(dx^2 + dy^2)$$

on the open disc $x^2 + y^2 < -\frac{4}{k}$. The other has appeared in disguise in exercise 6.10.6: If $k = -1$ it is defined by $ds^2 = y^{-2}(dx^2 + dy^2)$ on the half-plane $y > 0$, and is called the *upper-half-plane model*.

The unique simply connected riemannian surface of curvature -1 is called the *hyperbolic plane*.

11.2.3. Now let (X, g) be an arbitrary riemannian manifold (of any dimension) and $p : \tilde{X} \to X$ a covering space. Following 10.3.2, we can consider on $\tilde{X}$ the induced riemannian metric p^*g, which we denote by $\tilde{g}$. We call $(\tilde{X}, \tilde{g}) \to (X, g)$ a *riemannian covering space*.

11.2.4. It follows from 11.2.1 and the existence of universal covers that every surface (X, g) of constant curvature k is the quotient of $\mathbf{S}_k^2$ by a group of isometries, acting properly discontinuously without fixed points.

For $k > 0$, the classification of surfaces (4.2.25) implies that there are only two manifolds of constant curvature: S_k^2 itself and its quotient by the antipodal map, namely the real projective plane $P^2(\mathbf{R})$ (2.4.12.2). The projective plane inherits from S_1^2 a canonical geometry, called *elliptic*; in this geometry geodesics are all closed (cf. 11.10) and satisfy the incidence axioms for lines. For more information, see [Ber87, chapter 19], [Lel85] and [Gre80].

If $k = 0$, the possible quotients of properly discontinuous actions without fixed points are cylinders, Möbius strips, tori and Klein bottles (2.4.12). The first two are compact and the last two non-compact. Two tori are isometric if and only if the lattices they come from can be mapped into one another by a plane isometry; after normalization, we see there is a two-parameter family of flat tori. Parameters in the space of riemannian structures are traditionally known as *moduli*, so the moduli space for the torus is two-dimensional, and is in fact obtained from the shaded region shown in figure 11.2.4 by identifying boundary points that give the same riemannian structure. The bottom left corner of the shaded region corresponds to the square lattice $\mathbf{Z}^2$, and the bottom right corner to the hexagonal lattice. For more details, see [BGM71], [Cha84].

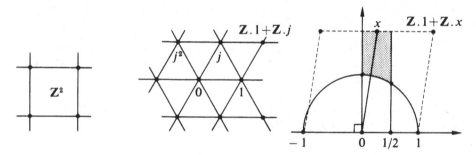

Figure 11.2.4.1

A Klein bottle is the quotient of a torus by an involution, and the existence of such an involution requires that the torus come from a rectangular lattice; thus, after normalization, we get only a one-parameter family of Klein bottles, the modulus being the width of the rectangle, for example. Cylinders and Möbius strips, too, have only one modulus.

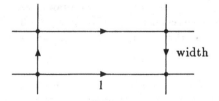

Figure 11.2.4.2

11.2.5. The case $k < 0$ is much more complex. First, any compact surface of genus $g \geq 2$ admits a metric of constant negative curvature: to see this, one can easily construct a regular polygon in the hyperbolic plane $\mathbf{S}_k^2$ whose vertex angles all equal $2\pi/4g$ [Ber87, 19.8.20], then glue its edges in the right pattern to obtain the surface. The complementary problem, that of studying the set of riemannian structures on such a surface, is much harder, and still an active research area; one important result is that the space of moduli has dimension $6g - 6$ if the surface is orientable and $3g - 3$ if not. For this huge subject, see [Wol85], [Thu79] and [BBL73] (non-orientable case).

The area of a compact surface of constant curvature $k \neq 0$, on the other hand, is very easily given by the Gauss–Bonnet formula (cf. 11.7.1); it equals $\text{area}(X) = 2\pi k^{-2}|\chi(X)|$.

The geometry of surfaces of constant curvature is extremely rich. In addition to the references mentioned above, the reader can consult [Wol72], [Spi79, vol. IV], and [Thu88].

11.3. The Two Variation Formulas

Consider a geodesic γ, parametrized by arclength (we assume this about all geodesics, unless we say otherwise), and a one-parameter family

11.3.1 $$C : [a, b] \times]-\eta, \eta[\to X$$

of curves $C_\alpha = C|_{[a,b] \times \{\alpha\}}$ such that $C_0 = \gamma$. We want to estimate the lengths $\text{leng}(C_\alpha) = l(\alpha)$. The first two derivatives of l are easily calculated:

11.3.2. Theorem (first variation formula). *If s_t denotes the transverse curve $\alpha \to C(t, \alpha)$, we have*

$$\frac{d(l(\alpha))}{d\alpha}(0) = \big(s_b'(0) \,|\, \gamma'(b)\big) - \big(s_a'(0) \,|\, \gamma'(a)\big). \qquad \square$$

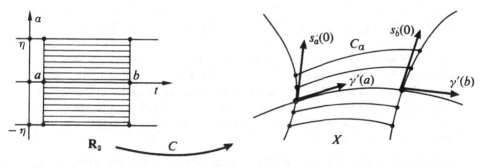

Figure 11.3.2.1

For example, this variation is zero if the endpoints are fixed (cf. 10.4.5). Another consequence is that if γ is a geodesic from m to n and δ is a geodesic from n to p, the strict triangle inequality $d(m, p) < d(m, n) + d(n, p)$ holds, unless $\gamma'(n) = \delta'(n)$.

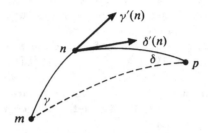

Figure 11.3.2.2

To express the second derivative, suppose, for simplicity, that all the transverse curves s_t are orthogonal to γ at 0. Orient X along γ. Then the transverse vector $s_t'(0)$ is of the form

$$s_t'(0) = y(t) \left(\gamma'(t) + \frac{\pi}{2} \right),$$

where y is a scalar function (cf. 10.4.8.3) and the expression in parentheses represents the vector γ' after a rotation of $\pi/2$.

11.3.3. Theorem (second variation formula). *If K is the total curvature and ρ is the geodesic curvature, we have*

$$\frac{d^2(l(\alpha))}{d\alpha^2}(0) = \left(\rho(b)y(b) - \rho(a)y(a) \right) + \int_a^b \left(y'^2 - K(\gamma(t))y^2 \right) dt. \qquad \Box$$

This formula, of which 11.5.5 is an integral version (via 11.5.3), is the most important tool in riemannian geometry at this time. We will use it many times in the sequel (cf. 11.6.2).

11.4. Shortest Paths and the Injectivity Radius

11.4.1. The cut locus. If ξ is a unit tangent vector and γ_ξ its geodesic, we know from 10.4.8.1 that $d(\gamma(0), \gamma(t)) = t$ for t small enough. This certainly can't be true if t gets too large: think of the diameter of a compact manifold! It makes sense to look at the (possibly infinite) infimum of the values of t for which $d(\gamma(0), \gamma(t)) < t$; this is called the *cut value* of ξ, and denoted by $\mathrm{cutval}(\xi)$.

11.4.1.1. Theorem. *If $t = \mathrm{cutval}(\xi)$, at least one of the two conditions must hold:*

(i)   *there exists another segment δ from $\gamma(0)$ to $\gamma(t)$, distinct from $\gamma_{[0,t]}$;*
      or

(ii) *the exponential map $\exp_{\gamma(0)} : T_{\gamma(0)}X \to X$ has rank < 2 at $t\xi$.*

In addition, if $\exp_{\gamma(0)}$ has rank < 2 at η, we have $\mathrm{cutval}\big(\eta/\|\eta\|\big) \leq \|\eta\|$. $\square$

11.4.1.2. Definition. *When $\exp_{\gamma(0)}$ has rank < 2 as in the theorem, we say that the points $m = \gamma(0)$ and $n = \gamma(t)$ are conjugate along the geodesic γ_ξ. The cut locus $\mathrm{cutloc}(m)$ of $m \in (X,g)$ is the set of points of the form $\exp_m\big(\mathrm{cutval}(\xi)\xi\big)$.*

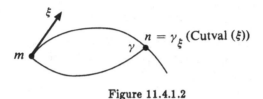

Figure 11.4.1.2

11.4.2. Examples of cut loci. We remarked in 10.4.9 that geodesics are hard to calculate explicitly; the cut locus can only be harder. The part coming from conjugate points is sometimes easier, and may help in the search.

In the sphere $\mathbf{S}_1^2$ the cut locus of any point is its antipodal point. In elliptic space, $\mathrm{cutloc}(m)$ is the projective line m^* dual to m.

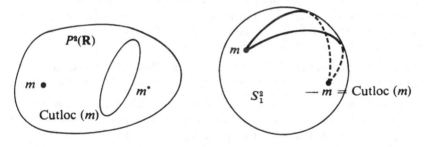

Figure 11.4.2.1

In a flat torus, the cut locus is the image under the exponential map of the boundary of the fundamental domain, or Voronoi diagram, of the underlying lattice [Ber87, 1.9.12].

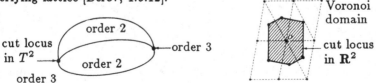

Figure 11.4.2.2

In a torus of revolution, the cut locus of a point m on the outer equator is formed by the inner equator, the opposite meridian and an arc of the outer equator, whose endpoints are conjugate to m.

numbers indicate how many segments join m to each point of its cut locus

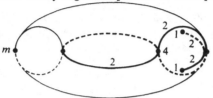

Figure 11.4.2.3

For an ellipsoid, the analysis in 10.4.9.5 allows one to show that the cut locus of an umbilic is the opposite umbilic, whereas the cut locus of any other point is an interval [Man81].

11.4.3. General results on the cut locus. The study of the cut locus was first undertaken by Poincaré, and other people to have obtained results include Myers, Warner, Gluck–Singer, Buchner and C. T. C. Wall. Presently we know the following (see [Buc78] and [Wal79]):

11.4.3.1. The cut locus of m is the closure of the set of points that can be connected to m by at least two distinct segments.

11.4.3.2. If (X, g) is real analytic, the cut locus of every point is a subanalytic set. In particular, in dimension two the cut locus is a graph. The number of segments joining a point to m is its order in the graph, and a point has order one when it's conjugate to m.

11.4.3.3. In general the cut locus can be wild. There exist riemannian metrics on S^2 for which the cut locus of almost all points is non-triangulable.

11.4.3.4. However, for *generic metrics* (we won't define this notion, but the idea is that almost all metrics are generic) the cut locus of every point is triangulable and has no points of order higher than three.

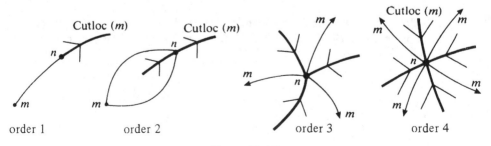

Figure 11.4.3

11.4.4. Injectivity radius. The *injectivity radius* of (X, g) at m is defined by

$$\text{inj}(m) = \text{inj}(X, m) = \inf \left\{ \text{cutval}(\xi) : \xi \in U_m X \right\},$$

where $U_m X$ is the unit tangent bundle at m. This number is always positive because the cut value is a continuous function on UM. The *injectivity radius* of (X, g) is then

$$\text{inj}(X, g) = \text{inj}(X) = \inf \left\{ \text{inj}(m) : m \in X \right\};$$

this number can be zero in general (figure 11.4.4) but is positive if X is compact.

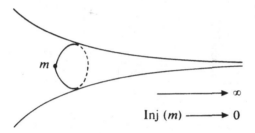

Figure 11.4.4

One reason why the injectivity radius is important is that every disc $B(m, r)$ with $r \le \text{inj}(m)$ is diffeomorphic to $\mathbf{R}^2$. Similarly, for every $m \in X$, the complement $X \setminus \text{cutloc}(m)$ is diffeomorphic to $\mathbf{R}^2$.

The injectivity radius also controls the area of (X, g). The *isoembolic inequality* says that

$$\text{area}(X, g) \ge \frac{4}{\pi} \text{inj}^2(g),$$

and equality only holds for $\mathbf{S}_k^2$, with $k > 0$. In fact, we can say even more [Cro84] if we know the area (or, in higher dimension, the volume) of the star-shaped set $E(x)$ determined on the tangent plane $T_x X$ at a point x by the cut value function $U_x X \to \,]0, \infty]$: there exists a universal constant $c(n)$, where n is the dimension of X, such that

$$\text{vol}^2(X, g) \ge c(n) \int_{x \in X} \text{vol}(E(x)) \, dx.$$

11.5. Manifolds with Curvature Bounded Below

We now consider surfaces (X, g) whose Gaussian curvature is $\geq k$, where k is a fixed real number. Here it is convenient to introduce a notion that was hiding behind geodesic coordinates (10.5.3.3):

11.5.1. Definition. *A Jacobi field on a geodesic γ of a surface is a scalar function f on γ such that $f'' + (K \circ \gamma)f = 0$, where the derivative is taken with respect to arclength.*

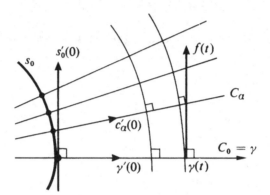

Figure 11.5.1

Every Jacobi field f on γ gives rise to a one-parameter family C of geodesics C_α, such that $C_0 = \gamma$ and f is the magnitude of the partial derivative of C with respect to α at C_0:

$$\frac{\partial C}{\partial \alpha}(t, 0) = f(t)\left(\gamma'(t) + \frac{\pi}{2}\right).$$

By 1.2.6 and 1.3.1 there exists a unique Jacobi field f on γ for any choice of $f(0)$ and $f'(0)$. These quantities represent the velocity and geodesic curvature ρ of the transverse curve s_0 of the family C at $t = 0$ (see 11.3 for notation). More precisely, we have

$$s_0'(0) = f(0)\left(\gamma'(0) + \frac{\pi}{2}\right) \qquad \text{and} \qquad \rho = \frac{f'(0)}{f(0)}.$$

11.5.2. Notation. In $\mathbf{S}_k^2$ the equation of a Jacobi field reduces to $f'' + kf = 0$. Denote by h_k the solution of that equation such that $h_k(0) = 0$ and $h_k'(0) = 1$, that is,

$$h_k(t) = \begin{cases} \frac{1}{\sqrt{k}}\sin\sqrt{k} & \text{if } k > 0, \\ t & \text{if } k = 0, \\ \frac{1}{\sqrt{-k}}\sinh\sqrt{-k} & \text{if } k < 0. \end{cases}$$

Also, denote by t_0 the first zero of h_k distinct from zero, namely $\pi/\sqrt{k}$ if $k > 0$ and $+\infty$ otherwise.

11.5.3. Fundamental lemma. *If (X, g) satisfies $K \geq k$, every Jacobi field such that $f(0) = 0$ and $f'(0) = 1$ satisfies $f(t) \leq h_k(t)$ until f has a zero. In particular, if $k > 0$, there always exists $t \in \,]0, \pi/\sqrt{k}]$ such that $f(t) = 0$.*   □

The conclusion may be false beyond t_0 (figure 11.5.3). This lemma follows from classical Sturm–Liouville theory, and has numerous consequences.

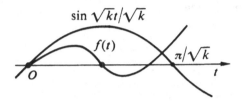

Figure 11.5.3

11.5.4. The case $k > 0$. For example, together with theorem 11.4.1.1 it shows that cutval$(\xi) \leq \pi/\sqrt{k}$ for every ξ. In particular, diam$(g) \leq \pi/\sqrt{k}$ for every (X, g) with $K \geq k$, and 11.1.4 implies that X is compact. (Notice that we're talking about the intrinsic diameter, even if the surface is embedded in $\mathbf{R}^3$.) By the Gauss–Bonnet formula below (11.7), this means that X can only be S^2 or $P^2(\mathbf{R})$. Finally, notice that $\pi/\sqrt{k}$ is exactly the diameter of $\mathbf{S}_k^2$.

The problem of surfaces (X, g) with $K > 0$ but not necessarily compact is more subtle. Cohn-Vossen [GM69] managed to show that all such surfaces are homeomorphic to $\mathbf{R}^2$.

11.5.5. Trigonometry. We denote by $\mathcal{T}^*$ a triangle on $\mathbf{S}_k^2$ having one angle α and adjacent sides b and c, and let $a^* = F(b, c, \alpha)$ be the length of the opposite side. We have [Ber87, chapters 18 and 19]:

$$a^2 = b^2 + c^2 - 2bc\cos\alpha \qquad \text{for } k = 0 \text{ (Euclidean plane)},$$
$$\cos a = \cos b \cos c - \sin b \sin c \cos\alpha \qquad \text{for } k = 1 \text{ (sphere)},$$
$$\cosh a = \cosh b \cosh c - \sinh b \sinh c \cos\alpha \quad \text{for } k = -1 \text{ (hyperbolic plane)}.$$

Now let $\mathcal{T}$ be a triangle on a surface (X, g) having angle α and sides b, c as above. If the opposite side, of length a, can be realized as a segment that does not intersect the cut locus of the vertex of angle α, we can integrate 11.5.3 to get the inequality $a \leq F_k(b, c, \alpha)$ if $K \geq k$. A. D. Alexandrov extended this and proved that $a \leq F_k(b, c, \alpha)$ for every triangle on such a manifold.

11.5.6. Areas. Integrating 11.5.3 immediately gives the inequality

$$\text{area}_X\big(B(m,r)\big) \leq \text{area}_{\mathbf{S}^2_k}\big(B(r)\big).$$

One can prove much more [Gro81, p. 65].

11.6. Manifolds with Curvature Bounded Above

The technique of Sturm–Liouville can be applied in the other direction:

11.6.1. Lemma. *If (X,g) satisfies $K \leq k$, every Jacobi field such that $f(0) = 0$ and $f'(0) = 1$ satisfies $f(t) \geq h_k(t)$ for $t \leq t_0$. In particular, f has no zeros before t_0, and no zeros at all if $k \leq 0$.*   □

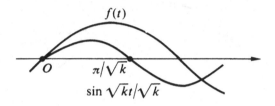

Figure 11.6.1

11.6.2. Consequences. For every disc of radius $r \leq t_0$ and $r \leq \text{inj}(m)$ we have

$$\text{area}_X\big(B(m,r)\big) \geq \text{area}_{\mathbf{S}^2_k}\big(B(r)\big).$$

Similarly, every sufficiently small triangle satisfies $a \geq F_k(b,c,\alpha)$. Alexandrov's result does not hold in general (figure 11.6.2), but it does hold if X is simply connected and $k \leq 0$; in this case we have $a \geq F_k(b,c,\alpha)$ for every triangle.

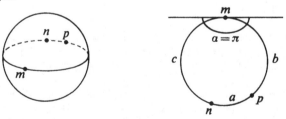

Figure 11.6.2

In particular, the celebrated *Hadamard's theorem* holds: in any manifold of non-positive curvature two arbitrary points are connected by a segment, which is unique if the manifold is simply connected (if not, there is a unique geodesic joining the two points for each homotopy class). This implies that a simply connected surface with $K \leq 0$ is diffeomorphic to $\mathbf{R}^2$.

11.6.3. Poles. A point in a riemannian surface is called a *pole* if its injectivity radius is infinite. A manifold having a pole is diffeomorphic to $\mathbf{R}^2$. The previous paragraph says that all points of a simply connected manifold of non-negative curvature are poles. But the vertex of an elliptic paraboloid is also a pole, although $K > 0$. The poles of quadrics in $\mathbf{R}^3$ have been studied in [Man81], which contains delicate results.

11.7. The Gauss–Bonnet and Hopf Formulas

Let's triangulate a compact riemannian surface (X, g) and apply formula 10.5.5.4 to each triangle. By definition, the Euler characteristic of X is $\chi(X) = V - E + F$, where V, E and F stand for the number of vertices, edges and faces, respectively. We immediately obtain the formula

11.7.1
$$\int_X K(m)dm = 2\pi\chi(X).$$

This is an all-important formula in the study of surfaces. It was extended to the non-compact case by Cohn-Vossen, after suitable modifications in the definitions [CG85]. For sharper results, having to do, for example, with the parts of X where K is positive and those where K is negative, see [BZ86].

11.7.2. Sticking to the compact case, we remark that 11.7.1 gives right away a matching of possible topological types with conditions on the curvature of a riemannian surface. If $K > 0$, we recover 11.5.4: X can only be S^2 or $P^2(\mathbf{R})$. If $K < 0$ we see that X must have at least one handle if it's oriented, two if not: this according to the list of surfaces in 4.2.25. If we allow $K \geq 0$ or $K \leq 0$, the possibilities remain the same, with the addition of manifolds of curvature identically zero; these, according to 11.2.4, are flat tori and flat Klein bottles. Finally, notice that if K is negative and constant, the last paragraph in 11.2.5 gives area$(X) = 4\pi K^{-2}(g-1)$, where g is the genus; thus the area increases with topological complexity, which is intuitive enough. But see also 11.9.

11.7.3. Application: minimal area of a surface. It is natural to study metrics g on X with *bounded geometry*, that is, those whose curvature satisfies $-1 \leq K \leq 1$. (We can also say that we control g by its acceleration.) The *minimal area* of a surface X, compact or not, is

$$\mathrm{minvol}(X) = \inf\{\mathrm{area}(X, g) : (X, g) \text{ is complete and } -1 \leq K_g \leq 1\}.$$

Clearly the minimal area of a compact surface is $2\pi|\chi(X)|$, by 11.7.1. If $\chi(X) \neq 0$ the minimal area is non-zero and is achieved by every metric of constant curvature ± 1, as the case may be; if $\chi(X) = 0$ (torus and Klein bottle) the infimum is not reached, but is the limit of flat metrics. We can say that the torus and the Klein bottle have lower-dimensional geometric limits, namely a circle or a point.

The non-compact case requires a much more delicate analysis. The answer can be found in [BP86]: the minimal area is zero, except for the plane $\mathbf{R}^2$, in which case it is $2\pi(1+\sqrt{2})$. This value is attained by only one manifold: a closed ball of radius $Arcsin\,2^{-3/2}$ in $\mathbf{S}_1^2$, sewn along its boundary to a piece of Beltrami surface whose boundary circle has matching length and geodesic curvature. This is a riemannian manifold of class C^1 only; the curvature jumps from $+1$ to -1 as we cross the seam.

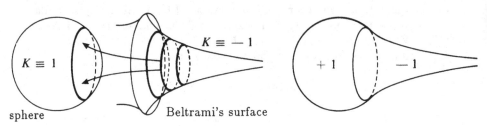

Figure 11.7.3

In higher dimension, the question of minimal volumes is far from being solved [Gro82].

11.7.4. Hopf's formula. We mentioned in 7.7.6.1 that

$$2\pi\chi(X) = \sum_i \mathrm{ind}_{x_i}\,\xi,$$

for every vector field ξ on a compact surface. This formula was not proved there; here it follows easily from 10.5.5.4 and its proof. We deduce a double formula:

11.7.4.1
$$\int_X K(m)\,dm = 2\pi\chi(X) = \sum_i \mathrm{ind}_{x_i}\,\xi.$$

An immediate corollary is that a surface immersed in $\mathbf{R}^3$ and homeomorphic to S^2 or $P^2(\mathbf{R})$ has at least one umbilic (10.6.4). Here's the reasoning for S^2: if it didn't, the principal directions would be well-defined and distinct at every point, and we could take a vector field on S^2 without singularities. The sum in 11.7.4.1 would be vacuous, and we'd have $\chi(X) = 0$, contradiction. For $P^2(\mathbf{R})$, take the oriented double cover.

In the case of the torus there may be no umbilics; take a torus or revolution, for example.

It is an old and open question [Tit73] whether a sphere embedded in $\mathbf{R}^3$ has at least two umbilical points.

11.8. The Isoperimetric Inequality on Surfaces

Is there on (X, g) a generalization of the classical isoperimetric inequality (section 9.3)? That is, given a submanifold-with-boundary Ω of X, with boundary C, is there an inequality tying leng(C) to area(Ω)?

11.8.1. The standard sphere. E. Schmidt found the complete answer for the round sphere: among all submanifolds-with-boundary of S_1^2 with a given area, closed balls (that is, spherical caps) are exactly the ones whose boundary has minimal length. His proof uses a method different from the one in 9.3; it consists in generalizing the idea of Steiner symmetrization [Ber87] to the sphere, which is possible because S^2 possesses symmetries (isometries) through every great circle.

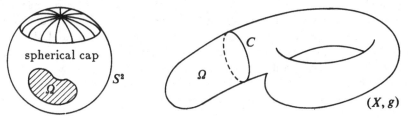

Figure 11.8.1

11.8.2. The general case. It is futile to hope for an inequality in full generality. For example, 11.8.1 clearly extends for every sphere S_k^2 ($k \geq 0$), but the constant involved depends on k. In addition, the dumbbells in figure 11.8.2 show that not only the curvature but also the diameter must appear in any isoperimetric inequality.

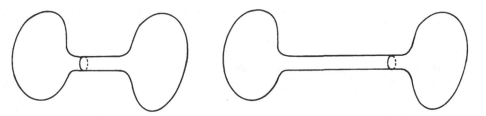

Figure 11.8.2

The best we can get is the following:

11.8.2.1. Theorem. *There exists a scalar function U of three variables such that, if (X, g) is an arbitrary compact riemannian surface with curvature $\geq k$ and Ω is a submanifold-with-boundary of X with boundary C, we have*

$$\text{leng}(C) \geq U\big(\text{area}(\Omega), \text{diam}(g), k\big). \qquad \square$$

This inequality is optimal. If $k > 0$ the limiting objects that optimize it are spheres, with spherical caps as submanifolds. If $k \leq 0$ the limiting objects are singular; they're obtained by gluing together two plane or hyperbolic discs.

The idea of the proof is to take a submanifold with a given area and the shortest possible boundary. This implies that the boundary is a "generalized circle," that is, a curve with constant geodesic curvature. Then one calculates the area of Ω based on this boundary, all of this controlled by a trivial generalization of 11.5.3. See [Ber85] and [BZ86].

11.8.3. The isoperimetric profile. To illustrate the difficulty of the isoperimetric problem when $k \leq 0$, we introduce the *isoperimetric profile* of a compact manifold (X, g), which is the function that associates to a real number $\alpha \in [0, \text{area}(X)]$ the infimum of the lengths of boundaries of submanifolds-with-boundary of (X, g) of area α. Figure 11.8.3 shows the isoperimetric profiles of the sphere, of the projective plane and of a flat torus. Here's what happens for a flat torus: if α is small, the best submanifolds-with-boundary are discs, but as α increases we get to a point where a strip is better (notice that the boundary of a strip consists of geodesics, hence of curves of constant geodesic curvature):

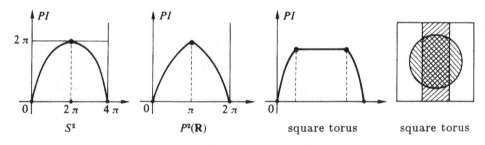

Figure 11.8.3

For more inequalities on the surfaces, see [Oss78].

11.9. Closed Geodesics and Isosystolic Inequalities

Take a surface in $\mathbf{R}^3$ with topology, and stretch an elastic band around some part of it. Letting go of the band you'll see that it shrinks to either a point or a closed geodesic on the surface; this follows from a general result:

11.9.0. Theorem. *Every free homotopy class of (X, g) contains a curve of minimal length, which is a closed geodesic.* □

By definition the *systole* of a non-simply connected riemannian manifold (X, g) is the infimum of the lengths of curves on X not homotopic to zero. This length, which we denote by $sys(X)$, is realized by a closed geodesic. We will only consider the compact case, otherwise there is no hope, as shown by figure 11.9.0.

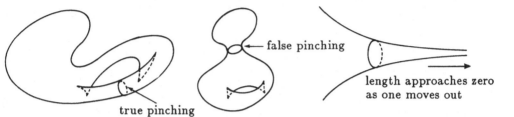

Figure 11.9.0

What the figure suggests is that, for a given value of $sys(g)$, the area of the manifold cannot be too small. In addition, the more holes there are, the larger the area should be; the area increases with topological complexity. Here's what we know at present [Gro83] about these *isosystolic inequalities* for surfaces (the higher-dimensional case is much harder):

11.9.1. Loewner's inequality. If X is the torus T^2, we have $area(g) \geq \frac{\sqrt{3}}{2} sys^2(g)$ for every g; equality obtains if and only if (X, g) is a flat torus coming from a hexagonal lattice (11.2.4).

11.9.2. Pu's inequality. If X is the projective plane $P^2(\mathbf{R})$, we have $area(g) \geq \frac{2}{\pi} sys^2(g)$ for every g, and equality holds if and only if g is the elliptic metric (11.2.4).

11.9.3. Gromov's inequality. For every surface X with γ holes and every metric g on X, we have $area(g) \geq c(\gamma) sys^2(g)$, where c is a positive function of γ that goes to infinity with γ.

11.9.4. Bavard's inequality. For the Klein bottle Bavard recently got [Bav86] the best possible inequality: $area(g) \geq \frac{2\sqrt{2}}{\pi} sys^2(g)$. He also proved that equality is only attained for a surface with singularities, obtain by gluing together two Möbius strips, each a quotient under the antipodal map of the spherical zone of S^2 contained between the parallels of latitude $\pi/4$ and $-\pi/4$.

11.9.5. Croke's inequality. Even if X is simply connected, that is, homeomorphic to S^2, we have an "intrinsic" isoperimetric inequality, not involving submanifolds-with-boundary. It says that

$$area(g) > L^2/961$$

for every g, where L is the length of the shortest non-trivial periodic geodesic of g. (Periodic geodesics exist on any surface, even if simply connected; see [Kli82].) The constant $\frac{1}{961}$ is not the best possible; it is conjectured that the best constant is $\frac{\sqrt{3}}{2}$, and that it is achieved by the singular surface consisting of two equilateral triangles glued along their boundaries in the obvious way. See [Cro].

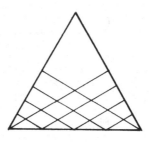

Figure 11.9.5

11.9.6. Note. For all these results the basic reference is [Gro83]. The case of dimension ≥ 3 is considerably harder.

11.10. Surfaces All of Whose Geodesics Are Closed

On the canonical sphere S^2 all geodesics are closed, and they all have the same (shortest) period 2π. In addition, S^2 is *antipodal*, in the sense that the cut locus of every point is again a point. A manifold (X, g) possessing the first property is said to be P1, and one possessing the second is said to be SP1.

There is no obvious reason why the fact that all geodesics are closed should imply that they have the same period; in fact, this is not the case in arbitrary dimension [Bes78, p. 185]. But Gromoll and Grove have recently shown [GG81] that for surfaces the first condition implies the second.

11.10.1. Possible surfaces. Besides S^2, we have seen only one example of a P1 metric: the elliptic metric on $P^2(\mathbf{R})$. Let's show that the sphere and the projective plane are the only surfaces that admit a P1 metric. By continuity, the homotopy class of all geodesics on X is a positive multiple of the same element of $\pi_1(X)$, which we call $[\gamma]$. By 11.9.0 this implies that every element of $\pi_1(X)$ is of the form $k[\gamma]$; but by looking at geodesics with same initial point and opposite directions we get $[\gamma] = -[\gamma]$. Thus $\pi_1(X)$ is either trivial or equal to $\mathbf{Z}_2$. Since X is compact (by the previous paragraph, for example), the only candidates from the list of surfaces (4.2.25) are the sphere and the projective plane.

Notice that SP1 structures on S^2 and P1 structures on $P^2(\mathbf{R})$ are in one-to-one correspondence given by the canonical double cover.

11.10.2. A double surprise. The first surprise is that there exist on S^2 many P1 riemannian metrics. R. Michel has shown that the P1 metrics of revolution are those of the form

$$ds^2 = (1 + h\cos r)dr^2 + \sin^2 r \, d\theta^2,$$

where h is any odd function from $]-1, 1[$ into itself.

But the space of P1 metrics on S^2 is much bigger than that: Guillemin showed that its tangent space at the point corresponding to the canonical metric is isomorphic to the set of odd functions on S^2, that is, functions such that $f(-x) = -f(x)$. Nothing else is known about this set, not even whether it is connected.

The second surprise is that the situation is radically different for $P^2(\mathbf{R})$: Green has shown that a P1 riemmanian structure on $P^2(\mathbf{R})$ must be the elliptic metric (with arbitrary curvature, of course).

The basic reference for this section is [Bes78].

11.11. Transition: Embedding and Immersion Problems

11.11.1. Topological problems. For completeness, we mention two non-metric results. The first is that the projective plane $P^2(\mathbf{R})$ cannot be embedded in $\mathbf{R}^3$, since it is non-orientable [Gre67]. But, as we have see in 10.2.4, it can be immersed in $\mathbf{R}^3$.

The second is a surprise: by a continuous sequence of immersions we can deform the standard embedding of S^2 into its opposite, turning the sphere inside out. This *eversion* has been described by Phillips in [Phi66] and, more simply and in more detail, by Morin and Petit in [MP78] and [MP80].

11.11.2. A metric problem. Let (X, g) be a riemannian surface, assumed compact for simplicity. Can (X, g) be isometrically embedded or immersed in $\mathbf{R}^3$? Certainly not in general; every immersed surface has at least one point where the curvature is strictly positive. To see this, enclose the surface in a large enough sphere, then shrink the sphere until it is tangent to the surface; at the tangency point K must be no less than the curvature of the sphere.

Apart from this, little is known about this very natural problem. The reason is our relative ignorance about an even more elementary question: Given a riemannian surface (X, g) and a point $m \in X$, is there an open neighborhood U of m in X such that $(U, g|_U)$ can be isometrically embedded in $\mathbf{R}^3$?

Presently we know that the answer is yes if $K(m) > 0$ or $K(m) < 0$ (this implies it is always yes if g is real analytic). Also, a positive answer has

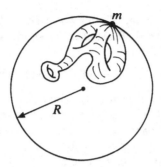

Figure 11.11.2

recently been given in the case $dK(m) \neq 0$, that is, when the derivative of K is non-zero at m.

See [GR70] and [Gro86] for more details on these problems.

11.11.3. The case of positive curvature. In one particular case the embedding problem has been completely solved:

11.11.3.1. Theorem. *If g is a riemannian metric on S^2 whose curvature is everywhere positive, the riemannian manifold (S^2, g) can be isometrically embedded in $\mathbf{R}^3$. Such an embedding is unique up to an isometry of $\mathbf{R}^3$.* $\square$

This result is due to the accumulated efforts of H. Weyl, Alexandrov, Nirenberg and Pogorelov. See 11.14.1 for the uniqueness part.

For the case of everywhere negative curvature, see section 11.15.

PART II: SURFACES IN $\mathbf{R}^3$

11.12. Surfaces of Zero Curvature

This is a typical example where global and local are in conflict, and global wins. We saw in 10.6.6.2 that every surface with $K = 0$ is locally developable, but globally such a surface doesn't even have to be ruled: we can make up wild surfaces by gluing pieces of cone or cylinder to a plane in C^∞ fashion (figure 11.12.0).

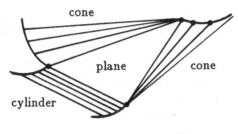

Figure 11.12.0

What's happening here is that the rank of the second fundamental form jumps between zero and one (it is never two when $K = 0$). But notice that this surface cannot be made complete: in fact, if a complete developable surface is not a cylinder it must develop somewhere a line of striction (10.2.3.7). This reasoning was made watertight by Hartman and Nirenberg:

11.12.1. Theorem. *The only complete surface $S \subset \mathbf{R}^3$ with zero curvature is the cylinder.* □

The proof is subtle; the fact that it was only carried out in 1959 illustrates well the difference in rigor between nineteenth-century geometers and those of our days. Notice that this result only came sixty years after Liebmann's (see the beginning of 11.14), which seems harder.

11.13. Surfaces of Non-Negative Curvature

What can one say about a compact surface X of curvature $K \geq 0$ and immersed in $\mathbf{R}^3$? By 11.7.1 X can only be the sphere, the projective plane, the torus or the Klein bottle. But in the last two cases the curvature must be zero everywhere, which is ruled out by 11.11.2. The case $K > 0$ was solved by Hadamard as early as 1898:

11.13.1. Theorem. *If X is compact, immersed in $\mathbf{R}^3$ and has positive curvature everywhere, then X is in fact embedded in $\mathbf{R}^3$ and is the boundary of a strictly convex set* (see [Ber87, chapter 11], if necessary).

Proof. This proof is too nice and simple to omit altogether. We do the orientable case, X homeomorphic to S^2. Since the curvature never vanishes, the Gauss map $\nu : X \to S^2$ (10.3.3) has maximal rank, namely two, everywhere. By 4.1.5 ν must be a covering map, hence a homeomorphism because S^2 is simply connected. This easily implies that the image of X in $\mathbf{R}^3$ lies all on one side of its tangent planes, which implies the conclusion by [Ber87, 11.5.5]. The non-orientable case is easy, and we leave it to the reader.   □

Notice that this argument works in any dimension, except one, since the circle is not simply connected. Thus there is no analog for 11.13.1 in the plane; compare our study of the global convexity of curves (9.6).

11.13.2. Generalization. One can try to extend Hadamard's theorem in two ways: by admitting non-compact surfaces, or surfaces with $K \geq 0$. Doing both at once doesn't work: think of a cylinder over a figure eight.

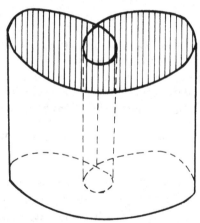

Figure 11.13.2

Thanks to the combined efforts of Chern–Lashof, Stoker and do Carmo–Lima, we have the best possible generalization of 11.13.1: the conclusion of 11.13.1 remains valid if X is compact and $K \geq 0$, or if X is arbitrary and $K \geq 0$ with $K(m) > 0$ for at least one point $m \in X$ [Car76, p. 387].

11.14. Uniqueness and Rigidity Results

We encountered in 10.2.3.6 and 10.5.3.9 local surfaces that are isometric but cannot be taken into one another by a Euclidean motion. In fact, there are numerous such local examples, since the system of partial differential equations that governs them is greatly underdetermined [Spi79, vol. III, chapter 12]. (Nevertheless, the general question of when a metric admits local deformations is still open.)

Now one can ask whether this kind of thing happens globally, particularly for compact surfaces. Historically, the first case to consider was the round sphere S^2: Liebmann was the first to show, in 1899, that every surface in $\mathbf{R}^3$ isometric to S^2 is a sphere of radius one. This is a particular case of a very beautiful result of Cohn-Vossen and Herglotz:

11.14.1. Theorem. *Two isometric surfaces of strictly positive curvature in $\mathbf{R}^3$ can be taken to one another by a Euclidean motion.*

Notice that this would be trivial in dimension higher than two (see 10.8).

Proof. Theorem 11.4.1 follows easily from Herglotz's formula (11.19.1.6), which, in turn, makes essential use of the Codazzi–Mainardi equation (10.7.1). □

The surfaces of revolution S and S' generated by the two meridians in figure 11.14.1 show that the condition $K > 0$ in 11.4.1 is essential. But no examples are known of isometric surfaces that are compact and real analytic [Spi79, vol. III, chapter 12].

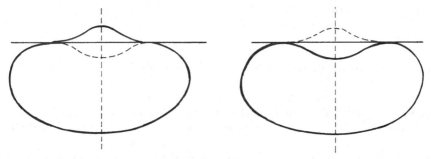

Figure 11.14.1

11.14.2. Deformations and rigidity. One can also ask whether a surface can be continuously deformed without changing the metric, as in 10.2.3.6. A weaker condition is flexibility: a surface is *flexible* if it can be continuously deformed in such a way that the derivative of the metric with respect to the parameter of the deformation vanishes at the original surface. Even for surfaces homeomorphic to S^2 (but non-convex, of course) the problem

has not been entirely solved. For all these questions and partial results, see
the very complete reference [Pog73] and [Spi79, vol. III].

However surprising this may seem, certain surfaces cannot be deformed
isometrically, even locally. A famous example, due to Efimov, is

$$(x, y, x^9 + \lambda x^7 y^2 + y^9),$$

where λ is transcendental. See [Jac82, p. 389], and also [Che85].

It is interesting to compare this problem with the analogous problem for
polyhedra in $\mathbf{R}^3$. A polyhedron is said to be *flexible* if it can be continuously deformed in such as way that only the dihedral angles between the
faces change. As early as 1813 Cauchy had demonstrated the analog of
11.14.1 for polyhedra: two isometric convex polyhedra can be taken to one
another by an isometry of ambient space. For non-convex polyhedra the
question remained open until 1978, when Connelly found a flexible polyhedron (see [Ber87, 12.8.4.2] and the references therein). But it is not known
whether flexible polyhedra necessarily have constant volume.

Another reason to study polyhedra is that there are numerous results
of Alexandrov and Pogorelov which use approximating polyhedra in the
study of convex surfaces (see [Ber87, section 12.9] and [Pog73]).

11.14.3. Punctured surfaces. Consider S^2 without its north and south
poles. Starting from the wedges in the middle of figure 10.5.3.10 Pogorelov
managed to construct an infinite number of immersions of this non-complete
manifold in $\mathbf{R}^3$, all isometric (that is, having curvature one). The same can
be done with the n-punctured sphere, where $n \geq 2$. But Green and Wu recently showed that these immersions can never be embeddings. Intuitively
one can see that the immersions must resemble the wedges in 10.5.3.10 but
covered several times. See [GW72].

11.15. Surfaces of Negative Curvature

A curious reader may have asked himself why we defined the hyperbolic
plane abstractly (11.2) instead of as a nice submanifold of $\mathbf{R}^3$. We have
seen surfaces of constant negative curvature, like Beltrami's surface and all
the others at the end of 10.5.3.10, but they only represent a portion of the
hyperbolic plane; they are not complete, and if you try to go too far you
hit a singularity. This is a pity, but it explains why hyperbolic geometry
had to wait until 1854 to be founded by Riemann, who introduces formula
11.2.2. It is also inevitable, due to the following result:

11.15.1. Theorem (Hilbert, 1901). *No embedded or immersed surface in*
$\mathbf{R}^3$ *can have constant negative curvature.*   □

This theorem was extended to the case of variable curvature by Efimov:

11.15.2. Theorem. *No immersed surface in* $\mathbf{R}^3$ *can have curvature* $K \leq k$, *where* $k < 0$ *is fixed.* □

The proof uses the completeness assumption to show that, starting from a point and going far enough, one necessarily hits a singularity, and that this happens in all directions, as exemplified by Beltrami's surface. For more details, see [Klo72]. □

11.16. Minimal Surfaces

We now get back to minimal surfaces (10.6.9.2), this time from the global point of view. One of the reasons to study local minimal surfaces is Plateau's problem: to find the surface of least area whose boundary is a given space curve. We will skip over this topic entirely; the interested reader can refer to the excellent and up-to-date survey [Mee81].

Here we just ask what are the surfaces in $\mathbf{R}^3$ whose mean curvature H is everywhere zero. Are there many of them? What is their geometry like? We will mention the more striking results about these problems; see [Mee81] for more.

First recall the examples of minimal surfaces we have already encountered: planes; Scherk's surface (10.2.1.3); the catenoid (example (3) in 10.6.6.6), the only minimal surface of revolution; the helicoid of pitch 2π (10.6.6.5), the only ruled minimal surface; Enneper's surface (10.6.6.3), which is an immersion only; and the Weierstrass-type surfaces in 10.2.3.6, which are fairly general.

11.16.1. Geometry. Every minimal surface has non-positive total curvature, because $H = \frac{1}{2}(k_1 + k_2) = 0$ implies $K = k_1 k_2 \leq 0$. Thus the results in 11.6 apply.

11.16.2. Compactness. There are no compact minimal surfaces: this follows from 11.16.1 and the reasoning in 11.11.2.

11.16.3. Theorem (Bernstein). *A minimal surface that is the graph of a function defined on the whole of* $\mathbf{R}^2$ *must be a plane.* □

This is in fact a result on the solutions of the partial differential equation

$$(1 + q^2)r - 2pqs + (1 + p^2)t = 0$$

from 10.6.9. Compare with Scherk's surface.

11.16.4. On the other hand, there exist *triply periodic* minimal surfaces, that is, surfaces invariant under the group $\mathbf{Z}^3$ of translations of $\mathbf{R}^3$. In fact there is a six-parameter family of them. See [Mee81, section 17], where Riemann surfaces, hyperelliptic curves and abelian integrals intervene.

11.16.5. Surfaces with given kinds of symmetry. Minimal surfaces can be topologically as complicated as desired. One simple example [JM83] consists in taking

$$f(Z) = \frac{1}{(Z^n - 1)^2} \qquad \text{and} \qquad g(Z) = Z^{n-1}$$

in the generalized Weierstrass formulas 11.16.6, where n is an arbitrary integer and Z ranges over $S^2 = \mathbf{C} \cup \{\infty\}$ minus the n-th roots of units. For $n = 2$ we get the catenoid (10.2.3.6, but the parametrization has changed) and for $n = 3$ the figure below:

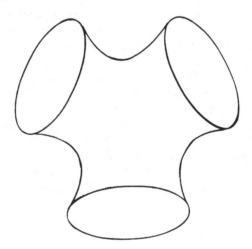

Figure 11.16.5

11.16.6. Generalized Weierstrass formulas. Formulas 10.2.3.6 can be generalized to generate all minimal surfaces. The starting point for the generalization is the idea of Riemann surfaces, that is, one-dimensional complex manifolds. Any orientable surface X can be given a complex manifold structure by covering it with conformal coordinate patches (10.4.2) and defining multiplication by i within each patch to be rotation by $\pi/2$, which is well-defined by metric and orientation together. If X is a minimal surface immersed in $\mathbf{R}^3$, the Gauss map $\nu : X \to S^2$ is conformal (10.6.9); with reference to the complex structure this says that ν is meromorphic, that is, complex analytic as a map into the Riemann sphere $S^2 = \mathbf{C} \cup \{\infty\}$ which its canonical complex structure (see [For81] for background).

One can work the other way around and start with an abstract Riemann surface X, a meromorphic function $\nu : X \rightarrow S^2$, and a meromorphic one-form $f\, dZ$ on X. Then the integrals

$$x(u,v) = \operatorname{Re} \int_{w_0}^{w} \left(1 - \nu^2(Z)\right) f(Z) dZ,$$

$$y(u,v) = \operatorname{Re} \int_{w_0}^{w} i\left(1 + \nu^2(Z)\right) f(Z) dZ,$$

$$z(u,v) = \operatorname{Re} \int_{w_0}^{w} 2\nu(Z) f(Z) dZ$$

from a fixed point w_0 to $w = (u,v) \in X$, define an immersion of X in $\mathbf{R}^3$ whose image is a minimal surface; and, conversely, every immersed minimal surface is obtained in this way.

This general formulation has numerous consequences. Here are a few:

11.16.7. Total curvature. For an immersed (or embedded) minimal surface X we consider the integral

$$\operatorname{totcurv}(X) = \int_X K(\nu) d\nu.$$

This integral is always defined (and possibly equal to $-\infty$) because K is non-positive (11.16.1); by 10.6.2.2 it equals the area of the image $\nu(X)$ of the Gauss map, taking multiplicities into account (cf. 7.4.3).

The following fundamental results are due to Osserman. The integral $\operatorname{totcurv}(X)$ is either infinite, or a positive integer multiple of -4π. The value -4π is only achieved for two surfaces: the catenoid and Enneper's surface. Moreover, $\nu(X)$ covers the whole of S^2 with the exception of at most six points (this is due to Xavier). If $\operatorname{totcurv}(X)$ is finite at most three points may be omitted. For more information, see [Mee81].

11.17. Surfaces of Constant Mean Curvature, or Soap Bubbles

We saw in 10.6.9 two motivations for the study of the surfaces of constant mean curvature. One was physical; the other was the proof of the isoperimetric inequality. Let's now consider problems of existence and uniqueness of global surfaces of constant mean curvature. We treat only the compact case; for non-compact examples see 10.6.9.6.

11.17.1. The sphere. The most obvious examples of surfaces of constant mean curvature are round spheres. Are there any others? Liebmann showed, as early as 1899 (cf. 11.14), that the answer is no if $K > 0$ everywhere. His proof is clinched by generously using the Codazzi–Mainardi equations (10.7) to show that all points are umbilics; this implies, even locally, that the surface is a round sphere (end of 10.6.4.2). This method was generalized by Hilbert and is a particular case of the one employed in the next section.

11.17.2. The theorems of Alexandrov and Hopf. What if one eliminates the condition $K > 0$ and allows all topological types? In 1955 Alexandrov showed tht every compact surface of constant mean curvature embedded in $\mathbb{R}^3$ is a round sphere. The proof is very difficult; it is a mixture of analysis and geometry, and concludes by showing that the surface has symmetries in all plane directions. See the excellent reference [Hop83] and [Spi79, vol. IV, chapter 9, addendum].

Soon after that, in 1956, H. Hopf demonstrated in [Hop83] that every immersion of S^2 with constant mean curvature is a round sphere. This proof is also very difficult; it uses the fact that S^2, as a Riemann surface, does not admit non-trivial quadratic differentials.

11.17.3. Wente immersions. What was still unknown, and remained so until 1984, was whether or not there existed immersions of compact surfaces with non-positive genus and constant mean curvature. Wente answered this question in the affirmative by constructing immersions of the torus T^2 with constant mean curvature. The general appearance of his immersions is shown in figure 10.2.4. For details, see [Wen85] or [Abr86].

11.17.4. Willmore's conjecture. We mention it here for its simplicity: consider, on a compact surface S, the integral $\int_S H^2(\nu)d\nu$. This integral is invariant not only under homotheties but also under inversions, that is, it is invariant under the conformal group of $\mathbb{R}^3$ [Ber87, chapters 10 and 18]. When does it have a minimum? If S is the sphere S^2 the answer is simple, by 11.7.4 and because $H^2 \geq K$: the minimum occurs if and only if S is a round sphere.

On the other hand, if S is the torus T^2, we are at sea. Willmore conjectured in 1965 that the integral is never less than $2\pi^2$ (the value it takes on a square torus). This conjecture remains open; a partial answer is given in [LY82]. Also, L. Simon recently showed that the minimum is indeed achieved for some compact C^∞ surface in $\mathbb{R}^3$.

11.17.5. Manifolds-with-boundary and Rellich's conjecture. By analogy with Plateau's problem (introduction to 11.16) we can ask what are the surfaces of given constant mean curvature having for boundary a fixed curve in space. We won't say much about this problem, but we mention that uniqueness is not to be expected here. For instance, if the curve is a circle in $\mathbb{R}^3$, the sphere of given curvature containing this circle is divided by the circle in two parts that are generally unequal; this gives two distinct solutions, a small and a large bubble. There have long been results on the existence of at least one solution, but Rellich's conjecture was that there existed at least two. This conjecture was recently solved by Brézis and Coron; see [BC84].

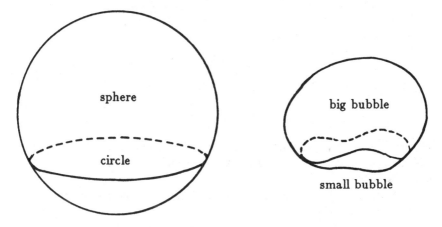

Figure 11.17.5

11.18. Weingarten Surfaces

We have see many examples of surfaces whose principal curvatures k_1 and k_2 seem to be intimately related: minimal surfaces ($k_1 = -k_2$), surfaces with constant mean curvature ($k_1 + k_2 = $ constant), surfaces all of whose points are umbilics ($k_1 = k_2$), ellipsoids of revolution (k_1^3/k_2; see example (2) in 10.6.6.6), boundaries of tubes ($k_1 = $ constant; see 2.7.6.2, 10.2.3.12 and 10.6.8.2.3). In all these cases k_1 and k_2 satisfy a universal relation.

11.18.1. Definition. *A Weingarten surface is a surface S for which there exists a C^∞ function U of two variables such that $U\big(k_1(v), k_2(v)\big) = 0$ for every $v \in S$.*

From 10.6.6.6 it follows that every surface of revolution is a Weingarten surface. But there are many Weingarten surfaces that are not of revolution, like boundaries of tubes, for instance. Using the preceding examples and a bit of cutting and pasting, one can construct C^∞ Weingarten surfaces with arbitrarily complicated topology. First, one can take an arbitrary knot and the associated tube. One can also form Weingarten surfaces of any genus if one knows how to attach a handle to a sphere preserving Weingartenness, so to speak. Figure 11.18.1 shows a way of doing this: glue a piece of round

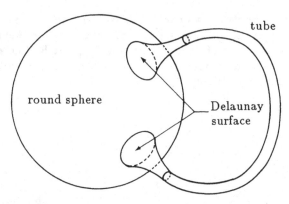

Figure 11.18.1

sphere, bounded by a circle, to a piece of Delaunay surface (10.6.9.6), then glue the latter to a tube, then glue the other end of the tube to the sphere in the same way.

Here's the only known global result on general Weingarten surfaces:

11.18.2. Theorem. *Let S be a Weingarten surface with positive curvature, and assume S is not the round sphere. Then no point of S can be at the same time a maximum for k_1 and a minimum for k_2.*   □

The proof make systematic use of the Codazzi–Mainardi equations, in the form 10.7.2.

11.18.2.1. Corollary. *A Weingarten surface with $K > 0$ and $k_1 = f(k_2)$, where f is a decreasing function, must be a round sphere.*   □

The results in 11.14 and 11.17.1 are particular cases of this.

11.18.3. If K is not assumed positive, we have virtually no global information: for instance, what are the real analytic Weingarten surfaces? As to local information, we saw a neat formula in 10.6.8.4.

One far out example: in [Dar72, vol. III, p. 322], one has reason to consider the family of surfaces that satisfy the equation

$$\frac{2}{k_1} - \frac{2}{k_2} = \sin\left(\frac{2}{k_1} + \frac{2}{k_2}\right)!$$

11.19. Envelopes of Families of Planes

We saw in 10.2.3.13 how to define a surface $S \subset \mathbf{R}^3$ as the envelope of a two-parameter family of planes. Here we work only with strictly convex surfaces, and we assume the origin is in the interior of S. The *support function p* of S assigns to $x \in S$ the distance from the origin to $T_x S$; we can consider it also as a function on the unit sphere S^2, via the Gauss map, which is a bijection because S is strictly convex. In 10.2.3.13 we stated formulas that give the contact point $v(\xi)$ of the tangent plane $P(\xi)$ to S as a function of $p : S^2 \to \mathbf{R}_+$. Similarly, the total curvature K and mean curvature H will be interchangeably considered as functions on S or on S^2.

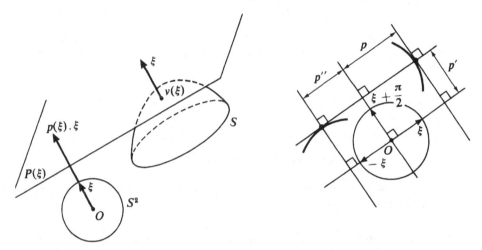

Figure 11.19

Euler's problem was to find a plane curve when its curvature as a function of the tangent is given (10.2.3.13). Here there are two natural problems, one for K (Minkowski) and one for H (Christoffel).

We start with some necessary formulas.

11.19.1. Formulary. The mean curvature is easy to compute. Extend $p : S^2 \to \mathbf{R}$ into a homogeneous function of degree one defined on the whole of $\mathbf{R}^3$: $p(t\xi) = tp(\xi)$. Then

11.19.1.1 $$2H(v(\xi)) = \left(\frac{\partial^2 p}{\partial x^2} + \frac{\partial^2 p}{\partial y^2} + \frac{\partial^2 p}{\partial z^2} \right)(\xi),$$

that is, H is proportional to the Laplacian Δp of p.

For $K(v(\xi))$ we have the following formula (where the indices denote derivatives):

$$\left(K(v(\xi)) \right)^{-1} = (p_{xx}p_{yy} - p_{xy}^2) + (p_{yy}p_{zz} - p_{yz}^2) + (p_{zz}p_{xx} - p_{zx}^2).$$

Using Stokes' formula and the Codazzi–Mainardi equations we obtain:

11.19.1.2 $\text{area}(S) = \displaystyle\int_{v \in S} p(v) H(v) dv$ (Minkowski),

11.19.1.3 $M = \displaystyle\int_S H(v) dv = \int_S p(v) K(v) dv$ (Minkowski),

11.19.1.4 $\displaystyle\int_{S^2} K\big(v(\xi)\big)^{-1} \xi \, d\xi = 0$

(where $d\xi$ is the canonical measure and both sides represent vectors), and

11.19.1.5 $\displaystyle\int_{S^2} H\big(v(\xi)\big) \xi \, d\xi = 0$

(ditto). Finally, we have the following difficult formula, due to Herglotz and involving two surfaces S and S' in $\mathbf{R}^3$, isometric under some map $\phi : S \to S'$:

11.19.1.6 $2 \displaystyle\int_S H(v) dv - 2 \int_s H'(\phi(v)) dv = \int_S k(S; S')(v) dv$,

where k is a function of the second fundamental forms of S and S', namely

$$k(S; S') = \det{}^{-1}(g)({}^t g)^{-1}\big(\phi^*(\mathrm{II}') - \mathrm{II}\big)g^{-1}$$

(here the first fundamental form can be expressed in any set of coordinates; the definition is invariant).

11.19.2. Applications. The reader can easily deduce by combining both of Minkowski's formulas that round spheres are the only ones that have K or H constant (cf. 11.14 and 11.17.1).

Herglotz's formula proves rigidity in 11.14.1; in fact, II and II', at corresponding points, have same determinant, since this determinant is the Gauss curvature and the surfaces are isometric. Since we're in dimension two, this implies an inequality on the traces via ϕ, that is, $k(S; S') \geq 0$. The rest is easy.

11.19.3. Minkowski's problem. The problem is to find a surface S such that the function $\xi \mapsto K(v(\xi))$ on S^2 is given (compare with 11.11.3). Uniqueness was shown by Minkowski, using generalizations of the formulas above.

Existence is a very difficult problem. Notice that 11.19.1.4 gives three necessary numerical conditions. Lewy and Nirenberg have shown that the problem is always solvable if these three conditions are satisfied; the proof is analytical. See [Spi79, vol. III, chapter 11] for references and proofs of the formulas above. See also the recent reference [Oli84].

11.19.4. Christoffel's problem. Here we're looking for S such that the function $\xi \mapsto H(v(\xi))$ on S^2 is given. The simplicity of 11.9.1.1 would make one think that the problem is simple, at least simpler than Minkowski's.

However, this is not the case. One can draw a parallel with the case of plane curves, where condition 9.5.2 on the curvature, which corresponds to our three conditions 11.9.1.5, is not sufficient to guarantee existence; for instance, the theorem of the four vertices 9.7.4 gives examples of curvature functions on S^1 that don't work. In higher dimension, the answer is even more complicated; after several partial solutions Firey managed, in 1968, to give necessary and sufficient conditions (in addition to 11.9.1.5) for the function H to come from a surface. These conditions are too complicated even to state here; we refer the reader to [Fir68].

Finally, we mention that the condition of strict positivity of the curvature (or the condition of strict convexity) is essential in both problems, as shown by the counterexample in figure 11.19.4:

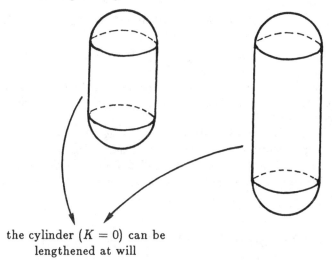

the cylinder $(K = 0)$ can be
lengthened at will

Figure 11.19.4

11.20. Isoperimetric Inequalities for Surfaces

Consider a compact surface S that is the boundary of a submanifold-with-boundary $\Omega \in \mathbf{R}^3$. Can one generalize the isoperimetric inequality 9.3 to this situation? The answer is yes, it can be done in any dimension. This is a classical result; see 6.6.9, [Ber87, section 12.11], or [BZ86].

11.20.1. Minkowski inequalities. Let V be the volume of Ω and A the area of S. We know that

$$A^3 \geq 36V^2,$$

and equality only holds for round spheres. Minkowski sharpened this inequality by introducing a third character, the integral of the mean curvature, which we called M in 11.19.1.3. He demonstrated the double inequality

$$\frac{A^4}{9V^2} \geq M^2 \geq 4\pi A.$$

As could be expected, each inequality only turns into an equality if S is a round sphere. For the proof, and generalizations in higher dimension, see [BZ86], [Wal78] and [Lei80].

11.20.2. Volume of tubular half-neighborhoods. Let S be a relatively compact surface in $\mathbf{R}^3$; we showed in 10.6.3 that the volume of the slab between two surfaces parallel to S and distant ε from it is

$$\mathrm{vol}(\mathrm{Tub}_\varepsilon S) = 2\varepsilon\,\mathrm{vol}(S) + \frac{2}{3}\varepsilon^3 \int_S K(v)\,dv.$$

And what is the volume between S and one of these parallel surfaces? It is equal to

$$\varepsilon\,\mathrm{vol}(S) + \varepsilon^2 M + \frac{1}{3}\varepsilon^3 \int_S K(v)\,dv,$$

which agrees with the formula above when we add the two volumes, since the sign of M depends on the side of S on which we're considering the parallel surface (cf. 10.3.3).

This formula is not hard to demonstrate using 10.6.6.7 and 10.6.9.1. It exists in all generality and the invariants generalizing the integrals of H and K are very important ones, and called *Lipschitz–Killing curvatures* [CMS84]. We've encountered the odd-indexed ones in 6.9.

11.21. A Pot-pourri of Characteristic Properties

We group here the results above concerning spheres and Dupin cyclids, and we add some new ones.

11.21.1. Theorem (local characterization of spheres). *Every local surface*
(i)   *all of whose points are umbilics;*
(ii)   *or, having at least one caustic reduced to a point;*
(iii) *or, invariant under a three-parameter local group of isometries of* $\mathbf{R}^3$
is necessarily a piece of round sphere.

Proof. Part (i) was discussed in 10.6.4.2. About part (ii) we just observe that the second caustic is also a point, coinciding with the first (the center of the sphere). Part (iii) is easy to check.   □

In a similar vein, it can be shown that pieces of cylinders of revolution are the only surfaces invariant under a two-parameter local group of isometries of $\mathbf{R}^3$. Invariant under one-parameter groups of isometries are all helicoidal surfaces, and in particular surfaces of revolution.

All this discussion has to do with extrinsic isometries. How about surfaces admitting intrinsic isometries, that is, maps from the surface into itself that preserve the first fundamental form $g = ds^2$? It is immediate that if a two-parameter group of such maps exists, the surface is locally homogeneous, and in particular its Gaussian curvature K is constant. We already know what such surfaces are, and in fact they admit a three-parameter local group of isometries. Thus we only have three cases: three, one and zero parameters, the latter being that of a generic surface. We now examine the case of one-parameter groups. By analyzing invariant curves and using the first variation formula (11.3.2) one can see that their orthogonal trajectories are necessarily geodesics, taken into one another by the group in question. Thus their expression in geodesic coordinates is

$$ds^2 = dr^2 + J^2(r)d\theta^2,$$

where J does not depend on θ. The technique of 10.5.3.10 allows to show immediately that such a metric can always be locally realized by a surface of revolution.

11.21.2. Theorem (global characterization of spheres). *Let $V \subset \mathbf{R}^3$ be a compact surface. If any one of the conditions below is satisfied, V is a round sphere:*
(i)   *all plane sections of V are circles;*
(ii)   *for all directions $\xi \in P^2(\mathbf{R})$ the cylinder parallel to ξ and circumscribed around V touches V along a circle (figure 11.21.2);*
(iii)   *all points in V have an antipodal point (combine 11.10 and 11.14);*
(iv)   *V has the least surface area among surfaces of equal volume;*
(v)   *V minimizes the integral of the mean curvature, among surfaces of equal area;*
(vi)   *V maximizes the integral of the mean curvature, among surfaces of equal area and volume;*
(vii)   *V is homeomorphic to S^2 and minimizes the integral of the square of the mean curvature (cf. 11.17.4);*
(viii) *the curvature of V is $\geq k$ everywhere and the intrinsic diameter of V is $\geq \pi/\sqrt{k}$ (cf. 11.5.4);*
(ix)   *V has the least extrinsic diameter among surfaces of equal volume;*
(x)   *the total curvature of V is a constant (more generally, see 11.18.2);*
(xi)   *V is embedded and its mean curvature is a constant.*

Proof. All of these results have been mentioned before, except for (i) and (viii), which we leave to reader; (ii), which we discuss below; and (ix), a result of Bieberbach that is proved (in one direction only) in [Ber87, 9.13.8]. For more details on all the results see [Bla56].   □

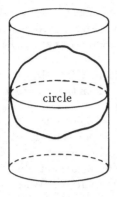

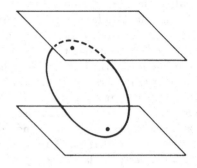

Figure 11.21.2   Figure 11.21.3.3

11.21.3. Global properties of spheres (non-characteristic)

11.21.3.1. A weaker version of 11.21.2(ii) is that all circumscribed cylinders touch a sphere along a plane curve; this property characterizes quadrics [Bla56].

11.21.3.2. All geodesics of a sphere are periodic. We saw in 11.10 that there are other surfaces with the same property.

11.21.3.3. The sphere has constant width. The *width* of V in a given direction is the distance between the two planes parallel to that direction that are tangent to V and as far apart as possible. Constant width means that one can move a sphere between two parallel planes in space with two degrees of freedom. This is actually a very weak condition; in the language of the support function p in 11.19, we're just saying that $p(\xi) + p(-\xi)$ is constant. For a recent discussion of this property, see [CG83].

11.21.3.4. The sphere also has constant *girth*, that is, the perimeter of the circumscribed cylinder is constant for every direction. Constant girth means that one can move the surface inside a paper cylinder (possibly deforming it without stretching) with two degrees of freedom. This is also a very weak condition, and turns out to be equivalent to constant width [Bla56].

11.21.3.5. The sphere also has a shadow of constant area. This, too, is a weak property. It is important to know that it is connected with important formulas of Cauchy about convex bodies; see [BZ86], [Lei80], [Die69, 16.24, problem 4], or yet [Ber87, 12.10.2]. A related result is that, for a surface whose shadow has constant area, the sum $K(m) + K(m')$ for points m, m' having parallel tangent planes is a constant. See [Bla56, p. 153].

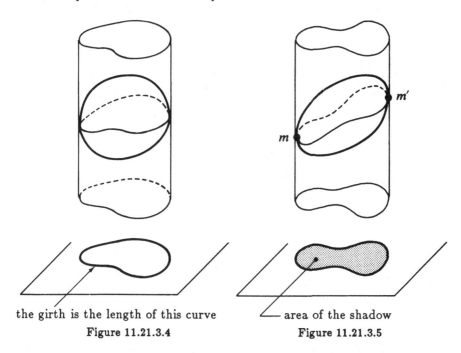

the girth is the length of this curve   area of the shadow
                Figure 11.21.3.4   Figure 11.21.3.5

11.21.4. Theorem (characterization of Dupin cyclids).

(i) *If V is a Dupin cyclid, its expression in pentaspheric coordinates [Ber87, 20.7] is that of a quadric (this property is shared by other surfaces; see 10.2.2.6).*

In addition, any of the conditions below is equivalent to V being a Dupin cyclid:

(ii) *V is obtained by inversion from a cylinder of revolution, or a cone of revolution, or a torus of revolution.*

(iii) *Both caustics of V degenerate into curves (this is a local property).*

(iv) *If V is homeomorphic to a torus, any round sphere in $\mathbf{R}^3$ cuts V into at most two connected components.*

Proof. Part (iv) is a very delicate result of Banchoff. For its proof, and the generalization of results such as (iii) to Dupin cyclids in higher dimension, see [Bla29], [Pin85] and references therein, [Kui84], [Heb81] and [Ban70].

□

Bibliography

[Abr86] U. Abresch. Constant mean curvature tori in terms of elliptic functions. 1986. Preprint. Cited in 10.2.4, 11.17.3.

[Ali84] J. Alias. *La voie férrée*. Eyrolles, Paris, 1984. Cited in 8.4.14.0.

[Ape86] François Apéry. La surface de Boy. *Adv. Math.*, 61:185–266, 1986. Cited in 10.2.4.

[AR67] Ralph Abraham and Joel Robbin. *Transversal Mappings and Flows*. Benjamin, New York, 1967. Cited in 9.2.9.

[Ban70] Thomas Banchoff. The spherical two-piece property and tight surfaces in spheres. *J. Diff. Geom.*, 4:193–205, 1970. Cited in 11.21.4.

[Bav86] C. Bavard. Inégalité isosystolique pour la bouteille de Klein. *Math. Annalen*, 274:439–441, 1986. Cited in 11.9.4.

[BBL73] L. Bérard-Bergery, J.-P. Bourguignon, and J. Lafontaine. Déformations localement triviales des métriques riemanniennes. In *Differential Geometry*, pages 3–32, AMS, Providence, R.I., 1973. Cited in 11.2.5.

[BC84] H. Brézis and J.-M. Coron. Multiple solutions of *H*-systems and Rellich's conjecture. *Comm. Pure and Appl. Math.*, 37:149–187, 1984. Cited in 11.17.5.

[Ben86] D. Bennequin. Caustique mystique. In *Séminaire Bourbaki 84/85*, pages 19–56, Société Mathématique de France, Paris, 1986. Cited in 10.6.

[Ber85] P. Bérard. *Lectures on Spectral Geometry. Lecture Notes in Mathematics, 1207*, Springer-Verlag, New York, 1986, or IMPA, Rio de Janeiro, 1985. Cited in 11, 11.8.2.1.

[Ber87] Marcel Berger. *Geometry.* Springer-Verlag, Berlin, 1987. Cited in 0.5.3.1–2, 2, 2.5.12.4, 6.6.9.2, 8.7.4, 8.7.16, 8.7.23, 9.9.3, 10.2.2.6, 10.2.3.3, 10.2.3.8,

10.2.3.13, 10.4.1.1, 10.4.2, 10.5.5.5, 10.6.2, 10.6.2.2, 10.6.8.3, 11.2.4–5, 11.4.2, 11.5.5, 11.8.1, 11.13.1, 11.14.2, 11.17.4, 11.20, 11.21.2, 11.21.3.5, 11.21.4.

[Bes78] Arthur Besse. *Manifolds All of Whose Geodesics are Closed. Ergebnisse der Mathematik, 93*, Springer-Verlag, Berlin, 1978. Cited in 11.10, 11.10.2.

[Bes86] Arthur Besse. *Einstein Manifolds. Modern Surveys in Mathematics*, Springer-Verlag, New York, 1986. Cited in 3.5.15.5.

[BF58] M. Barner and F. Flohr. Die Vierscheitelsatz und seine Verallgemeinerungen. *Der Mathematikunterricht*, 4:43–73, 1958. Cited in 9.7.5.

[BGM71] M. Berger, P. Gauduchon, and E. Mazet. *Le spectre d'une variété riemannienne. Lecture Notes in Mathematics, 194*, Springer-Verlag, New York, 1971. Cited in 11.2.4.

[Bla29] Wilhelm Blaschke. *Vorlesungen über Differentialgeometrie, III: Differentialgeometrie der Kreise und Kugeln. Grundlehren der Mathematischen Wissenschaften, 29*, Springer-Verlag, Berlin, 1929. Cited in 11.21.4.

[Bla56] Wilhelm Blaschke. *Kreis und Kugel*. de Gruyter, Berlin, second edition, 1956. Cited in 11.21.2, 11.21.3.1, 11.21.3.4–5.

[BM70] Leonard M. Blumenthal and Karl Menger. *Studies in Geometry*. W. H. Freeman, San Francisco, 1970. Cited in 8.0.2.1.

[Bou74] Nicolas Bourbaki. *Algebra* (chapters I–III). *Elements of Mathematics, 2*, Addison-Wesley, Reading, Mass., 1974. Cited in 0.1.15.1, 0.1.15.7.

[BP86] C. Bavard and P. Pansu. Sur le volume minimal de $\mathbf{R}^2$. *Ann. Sci. Ec. Norm. Sup. (ser. 4)*, 19, 1986. Cited in 11.7.3.

[Buc78] M. Buchner. The structure of the cut-locus in dim $\leq$ 6. *Compositio Math.*, 37:103–119, 1978. Cited in 11.4.3.

[BW75] Max Born and Emil Wolf. *Principles of Optics*. Pergamon, New York, fifth edition, 1975. Cited in 10.6.

[BZ86] Yu. D. Burago and V. A. Zalgaller. *Geometric Inequalities. Grundlehren der Mathematischen Wissenschaften, 285*, Springer-Verlag, Berlin, 1986. Cited in 10.8, 11.7.1, 11.8.2.1, 11.20, 11.20.1, 11.21.3.5.

[Car37a] C. Carathéodory. *Geometrische Optik. Ergebnisse der Mathematik, 4, pt.5*, Springer-Verlag, Berlin, 1937. Cited in 10.6.

[Car37b] Elie Cartan. *Leçons sur la théorie des espaces à connexion projective*. Gauthier-Villars, Paris, 1937. Cited in 10.4.9.6.

[Car63] Henri Cartan. *Elementary Theory of Analytic Functions of One or More Complex Variables*. Hermann, Paris, 1963. Cited in 9.1.8.4.

[Car70] Henri Cartan. *Differential Forms*. Hermann, Paris, 1970. Cited in 0.3.7.1, 0.3.13, 5.6.1.

[Car71] Henri Cartan. *Differential Calculus*. Hermann, Paris, 1971. Cited in 0, 0.0.8–10, 0.0.12–13, 0.2.8.1, 0.2.8.3–4, 0.2.8.7, 0.2.9.2, 0.2.12, 0.2.13.1, 0.2.17, 0.2.22, 0.2.26, 1.6.0, 1.6.2–3, 1.6.10, 6.8.11–12.

[Car76] Manfredo P. do Carmo. *Differential Geometry of Curves and Surfaces.* Prentice-Hall, Englewood Cliffs, N.J., 1976. Cited in 10, fig. 10.2.1.3.3, fig. 10.2.2.5, fig. 10.2.2.7.2, fig. 10.2.2.9, fig. 10.2.3.5, 10.4.9.5, 10.4.9.6, 10.7.3, 11.13.2.

[Car84] Elie Cartan. Les problèmes d'équivalence. In *Oeuvres complètes*, pages 1311–1334, CNRS, Paris, 1984. Cited in 10.5.4.

[CG83] G. D. Chakerian and H. Groemer. Convex bodies of constant width. In P. M. Gruber and J. M. Wills, editors, *Convexity and Its Applications*, Birkhäuser, Basel, 1983. Cited in 11.21.3.3.

[CG85] Jeff Cheeger and Mikhail Gromov. On the characteristic numbers of complete manifolds of bounded curvature and finite volume. In *Differential Geometry and Complex Analysis*, pages 115–154, Springer-Verlag, Berlin, 1985. Cited in 11.7.1.

[Cha53] J. Chazy. *Dynamique des systèmes matériels.* Gauthier-Villars, Paris, 1953. Cited in 10.6.

[Cha84] I. Chavel. *Eigenvalues in Riemannian Geometry.* Academic Press, Orlando, Fla., 1984. Cited in 11, 11.2.4.

[Che85] S.-S. Chern. Deformation of surfaces preserving principal curvatures. In *Differential Geometry and Complex Analysis*, pages 155–164, Springer-Verlag, Berlin, 1985. Cited in 11.14.2.

[Cho68] Y. Choquet-Bruhat. *Géométrie différentielle et systèmes extérieurs.* Dunod, Paris, 1968. Cited in 3.5.15.1.

[CMS84] J. Cheeger, W. Müller, and R. Schrader. On the curvature of piecewise flat manifolds. *Comm. Math. Phys*, 92:405–454, 1984. Cited in 6.9.8, 10.5.5.7, 11.20.2.

[Coo68] Julian L. Coolidge. *A History of the Conic Sections and Quadric Surfaces.* Dover, New York, 1968. Cited in 10.2.3.14.

[Cro84] Christopher Croke. Curvature free volume estimates. *Inventiones Math.*, 76:515–521, 1984. Cited in 11.4.4.

[Cro] Christopher Croke. Area and the length of the shortest closed geodesic. Preprint. Cited in 11.9.5.

[Dar17] Gaston Darboux. *Principes de Géométrie Analytique.* Gauthier-Villars, Paris, 1917. Cited in 10.2.2.6.

[Dar72] Gaston Darboux. *Leçons sur la Théorie Générale des Surfaces.* Chelsea, Bronx, N.Y., 1972. Cited in 10, 10.2.1.3–4, 10.2.2.7, 10.2.3.1, 10.2.3.6, 10.2.3.10, 10.4.2, 10.4.9.5–6, 10.5.5.7, 10.6.6.8, 10.6.7.1, 10.6.8.4, 10.8, 11.18.3.

[Die69] Jean Dieudonné. *Treatise on Analysis. Pure and Applied Mathematics, 10*, Academic Press, New York, 1969–. Cited in the preface, 1.6.1–2, 3.5.15.1, 4.3, 6.7.20, 9.2.3.1, 11.21.3.5.

[Dix67] Jacques Dixmier. *Cours de Mathématiques du premier cycle, première année.* Gauthier-Villars, Paris, 1967. Cited in 0.2.6, 0.2.12.

[Dix68] Jacques Dixmier. *Cours de Mathématiques du premier cycle, deuxième année*. Gauthier-Villars, Paris, 1968. Cited in 0, 0.1.1, 0.1.3, 0.1.5, 0.1.15.1, 0.2.8.1, 0.2.17, 0.2.24, 0.2.26, 2.1.6.4, 4.2.6, 6.5.3, 7.5.2, 8.3.6.

[Dom79] P. Dombrowski. *Fifty Years after Gauss's "Disquisitiones Generales Circa Superficies Curvas"*. Astérisque, *62*, Société Mathématique de France, Paris, 1979. Cited in 10.5.5.7.

[Don83] S. Donaldson. An application of gauge theory to four-dimensional topology. *J. of Diff. Geometry*, 18:279–316, 1983. Cited in 4.2.26.

[Eel87] James Eells. The surfaces of Delaunay. *Math. Intelligencer*, 9:53–57, 1987. Cited in 10.6.9.6.

[Eis49] L. P. Eisenhart. *Riemannian Geometry*. Princeton University Press, Princeton, 1949. Cited in 10.4.9.6.

[Eis62] L. P. Eisenhart. *Transformations of Surfaces*. Chelsea, New York, 1962. Cited in 10.8.

[Fed69] Herbert Federer. *Geometric Measure Theory. Grundlehren der Mathematischen Wissenschaften, 153*, Springer-Verlag, Berlin, 1969. Cited in 9.3.4.1.

[Fel68] E. A. Feldman. Deformations of closed space curves. *J. of Diff. Geometry*, 2:67–75, 1968. Cited in 9.4.17.2.

[Fir68] W. Firey. Christoffel's problem for general convex bodies. *Mathematika*, 15:7–21, 1968. Cited in 11.19.4.

[For81] Otto Forster. *Lectures on Riemann Surfaces. Graduate Texts in Mathematics, 81*, Springer-Verlag, New York, 1981. Cited in 11.16.6.

[GG81] D. Gromoll and K. Grove. On metrics on S^2 all of whose geodesics are closed. *Inventiones Math.*, 65:175–177, 1981. Cited in 11.10.

[Glu71] Herman Gluck. The converse to the four vertex theorem. *L'Enseignement Mathématique (sér. 2)*, 17:295–309, 1971. Cited in 9.7.10.4.

[GM69] D. Gromoll and W. Meyer. On complete open manifolds of positive curvature. *Annals of Math.*, 90:75–90, 1969. Cited in 11.5.4.

[GR70] M. Gromov and V. Rokhlin. Embeddings and immersions in Riemannian geometry. *Russian Math. Surveys*, 25(5):1–57, 1970. Cited in 11.11.2.

[Gra71] André Gramain. *Topologie des Surfaces*. Presses Universitaires de France, Paris, 1971. Cited in 4.2.25.

[Gre67] Marvin J. Greenberg. *Lectures in Algebraic Topology*. W. A. Benjamin, Reading, Mass., 1967. Cited in 4.2.24.2, 5.4.14, 6.10.2.2, 7.6, 7.6.2, 7.6.9, 9.3.4.2, 11.11.1.

[Gre80] Marvin J. Greenberg. *Euclidean and Non-Euclidean Goemetry, Development and History*. W. H. Freeman, San Francisco, 1980. Cited in 11.2.4.

[Gro81] Mikhail Gromov. *Structures métriques pour les variétés riemanniennes.* CEDIC–Nathan, Paris, 1981. Cited in 11.5.6.

[Gro82] Mikhail Gromov. Volume and bounded cohomology. *Publications Mathématiques de l'IHES*, 56:5–100, 1982. Cited in 11.7.3.

[Gro83] Mikhail Gromov. Filling riemannian manifolds. *J. of Diff. Geometry*, 18:1–147, 1983. Cited in 11.9.0, 11.9.6.

[Gro86] Mikhail Gromov. *Partial Differential Relations. Ergebnisse der Mathematik, 3. Folge, 9*, Springer, Berlin, 1986. Cited in 11.11.2.

[Gui69] A. Guichardet. *Calcul intégral.* Armand Colin, Paris, 1969. Cited in 0, 0.4, 0.4.3.1–2, 0.4.4, 0.4.4.1, 0.4.4.3, 0.4.5–6, 0.4.8.0, 3.3.11.1, 3.3.11.5, 3.3.16, 6.1.4.9.

[Gun62] Robert Gunning. *Lectures on Riemann Surfaces. Mathematical Notes, 2*, Princeton University Press, Princeton, 1962. Cited in 11.

[GW72] R. Greene and H. Wu. On the rigidity of punctured ovaloids, II. *J. of Diff. Geometry*, 6:459–472, 1972. Cited in 11.14.3.

[Hal77] Weaver Halpern. Inverting a cylinder through isometric immersions and isometric embeddings. *Trans. Amer. Math. Soc.*, 230:41–70, 1977. Cited in 10.2.3.8.

[Hal88] G.-H. Halphen. *Traité des Fonctions Elliptiques et de leurs applications.* Gauthier-Villars, Paris, 1888. Cited in 10.5.5.7.

[HC52] D. Hilbert and S. Cohn-Vossen. *Geometry and the Imagination.* Chelsea, New York, 1952. Cited in fig. 10.2.2.3.2, fig. 10.2.2.6, fig. 10.2.3.14.

[Heb81] J. Hebda. Manifolds admitting taut hypersurfaces. *Pacific J. Math.*, 97:119–124, 1981. Cited in 11.21.4.

[Hic65] Noel Hicks. *Notes on Differential Geometry.* Van Nostrand, Princeton, 1965. Cited in the preface, 10.

[Hil] J. N. Hilliard. *La Prestidigitation au XX⁰ Siècle: Tours Divers.* Payot. Cited in 7.4.14, 10.2.3.8.

[Hop83] H. Hopf. *Differential Geometry in the Large. Lecture Notes in Mathematics, 1000*, Springer-Verlag, Berlin, 1983. Cited in 10, 11.17.2.

[Hu69] Sze-Tsen Hu. *Differentiable Manifolds.* Holt, Rinehart and Winston, New York, 1969. Cited in the preface, 5.4.10.

[Jac82] H. Jacobowitz. Local isometric embeddings. In S.-T. Yau, editor, *Seminar on Differential Geometry*, pages 381–394, Princeton University Press, Princeton, 1982. Cited in 11.14.2.

[JM83] L. P. de Melo Jorge and W. H. Meeks. The topology of complete minimal surfaces of finite total gaussian curvature. *Topology*, 22(2):203–221, 1983. Cited in 11.16.5.

[Kli78] Wilhelm Klingenberg. *A Course in Differential Geometry. Graduate Texts in Mathematics, 51*, Springer-Verlag, New York, 1978. Cited in 10, fig. 10.2.3.9.

[Kli82] Wilhelm Klingenberg. *Riemannian Geometry*. de Gruyter, Berlin, 1982. Cited in 10.4.9.5, 11.9.5.

[Klo72] Tilla Klotz-Milnor. Efimov's theorem about complete immersed surfaces of negative curvature. *Advances in Math.*, 8:474–543, 1972. Cited in 11.15.2.

[KN69] S. Kobayashi and K. Nomizu. *Foundations of Differential Geometry.* Tracts in Mathematics, 15, Interscience, New York, 1963–1969. Cited in the preface, 3.5.15.5, 7.5.2, 7.5.7, 10, 10.7.3, 10.8.

[Kno80] H. Knörrer. Geodesics on the ellipsoid. *Inventiones Math.*, 59:119–143, 1980. Cited in 10.4.9.5.

[Kow80] O. Kowalski. Additive volume invariants of Riemannian manifolds. *Acta Mathematica*, 145:205–225, 1980. Cited in 6.9.8.

[Kui70] Nicholas Kuiper. Minimal total absolute curvature for immersions. *Inventiones Math.*, 10:209–238, 1970. Cited in 9.6.5.1, 9.6.5.2.

[Kui84] Nicholas Kuiper. Geometry in total absolute curvature theory. In *Perspectives in Mathematics*, Birkhäuser, 1984. Cited in 11.21.4.

[LA74] J. Lelong-Ferrand and J.-M. Arnaudiès. *Géometrie et Cinématique. Cours de mathématiques, vol. 3*, Dunod, Paris, 1974. Cited in 8.7.17.5.

[Lan68] Serge Lang. *Analysis I*. Addison-Wesley, Reading, Mass., 1968. Cited in 1.6.1.

[Lan69] Serge Lang. *Analysis II*. Addison-Wesley, Reading, Mass., 1969. Cited in 1.2.7, 1.6.2, 2.2.2, 6.2.2.

[Lei80] K. Leichtweiss. *Konveze Mengen*. Springer-Verlag, Berlin, 1980. Cited in 10.8, 11.20.1, 11.21.3.5.

[Lel63] Jacqueline Lelong-Ferrand. *Géométrie Différentielle*. Masson, Paris, 1963. Cited in 10, 10.4.7. Cited in 9.7.10.1.

[Lel82] Jacqueline Lelong-Ferrand. Les géodésiques des structures conformes. *Comptes Rendus Acad. Sci. Paris*, 294:629–632, 1982. Cited in 10.4.9.6.

[Lel85] Jacqueline Lelong-Ferrand. *Les fondements de la géometrie*. Presses Universitaires de France, Paris, 1985. Cited in 11.2.4.

[Lem67] J. Lemaire. *Hypocycloïdes et Epicycloïdes*. Blanchard, Paris, 1967. Cited in 8.7.17.5.

[LS82] D. Lehman and C. Sacré. *Géométrie et topologie des surfaces*. Presses Universitaires de France, Paris, 1982. Cited in 10.

[LY82] P. Li and S.-T. Yau. A new conformal invariant and its applications to the Willmore conjecture and the first eigenvalue of compact surfaces. *Inventiones Math.*, 69:269–291, 1982. Cited in 11.17.4.

[Mac49] E. Mach. *The Principles of Physical Optics*. Dover, New York, 1949. Cited in 10.6.

[Man81] H. von Mangoldt. Über diejenigen Punkte auf positiv gekrümten Flächen, welche die Eigenschaft haben, daß die von ihnen ausgehenden geodätischen Linien nie aufhören, kürzeste Linien zu sein. *Crelle's J.*, 91:23–52, 1881. Cited in 11.4.2, 11.6.3.

[Mas77] William S. Massey. *Algebraic Topology: an Introduction. Graduate Texts in Mathematics, 56*, Springer-Verlag, New York, 1977. Cited in 4.2.25–26.

[Mee81] William Meeks III. A survey of the geometric results in the classical theory of minimal surfaces. *Bol. Soc. Bras. Mat.*, 12:29–86, 1981. Cited in 11.16, 11.16.4, 11.16.7.

[Mil63] John Milnor. *Morse Theory. Annals of Mathematical Studies, 51*, Princeton University Press, Princeton, 1963. Cited in the preface, 4.2.24, 4.2.24.1, 4.2.24.4.

[Mil69] John Milnor. *Topology from the Differentiable Viewpoint.* University Press of Virginia, Charlottesville, 1969. Cited in the preface, 7.7.6.1.

[MP78] Bernard Morin and Jean-Pierre Petit. Le retournement de la sphère. *Comptes Rendus Acad. Sci. Paris*, 287:767–770, 791–794, 879–882, 1978. Cited in 11.11.1.

[MP80] Bernard Morin and Jean-Pierre Petit. Le retournement de la sphère. In *Les Progrès des Mathématiques*, pages 32–45, Pour la Science/Belin, Paris, 1980. Cited in 11.11.1.

[Nit75] Johannes C. C. Nitsche. *Volesungen über Minimalflächen. Grundlehren der Mathematischen Wissenschaften, 199*, Springer-Verlag, Berlin, 1975. Cited in 10.2.3.6.

[Oli84] V. Oliker. Hypersurfaces in R^{n+1} with prescribed curvature and related equations of Monge–Ampère type. *Communic. in Partial Diff. Equations*, 9:807–838, 1984. Cited in 10.8, 11.19.3.

[ONe66] B. O'Neill. *Elementary Differential Geometry.* Academic Press, New York, 1966.

[Oss69] Robert Osserman. *A Survey of Minimal Surfaces. Mathematical Studies, 25*, Van Nostrand–Reinhold, New York, 1969. Cited in 10.2.3.6.

[Oss78] Robert Osserman. The isoperimetric inequality. *Bull. Amer. Math. Soc.*, 84:1182–1238, 1978. Cited in 11.8.3.

[Oss85] Robert Osserman. The four-or-more vertex theorem. *Amer. Math. Monthly*, 92:332–337, 1985. Cited in 9.9.7.

[Oza84] T. Ozawa. On Halpern's conjecture for closed plane curves. *Proc. Amer. Math. Soc.*, 92:554–560, 1984. Cited in 9.8.1.1.

[Oza85] T. Ozawa. The numbers of triple tangencies of smooth plane curves. *Topology*, 24:1–13, 1985. Cited in 9.8.1.1.

[Pal57] Richard Palais. On the differentiability of isometries. *Proc. Amer. Math. Soc.*, 8:805–807, 1957. Cited in 10.4.4.

[Phi66] Anthony Phillips. Turning a surface inside out. *Scientific American*, 214:112–120, May 1966. Cited in 11.11.1.

[Pin85] U. Pinkall. Dupin hypersurfaces. *Math. Annalen*, 270:427–440, 85. Cited in fig. 10.2.4.1–3, 11.21.4.

[Pog73] A. V. Pogorelov. *Extrinsic Geometry of Convex Surfaces*. American Math. Society, Providence, R.I., 1973. Cited in 11.14.2.

[Poh68] William F. Pohl. The self-linking number of a closed space curve. *J. of Math. and Mechanics*, 17:975–986, 1968. Cited in 9.6.5.4.

[RdC22] Eugène Rouché and Charles de Comberousse. *Traité de Géométrie*. Gauthier-Villars, Paris, 1922. Cited in fig. 10.2.1.2, fig. 10.2.2.3.1, 10.2.2.7, fig. 10.2.2.7.2.

[Rud74] Walter Rudin. *Real and Complex Analysis*. McGraw-Hill, New York, second edition, 1974. Cited in 9.3.2, 10.2.3.6.

[Sal74] G. Salmon. *A Treatise on the Analytic Geometry of Three Dimensions*. Hodges, Dublin, third edition, 1874. Cited in 10.2.2.7, 10.2.3.14, 10.4.9.5, 10.6.8.3.

[Ser69] J. Serrin. On surfaces of constant curvature that span a given curve. *Math. Zeitung*, 112:77–88, 1969. Cited in 10.6.9.5.

[SG82] J. Sotomayor and C. Gutierrez. Structurally stable configurations of lines of principal curvature. *Astérisque*, 98–99:195–215, 1982. Cited in 10.6.8.1. Cited in the preface.

[Spi65] Michael Spivak. *Calculus on Manifolds*. Benjamin, New York, 1965. Cited in the preface.

[Spi79] Michael Spivak. *A Comprehensive Introduction to Differential Geometry*. Publish or Perish, Berkeley, Ca., second edition, 1979. Cited in the preface, 2.2.10.6, 3.5.15.1, 6.2.2, 8.6.17, 10, fig. 10.2.2.10, fig. 10.2.3.6, 10.4.2, 10.4.9.6, 10.5.5.4, 10.6.6.2, 10.6.6.4, 10.7.3, 10.8, 11.2.5, 11.14, 11.14.1–2, 11.17.2, 11.19.3.

[ST67] I. M. Singer and John A. Thorpe. *Lecture Notes on Elementary Topology and Geometry*. Scott, Foresman, Glenville, Ill., 1967. Cited in 5.4.10, 10.

[Ste64] Shlomo Sternberg. *Lectures on Differential Geometry*. Prentice-Hall, Englewood Cliffs, N.J., 1964. Cited in the preface, 3.1.5, 3.5.15.1, 3.5.15.5, 10, 10.5.5.4.

[Ste??] Ian Sterling. A generalization of a theorem of Delaunay to rotational w-hypersurfaces of σ_e-type in H^{n+1} and S^{n+1}. Preprint. Cited in 10.6.9.6.

[Sto69] J. Stoker. *Differential Geometry*. Wiley, New York, 1969. Cited in 10.6.2.2, 10.7.3.

[Str61] D. Struik. *Lectures on Classical Differential Geometry*. Addison-Wesley, Reading, Mass., 1961. Cited in 10, 10.4.7, 10.5.3.2.

[Syn37] J. L. Synge. *Geometrical Optics*. Cambridge University Press, London, 1937. Cited in 10.6.

[Tho79] John A. Thorpe. *Elementary Topics in Differential Geometry*. Springer-Verlag, New York, 1979. Cited in 10.

[Thu79] William P. Thurston. *Travaux de Thurston sur les Surfaces: Séminaire Orsay. Astérisque, 66–67*, Société Mathématique de France, Paris, 1979. Cited in 11.2.5.

[Thu82] William P. Thurston. Three-dimensional manifolds, Kleinian groups and hyperbolic geometry. *Bull. Amer. Math. Soc. (new series)*, 6:357–381, 1982. Cited in 4.2.26.

[Thu88] William P. Thurston. *The Geometry and Topology of Three-manifolds*. Princeton University Press, Princeton, 1988. Cited in 4.2.26, 11.2.5.

[Tit73] C. Titus. A proof of a conjecture of Loewner and of the conjecture of Carathéodory on umbilic points. *Acta Mathematica*, 131:43–47, 1973. Cited in 11.7.4.1.

[Val84] Georges Valiron. *The Geometric Theory of Ordinary Differential Equations and Algebraic Functions. Lie Groups: History, Frontiers and Applications, XIV*, Math Sci Press, Brookline, Mass., 1984. Cited in 10, 10.2.3.10.

[Ven79] P. Venzi. Geodätische Abbildungen riemannscher Mannigfaltigkeiten. *Tensor*, 33:313–321, 1979. Cited in 10.4.9.6.

[Wal78] R. Walter. *Differentialgeometrie*. BI.-Wissenschaftsverlag, Mannheim, 1978. Cited in 10, 11.20.1.

[Wal79] C. T. C. Wall. Geometric properties of generic differentiable manifolds. In *Geometry and Topology*, Springer-Verlag, Berlin, 1979. Cited in 11.4.3.

[War71] Frank Warner. *Foundations of Differentiable Manifolds and Lie Groups*. Scott, Foresman, Greenville, Ill., 1971. Cited in the preface, 3.5.15.1, 3.5.15.5, 5.4.10.

[Wen85] H. Wente. A counter-example in 3-space to a conjecture of H. Hopf. In *Arbeitstagung Bonn 1984*, pages 421–428, Springer-Verlag, Berlin, 1985. Cited in 11.17.3.

[Wey39] Hermann Weyl. On the volume of tubes. *Amer. J. Math*, 61:461–472, 1939. Cited in 6.9.8.

[Wol72] Joseph A. Wolf. *Spaces of Constant Curvature*. Berkeley, J. A. Wolf, 1972. Cited in 11.2.5.

[Wol85] S. Wolpert. The topology and geometry of the moduli space of Riemann surfaces. In *Arbeitstagung Bonn 1984*, pages 431–451, Springer-Verlag, Berlin, 1985. Cited in 11, 11.2.5.

[Won72] Yung-Chow Wong. On an explicit characterization of spherical curves. *Proc. Amer. Math. Soc.*, 34:239–242, 1972. Cited in 8.7.12.

[Wun62] W. Wunderlich. Über eine abwickelbare Möbiusband. *Monatshefte Math.*, 66:276–289, 1962. Cited in 10.2.3.8.

[Zwi63] C. Zwikker. *The Advanced Geometry of Plane Curves and their Applications.* Dover, New York, 1963. Cited in 8.7.17.5.

Index of Symbols and Notations

Non-alphabetical symbols are listed first, grouped according to type.
Greek letters are entered as if spelled out.

E^* 0.1

f^* 0.1.8, 3.6.3

$f^*\delta$ 0.1.29.3, 0.3.11.1, 3.3.2

$f^*\omega$ 5.2.4

$f_*\xi$ 7.7.2

$*\alpha$ 0.5.1

$f'(x)$ 0.2.1

f'_{E_i}, f'_{x_i} 0.2.8.5

$f'(a)$ 0.2.8.8

$f''(x)$ 0.2.11

$\alpha'(t)$ 2.5.17.3

$\tilde{X}$ 5.3.25

$\tilde{g}$ 11.2.3

$\int_a^b \alpha_t\, dt$ 0.3.15.4, 5.2.10.4

$\int_X f\mu$ 0.4.1, 0.4.7.1

$\int_X f\delta, \int_X f\, d\delta$ 3.3.14

$\int_D \alpha$ 6.1.4.10

$\int_D f\omega_0$ 6.5.12

$\int_C K$ 9.6.1

$\partial f/\partial\lambda$ 0.4.8.0

$\partial f/\partial x_i$ 0.2.8.5, 5.3.13

∂D 5.3.34

$(\cdot\,|\,\cdot)$ 0.1.15.1

(x, y, z) 0.1.16

$\|\cdot\|$ 0.1.15.1

$|\omega|$ 6.1

$[\xi, \eta]$ 2.8.17.2

$\{f, g\}, \{F, G\}$ 7.4.8

$-X$ 6.1.4.3

$x \times y$ 0.1.17

X/G 2.4.8

$(U, \phi, u) \sim (V, \psi, v)$ 2.5.8

$\mu \otimes \nu$ 0.4.5

$\delta \otimes \epsilon$ 3.3.18.1

$\otimes^r TX, \otimes^r T^*X$ 5.1.6

$f_1 \wedge \cdots \wedge f_r$ 0.1.1

$\alpha \wedge \beta$ 0.1.4

$\flat, \sharp$ 0.1.15.1

$\boxed{\phi}$ 5.2.2.2

$\circledm$ 8.1.7

1_t 2.5.17.2

α_x 1.3.1.1

$B(a,r), \overline{B}(a,r), B_d(0,1)$ 0.0.3
$B(E)$ 3.6.3
$b_k(X)$ 4.2.24.2
$B^r(X)$ 5.4.2
Bilsym(E) 4.2.6

$C^0(X)$ 0.0.6
$C^0(X;Y)$ 0.0.5
$C^1(U), C^1(U;F)$ 0.2.5
$C^p(U), C^p(U;F)$ 0.2.14
$C^p(X;Y)$ 2.3.1
$C^\infty(U;F)$ 0.2.14
$C^\infty(X)$ 2.8.17
$C_\mu^{\rm int}(X), C_\mu^{\rm int}(X;E)$ 0.4.7
$C_\delta^{\rm int}(X)$ 3.3.14
can 2.7.4
$\chi(X)$ 4.2.24.2
cont$(\xi)\alpha$ 5.3.12
cutloc(m) 11.4.1.2
cutval(ξ) 11.4.1

$D_i f$ 0.2.8.5
$\mathcal{D}(f)$ 1.3.3, 1.5.1
$df(x)$ 2.5.23
$df(\xi)$ 2.8.17.1
$\mathcal{D}(\xi)$ 3.5.6
df 5.2.6
$D_F\alpha$ 5.5.3
$D_{f'}Z$ 10.4.7
$d(v,w)$ 10.4.3
$D_\xi\eta$ 10.4.7
dm 10.5.5.2
deg α 0.1.6
deg(f) 7.3.1
$\Delta_p(U)$ 0.3.11.1
$\Delta(X)$ 3.3.1
$\Delta_q(X)$ 3.3.8
Dens(E) 0.1.28
Dens(X) 3.3.1
der$(C^\infty(X))$ 2.8.17
Diff$^+(I;J)$, Diff$^-(I;J)$ 8.1.10
Diff$^p(X)$, Diff$^p(X;Y)$ 2.3.5
Diff(X), Diff$(X;Y)$ 2.3.5
div f 6.6.9.2

e_I^* 0.1.3
E 10.4.1.1
End(E) 2.8.10
End(TX) 5.1.6
ε_σ 0.1.2
exp u 2.8.11
exp$_x$ 11.1.1

F 10.4.1.1
$F^r(X)$ 5.4.2

G 10.4.1.1
$G_t x$ 1.3.3
G_t^s 1.5.3
$G(x)$ 2.4.5
$G_{k,d}$ 2.8.8
$G_t x$ 3.5.6
$\gamma(f)$ 6.10.2.2
GL(E) 2.8.10

H 10.6.2
$H^k(X)$ 4.2.24.2
h_k 11.5.2

i 8.5.1
ind$_m \xi$ 7.4.16
Inv^r 5.8.1, 5.8.8
Isom$(E;F)$ 0.0.11

$J(f)$ 0.2.8.9
$J(x)$ 1.3.1.1, 1.5.1

K 2.4.12.4
$K(X)$ 0.4
k_1, k_2 .... 10.6.2
$K_m C$ 8.4.1

$L^1(X)$ 0.4
$L(E;F)$ 0.0.4
$L(E,F;G)$ 0.2.8.3
$L^2([0,2\pi])$ 9.3.2
L_ξ 5.5.7.3
λ_E 0.1.15.4
$\Lambda^0 E^*, \Lambda^r E^*$ 0.1.1
ΛE^* 0.1.7
$\Lambda^r T^* X$ 5.1.2
leng(f) 3.6.3, 10.4.3
link(f,g) 7.4.9
μ_E 0.1.26

$N_X, N^e X, \overline{N^e X}, N_x X$ 2.7.2

ν 6.4.3

$NU^e X, NUX$ 2.7.2

$O(X), O_x(X), O_A(X)$ 0.0.2

$o(\|h\|)$ 0.2.1

$O(n)$ 2.1.6.4

$O(E)$ 2.8.10

$\underline{\Omega}_p^r(U), \underline{\Omega}^r(U)$ 0.3.1

$\underline{\Omega}_p^*(U)$ 0.3.2

$\Omega_q^r(X)$ 5.2.1

$\Omega_q(X)$ 5.2.3.1

ω_i 5.3.10.1

$\Omega^r(X)$ 5.4

$\mathcal{O}(E)$ 0.1.13

$P^d(\mathbf{R})$ 2.4.12.2

$P^d(\mathbf{C})$ 2.8.26

$P(x)$ 3.5.15.3

P1 ....11.10

r......................4.2.16

$R^r(X)$ 5.4.4

s 4.2.16

S_r 0.1.2

$\mathbf{S}_k^2$ 11.2

S^d 2.1.6.2

σ 5.3.17.2

$SL(E)$ 2.8.10

SP1 ....11.10

$sys(X)$ 11.9.0

t........................4.2.16

$^t A$ 0.1.15.6

T^d 2.1.6.3

$T^{(p,q)}$ 10.3.2

$T_x f$ 2.5.14

TX 2.5.24

Tf 2.5.26

$T_x V$ 2.5.3

$T_x X$ 2.5.9

$T_x^* X$ 5.1.1

t_0 ....11.5.2

$\tan g_m C$ 8.2.1.4

θ_x......................2.5.10, 2.5.22

$totcurv(X)$ 11.16.7

$Tub^e X$ 2.7.6

$V_{k,d}$ 2.8.8

$V(X)$ 2.8.17

V_f 10.6.9

$vol(V)$ 6.6.3

$vol(X, \delta)$ 6.6.7

Index

Numbers refer to the lowest-level subdivision yet started: for example, 4 is the introduction to chapter 4, before the beginning of section 4.1. Italics indicate a definition, or the statement of a result. Entries in brackets can be looked up in the Bibliography, where they are accompanied by a list of citations.

abelian integral, 10.2.3.1
Abraham, R., [AR67]
Abresch, U., [Abr86]
abstract manifold, 2.2
acceleration, *8.4.10*
action of group, *2.4.8*–10, 5.3.9–10
Alexandrov, 11.5.5, 11.6.2, 11.11.3.1,
    11.14.2, 11.17.2
algebra structure 2.3.4, 5.2.3.1, 5.2.4.2
algebraic curve, 10.2.3.1
  – surface, *10.2.2.4*
  – topology, *see* de Rham groups,
    homotopy
Alias, J., [Ali84]
Allendoerfer–Weyl–Fenchel–
    Gauss–Bonnet–Chern
    theorem, 6.9.8
almost everywhere, *0.4.4.4*
alternating form, *see also* exterior
    algebra
  – –, degree of, *0.1.6*
  – –, positive, *0.1.14*
  – –, pullback of, *0.1.8*

– –s, space of, *0.1.15.1*-3, 0.5.2
Ampère's theorem, 7.8.15
angle, right, 8.5.1
annulus, 5.6.4
antiderivation, *0.1.19*, *5.5.6*, 5.5.7.1-5
antipodal manifold, *11.10*
  – map on orientable double cover,
    5.3.31.1
  – – on $\mathbf{R}^2$, 2.8.14
  – – on the sphere, *2.4.7.2*
  – points, *10.4.9.1*, 11.1.1, 11.21.2(c)
Apéry, F., [Ape86]
apsidal surface, 10.2.2.7
arc, *6.5.9*
  –, biregular, *8.2.2.1*, *8.2.2.7*, 8.4.3
  –, geometric, *8.1.4*-7
  –, orientation of, 8.1.10, 8.2.1.6
  –, parametrized, *8.1.1*, 8.4.8
Archimedes's theorem, 6, 6.5.15
arclength, *see also* parametrization by
    arclength, 8.3.5
area of a surface, *6.5.1*, *6.6.3*, *6.6.7*,
    11.5.6

area *continued*:
  - of a plane curve, *9.3.1*
  - of surface of constant curvature, 11.2.5
  - of ball, 11.5.6, 11.6.2
area-preserving coordinates, *10.4.2*
Arnaudiès, J.-M., [LA74]
around x, chart, *2.2.3*
astigmatism, *10.6.8.1*
astroid, *8.7.17.3*
asymptotic direction, *10.6.4.1*
at x, chart, *2.2.3*
atlas, *2.2.1*
  - and canonical topology, 2.2.8
  -es, compatible, *2.2.4*
  -, maximal, *2.2.7*
  - on sphere, 2.8.7
  - on projective space, 2.8.30

Bacry, D., preface
ball, 0.0.3, 5.3.37.1
  -, volume of, 6.5.5
  -, area of, 11.5.6, 11.6.2
Banach space, 0.0.10, 0.2.1, 0.2.5–6, 0.2.8.1, 0.2.8.4–5, 0.2.9.2, 0.2.10, 0.2.12, 0.2.18, 0.2.22–23, 1.1, 1.1.1–2, 1.4.5, 2.2.2
Banchoff, T., 11.21.4, [Ban70]
Barner, M., [BF58]
base, countable, 3.1.3, 3.2.5
basis, positive, *0.1.14*
Bavard, C., [BP86], [Bav86]
  -'s inequality, 11.9.4
behavior, *see* local behavior
Beltrami, 10.4.9.6; *see also* Laplace–Beltrami
Beltrami's surface, *10.2.3.5*, 10.4.1.4, 10.5.5.5, 10.6.6.6, 11.7.3, 11.15, 11.15.2
Bennequin, D., [Ben86]
Bérard, P., [Ber86]
Bérard-Bergery, L., [BBL73]
Berger, M., [BGM71], [Ber87]
Bernoulli's lemniscate, *8.7.18*
Bernstein's theorem, 11.16.3
Bertrand curve, *8.7.23*
Besse, A., [Bes78], [Bes86]
Betti number, *4.2.24.2*, 5.4.11
Bieberbach, 11.21.2

binormal, *8.6.2*
birefringent medium, 10.2.2.7
biregular arc, *8.2.2.1*, *8.2.2.7*, 8.4.3
Bjerre, *see* Fabricius–Bjerre–Halpern
Blaschke, W., [Bla29], [Bla56]
blow-up, *2.8.16*
Blumenthal, L., [BM70]
Bonnesen's inequality, 9.9.3
Born, M., [BW75]
bound vector, 2.7.6
boundary of form, *6.2.2*
  - of manifold-with-boundary, *5.3.34*–37
  - of normal and bundle, 5.3.37.2
  - of tube, 10.2.3.10–12
bounded curvature, 11.5–6
  - geometry, *11.7.3*
  - topology, 4.2.26
Bourbaki, N., [Bou74]
Bourguignon, J.-P., preface, [BBL73]
Boy's surface, 10.2.4
bracket of two vector fields, *2.8.17.2*, 3.5.15.2
branch, *8.1.8.3*
Brézis, H., 11.17.5, [BC84]
Brouwer, 6.3.5–6
bubbles, 11.17
Buchner, M., 11.4.3, [Buc78]
bump function, 0.2.16, 2.3.7.1, 3.1.2, 3.2.7, 5.5.9.1, 5.7.1.4
bundle, density, *3.3.1*
  -, normal, *2.7.4*–10, 5.3.37.2, 6.7
  -, tangent, *2.5.25*, 3.1.7, 4.2.26, 5.9.9
  -, unitary normal, *2.7.4*, 5.3.37.2, 6.7.17, 7.5
Burago, Yu., [BZ86]

calculus, 0.2, 5.4.11
  -, fundamental theorem of, 3.4.4
canonical density, *0.1.26*, *0.3.11.1*, 3.3.18.2, *6.6.1*, *6.7.9*, 6.7.12
  - differentiable structure on $\mathbf{R}^n$, 2.2.10
  - involution of orientable double cover, 5.3.31.1
  - isomorphism, 2.5.12.3, 2.5.22
  - manifold structure of a submanifold, 2.6.2

canonical *continued*:
- map on normal bundle, *2.7.5*, 2.7.10
- measure, *10.5.5.2*
- norm on $L(E; F)$, *0.0.10*
- normal vector field, *6.4.3*
- orientation of $\mathbf{R}^n$, 5.3.8.1
- - of sphere, 5.3.37.1
- projection, 2.3.3.1, 3.3.1
- topology, 0.0.9, *2.2.6–7*
- vector field, *2.5.17.2*, 3.5.1
- volume form, *0.1.15.5*
- - - on orientable double cover, 5.3.31.2
- - - on $\mathbf{R}^n$ and submanifolds thereof, 5.3.17.1, 6.4.1
- - - on torus, 5.3.10.1, 6.5.11
- - - on sphere, 5.3.17.2, 6.4.6
Cantor set, 4.2.26
Carathéodory, C., [Car37a]
cardioid, *8.7.17.3*
Carmo, M. do, 11.13.2, [Car76]
Cartan, E., [Car37b], [Car84]
Cartan, H., [Car63], [Car70], [Car71]
category theory, 5.1.6, 5.4.6
catenary, 8.7.20, 10.2.2.9, 10.2.3.5
catenoid, *10.2.2.9*, 10.2.3.5–6, 10.6.6.6, 10.6.9.6
Cauchy, 11.14.2, 11.21.3.5
caustic, 8.7.17.3, 10.6, *10.6.8.1*
celestial mechanics, 10.4.9.3
center of curvature, *8.4.15*, 8.7.20, 10.2.2.9
- of hypo- or epicycloid, *8.7.17.3*
- of mass, *6.5.14*, 6.5.15, *6.10.16*, 6.10.25
centered at x, chart, *2.2.3*
chain, *6.2.2*
- rule, 2.5.15, 2.5.23.3, 2.5.27, 2.5.29, 4.2.13
Chakerian, G., [CG83]
change of variable formula, 0.4.6
chart, *2.2.3*, *2.2.5*, 2.2.9, 2.3.6.2, 2.8.20
-, complete, 10.1.2
-, latitude-longitude, 6.1.6, 10.2.3.3, 10.4.1.3
- on S^2, 6.1.6, 10.2.3.3
-, positively oriented, *5.3.22*
Chavel, I., [Cha84]

Chazy, J., [Cha53]
Chebyshev coordinates, *10.4.2*
Cheeger, J., [CG85], [CMS84]
Chern, S.-S., 6.9.8, [Che85]
Chinese lantern, 6.5.4
Choquet-Bruhat, Y., [Cho68]
Christoffel symbols, *10.4.7*, 10.7.1
-'s problem, 11.19.4
circle, *2.1.6.2*, 2.4.3, 2.4.12.1, 2.6.13.1, 2.8.23, 5.3.17.4
-, de Rham group of, 5.4.9.2
-, topological, 3.4.1, 7.6
Clairault's relation, *10.4.9.3*, 10.4.9.5
class, *see* differentiability class
classification of higher-dimensional manifolds, 4.2.26
- of one-dimensional manifolds, 3.4.1
- of simply connected compact manifolds, 4.2.26
- of surfaces, 4.2.25–26
closed curve, *9.1.4*
- form, 3.5.15.5, *5.4.2*, 5.4.11
Codazzi–Mainardi equation, *10.7*, 10.7.2, 10.7.3, 11.14.1, 11.17.1, 11.18.2, 11.19.1.1
codimension, *2.1.1*, 3.3.17.2
cogwheel design, 8.7.17.5
-, Lahire's, *8.7.17.3*
Cohn-Vossen, S., 11.5.4, 11.7.1, 11.14.1, [HC52]
cohomology groups, 4.2.24.2, 5.4.9
- ring, 5.9.15
Comberousse, C. de, [RdC22]
commutativity of flows, 3.5.15.2
compact Lie group, 5.8.9
-, locally, 2.2.11
- manifold, 3.5.9
-, relatively, *3.2.5*
- subsets, exhaustion by, 3.2.6
- support, 5.6.2, 7.1.1–10
comparison tests, 1.6.0
- theorem, Schur's, 8.7.22
compatible atlases, *2.2.4*
- coordinate system, *2.1.9*, 2.5.7.2
complement of submanifold, 2.8.24
-, orthogonal, *2.7.1*
complete chart, 10.1.2
- metric space, 0.0.13.2, 11.1.1

complete *continued*:
 - surface, 11.1.3
complex integration, 10.2.3.6
 - numbers, 2.8.5(d), 2.8.13,
   2.8.25–27, 4.4.6(a), 5.9.17,
   10.2.3.6
 - projective space, *2.8.26*–27(a),
   4.4.6(a)
components of acceleration, 8.4.12
composition of maps, 0.2.13.1, 2.3.3.2,
   5.3.21; *see also* chain rule
concavity, *8.2.2.11*
cone, *10.2.2.4*, 10.2.3.8, 10.2.3.10,
   10.4.9.2, 11.12
confocal quadrics, *10.2.2.3*
conformal coordinates, *10.4.2*
 - map, *0.5.3.1*
conic, 10.2.1.4
 -, focal, *10.2.3.12*, 10.2.3.14
conjecture, Rellich's, 11.17.5
 -, Willmore's, 11.17.4
conjugate points, *11.4.1.2*
connectedness, 2.2.13
 - of the complement of a
   submanifold, 2.8.24
 - of the orientable double cover,
   5.3.31.2
conoid, Plücker's, 10.2.1.4, 10.2.2.4,
   10.2.3.8
constant curvature, 10.4.9.6, 10.5.2,
   10.5.3.10, 11.2
 - mean curvature, 10.6.9.6, 11.17
contact, *8.7.11*, 9.7.10.1
continuity of integral, *0.4.8.2*, 6.1.4.11,
   9.5.4
continuous function, 0.0.6
 - linear map, 0.0.4
 - map, 0.0.5
continuously differentiable map, *0.2.5*
contracting map, *0.0.13.1*
contraction, *0.1.18*, *5.3.12*, 5.5.7.2
contravariant functor, *5.4.6*
convention on Hausdorff, 2.2.10.7
convex curve, *9.6.3*
 -, globally, *9.6.2*
 - polyhedron, 11.14.2
 - submanifold-with-boundary, 9.6.2
 - surface, 11.19
Coolidge, J., [Coo68]

coordinate change, *2.1.9*
 - system, *2.1.8*–9, 2.5.7.2, 10.4.2
coordinates, area-preserving, *10.4.2*
 -, Chebyshev, *10.4.2*
 -, conformal, *10.4.2*
 -, cylindrical, *6.10.15*
 -, elliptic, *10.2.2.3*
 -, geodesic, *10.4.2*, 10.4.9.4, 10.5.1,
   10.5.3.3
 -, homogeneous, *2.8.26*
 -, Liouville, *10.4.2*, 10.4.9.5–6
 -, local, *2.2.3*
 -, normal, *10.5.1*
 -, orthogonal, *10.4.2*, 10.4.7
 -, polar, 0.2.21, 8.2.2.13, 8.4.14.2,
   10.2.1.3; *see also* geodesic
   coordinates
 -, spherical, *6.5.8*
Coron, J.-M., 11.17.5, [BC84]
countable base, 3.1.3, 3.2.5
covariant derivative, *10.4.7*, 10.7
covering map, *2.4.1*, 3.3.7, 3.6.1, 4.1.5
 - space, *see* covering map
criterion for differentiability, 0.2.8.6,
   5.2.2.3
 - for orientability, 5.3.24, 5.3.27
critical point, *4.1.1*, 4.2.8–12
 - -, characterization of, 4.4.1
critical value, *4.1.1*, 4.3.1
Croke, C., [Cro84], [Cro]
 -'s inequality, 11.9.5
cross product, *0.1.17*
cross-ratio, *8.7.23*, 10.2.3.8
cultural digression, 1.6, 3.5.15,
   4.2.24–26, 6.9.8, 7.7.6
curvature bounded above, 11.6
 - - below, 11.5
 -, center of, *8.4.15*, 8.7.20, 10.2.2.9
 -, Gaussian, 6.9.7, 10.4.3, 10.5,
   10.5.1.2, 10.5.2–5, 10.5.3.6,
   10.5.5, 10.6.2.2, 11, 11.2, 11.5,
   11.19.2, 11.21.1
 -, geodesic, *10.4.7*, 10.6.1, 11.3.3
 -, line of, 10.2.3.12, *10.6.4.2*, 10.7.2
 -s, Lipschitz–Killing, *11.20.2*
 -, mean, *10.6.2*, 10.6.9
 -, Menger, 8.7.13
 -, negative, 11.15
 -, non-negative, 11.13

curvature *continued*:
- –, normal, *10.6.1.1*
- – of circle, 8.4.14.1
- – of curve, *8.4.1*, 8.4.14.2
- – of parametrized arc, *8.4.8*
- –, positive, 11.5.4, 11.11.3; *see also* non-negative curvature
- –, principal, *10.6.2*
- –, radius of, *8.4.15*, 8.7.19
- –, sectional, 10.8
- –, signed, *8.5.2*
- – tensor, 10.4.3
- –, total, *9.4.10*, *9.6.1*, 10.5.1, *10.6.2*, 11.16.7; *see also* Gaussian curvature
- –s, Weyl, *6.9.6*
- –, zero, 11.12
curve, *0.2.9.1*, 0.2.25, 2.1.3, *2.5.17.1*, *3.4.2*, 4.4.2, 5.3.17.4, *7.4.8*
- –, algebraic, 10.2.3.1
- – and Frenet frame, 8.6.12.2
- –, arc of, *6.5.3*
- –, area of, *9.3.1*
- –, Bertrand, *8.7.23*
- –, closed, *9.1.4*
- –, convex, *9.6.3*
- –, integral, *1.2.2*, *1.4.1*, *3.5.3*, *3.5.11*
- –, length of, *3.6.3*, 6.5.3
- –s, linked, *7.4.11*
- –, local behavior of, 8.5.3
- – on surface, 10.6.1
- – on sphere, 9.9.6
- –s, pair of, *7.4.8*
- –, Peano, 8.0.2.1
cuspidal edge, *see* line of striction
cut locus, *11.4.1*–3
cut value, *11.4.1*
cyclid, *10.2.2.6*
cylinder, 5.6.4, *10.2.2.1*, 10.2.3.8, 10.2.3.10, 11.12
- –, elliptic, 10.2.2.1
- –, hyperbolic, 10.2.2.1
- –, parabolic, 10.2.1.2
cylindrical coordinates, *6.10.15*
- – wedge, *6.10.30*

Darboux, G., 10, [Dar17], [Dar72]
- –'s theorem, 3.5.15.5

de Rham group, 2, *4.2.24.2*, 5, *5.4.4*, 5.4.7, 5.4.9.1–2, 5.4.10–13, 5.5.11, 5.6.3–4, 5.7–8, 5.9.15–18, 6.3.3, 7, 7.1.9, 7.2.1, 7.3.2, 7.5.6.1, 7.6
- – –'s theorem, 5.4.10
deformation of surface, 11.14.2
degree of form, *0.1.6*, *0.3.1*, *5.2.1*
- – of Gauss map, 7, 7.5, 7.5.2, 7.5.4
- – of map, 5.4.13, *7.3.1*, *7.6.4*, 10.2.3.8
- – of tangent map, 9.4.1
Delaunay, 10.6.9.6, 11.18.1
density, *0.1.25*, 0.1.29, *3.3.1*–8, *3.4.2*–4, 3.6.5, 5.9.4, *6.5.14.1*
- – bundle, *3.3.1*
- –, canonical, *0.1.26*, *0.3.11.1*, 3.3.18.2, *6.6.1*, *6.7.9*, 6.7.12
- –, integral with respect to, 3.3.14
- –, measure associated with, 3.3.11
- –, product, *3.3.18.1*–5
- –, pullback of, 0.3.11.1, *3.3.2*, 3.3.16
derivation, *2.8.17*, *5.5.6*
derivative, *0.2.3*, *2.5.29*, *5.2.6*
- – at a point, *0.2.2*
- –, elementary notion of, 0.2.4
- –, partial, 0.2.8.8, 0.2.15.5
- –, second, *0.2.11*–13, 4.2, 4.2.4
- – of solution of differential equation, 1.6.2
- –s, uniform convergence of, 4.4.10
determinant form, 0.1.12.1
developable surface, *10.2.3.9*, 10.2.3.12, 10.4.1.8, 10.4.9.2, 10.6.6.2
diagonalization, 4.2.16–17
diameter, *11.1.4*
Dieudonné, J., [Die69]
diffeomorphism, *0.2.18*
differentiability class, 0.2.5, 0.2.11, 0.2.14, 2.1.1, 2.2.1, 2.2.5, 2.3.1, 3.3.5, 5.2.1
- – criterion, 0.2.8.6, 5.2.2.3
differentiable manifold, *2.2.5*
- – map, *0.2.1*
- – – on manifold, *2.3.1*–3
- – structure, *2.2.5*
differential, *2.5.23*, *5.2.6*
- – calculus, 0.2, 5.4.11

differential *continued*:
- equation, *1.1.1*, *3.5.11*, 3.5.15.3;
  see also vector field
- –s, comparison tests for, 1.6.0
- –, linear, *1.6.3–4*
- –, order of, *1.1.1–2*
- –s, system of, *1.1.1*
- form, *0.3.1*, 3.5.15.6, *5.2.1*, 5.2.7
- –, boundary of, *6.2.2*
- –, closed, 3.5.15.5, *5.4.2*, 5.4.11
- –, degree of, *0.3.1*
- –, exact, *5.4.2*
- –s, family of, *0.3.15.1–3*,
  5.2.10.1–4
- – in coordinates, 0.3.5, 5.2.8
- –, integrable, *6.1*
- –, invariant, 5.3.10.2, 5.9.1–3
- –, pullback of, *0.3.7*, *5.2.4.1*
- –, rank of, *3.5.15.5*
- –, restriction of, *5.2.5.1*
differentiation under integral sign,
  0.4.8, 7.1.5, 5.2.10.5
digression, *see* cultural digression
dimension argument, 5.2.5.2
- of manifold, *2.2.5*
- of submanifold of $\mathbf{R}^n$, *2.1.1*
- of tangent space to submanifold,
  2.5.6
Diquet's formula, 10.5.1.3
direction, asymptotic, *10.6.4.1*
- , principal, *10.6.2*, 10.8
distance, *3.6.3*, *10.4.3*; *see also* length
distribution, parameter of, *10.2.3.8*
divergence, *6.6.9.2*
Dixmier, J., [Dix67], [Dix68]
Dombrowski, P., [Dom79]
Donaldson, S., [Don83]
double cover, *see* orientable double
  cover
- point, *8.1.8*, *9.3.4*
- tangent, *9.3.3*
dual of a vector space, *0.0.4*
Dupin cyclid, *10.2.2.6*, 10.2.3.12,
  10.6.8.2, 11.21.4
- indicatrix, *10.3.3*

Eells, J., [Eel87]
Efimov, 11.14.2, 11.15.1
Eisenhart, L., [Eis49], [Eis62]

elastic band, 10.4.5; *see also* string
electricity, 8.7.17.5
ellipse, 8.7.19, 10.2.2.7
ellipsoid, 10.2.2.3, 10.2.3.4, 10.4.1.6,
  10.4.9.5, 10.5.3.7, 10.5.5.7,
  10.6.6.6, 10.6.6.8, 10.6.8.3,
  11.4.2.3, 11.8
- , mechanical generation of,
  10.2.2.7, 10.2.3.14
elliptic coordinates, *10.2.2.3*
- cylinder, 10.2.2.1
- geometry, *11.2.4*
- paraboloid, *6.10.12*, 10.2.1.2
- point, *10.6.4.1*
- space, *11.2.4*
embedding, *2.6.9–12*, 3.1, 9.6.5.3,
  11.11
Enneper's surface, 10.2.2.5, 10.2.3.6,
  10.2.3.13, 10.4.1.5, 10.5.3.5,
  10.6.6.3
envelope, 10.2.3.5, 10.2.3.12–13,
  10.6.8.2(3), 11.19
epicycloid, *8.7.17.1–5*
equation, Codazzi–Mainardi, *10.7*,
  10.7.2, 10.7.3, 11.14.1, 11.17.1,
  11.18.2, 11.19.1.1
- , differential, *1.1.1–2*, 1.6.1–2,
  3.5.11, 3.5.15.3; *see also* vector
  field
- , Euler, *see* Euler form
- , local, 2.1.3.1, 2.6.15
equilibrium, 7.8.14
equivalent parametrized arcs, *8.1.4*
- – –, strictly, *8.1.11*
- volume forms, *5.3.4*
Euclidean norm, 0.0.9
Euclidean structure on space of
  alternating forms, 0.1.15.1–3,
  0.5.2
Euler characteristic, *4.2.24.2–4*,
  4.4.6(c), 6.9.16, 7, 7.5.4, 7.5.6.3,
  7.7.6.1, 11.7
- equation, *see* Euler form
- form, 8.7.21, 9.9.3, 9.9.7,
  10.2.3.13, 11.19
eversion, *11.11.1*
evolute, *8.7.5*, 8.7.16, 8.7.17.4
exact form, *5.4.2*
exhaustion by compact subsets, 3.2.6

exponential map, *11.1.1*
exterior, *9.2.2*
   - algebra, 0.1, 0.3.2; *see also*
      exterior derivative
   - - on Euclidean spaces, 0.1.15
   - - on manifolds, 5.2
   - product, *0.1.4*
   - power, *0.1.15.7*
   - derivative, *0.3.12.0*, *5.2.9.1*, 5.4.2,
      5.5.7.1
   - -, in coordinates, 5.2.9.7

Fabricius–Bjerre–Halpern, 9.8
family of diffeomorphisms, *see*
      one-parameter group
   - of differential forms, *0.3.15.1–3*,
      5.2.10.1–4
   - of maps, pullback under, 5.2.10.6
   - of normals to a surface, 3.5.15.5
   -, triply orthogonal, *10.2.2.3*,
      10.2.2.6, 10.6.8.3
Federer, H., [Fed69]
Feldman, E., 9.4.17.2, [Fel68]
Fenchel, 6.9.8
Ferrand, J., preface; *see also*
      Lelong-Ferrand
field, *see* vector field, Jacobi field
Firey, W., 11.19.4, [Fir68]
first fundamental form, *10.3.1–2*, 10.4,
      10.7–8, 11.19
   - - - in coordinates, 10.4.1
   - integral, *10.4.9.3*
fixed point, 0.0.13.2, 6.3.5
flat map (♭), *0.1.15.1*
flexible, *11.14.2*
Flohr, F., [BF58]
flow, *see also* one-parameter family of
      diffeomorphisms
   -s, commutativity of, 3.5.15.2
   - defined everywhere, 3.5.9
   -, domain of, *1.3.3*, 1.3.6, 3.5.6
   -, global, *1.3.3*, *3.5.6*, 3.5.11
   -, local, *1.2.3*, *3.5.3*, 3.5.11
   -, uniqueness of, 1.3.1, 3.5.4, 3.5.11
focal conic, *10.2.3.12*, 10.2.3.14
   - point, *2.7.11*
   - surface, *10.6.8.1*
form, alternating, *see* alternating form
   - -, degree of, *5.2.1*

-, determinant, 0.1.12.1
-, differential, *see* differential form
-, Euler, 8.7.21, 9.9.3, 9.9.7,
   10.2.3.13, 11.19
-, first fundamental, *10.3.1–2*, 10.4,
   10.7–8, 11.19
-, linear, *0.0.4*
-, second fundamental, *10.3.3*,
   10.6–8
-, symmetric bilinear, 0.2.8.3, 3.6.3,
   4.2, 4.2.6
-, third fundamental, *10.3.3*
-, volume, *0.1.15.5*, *5.3.2–17*
formula, change of variable, 0.4.6
-, Diquet's, 10.5.1.3
-s, Frenet, *8.6.6*, 10.2.3.10
-, Gauss's, 10.5.3.2
-, Girard's, *10.5.5.5*, 10.6.2.2
-, Hopf's, 11.7.4
- of the three levels, 6.10.32
-, Puiseux's, *10.5.1.3*
-s, variation, 11.3.2, 11.3.3
-, Weierstrass's, 10.2.3.6, 11.16.5–6
formulary for second fundamental
   form, 10.6.5
Forster, O., [For81]
four-vertex theorem, 9.7.4
France, teaching in, preface
Frenet formulas, *8.6.6*, 10.2.3.10
   - frame, *8.6.6*, 8.6.10–13, 9.9.4,
      10.6.7, 10.7.3
Fresnel, 10.2.2.7
Frobenius' theorem, 2.8.17.2,
   3.5.15.3–6, 10.7.3
Fubini's theorem, *0.4.5.1*, 3.3.18.6,
   3.3.18.7, 6.2.1.3, 6.5.9, 6.5.10,
   6.6.9.2, 6.7.16, 7.1.8
function, *see also* map
   -, bump, 0.2.16, 2.3.7.1, 3.1.2, 3.2.7,
      5.5.9.1, 5.7.1.4
   -, continuous, 0.0.6
   -, height, 4.1.4.2, 6.5.15
   -, holomorphic, 10.2.3.6
   -, integrable, *0.4*, 3.3.14
   -, multi-valued, 10.2.3.6
   -, periodic, 9.1.7
   -, support, *11.19*
   -, triply periodic, 11.16.4
functor, 5.1.6, 5.4.6

fundamental form, *see* first, second, third
- --s, links between the two, 10.7–8
- group, 4.2.26, 5.4.13; *see also* simply connected
- theorem of calculus, 3.4.4
funnel, 6.10.14

Gauduchon, P., [BGM71]
Gauss, 10.4.1.7, 10.5.3.2, 10.5.5.7, 10.6.2
- -'s formula, 10.5.3.2
- map, *6.8.13*, 6.9.15, *10.9.9*, 10.6.2.2, 10.6.9.3, 10.8, 11.13.1, 11.16.6–7, 11.19
- - -, degree of, 7, 7.5, 7.5.2, 7.5.4
- -'s *Theorema egregium*, *10.5.9.2*, 10.6.2.1
Gauss–Bonnet theorem, preface, 6.9.8, 7, *7.5.4*, 7.5.7, 10.5.5.4–5, 11, 11.2.5, 11.5.4, 11.7
Gaussian curvature, 6.9.7, 10.4.3, 10.5, *10.5.1.2*, 10.5.2–5, 10.5.3.6, 10.5.5, 10.6.2.2, 11, 11.2, 11.5, 11.19.2, 11.21.1
generic metrics, 10.6.8.1, 11.4.3.4
geodesic, 10.4.5, 10.4.8–9
- , closed, 11.9–10
- coordinates, *10.4.2*, 10.4.9.4, 10.5.1, 10.5.3.3
- curvature, *10.4.7*, 10.6.1, 11.3.3
- map, *10.4.9.6*
- torsion, *10.6.7*
geodesy, 10.5.5.6
geometric arc, *8.1.4*
- - , oriented, *8.1.11*
Géométrie Différentielle, preface
Girard's formula, *10.5.5.5*, 10.6.2.2
girth, *11.21.9.4*
global flow, *1.9.9*, *9.5.6*, 3.5.11
- surface, *10.1.2*
- uniqueness of flow, 1.3.1, 3.5.4, 3.5.11
globalization, 2.1.6.5, 8.2.2.16
globally convex, *9.6.2*
Gluck, H., [Glu71]
Gluck–Singer, 11.4.3
graded algebra, 5.5.6, 5.9.15
gradient, 10.2.2.12

Gram–Schmidt process, 6.7.10
Gramain, A., [Gra71]
graph, 2.1.2, 2.1.3.1, 4.4.9, 10.2.1, 10.4.1.2, 10.5.3.4, 10.6.6.1
grassmannian, *2.8.8*
Grauenstein, *see* Whitney–Grauenstein
Green, 11.10.2, 11.14.3
Greenberg, M., [Gre67], [Gre80]
Greene, R., [GW72]
Greenwich meridian, 6.1.6
Groemer, H., [CG83]
Gromoll, D., 11.10, [GG81], [GM69]
Gromov, M., 6.6.9.2, [CG85], [GR70], [Gro81], [Gro82], [Gro83], [Gro86]
- -'s inequality, 11.9.3
group action, *2.4.8*–10, 5.3.9–10
- , cohomology, 4.2.24.2, 5.4.9
- , de Rham, 2, *4.2.24.2*, 5, *5.4.4*, 5.4.7, 5.4.9.1–2, 5.4.10–13, 5.5.11, 5.6.3–4, 5.7–8, 5.9.15–18, 6.3.3, 7, 7.1.9, 7.2.1, 7.3.2, 7.5.6.1, 7.6
- , fundamental, 4.2.26, 5.4.13; *see also* simply connected
- , holonomy, 3.5.15.5
- , Lie, preface, 3.5.15.5, *3.6.5*, 5.8.9
- , linear, 2.8.10–11, 3.6.5
- , one-parameter, *1.9.5*, 3.5.10, 5.5.7.4
- , orthogonal, 2.1.6.4, 2.8.10–11, 5.8.9
Grove, K., 11.10, [GG81]
Guichardet, A., [Gui69]
Guillemin, 11.10.2
Guldin theorems, 6.10.15–16
Gunning, R., [Gun62]
Gutierrez, C., [SG82]

Haar measure, *9.6.5*, 5.8.9
Hadamard's theorem, *11.6.2*, 11.13.1–2
Halpern, W., [Hal77]; *see also* Fabricius–Bjerre–Halpern
Halphen, G.-H., [Hal88]
Hartman, 11.12
Hausdorff, 2.2.10.4, 2.2.10.7, 2.2.11, 2.3.7.3, 2.4.9–10, 2.4.12.4, 2.5.25.1, 2.6.12, 2.8.8, 3.1.2, 3.2.4, 3.5.4–5, 3.6.4, 4.1.5, 5.1.5, 5.3.27

Hebda, J., [Heb81]
height function, 4.1.4.2, 6.5.15
helicoid, 2.8.22, *10.2.1.3*, 10.2.2.10, 10.2.3.6, 10.2.3.8
helix, *8.7.10*
  –, circular, *8.4.14.4*, 8.6.11.1, 8.6.16
  –, spherical, *8.7.17.3*
Herglotz, 11.14.1, 11.19.1.6, 11.19.2
Hermann, 6.9.8
Hessian, 0.5.3.2, *4.2.2*, 4.2.16, 4.4.2
Hicks, N., [Hic65]
Hilbert, D., 9.3.2, 11.15.1, 11.17.1, [HC52]
Hilliard, J., [Hil]
holomorphic function, 10.2.3.6
holonomy groups, 3.5.15.5
homogeneous coordinates, *2.8.26*
homologous, *5.4.5*, 5.5.11
homotopy, 4.2.26, *7.4.1*, *7.4.8*, *9.4.6*
Hopf, H., [Hop83]
  – fibration, *2.8.25*
  –'s formula, 11.7.4
  – invariant, *6.10.2.2*
  –'s theorem, 11.17.2
Hopf–Rinow, theorem of, 11.1.1–4
horizontal tangent space, 4.1.4.2
Hu, S.-T., [Hu69]
hydrostatics, 10.6
hyperbolic cylinder, 10.2.2.1
  – paraboloid, *6.10.12*, 10.2.1.2, 10.2.3.8
  – plane, *11.2.2*, 11.4.3
  – point, *10.6.4.1*
hyperboloid, 10.2.2.3, 10.2.3.8
hyperplane reflection, *0.5.3.1*
hyperquadric, *2.8.9*
hypersurface, *2.1.6.5*, 2.5.7.1, 2.6.15, 10.8
hypocycloid, *8.7.17.1*-5

I.H.E.S, preface
image of geometric arc, *8.1.6*
  – of parametrized arc, *8.1.3*
  – of point, *8.1.7*, *9.1.9*
immersed surface, *10.1.4*, 10.2.4
immersion, *0.2.23*, 2.1.3.1, *2.6.9*–12, 2.7, 11.11, 11.17.3
implicit function theorem, *0.2.26*

– characterization of submanifolds, 2.1.2
implicitly defined surface, 10.2.2, 10.5.3.8, 10.6.6.4
index of critical point, *4.2.8*, *4.2.11*
  – of point, *7.6.8*, *9.1.11*
  – of singularity, *7.4.16*, 7.7.4
indicatrix of Dupin, *10.3.3*
inequality, Bavard's, 11.9.4
  –, Bonnesen's, 9.9.3
  –, Croke's, 11.9.5
  –, Gromov's, 11.9.3
  –, isoembolic, *11.4.4*
  –, isoperimetric, 6.6.9, 9.3, 10.6.9.7, 11.8, 11.20
  –, isosystolic, *11.9.0*
  –, Loewner's, 11.9.1
  –, Pu's, 11.9.2
  –, strict triangle, 10.3.2
  –, Wirtinger's, 9.3.2
infinite-dimensional manifold, 2.2.2
inflection point, *9.8.5*
initial condition, *1.2.2*, *3.5.3*
injectivity radius, *11.4.4*
inner product, 0.1.15.1
inside, *9.2.2*
integrable form, *6.1*
  – function, *0.4*, 3.3.14
  – vector field, 5
integral, 0.4.2, 3.3.14–16, *6.1.3*, *6.2.2*
  –, abelian, 10.2.3.1
  –, continuity of, *0.4.8.2*, 6.1.4.11, 9.5.4
  – curve, *1.2.2*, *1.4.1*, *3.5.3*, *3.5.11*
  –, first *10.4.9.3*
integration, 0.4, 3.3.11.5, 10.2.3.6
  – of a family of differential forms, 0.3.15.3, 5.2.10.4
interior, *9.2.2*
intrinsic components of acceleration, 8.4.12
  – metric, *3.6.3*, *10.4.3*
invariance of degree under homotopy, 7.4.3, 7.6.5
  – – index under diffeomorphism, 7.7.3
invariant forms, 5.9.1–3
inverse function theorem, 0.2.22, 2.3.7.1, 2.5.20, 2.8.11

inverse *continued*:
- image, *see also* pullback
- - of point, 5.9.7
inversion, *0.5.9.1*, 8.7.4, 10.2.3.12
isoembolic inequality, *11.4.4*
isolated singularity, *7.4.15*, *7.7.2.1*
isometry, 6.9.5, 8.3.4, 8.3.12.3, 10,
    10.2.3.6-8, 10.3.2, 10.4.1.1,
    10.4.1.7-8, 10.4.4, 10.4.9.2,
    10.5.1.2, 10.5.2 10.5.3.9, 10.5.4,
    10.8, 11.11.2-3, 11.14, 11.19.1.5,
    11.19.2, 11.2.1, 11.2.4, 11.8.1,
    11.21.1
isomorphism $T_p X \sim \mathbf{R}^d$, 2.5.10,
    2.5.12.3
isoperimetric inequality, 6.6.9, 9.3,
    10.6.9.7, 11.8, 11.20
- profile, *11.8.9*
isosystolic inequality, *11.9.0*
isotopy, *7.7.1*

Jabœuf, F., preface
Jacobi field, 10.5.3.3, *11.5.1*-3, 11.6.1
jacobian (determinant), *0.2.8.9*,
    0.3.10.2, 3.3.6, *4.2.17*, 5.3.36,
    6.6.9.2, 6.7.11, 6.8.4, 7.7.8, 9.8.8,
    10.4.1.1
- matrix, *0.2.8.8*, 0.2.21, 2.3.7.2,
    2.6.13.3
Jacobowitz, H., [Jac82]
Jordan's theorem, 7.6.8, 9, 9.2, 9.3.4.2,
    9.5.1-3, 9.6.4
Jorge, L., [JM83]

Kepler's laws, 10.4.9.3
k-fold cover, *2.4.4*
Killing, *see* Lipschitz-Killing
kinematics, 8.4.5, 8.4.10
Klein bottle, *2.4.12.4*, 4.2.24.3, 4.4.4,
    5.3.19, 5.9.10, 5.9.18, 11.2.4,
    11.7.2-3, 11.9.4, 11.13
Klingenberg, W., [Kli78], [Kli82]
Klotz-Milnor, T., [Klo72]
Knörrer, H., [Kno80]
Kobayashi, S., [KN69]
Kowalski, O., [Kow80]
Kuiper, N., [Kui70], [Kui84]

Lafontaine, J., preface, [BBL73]
Lahire's cogwheel, *8.7.17.9*
Lang, S., preface, [Lan68], [Lan69]
lantern, Chinese, 6.5.4
Laplace-Beltrami operator, 11
Lashof, 11.13.2
latitude, *10.4.1.4*
-, parallels of, *9.9.5*
latitude-longitude chart, 6.1.6,
    10.2.3.3, 10.4.1.3
lattice, 2.4.7.1
laws, Kepler's, 10.4.9.3
Lebesgue measure, *0.4.3.1*, 0.4.4-6,
    0.4.8.0, 3, 3.3.11.5, *3.3.12*-13,
    3.3.17.3-5, 6.1, 6.1.4.2, 6.1.6,
    6.6.9.2
Legendre, 10.5.5.6
Lehman, D., [LS82]
Leichtweiss, K., [Lei80]
Lelong-Ferrand, J., [LA74], [Lel63],
    [Lel82], [Lel85]; *see also* Ferrand
Lemaire, J., [Lem67]
lemma, Poincaré, 5.4.12-13, 5.6.1-2,
    5.7.1.4, 7.1.2.1
-, Zorn's, 3.4.5.2
lemniscate, *8.7.18*
length, *3.6.9*, *6.5.1*-3, *6.6.9*, *6.6.7*,
    8.9.7, *8.9.9*, 9.3.1, *10.4.9*
Levy, S., preface
Lewy, 11.19.3
Li, P., [LY82]
Lie, Sophus, 10.2.3.1
- algebra, 3.5.15.5
- derivative, 5.4.11, *5.5.9*
- group, preface, 3.5.15.5, *3.6.5*,
    5.8.9
Liebmann, 11.12.1, 11.14, 11.17.1
lifting, *7.6.2*
Lima, *see* Carmo-Lima
limaçon of Pascal, *8.4.14.9*, 8.7.17.3
line, long 2.2.10.6
- of curvature, 10.6.4, *10.6.8.1*-3
- of striction, *10.2.9.8*-9, 10.6.8.3
linear differential equation, *1.6.9*-*4*
- form, *0.0.4*
- group, 2.8.10-11, 3.6.5
- map, 0.0.4
linked curves, *7.4.11*
linking number, *7.4.9*

links between the two fundamental
forms, 10.7–8
Liouville, *see also* Sturm–Liouville
 – coordinates, *10.4.2*, 10.4.9.5–6
 –'s theorem, 0.5.3
Lipschitz map, *0.0.13.1*, 0.2.6, 0.4.4.5
 – vector field, 1.2.6, 1.2.7, 1.3.1,
   1.6.0
Lipschitz–Killing curvatures, *11.20.2*
Lissajous figure, *9.1.8.3*
local behavior of a curve, 8.5.3
 – – – a map, 0.2.23–26
 – – – a surface, 4.2.20
local connectedness, 2.2.12
 – coordinates, *2.2.3*
 – equations, 2.1.3.1, 2.6.15
 – flow, *1.2.3*, *3.5.3*, 3.5.11
 – –, continuity of, 1.2.6
 – –, differentiability of, 1.2.7
 – –, existence and uniqueness of,
   1.2.6, 1.4.5, 1.4.7, 3.5.3, 3.5.11
local surface, *10.1.2*
locally compact, 2.2.11
 – constant, 2.4.4
 – convex, 8.2.2.15
 – finite, *3.2.1*
 – Lipschitz map, *0.0.13.1*, 0.2.7
Loewner's inequality, 11.9.1
logarithmic spiral, *8.7.16*
long line, 2.2.10.6
loop, *7.6.7*

Mach, E., [Mac49]
magic tricks, 7.4.14, 7.8.11, 10.2.3.8
Mainardi, *see* Codazzi-Mainardi
Mangoldt, H. von, [Man81]
manifold, *see also* curve, surface,
   riemannian manifold
 –, abstract, 2.2
 –, antipodal, *11.10*
 –, classification of, 4.2.26
 –, compact, 3.5.9
 –, differentiable, *2.2.5*
 –, dimension of, *2.2.5*
 –, infinite-dimensional, 2.2.2
 –, non-Hausdorff, 2.2.10.4, 3.5.5
 –, one-dimensional, 3.4.1, 10.1.4; *see
   also* curve
 –, orientation of, *5.3.5*

 –s, product of, *2.2.10.3*–5, 2.3.3.1–3
   2.8.18, 5.6.3
 – structure, 2.6.2
 –, topological, *2.2.5*
 –, two-dimensional, *see* surface
 –, unreasonable, *2.2.10.5*
manifold-with-boundary, *5.3.33*–37,
   10.5.5.4 11.17.5; *see also*
   submanifold-with-boundary
 –, boundary of, *5.3.34*–37
manifolds-with-boundary, product of,
   5.9.12
map, *see also* function, isomorphism,
   isometry
 –, antipodal, *2.4.7.2*, 2.8.14, 5.3.31.1
 –, canonical, 2.5.12.3, 2.5.22, *2.7.5*,
   2.7.10
 –, conformal, *0.5.3.1*
 –, contracting, *0.0.13.1*
 –, covering, *2.4.1*, 3.3.7, 3.6.1, 4.1.5
 –, degree of, 5.4.13, *7.3.1*, *7.6.4*,
   10.2.3.8
 –, differentiable, *0.2.1*, *2.3.1*
 –, exponential, *11.1.1*
 –, flat (♭), *0.1.15.1*
 –, Gauss, *see* Gauss map
 –, geodesic, *10.4.9.6*
 –, linear, 0.0.4
 –, Lipschitz, *0.0.13.1*, 0.2.6, 0.4.4.5
 –, local behavior of, 0.2.23–26
 –, regular, *0.2.20*, *2.6.9*, 2.7.10, 3.3.7
 –, restriction of, 2.3.3.4, 2.6.6–7
 –, section of, *0.2.8.5*
 –, sharp (♯), *0.1.15.1*
 –, symmetric bilinear, *see*
   symmetric bilinear form
 –, tangent, *2.5.14*–20
 –, unit tangent, *9.4*
Massey, W., [Mas77]
maximal atlas, *2.2.7*
 – integral curve, *1.3.1.1*, 1.5.1, 3.5.5
Mazet, E., [BGM71]
mean curvature, *10.6.2*, 10.6.9
measure associated with a density,
   3.3.11
 –, canonical, *10.5.5.2*
 –, Haar, *3.6.5*, 5.8.9
 –, Lebesgue, *0.4.3.1*, 0.4.4–6,
   0.4.8.0, 3, 3.3.11.5, *3.3.12*–13,

measure *continued*:
    3.3.17.3–5, 6.1, 6.1.4.2, 6.1.6,
      6.6.9.2
    – of image, 3.3.17.1–4
    –, product, *0.4.5*, 3.3.18.5
    –, Radon, *0.4*
    – zero, *0.4.4.0*, *3.3.13*, 3.3.17.2,
      3.3.17.6
mechanics, 8.7.17.5, 10.6; *see also*
    kinematics, center of mass
    –, celestial, 10.4.9.3
Meeks, W., [JM83], [Mee81]
Menger, K., [BM70]
    – curvature, 8.7.13
Mercator projection, 10.2.3.3, 10.4.1.3
metacenter, 10.6, 10.6.8.1
metric space, 0.0.3
    – –, complete, 0.0.13.2, 11.1.1
metrizability, 0.4, 3.3.11.1, 3.6.3
Meyer, W., [GM69]
Michel, 11.10.2
Milnor, J., [Mil63], [Mil69]
Minding, 10.2.3.5
minimal area, *11.7.3*
    – surface, 10.2.1.3, 10.2.3.6,
      10.6.6.3, 10.6.9.1, 11.16
    – – of revolution, 10.2.2.9
    – – of translation, 10.2.1.3
Minkowski, 11.19.1.2–3
    –'s problem, 11.19.3
    – inequalities, 11, 11.20.1
mirror, *see* caustic
mixed problems, 10.8
    – product, *0.1.16*, 10.2.3.8
Möbius strip, 5.3.6, 5.3.17.1, 5.9.11,
      6.10.20, 7.8.11, 10.2.3.8, 11.2.4,
      11.9.4
models for the hyperbolic plane,
    11.2.2, 11.4.3
moduli, *11.2.4*
molding surface, *10.2.3.10*, 10.6.8.2
monkey saddle, 4.2.22
Morin, B., [MP78], [MP80]
morphism, *2.3.1*; *see also* differentiable
    map
Morse reduction, 4.2.13, 9.5.4
    – theory, preface, 4, *4.2.24*–25,
      7.5.1, 7.5.4
Moser's theorem, 3, 7, 7.2.3

Müller, W., [CMS84]
multiple point, *8.1.8*, *9.1.9*
multiplicity, *2.4.4*, *8.1.8*, *9.1.9*
multi-valued function, 10.2.3.6
Myers, 11.4.3

naval architecture, 10.6
negative curvature, 11.15
nephroid, *8.7.17.3*
Nirenberg, 11.11.3.1, 11.12, 11.19.3
Nitsche, J., [Nit75]
Nomizu, K., [KN69]
non-degenerate critical point, *4.2.7*,
    4.2.11, 4.2.18, *4.2.21*, 4.4.10
non-Hausdorff manifold, 2.2.10.4, 3.5.5
non-negative curvature, 11.13
non-orientable, *5.3.5*
    – at infinity, 4.2.26
non-separable space, 3.6.4
norm, 0.0.9–10, 0.1.29.6, 0.4.8.0
    – of uniform convergence, 1.2.6.2
normal bundle, *2.7.4*, 6.7
    – coordinates, *10.5.1*
    – curvature, *10.6.1.1*
    –s, family of, 3.5.15.5
    – space, *2.7.2*
normed vector space, 0.0.10, 0.4.8.0
number, Betti, *4.2.24.2*, 5.4.11
    –, linking, *7.4.9*
    –, turning, *9.4.2*
    –, winding, *7.3.7*, *7.6.8*

Oliker, V., [Oli84]
one-dimensional manifolds, 3.4.1,
    10.1.4; *see also* curve
O'Neill, B., [ONe66]
one-parameter group of
    diffeomorphisms, 3.5.10, 5.5.7.4
    – – of homeomorphisms, *1.3.5*
optics, 8.7.17.5, 10.2.2.7, 10.6, 10.6.8.1
orbit, *2.4.5*
order of differential equation, *1.1.1–2*
orientability, *5.3.5*, 6.7.26
    – criterion, 5.3.24, 5.3.27
orientable, *5.3.5*
    – double cover, 5.3.27–29
orientation of arc, 8.1.10, 8.2.1.6
    – of boundary, 5.3.36
    – of manifold, *5.3.5*

orientation *continued*:
  – of product, 5.3.8.4, 5.9.8
  – of simple closed curve, 9.2.6
  – of sphere, 5.3.17.2, 5.3.37.1
  – of submanifold, 5.3.8.2
  – of surface, 10.3.3
  – of tangent bundle, 5.9.9
  – of torus, 5.3.10.1
  – of vector space, *0.1.13*, 5.3.8.1
orientation-preserving, *0.1.14*, *5.9.20*
orientation-reversing, *0.1.14*, *5.9.20*
oriented closed curve, *9.1.5*
  – simple closed curve, *9.1.1*
orthogonal complement, *2.7.1*
  – coordinates, *10.4.2*, 10.4.7
  – group, 2.1.6.4, 2.8.10–11, 5.8.9
  – subspace, *see* normal subspace
  – vectors, *4.2.6*
osculating circle, *8.4.15*, 8.7.4
  – paraboloid, *10.6.4*
  – plane, *8.2.2.2*, *8.2.2.*7-9, 10.2.3.9
  – sphere, *8.7.9*
Osserman, R., 9.7.5, 9.9.7, 11.16.7,
  [Oss69], [Oss78], [Oss85]
outside, *9.2.2*
Ozawa, T., 9.8.1.1, [Oza84], [Oza85]

P1, 11.10
Palais, R., [Pal57]
Pansu, P., [BP86]
paper strip, 10.2.3.8
parabola, 10.6.9.6
parabolic cylinder, 10.2.1.2
  – point, *10.6.4.1*
paradox of the funnel, 6.10.14
parallel, *10.4.6*
  – of latitude, *9.9.5*
  – surface, 10.2.2.12, 10.2.3.11,
    10.6.6.7, 10.6.8, 10.6.9.1
  – transport, *10.4.6.1*, 10.5.5
parameter of distribution, *10.2.3.8*
parameter-dependent vector field,
  1.4.6–7
parametrically defined surface, 10.2.3
parametrization, *8.1.4*
  – by arclength, *3.4.3*, *8.3.1*
  – – –, existence of, 3.4.6
  – – –, invariance under isometries,
    8.3.4

  – of ruled surface, standard,
    10.2.3.8
parametrized arc, *8.1.1*
Paris, preface
Parseval's theorem, 9.3.2
partial derivative, 0.2.8.8, 0.2.15.5
partition of unity, *3.2.2*, 3.2.4, 5.3.24,
  5.7.1.4
Pascal limaçon, *8.4.14.3*, 8.7.17.3
path-connectedness, 2.2.13
Peano curve, 8.0.2.1
period, 8.1.9, 9.1.7, 9.1.8.1
periodic function, 9.1.7
Petit, J.-P., [MP78], [MP80]
Phillips, A., [Phi66]
physics, 5, 6.5.16; *see also* mechanics,
  kinematics, optics
Pinkall, U., [Pin85]
pitch, *10.2.2.10*
planar point, *10.6.4.1*
plane arc, *8.2.2.3*
  – curve, *see* curve
plane, equation of, 10.2.2.2
Plateau's problem, 11.16, 11.17.5
Plücker's conoid, 10.2.1.4, 10.2.2.4,
  10.2.3.8
Pogorelov, A., 11.11.3.1, 11.14.2–3,
  [Pog73]
Pohl, W., [Poh68]
Poincaré lemma, 5.4.12–13, 5.6.1–2,
  5.7.1.4, 7.1.2.1
  – –, generalization of, 5.6.3
  – model, *11.2.2*, 11.4.3
point, antipodal, *10.4.9.1*, 11.1.1,
  11.21.2(c)
  –, conjugate, *11.4.1.2*
  –, critical, *4.1.1*, 4.2.7–12, 4.2.21,
    4.4.10
  –, double, *8.1.8*, *9.8.4*
  –, elliptic, *10.6.4.1*
  –, fixed point, 0.0.13.2, 6.3.5
  –, focal, *2.7.11*
  –, hyperbolic, *10.6.4.1*
  –, image of, *8.1.7*, *9.1.9*
  –, index of, *7.6.8*, *9.1.11*
  –, index of critical, *4.2.8*, *4.2.11*
  –, inflection, *9.8.5*
  –, inverse image of, 5.9.7
  –, multiple, *8.1.8*, *9.1.9*

point *continued*:
- of geometric arc, *8.1.7*
- on curve, *9.1.9*
-, parabolic, *10.6.4.1*
-, planar, *10.6.4.1*
-, regular, *4.1.1*
-, regular double, *9.8.4*
-, regular inflection, *9.8.5*
-, simple, *8.1.8*
-, triple, *8.1.8*
pointing inward, *6.9.7*, 7.4.18
- outward, *6.4.4*
polar coordinates, 0.2.21, 8.2.2.13,
    8.4.14.2, 10.2.1.3; *see also*
    geodesic coordinates
pole, *0.5.9.1*, *11.6.9*
polygonal approximation, 8.3.6
polyhedron, 11.14.2
polynomial, 0.2.8.2, 10.2.2.4
positive basis, *0.1.14*
- form, *0.1.14*
- curvature, 11.5.4, 11.11.3; *see also*
    non-negative curvature
power of an inversion, *0.5.9.1*
principal curvature, *10.6.2*
- direction, *10.6.2*, 10.8
- normal, *8.4.11*, *8.6.2*
problem, Christoffel's, 11.19.4
-, Minkowski's, 11.19.3
-, Plateau's, 11.16, 11.17.5
product density, *9.9.18.1*–5
- measure, *0.4.5*, 3.3.18.5
- of manifolds, *2.2.10.9*–5, 2.3.3.1–3
    2.8.18, 5.6.3
- topology, *5.2.10.6*
projective differential geometry,
    10.4.9.6
- plane, 2.6.13.3, 10.2.4; *see also*
    projective space
- –s, product of, 5.9.6
- space, *2.4.12.2*, 2.6.13.3, 2.8.16,
    4.2.24.3, 4.4.3, 4.4.5, 4.4.6(d),
    5.3.18, 5.4.12–13, 5.7.2, 5.9.5; *see*
    also projective plane
- –, complex, *2.8.26*–27, 4.4.6(a)
- –, quaternionic, *2.8.26*–27,
    4.4.6(b)
properly discontinuously without fixed
    points, *2.4.5*, 2.8.12, 3.1.7.2,

5.9.17, 6.6.8, 11.2.4; *see also*
    torus, projective space
Prüfer's surface, 3.1.4.3, *9.6.4*
pseudosphere, *see* Beltrami's surface
Pu's inequality, 11.9.2
Puiseux's formula, *10.5.1.9*
pullback of alternating form, *0.1.8*
- of density, 0.3.11.1, *9.9.2*, 3.3.16
- of differential form, *5.2.4.1*
- – – under a family of maps,
    5.2.10.6
- – – under covering map, 5.3.8.3
- of riemannian structure, *10.9.2*
punctured surface, 11.14.3
push-forward, *7.7.2*

quadric, 2.8.9, 10.2.1.2; *see also*
    ellipsoid, etc.
-, central, *2.8.9*
-s, confocal, *10.2.2.9*
-, homofocal, 10.2.3.14
-, proper, 10.2.2.3
quaternionic projective space,
    2.8.26–27, 4.4.6(b)
quaternions, 2.8.5(e), 2.8.25
quotient by an action, *2.4.8*–10,
    5.3.9–10

Radon measure, *0.4*
rank of a form, *9.5.15.5*
recipes for torus, 2.1.6.3, 2.4.12.1
reflection, *0.5.9.1*
refraction, 10.2.2.7
regular double point, *9.8.4*
- – tangent, *9.8.9*
- inflection point, *9.8.5*
- map, *0.2.20*, *2.6.9*, 2.7.10, 3.3.7
- point, *4.1.1*
- value, *4.1.1*
- –s, abundance of, 4.1.5, 4.3.6; *see*
    also Sard's theorem
relation, Clairault's, *10.4.9.9*, 10.4.9.5
relatively compact, 3.2.5
Rellich's conjecture, 11.17.5
restriction of form, *5.2.5.1*
- of map, 2.3.3.4, 2.6.6–7
Riemann, 11.15; *see also* riemannian
- integral, 0.4.3.1

Riemann *continued*:
- surface, 11, 11.16.4, 11.16.6, 11.17.2
riemannian covering space, *11.2.9*
- geometry, 3.5.15.5, 10, 10.4.3, 11.3.3
- manifold, 3.6.3, 6.9.8, 7.5.7, 10.3.2, 10.4.3–4, 10.4.7, 10.6.8.1, 11, 11.1.3, 11.2.3, 11.7.3, 11.11.3.1
- metric, 6.9.7, 6.9.8, 10.5.4, 11.2.3, 11.11.3.1; *see also* riemannian manifold, riemannian surface
- structure, *9.6.9*, 10, *10.9.1-2*, 10.4.1.1, 10.4.4, 11.2.4–5
- surface, 10.4.7, 10.4.9.6, 10.5.4, 11, 11.2.1–2, 11.6.3, 11.7, 11.7.2, 11.8.2.1, 11.11.2
right angle, 8.5.1
rigid motion, *8.5.4.2*, 10.2.1.4, 10.2.3.10
rigidity, 11.14.2
Rinow, *see* Hopf-Rinow
Robbin, J., [AR67]
Rokhlin, V., [GR70]
rotation by $\pi/2$, 8.5.1
Rouché, E., [RdC22]
Rudin, W., [Rud74]
ruled surface, 10.2.1.4, *10.2.9.7*, 10.2.3.10, 10.4.1.7, 10.5.3.6, 10.6.6.5
  - -, standard parametrization of, *10.2.9.8*
ruling, *10.2.9.7*

Sacré, C., [LS82]
Salmon, G., [Sal74]
Sard's theorem, 3.3.17.6, 4, 4.3, *4.9.1*, 7.3.2.1, 7.5.4, 9.2.7, 9.2.9
scalar product, 0.1.15.1
Scherk's surface, *10.2.1.9*, 11.16, 11.16.3
Schmidt, 11.8.1
Schrader, R., [CMS84]
Schur's comparison theorem, 8.7.22
Schwarz's theorem, 0.2.13, 0.3.12.1
screw motion, *10.2.2.10*
second derivative, *0.2.11*–13, 4.2, 4.2.4
- fundamental form, *10.9.9*, 10.6–8

- - -, formulary for, 10.6.5
section of map, *0.2.8.5*
sectional curvature, 10.8
segment, *10.4.4*
semigroup of local homeomorphisms, *1.9.5*, 3.5.10
separability, *9.1.9*–5, 3.1.7.1
Serrin, J., [Ser69]
shadow, 11.21.3.5
sharp map ($\sharp$), *0.1.15.1*
ship hulls, 10.6
shortest paths, *10.4.4*, 10.4.8, 11.1.1, 11.4
- period, 9.1.8.1
signature, *2.8.9*
signed curvature, *8.5.2*
similarity, *0.5.9.1*
Simon, L., 11.17.4
simple closed curve, *9.1.1*
  - - -, orientation of, 9.2.6
- point, *8.1.8*
simply connected, 4.2.26, 11.2.1; *see also* fundamental group
Singer, I., [ST67]
singularity, *7.4.15*, *7.7.2.1*
slab, 11.20.2; *see also* tube
smokestacks, 10.6
soap bubbles, 11.17
solid torus, 6.9.11.2
solution of a differential equation, *1.1.1-2*, 1.6.1; *see also* integral curve
Sotomayor, J., [SG82]
SP1, 11.10
space, *see also* Banach space, vector space, projective space
-, covering, *see* covering map
-, elliptic, *11.2.4*
-, metric, 0.0.3, 0.0.13.2, 11.1.1
-, normal, *2.7.2*
-, symmetric, 3.5.15.5
-, tangent, *see* tangent space
-, topological, 0.0.2; *see also* Hausdorff
- -, non-separable, 3.6.4
speed, *8.4.6*; *see also* velocity
sphere, *2.1.6.2*, 2.8.2, 4.1.4.3, 4.2.24.3, 4.4.3, 4.4.6(d), 5.3.17.2, 5.4.12–13, 5.9.17, 10.2.1.1,

sphere *continued*:
    10.2.3.3, 10.4.1.3, 10.4.9.1, 11.8.1,
    11.17.1, 11.21.1–3
    –, canonical orientation of, 5.3.17.2,
        5.3.37.1
    –, eversion of, *11.11.1*
    –s, product of, 5.9.6; *see also* torus
    –, volume of, 6.5.5
spheric mirror, 8.7.17.3
spherical coordinates, *6.5.8*
    – zone, *6.10.27*
Spivak, M., [Spi79]
Springer-Verlag, preface
standard parametrization of ruled
    surface, *10.2.3.8*
star-shaped, 4.2.13, 5.4.11, 5.6.1
Steiner, 11.8.1
stereographic projection, 2.8.7, 2.8.25,
    5.7.1, 6.10.8, 7.8.10, 9.7.9,
    10.2.3.3, 10.4.1.3
Sterling, I., [Ste]
Sternberg, S., [Ste64]
Stoker, J., 11.13.2, [Sto69]
Stokes' theorem, preface, 5, 5.4.14,
    6, 6.2, *6.2.1–2*, 6.3, 6.3.4, 6.3.7,
    6.5.5–6, 6.5.16, 6.6.9.2, 7.1.2.2,
    7.2.1, 7.4.18, 7.8.13, 9.3.3, 9.8.9,
    10.5.5.4, 11.19.1.1
strict triangle inequality, 10.3.2
striction, line of, *10.2.3.8*–9, 10.6.8.3
strictly equivalent parametrized arcs,
    8.1.11
string, 10.2.3.14; *see also* elastic band
Struik, D., [Str61]
Sturm–Liouville theory, 11.5.3, 11.6
submanifold, 2.1.3.1, *2.6.1*–14
    –s, counterexamples of, 2.1.5, 2.8.4
submersion, *0.2.23*, 2.1.3.1, *2.6.9*,
    2.6.14, 5.9.7
subordinate, *3.2.1*
support, compact, 5.6.2, 7.1.1–10
    – function, *11.19*
surface, *4.2.25*, 10, 11
    –, algebraic, *10.2.2.4*
    –, apsidal, 10.2.2.7
    –, area of, *6.5.1*, *6.6.3*, *6.6.7*, 11.5.6
    –, Beltrami's, *10.2.3.5*, 10.4.1.4,
        10.5.5.5, 10.6.6.6, 11.7.3, 11.15,
        11.15.2

–, Boy's, 10.2.4
–s, classification of, 4.2.25–26
–, complete, 11.1.3
–, convex, 11.19
–, deformation of, 11.14.2
–, developable, *10.2.3.9*, 10.2.3.12,
    10.4.1.8, 10.4.9.2, 10.6.6.2
–, Enneper's, 10.2.2.5, 10.2.3.6,
    10.2.3.13, 10.4.1.5, 10.5.3.5,
    10.6.6.3
–, focal, *10.6.8.1*
–, global, *10.1.2*
–, immersed, *10.1.4*, 10.2.4
–, implicitly defined, 10.2.2,
    10.5.3.8, 10.6.6.4
–, local, *10.1.2*
–, local behavior of, 4.2.20
–, minimal, 10.2.1.3, 10.2.3.6,
    10.6.6.3, 10.6.9.1, 11.16
–, molding, *10.2.3.10*, 10.6.8.2
–, normals to a, 3.5.15.5
– of constant curvature, *see*
    constant curvature
– of translation, 10.2.1.3, 10.2.3.1
– of minimal area, 10.6.9.2, 10.6.9.7;
    see also minimal surface
– of revolution, 6.10.4, 10.2.2.9,
    10.2.3.5, 10.2.3.10, 10.2.3.12,
    10.4.1.4, 10.4.9.3, 10.5.3.9,
    10.6.6.6, 10.6.9.6
–, orientation of, 10.3.3
–s, parallel, 10.2.2.12, 10.2.3.11,
    10.6.6.7, 10.6.8, 10.6.9.1
–, parametrically defined, 10.2.3
–, Prüfer's, 3.1.4.3, *3.6.4*
–, punctured, 11.14.3
–, Riemann, 11, 11.16.4, 11.16.6,
    11.17.2
–, riemannian, 10.4.7, 10.4.9.6,
    10.5.4, 11, 11.2.1–2, 11.6.3, 11.7,
    11.7.2, 11.8.2.1, 11.11.2
–, ruled, 10.2.1.4, *10.2.3.7*,
    10.2.3.10, 10.4.1.7, 10.5.3.6,
    10.6.6.5
– Scherk's, *10.2.1.3*, 11.16, 11.16.3
– tetrahedral, *10.2.2.8*
– Veronese's, *2.1.6.8*–9, 2.8.5, 7.8.17
– wave, 10.2.3.13, *10.2.2.7*
symbol, Christoffel, *10.4.7*, 10.7.1

symmetric bilinear form, 0.2.8.3, 3.6.3,
  4.2, 4.2.6
  – bilinear map, *see* symmetric
    bilinear form
  – spaces, 3.5.15.5
symmetry, minimal surfaces with,
  11.16.5
Synge, J., [Syn37]
system, coordinate, *2.1.8*–9, 2.5.7.2,
  10.4.2
  – of differential equations, *1.1.1*
systole, *11.9.0*

tangent bundle, *2.5.25*, 3.1.7, 4.2.26
tangent map, *2.5.14*–20
  – –, degree of, 9.4.1
  – plane, 4.2.20
  – space, *2.5.3*, *2.5.9*, 2.5.12
  – –, characterization of, 2.5.7
  – – of product, 2.5.18
  – to curve, *8.2.1.1*, *8.2.1.4*
  – vector, *2.5.1*, *2.5.9*
Taylor series, 4.2.16, 4.2.23, 4.4.7,
  8.2.2.15, 8.6.12.2, 8.7.11
tensor, *10.3.2*, 10.4.3
tetrahedral surface, *10.2.2.8*
theorem, of Allendoerfer–Weyl–
    Fenchel–Gauss–Bonnet–Chern,
    6.9.8
  –, Ampère's, 7.8.15
  –, Archimedes's, 6, 6.5.15
  –, Bernstein's, 11.16.3
  –, Darboux's, 3.5.15.5
  –, de Rham's, 5.4.10
  –, four-vertex, 9.7.4
  –, Frobenius', 2.8.17.2, 3.5.15.3–6,
    10.7.3
  –, Fubini's, *0.4.5.1*, 3.3.18.6,
    3.3.18.7, 6.2.1.3, 6.5.9, 6.5.10,
    6.6.9.2, 6.7.16, 7.1.8
  –, Gauss–Bonnet, preface, 6.9.8, 7,
    7.5.4, 7.5.7, 10.5.5.4–5, 11, 11.2.5,
    11.5.4, 11.7
  –s, Guldin, 6.10.15–16
  – Hadamard's, *11.6.2*, 11.13.1–2
  –, Hopf's, 11.17.2
  – of Hopf–Rinow, 11.1.1–4
  –, implicit function, *0.2.26*

–, inverse function, 0.2.22, 2.3.7.1,
  2.5.20, 2.8.11
–, Jordan's, 7.6.8, 9, 9.2, 9.3.4.2,
  9.5.1–3, 9.6.4
–, Liouville's, 0.5.3
–, Moser's, 3, 7, 7.2.3
–, Parseval's, 9.3.2
–, Sard's, 3.3.17.6, 4, 4.3, *4.3.1*,
  7.3.2.1, 7.5.4, 9.2.7, 9.2.9
–, Schur's comparison, 8.7.22
–, Schwarz's, 0.2.13, 0.3.12.1
–, Stokes', preface, 5, 5.4.14, 6,
  6.2, *6.2.1*–2, 6.3, 6.3.4, 6.3.7,
  6.5.5–6, 6.5.16, 6.6.9.2, 7.1.2.2,
  7.2.1, 7.4.18, 7.8.13, 9.3.3, 9.8.9,
  10.5.5.4, 11.19.1.1
–, turning tangent, 9.5
–, Weierstrass's, 6.10.4
–, Whitney–Grauenstein, 9, 9.4.8
Theorema egregium, *10.5.3.2*, 10.6.2.1
third fundamental form, *10.3.3*
Thorpe, J., [ST67], [Tho79]
three levels, formula of, 6.10.32
Thurston, W., [Thu79], [Thu82],
    [Thu88]
Titus, C., [Tit73]
topological manifolds, *2.2.5*
  – space, 0.0.2; *see also* Hausdorff
  – –, non-separable, 3.6.4
topology, algebraic, *see* fundamental
    group, homotopy
  –, bounded, 4.2.26
  –, canonical, 0.0.9, *2.2.6*–8
  –, product, *5.2.10.6*
toroidal coil, *8.6.11.3*
torsion, *8.6.5*–9, 8.6.12, 9.7.9, 10.2.3.9
  –, geodesic, *10.6.7*
torus, *2.1.6.3*, 2.4.12.4, 2.8.4, 4.1.4.4,
  4.2.5, 4.2.9.1, 4.2.24.3, 4.4.4,
  5.3.10.1, 5.4.12–13, 5.9.13,
  10.2.3.12, 10.2.4
  –, solid, 6.9.11.2
total curvature of curve, *see* total
    signed curvature
  – – of surface, 10.5.1, *10.6.2*,
    11.16.7; *see also* Gaussian
    curvature
  – differential, *5.4.11*
  – signed curvature, *9.4.10*

total *continued*:
  – unsigned curvature, *9.6.1*
totally geodesic submanifolds, 3.5.15.5
tractrix, *10.2.3.5*
transversality, *9.2.9*
traveler, 7.8.14
trigonometry, 11.5.5, 11.6.2
triple point, *8.1.8*
triply orthogonal family, *10.2.2.3*,
  10.2.2.6, 10.6.8.3
  – periodic function, 11.16.4
tube, *2.7.6*, 2.8.29
  –, boundary of, 10.2.3.10–12
  –, volume of, 6.7–9, 7.5, 10.6.3
tubular neighborhood, *see* tube
turning number, *9.4.2*
  – tangent theorem, 9.5
  – the sphere inside out, *11.11.1*
  – towards the origin, *8.2.2.13*
twice differentiable map, *0.2.11*

umbilic, *9.7.10.5*, 10.2.3.6, 10.4.9.5,
  10.6.4.1, 10.6.8.1, 10.6.8.3, 11.4.2,
  11.7.4, 11.21.1
Umlaufsatz, 9.5
unbounded, *see* bounded
unit tangent map, *9.4*
  – – vector, *8.3.11*
unitary normal bundle, *2.7.4*, 5.3.37.2,
  6.7.17, 7.5
unlinked, *see* linked
unreasonable manifold, *2.2.10.5*
upper-half-plane model, *11.2.2*

Valiron, G., [Val84]
variation formulas, 11.3.2, 11.3.3
vector, bound, 2.7.6
  – bundle, *see* bundle
  – field, *1.2.1*, *3.5.1*
  – – associated with family of
    diffeomorphisms, 3.5.14, 5.5.5
  – –s, bracket of, *2.8.17.2*, 3.5.15.2
  – –, canonical, *2.5.17.2*, 3.5.1
  – –, canonical normal, *6.4.3*
  – –, integrable, 5
  – –, Lipschitz, 1.2.6, 1.2.7, 1.3.1,
    1.6.0
  – – on sphere, 7.4.5
  – – on compact manifold, 3.5.9

  – –, parameter-dependent, 1.4.6–7
  – –, time-dependent, *1.4.1*–5, 1.5.1,
    3.5.11
  – space, 0.0.4
  – –, dual of, *0.0.4*
  – – isomorphism, 0.0.11
  – –, normed, 0.0.10, 0.4.8.0
  – –, orientation of, *0.1.13*, 5.3.8.1
  –, tangent, *2.5.1*, *2.5.9*
  –, unit tangent, *8.3.11*
vector-valued integral, 0.4.7
velocity, *0.2.9.1*, *2.5.17.3*, *2.5.28*, *8.4.6*
  –, scalar, *8.4.6*
Venzi, P., [Ven79]
Veronese's surface, *2.1.6.8*–9, 2.8.5,
  7.8.17
vertex, *9.7.1*, 9.7.4, 9.7.9
Viviani's window, *6.10.31*
volume, 2, *6.5.1*, *6.6.3*, *6.6.7*
  – form, *5.3.2*, 5.3.17
  – – invariant under a group, 5.3.10.2
Voronoi diagram, 11.4.2

Wall, C., 11.4.3, [Wal79]
Walter, R., [Wal78]
Wankel engine, 8.7.17.5
Warner, F., 11.4.3, [War71]
wave surface, 10.2.3.13, *10.2.2.7*
wedge, cylindrical, *6.10.30*
  – product, 5.2.3
Weierstrass's theorem, 6.10.4
  –'s formula, 10.2.3.6, 11.16.5–6
Weingarten endomorphism, *10.3.3*,
  10.6.2, 10.7
  – surface, preface, 10.6.6.6,
    11.18.1–3
well-ordered set, 2.2.10.6
Wente, H., [Wen85]
  – immersion, 11.17.3
Weyl, H., 6, 6.9.8, 11.11.3.1, [Wey39]
  – curvatures, *6.9.6*
Whitney, 3.1.5
Whitney–Grauenstein theorem, 9,
  9.4.8
width, *11.21.3.3*
Willmore's conjecture, 11.17.4
winding number, *7.3.7*, *7.6.8*
window, Viviani's, *6.10.31*
Wirtinger's inequality, 9.3.2

without fixed points, *see* properly
    discontinuous
Wolf, E., [BW75]
Wolf, J., [Wol72]
Wolpert, S., [Wol85]
Wong, Y.-C., [Won72]
Wu, H., 11.14.3, [GW72]
Wunderlich, W., [Wun62]

Xavier, 11.16.7

Zalgaller, V., [BZ86]
zero curvature, 11.12
   – section, *5.9.4*
   – torsion, 9.7.9
Zorn's lemma, 3.4.5.2
Zwikker, C., [Zwi63]

Graduate Texts in Mathematics

continued from page ii

48 SACHS/WU. General Relativity for Mathematicians.
49 GRUENBERG/WEIR. Linear Geometry. 2nd ed.
50 EDWARDS. Fermat's Last Theorem.
51 KLINGENBERG. A Course in Differential Geometry.
52 HARTSHORNE. Algebraic Geometry.
53 MANIN. A Course in Mathematical Logic.
54 GRAVER/WATKINS. Combinatorics with Emphasis on the Theory of Graphs.
55 BROWN/PEARCY. Introduction to Operator Theory I: Elements of Functional Analysis.
56 MASSEY. Algebraic Topology: An Introduction.
57 CROWELL/FOX. Introduction to Knot Theory.
58 KOBLITZ. *p*-adic Numbers, *p*-adic Analysis, and Zeta-Functions. 2nd ed.
59 LANG. Cyclotomic Fields.
60 ARNOLD. Mathematical Methods in Classical Mechanics.
61 WHITEHEAD. Elements of Homotopy Theory.
62 KARGAPOLOV/MERZLJAKOV. Fundamentals of the Theory of Groups.
63 BOLLABÁS. Graph Theory.
64 EDWARDS. Fourier Series. Vol. I. 2nd ed.
65 WELLS. Differential Analysis on Complex Manifolds. 2nd ed.
66 WATERHOUSE. Introduction to Affine Group Schemes.
67 SERRE. Local Fields.
68 WEIDMANN. Linear Operators in Hilbert Spaces.
69 LANG. Cyclotomic Fields II.
70 MASSEY. Singular Homology Theory.
71 FARKAS/KRA. Riemann Surfaces.
72 STILLWELL. Classical Topology and Combinatorial Group Theory.
73 HUNGERFORD. Algebra.
74 DAVENPORT. Multiplicative Number Theory. 2nd ed.
75 HOCHSCHILD. Basic Theory of Algebraic Groups and Lie Algebras.
76 IITAKA. Algebraic Geometry.
77 HECKE. Lectures on the Theory of Algebraic Numbers.
78 BURRIS/SANKAPPANAVAR. A Course in Universal Algebra.
79 WALTERS. An Introduction to Ergodic Theory.
80 ROBINSON. A Course in the Theory of Groups.
81 FORSTER. Lectures on Riemann Surfaces.
82 BOTT/TU. Differential Forms in Algebraic Topology.
83 WASHINGTON. Introduction to Cyclotomic Fields.
84 IRELAND/ROSEN. A Classical Introduction to Modern Number Theory.
85 EDWARDS. Fourier Series: Vol. II. 2nd ed.
86 VAN LINT. Introduction to Coding Theory.
87 BROWN. Cohomology of Groups.
88 PIERCE. Associative Algebras.
89 LANG. Introduction to Algrebraic and Abelian Functions. 2nd ed.
90 BRØNDSTED. An Introduction to Convex Polytopes.
91 BEARDON. On the Geometry of Discrete Groups.
92 DIESTEL. Sequences and Series in Banach Spaces.

93 DUBROVIN/FOMENKO/NOVIKOV. Modern Geometry — Methods and Applications Vol. I.

94 WARNER. Foundations of Differentiable Manifolds and Lie Groups.

95 SHIRYAYEV. Probability, Statistics, and Random Processes.

96 CONWAY. A Course in Functional Analysis.

97 KOBLITZ. Introduction in Elliptic Curves and Modular Forms.

98 BRÖCKER/TOM DIECK. Representations of Compact Lie Groups.

99 GROVE/BENSON. Finite Reflection Groups. 2nd ed.

100 BERG/CHRISTENSEN/RESSEL. Harmonic Analysis on Semigroups: Theory of Positive Definite and Related Functions.

101 EDWARDS. Galois Theory.

102 VARADARAJAN. Lie Groups, Lie Algebras and Their Representations.

103 LANG. Complex Analysis. 2nd ed.

104 DUBROVIN/FOMENKO/NOVIKOV. Modern Geometry — Methods and Applications Vol. II.

105 LANG. $SL_2(\mathbf{R})$.

106 SILVERMAN. The Arithmetic of Elliptic Curves.

107 OLVER. Applications of Lie Groups to Differential Equations.

108 RANGE. Holomorphic Functions and Integral Representations in Several Complex Variables.

109 LEHTO. Univalent Functions and Teichmüller Spaces.

110 LANG. Algebraic Number Theory.

111 HUSEMÖLLER. Elliptic Curves.

112 LANG. Elliptic Functions.

113 KARATZAS/SHREVE. Brownian Motion and Stochastic Calculus.

114 KOBLITZ. A Course in Number Theory and Cryptography.

115 BERGER/GOSTIAUX. Differential Geometry: Manifolds, Curves, and Surfaces.

116 KELLEY/SRINIVASAN. Measure and Integral, Volume 1.

117 SERRE. Algebraic Groups and Class Fields.

118 LANG. Cyclotomic Fields.